ROUTLEDGE HANDBOOK OF MOTOR CONTROL AND MOTOR LEARNING

The *Routledge Handbook of Motor Control and Motor Learning* is the first book to offer a comprehensive survey of neurophysiological, behavioural and biomechanical aspects of motor function. Adopting an integrative approach, it examines the full range of key topics in contemporary human movement studies, explaining motor behaviour in depth from the molecular level to behavioural consequences.

The book contains contributions from the world's leading experts in motor control and motor learning, and is composed of five thematic parts:

- Theories and models
- Basic aspects of motor control and learning
- Motor control and learning in locomotion and posture
- Motor control and learning in voluntary actions
- Challenges in motor control and learning

Mastering and improving motor control may be important in sports, but it becomes even more relevant in rehabilitation and clinical settings, where the prime aim is to regain motor function. Therefore, the book addresses not only basic and theoretical aspects of motor control and learning but also applied areas like robotics, modelling and complex human movements. This book is both a definitive subject guide and an important contribution to the contemporary research agenda. It is, therefore, important reading for students, scholars and researchers working in sports and exercise science, kinesiology, physical therapy, medicine and neuroscience.

Albert Gollhofer is Professor and Head of the Department of Sport Sciences at the University of Freiburg, Germany. He is former President of the European College of Sport Science and of the German Society of Biomechanics.

Wolfgang Taube is Professor and Head of the Movement and Sport Science Department at the University of Fribourg in Switzerland. He is a member of the national research council Switzerland and executive member of the Swiss Society of Sport.

Jens Bo Nielsen is Professor of Human Motor Control at the Institute of Exercise and Sport Sciences & Institute of Neuroscience and Pharmacology, University of Copenhagen, Denmark. He is head of the research group Copenhagen Neural Control of Movement.

ROUTLEDGE HANDBOOK OF MOTOR CONTROL AND MOTOR LEARNING

Edited by Albert Gollhofer, Wolfgang Taube and Jens Bo Nielsen

Routledge
Taylor & Francis Group

LONDON AND NEW YORK

First published 2012
by Routledge
2 Park Square, Milton Park, Abingdon, Oxfordshire OX14 4RN

Simultaneously published in the USA and Canada
by Routledge
711 Third Avenue, New York, NY 10017

First issued in paperback 2014

Routledge is an imprint of the Taylor & Francis Group, an informa business

British Library Cataloguing in Publication Data
A catalogue record for this book is available from the British Library

Library of Congress Cataloging in Publication Data
Library of Congress Cataloging-in-Publication Data
Routledge handbook of motor control and motor learning / edited by
Albert Gollhofer Wolfgang Taube and Jens Bo Nielsen.
p. cm.
ISBN 978–0–415–66960–3 (hbk)
ISBN 978–1–138–86042–1 (pbk)
ISBN 978–0–203–13274–6 (ebk)

1. Motor learning—Handbooks, manuals, etc.   2. Motor control—Handbooks,
manuals, etc.   I. Gollhofer, Albert.   II. Taube, Wolfgang.   III. Nielsen, Jens Bo.
BF295.R68 2013
152.3'34—dc23
2012019652

ISBN 978–0–415–66960–3 (hbk)
ISBN 978–1–138–86042–1 (pbk)
ISBN 978–0–203–13274–6 (ebk)

Typeset in Bembo
by RefineCatch Limited, Bungay, Suffolk

CONTENTS

Contents

Contents

LIST OF FIGURES AND TABLES

Figures

Tables

ACKNOWLEDGEMENTS

A book of this type can only be realized when the best scholars in the field – motor control and motor learning – are committed to contribute and to enrich the diversity with their perspectives. We therefore thank all our colleagues with their teams who participated with their valuable and state-of-the-art scientific manuscripts. Many thanks to all of you who devoted your valuable time in order to get this challenging project realized.

Multidisciplinary book projects need expertise from various scientific fields. We greatly acknowledge the effort and enthusiasm of each individual in our own research groups who has been in charge of reading and editing each individual chapter, communicating with the authors and finally getting organized what we can hold in our hands today – the final product of the project: the *Routledge Handbook of Motor Control and Motor Learning*.

INTRODUCTION

Imagine that you have to cross the busy Fifth Avenue in New York City. In order to safely cross the street you need to detect and anticipate the movement of other pedestrians in order to avoid walking into them; when and how much will they accelerate or decelerate, how will they react in response to external signs like street lights or obstacles and which way will they go in order to avoid walking into you? Crossing Fifth Avenue is clearly not as simple a task as it first seems when we start thinking of the complex calculations that are necessary in order to navigate through a mass of independently acting subjects. Our brain handles tasks like this continuously without us giving it much thought, yet even the most advanced computers and robots are far from accomplishing anything like it.

And this is not all: The maintenance of balance and the execution of gait are things that we normally take for granted, but walking safely across the street involves the precise coordinated control of dozens of muscles throughout the body based on the processing of enormous amounts of information from our body and the environment in order to orient the body in space and in relation to gravity with each step. This is no simple achievement.

It is easy to think of many other examples of the impressive capacity of our central nervous system to manage everyday tasks. But do you know how you and others have acquired these complex motor skills? Have you ever wondered why cycling is so easy now and has been so extremely difficult at the beginning? Have you ever questioned yourself why your brain forgets rapidly a recently learned French word but remembers easily how to ride a bike or to ski after a ten-year break? Thus there are obviously differences between learning a language and acquiring a motor task.

Our nervous system, based on 100 billions of highly interconnected nerve cells, represents one of the most complex structures in the entire universe allowing not only complex cognitive but also motor tasks. The attempts to reach a similar level of performance artificially are relatively advanced in certain areas but lack deep foundation in others: When considering cognitive tasks like playing chess, the world's fastest computers are nowadays able to perform 10 petaflops ($10 + 10^{15}$ bit$\star$sec^{-1}) per second and this capacity is more than sufficient to outperform the world's best human chess players. In contrast, when it comes to motor control and/or motor learning, even the performance of a young child is by far superior to any mechanistically scaled action of a robot with respect to elegance and movement repertoire.

Our motor system needs to select and apply appropriate activation patterns to achieve an intended task, which is complicated by the fact that for every given movement, numerous possibilities of trajectories exist, as a consequence of the redundancy of optimal solutions. The human brain is amazingly good at coping with the broad range of possible solutions for one and the same mechanical task. This can best be seen in highly skilled athletes such as basketball players, who are able to hit the basket despite continuously changing conditions (distance, opponent, velocity, etc.) due to their ability to rapidly select the right option meeting perfectly the requirements of the task. Furthermore, if task execution was not successful, our motor system has the capacity to learn from its own errors. In this respect it has been recognized that just the small 'errors' between the desired motor action and the actual performance are the essential fuel to drive motor learning. Our central nervous system is adaptable in the short term to master new and recently non-experienced situations but also allows outlasting plasticity to increase performance and to be competitive in the long term. Appropriate motor activity as well as adaptation of existing motor patterns and learning of new tasks is essential to eat, communicate, relocate and, thus, survive. Fortunately, our motor control system has the fundamental ability of learning throughout the lifespan.

Mastering and improving motor control may be important in sports but seems even more relevant in rehabilitation and in clinical settings, where the prime aim is to regain motor function. For this purpose, specific motor circuits have to be restored or have to be compensated by other structures. Those rehabilitative learning processes of previously known motor actions seem to follow many of the same, but also to some extent different patterns than the learning in the healthy system. Thus, (clinical) research in this context is of great interest.

To enclose all these fascinating aspects of motor control and motor learning, the editorial team has tried to allocate the complexity of the material and the diversity of possible applications in a modular fashion. In the first module, the reader is guided through the basic theories and models of motor control and motor learning. Questions which are addressed in this part are, for instance, 'what can be learned from animal models?' or 'how can we bridge the gap between biological systems and technical applications in order to realize brain–machine interfaces, prosthesis and robots?'.

The second module concentrates on the elements of motor learning. Our knowledge about the role of feedback has increased drastically in recent years and novel experiments narrate the story of the potential possibilities implementing feedback for boosting learning and adaptation. Theories about explicit and implicit types of learning have been substantiated recently in experimental studies using functional imaging. Here the transition from explicit to implicit motor control is a potential area to study altered motor control while learning proceeds.

Neuromuscular function and adaptation of whole body movements are described in the third module. Specific emphasise is laid on locomotion, postural control and jumping movements as they are important elements of daily life and sports. Moreover, in pathological situations when locomotor and/or postural functions are compromised, basic elements in motor competence are impaired and functional limitations of daily activities are the consequence.

The next module describes the role of body representation in regard to different aspects of body awareness, analyzes the physiological aspects of movement and reports about the functional adaptations in primarily voluntarily executed muscle actions. In the last module the role of fatigue is addressed. Already at the early start of an ongoing movement, both neural and muscular systems are affected by fatigue consequently compromising motor control and motor learning. The interesting question, whether fatigue also affects motor learning is also discussed.

The book addresses people with a general understanding of neuroscience who are interested in gaining a deeper insight into integrative neurophysiology. Postgraduate and PhD students in the areas of human movement and sports sciences as well as occupational therapists, neurophysiologists and people who are involved in clinical rehabilitation are especially addressed. The authors have attempted to couple neurophysiology and biomechanics whenever and wherever possible. It was a distinct challenge to supplement behavioural and biomechanical aspects with explanations about the underlying neurophysiology in order to put neurophysiological mechanisms into a behavioral perspective.

PART I

Theories and models

1

THEORETICAL MODELS OF MOTOR CONTROL AND MOTOR LEARNING

Adrian M. Haith[1] and John W. Krakauer[2]

[1]NEUROLOGY / BIOMEDICAL ENGINEERING, JOHNS HOPKINS UNIVERSITY, BALTIMORE, USA
[2]DEPARTMENTS OF NEUROLOGY AND NEUROSCIENCE, JOHNS HOPKINS UNIVERSITY, BALTIMORE, USA

Introduction

The ease with which one is able to walk, talk, manipulate objects and play sports belies the fact that generating coordinated movement is a tremendously complex task. We have to control our bodies through a muscular system that is highly redundant, nonlinear and unreliable. Furthermore, we are reliant on sensory feedback that is also unreliable and substantially delayed. Yet many tasks that robotic systems achieve either cumbersomely or not at all are routine to us. Expert performers push the limits of performance even further. Our advantage over synthetic manipulators and – arguably – a professional sportsperson's advantage over a rookie, lies not so much in the hardware performing the task, but in the way it is controlled.

A theoretical approach to motor control and motor learning seeks to explain regularities in behavior in terms of underlying principles. This typically entails formulating mathematical models that describe the mechanics of the body or task, the way in which appropriate motor commands are selected, or the way in which prior experience influences future behavior. Many theories are mechanstic in nature – appealing to computations or plasticity occurring at the level of individual neurons or synapses in order to explain observations at the behavioral level. More abstract theories may not necessarily refer to any specific neural substrate but instead seek to explain behavior in terms of the way in which information can be represented and transformed. In both approaches, predictions about behavior stem largely from constraints imposed by the assumed circuitry or algorithm. Note that these two modelling approaches are similar to the implementational and algorithmic levels of analysis discussed by Marr (1982).

An alternative approach is to set aside questions about mechanism or algorithms and attempt to characterize and understand motor system function purely at the behavioral level. The sheer flexibility of the motor system makes it seem unlikely that underlying mechanisms place a significant constraint on the kinds of movement that can be generated. Instead, it seems that regularities in behavior are mostly dictated by ... mostly dictated by features of the task at hand rather than by features of the underlying implementational mechanism. A *normative* modeling approach seeks to explain behavior by first understanding the precise computational problem that the brain faces, and then asking what, theoretically, is the best possible way to solve it (akin to Marr's computational level of analysis). Finding solutions to such problems typically leverages ideas from control theory or machine learning. Mechanistic and normative approaches are far from mutually exclusive endeavors – breakthroughs in normative models of behavior often inspire and help guide mechanistic models. A deeper mechanistic understanding can help to constrain normative models. The normative point of view effectively assumes that the underlying neural mechanisms have omnipotent capacity. Consequently, aspects of the task itself, rather than the underlying mechanisms responsible for implementing the solution, are what primarily dictate our patterns of behavior.

In this chapter, we provide an introduction to the core concepts that underlie most recent theoretical models of motor control, state estimation and motor learning. We examine the assumptions – many of which often go unchallenged – underlying these models and discuss common pitfalls in their application. Finally, we discuss important unanswered questions and consider possible future directions for research.

Theoretical models of control

The fundamental problem the motor system faces is to decide upon appropriate motor commands to bring about a desired outcome in the environment. For example, suppose you want to move your hand to push a button to call an elevator. What makes this problem difficult is that it is not enough to simply know the location of the elevator button in space. Changing the position of the arm can only be done very indirectly by using the muscles to generate forces that cause acceleration about the joints of the arm. Thus the dynamics of our bodies place a fundamental constraint on how we are able to move. Furthermore, these dynamics are highly nonlinear – the exact same motor commands may lead to a very different acceleration depending on the state of the arm and muscles. As a result, even an apparently simple task like a point-to-point movement actually requires a complex sequence of motor commands to achieve success.

Compounding the fact that task goals are distally and nonlinearly related to motor commands, motor execution itself is highly unreliable – forces generated by a muscle are inherently variable. Though there are many potential mechanisms that contribute to this variability (Faisal et al., 2008), the net effect appears to be that force variability grows linearly with force magnitude (Jones et al., 2002) – a phenomenon known as *signal-dependent* or multiplicative noise. Managing this noise to minimize its negative impact is a major theme in models of the motor system. Since execution noise acts at every instant, its effects will, if unchecked, accumulate over the course of a movement so that even moderate variability can end up significantly interfering with task goals.

In addition to the problems of acting through potentially complex dynamics and noise, a further factor that complicates the control problem is *redundancy*. Although a movement goal might be specified by a unique point in space, there is no such unique set of controls to get there. The movement may take many potential paths through space, may take any amount of time and vary in speed during movement in infinitely many ways, all of which must be achieved using very different muscle activations. Even for a given trajectory, many different combinations of muscle activations can lead to exactly the same kinematic outcome, just with varying degrees of co-contraction. The importance of redundancy in the motor system has long been recognized. Since the time of Bernstein (1967), who dubbed the need to resolve redundancy as 'the degrees-of-freedom problem', it has been common to regard redundancy as a nuisance that the motor system has to deal with on top of all the other factors that complicate control. However, far from viewing redundancy as a problem, redundancy should actually be regarded as a positive thing. It makes it easier to find solutions to a given task and allows goals to be achieved more flexibly and robustly. Redundancy, therefore, makes life easier for the motor system to develop adequate means of control and in general enables superior control strategies. However, redundancy complicates life for the motor system in the sense that it leads to a more complex and challenging control problem if one wants to exploit it intelligently.

The considerable redundancy in a task such as a point-to-point reaching movement means that, in principle, the same task could be successfully performed by two different people in totally different ways. Yet experimental data show that point-to-point reaching movements tend to have highly consistent characteristics across individuals. For example, Morasso (1981) found that kinematics of point-to-point movements were similar across individuals as well as across different directions, different amplitudes and in different parts of one's workspace. All movements followed a more or less Cartesian straight path and showed a characteristic bell-shaped velocity profile. One possible explanation for this regularity is that it is a consequence of a particular, idiosyncratic control mechanism that is common across individuals. In other words, regularities across the population may be arbitrary and purely a consequence of a shared motor heritage. For example, some models have attempted to explain regularities in the way we move as emerging from a simplistic controller coupled to the intrinsic dynamical properties of the musculoskeletal system (Gribble et al., 1998). An alternative view, and one that is adopted by the majority of recent theories of motor control, is that we possess sophisticated controllers that select the particular movements we make because they optimize some aspect of our behavior. Regularity across individuals emerges due to common properties of the underlying task. Explaining features of behavior as being the result of an optimization process has the advantage that, in principle, a range of behaviors can be explained through a single set of principles. Exactly what aspect of behavior should be optimized in such models is difficult to say, since it is something that the motor system – or, at least, evolutionary pressures – dictates. In most cases, one assumes that the motor system aims to minimize a cost function that reflects some combination of effort, variability, or the satisfaction of task goals. As we will see, in many cases it can be shown that a rational underlying principle can offer a parsimonious explanation for observed features of movement and generate novel predictions about features that our movements should possess.

A fundamental concern with normative models of control is that it might be possible to frame any regular behavior as optimizing something. Adhering to cost functions that make some ecological sense provides some protection against this concern. However, it is impossible to say in any principled way what kinds of cost functions should and should not be allowable. And, in any case, this cannot completely avoid the possibility of inferring spurious cost functions from behavior that may not truly be optimizing anything. It is therefore important to – where possible – specify cost functions *a priori* rather than reverse-engineering a cost function based on observed behavior. That said, there is reason to believe that behavior truly does reflect a process of optimization. When subjects are asked to control an unfamiliar object with complex dynamics, each individual initially adopts an idiosyncratic way of solving the task. With extended practice, however, all subjects gradually converged on almost identical patterns of behavior (Nagengast et al., 2009). This convergence is naturally explained by the idea of an optimization process.

Optimal control

The motor system can be modeled mathematically at many different levels of detail. Most simply, one can model the end-effector as a point mass subject to accelerations in one-, two- or three-dimensional Cartesian space. A more realistic model replaces the point mass dynamics with a multi-link rigid body subject to torques around each joint. More detailed models still replace torques with the combined action of individual muscles generating forces across joints and may encompass intrinsic properties of the muscles themselves, such as the nonlinear relationship between muscle length, muscle velocity and muscle force. There is no single 'correct' level of detail to adopt. A point mass serves as an excellent model of the oculo-motor system (Robinson et al., 1986), but may be overly simplistic in other settings. A more detailed musculoskeletal model may be unnecessarily cumbersome for modeling some behaviors, but can in certain cases prove to enlighten our understanding of how a task is performed (Todorov, 2000). The appropriate level of modeling detail is largely a matter of judgment. However, for the sake of both parsimony and transparency, it is generally best to work with the simplest model possible that is able to explain a particular phenomenon of interest.

Mathematically, regardless of the level of detail employed, we can represent the state of the body at time t by the vector $\mathbf{x}_t$. This will typically contain the position and velocity of an end-effector or set of joints (e.g. the shoulder and elbow angles of the arm), but may also include things such as intrinsic states of each muscle. The motor commands themselves, which we denote by the time-varying vector $\mathbf{u}_t$, may correspond to joint torques, muscle forces or motor neuronal activity that only indirectly leads to changes in muscle force. The dynamics of the system – the way in which motor commands change the state – can be expressed in terms of a forward dynamics equation:

$$\dot{\mathbf{x}}_t = f(\mathbf{x}_t, \mathbf{u}_t). \tag{1.1}$$

This describes how changes in the state, represented as a derivative, depend in a particular way on the current state and on the current outgoing motor commands. The change in state is represented as a derivative – though note that this is the change in state of *all* components of the state. If the vector $\mathbf{x}_t$ contains position and velocity, then $\dot{\mathbf{x}}_t$ contains velocity and acceleration. This equation describes mathematically the properties of the apparatus under control.

The role of the controller is to specify the motor commands $\mathbf{u}_t$. Typically, the process of motor command selection is described mathematically in terms of a *control policy* – a mapping π from some relevant variable, such as time or state of the body, to controls $\mathbf{u}_t$. A control policy can be either purely feedforward (open-loop), in which case the control policy is a mapping from time to motor commands

$$\mathbf{u}_t = \pi(t), \tag{1.2}$$

or feedback (closed-loop), in which case the motor commands may depend also on the current state of the plant

$$\mathbf{u}_t = \pi(\mathbf{x}_t, t). \tag{1.3}$$

More generally, feedback control commands may depend directly on sensory feedback rather than on the state of the plant *per se*. The control policy, coupled with the dynamics of the motor apparatus, fully determines the course of behavior (barring the effects of execution noise). The job of the motor system, therefore, is to employ a control policy that will ensure completion of the task while minimizing some cost J that will in general depend on both the states and the outgoing motor commands.

An early theory, put forward to explain the kind of kinematic regularities observed by Morasso (1981), is that movements are selected to be as smooth as possible. Flash and Hogan (1985) suggested squared jerk – the derivative of acceleration – as an appropriate measure. If the scalar position of the effector at time t is x_t, then the minimum jerk hypothesis states that among all possible ways of moving from $x_0 = 0$ to a goal located at g in time T (i.e. x_t) = g), the best way to move is the one that minimizes a cost J that is given by the summed squared jerk (the derivative of acceleration):

$$J = \int_0^T \left| \dddot{x}_t \right|^2 dt..$$ (1.4)

In this case, it is possible to analytically find the unique best sequence of positions (and therefore unique motor commands) that minimize this cost – see Shadmehr and Wise (2005) for details. The result is a smooth trajectory that is best characterized by examining the velocity, which follows a symmetric, bell-shaped profile, matching human movement data quite well.

A more dynamical version of this idea is to replace jerk with the rate of change of torque about a joint (Uno et al., 1989):

$$J = \int_0^T \left| \dot{\tau}_t \right|^2 dt.$$ (1.5)

Note, since forces and torques are linearly related to accelerations (according to Newton's second law of motion), rate of change of torque is qualitatively similar to jerk. However, nonlinearities in the mapping from joint angles to end-effector location, and the fact that limb dynamics are different in different directions, mean that these two costs do not generate exactly the same predictions.

This pair of models (minimum jerk and minimum torque change) represents the first time that movement regularity was formally described in terms of an optimization of some feature of the movement. These two theories have been superseded by more recent frameworks and we therefore won't dwell on the pros and cons of these models. However, they do largely capture many of the important features of more contemporary models. Both minimum jerk and minimum torque change essentially penalize some kind of third derivative of position. More recent formulations mostly penalize motor commands quadratically:

$$J = \int_0^T \left| u_t \right|^2 dt.$$ (1.6)

In these models motor commands are typically assumed to influence the rate of change of muscle force (Todorov & Jordan, 2002; Diedrichsen, 2007; Izawa et al., 2008). Therefore, these models also effectively penalize the squared third derivative of position.

Why should the squared motor command matter? The quadratic form of this cost is partly chosen for mathematical convenience. Many problems turn out to be straightforward to solve, provided one assumes such a quadratic penalty on motor commands. But the main reason the use of such a cost function persists is that it has proven successful in leading to models that offer a faithful description of behavior. It is important to note that it is not actually chosen based on any principled theoretical rationale. Energy consumption, which can be quantified in terms of ATP utilization, appears to scale linearly with muscle force (Szentesi et al., 2001), not quadratically. So even if the motor commands themselves represent or are at least well-correlated with muscle force, this cost does not appear to reflect expended energy. Instead, this quadratic motor command penalty should be viewed as a more abstract notion of 'effort', which appears to be successful in describing certain aspects of behavior, despite having no clear theoretical foundation.

There is some independent empirical support for the notion that the motor system attempts to minimize some quadratic cost function. In static force generation tasks, subjects must generate a sustained force of a specific amplitude in a specific direction. These tasks are particularly useful for studying how the motor system resolves redundancy. They eliminate the dynamical complexities and kinematic redundancies associated with point-to-point movement, thereby isolating the redundancy associated with coordinating multiple muscles to generate a prescribed force. Because there are multiple muscles spanning each joint, generating a force in a particular direction – e.g. at the wrist – can be achieved through an infinite number of combinations of individual muscle activity (Hoffman & Strick, 1999). It is possible, for instance, to generate the same net force with varying degrees of co-contraction of an agonist/antagonist pair. In reality, muscles at joints with multiple degrees of freedom are not organized as simple agonist/antagonist pairs, but simply as a collection of muscles that can each generate force in different directions. At the wrist, for instance, there are five muscles that can each generate force along a single line of action. Generating a force in a direction that is not aligned with the line of action of any one muscle requires multiple muscles to be recruited. Hoffman and Strick (1999) measured the contributions of individual muscles during such a static force generation task at the wrist and found that individual muscles possessed a tuning curve that was in fact not centered on their anatomical line of action. The preferred pulling directions – i.e. the force direction for which each muscle was most active – were found to be more uniformly distributed than the anatomical pulling directions. Fagg and colleagues (2002) showed that these patterns of muscle activations can be explained through a model in which muscle forces are chosen to minimize the squared force in each muscle. This finding is not limited to the wrist; a similar result was described by Kurtzer et al. (2006) for force generation using the elbow and shoulder. In this case, the force is shared across mono-articular (spanning one joint) and bi-articular (spanning two joints) muscles of the arm. Alternative versions of this cost with exponents other than two can be considered (e.g. penalizing $|u|^3$). Powers below two predict broader tuning with greater sharing of load across muscles, while higher powers predict recruitment of fewer muscles where possible. A quadratic cost appears to describe data about as well as any alternative. In practice, however, dissociating these variants on the notion of an effort cost relies on quantitatively precise predictions that may be more sensitive to other parameters of the model than those in the cost function (Kurtzer et al., 2006).

Minimum endpoint variance

In the above examples, movements are considered to be deterministic – i.e. there is no noise associated with executing a movement. However, movements are inherently variable. Importantly, the variability introduced during a movement appears to grow linearly with the size of the motor commands (Jones et al., 2002; Faisal et al., 2008). This leads to faster movements being more variable than slower movements. Harris and Wolpert (1998) proposed that a natural cost on candidate control policies is the endpoint variability. Endpoint variability reflects an accumulation of instantaneous signal-dependent noise that is introduced throughout the course of the movement. The endpoint variance can be expressed as a weighted sum of squared motor commands (Shadmehr et al., 2010). This quadratic structure is similar to the effort costs considered above, and solution can be found through similar methods. Applying this principle to model fast point-to-point movements of the eye (saccades) leads to saccade velocity profiles that are remarkably similar to those measured experimentally.

Effort versus variability

The fact that endpoint variance is a quadratic function of motor commands means that minimizing endpoint variance can be considered equivalent to minimizing a time-varying cost on squared motor commands, i.e.

$$J = \int_{0}^{T} w(t)u_t^2\, dt, \tag{1.7}$$

where $w(t)$, illustrated in Figure 1.1c, is a time-varying cost weight that depends on the dynamics of the eye. Motor commands early in the movement are penalized far less heavily than motor commands late in the movement. The reason for this is that the elasticity of the eye acts to dampen out variability over time, making it advantageous to introduce noise earlier in the movement than later on. The trajectories of saccades that minimize endpoint variability have a velocity profile that is skewed towards the beginning of the movement (Figure 1.1a), when the effective cost of motor commands is relatively cheap. The time-varying cost implicit in the minimum endpoint framework can be contrasted with an effort cost that is uniform in

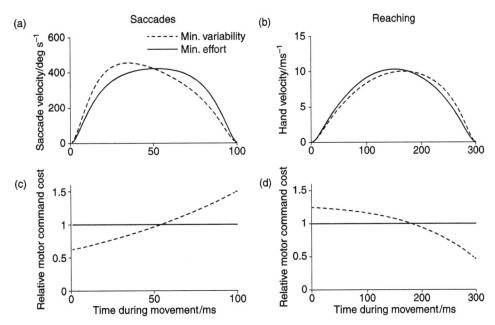

Figure 1.1 Comparison of effort and endpoint variability costs for open-loop arm and eye movements. (a) velocity profile for a 30 degree saccade lasting 100ms, as predicted by effort (solid line) and endpoint variability (dashed line) costs; (c) relative weight of quadratic motor command cost as a function of time within movement; (b) and (d) as (a) and (c) but for a point-to-point reaching movement of 10cm lasting 300ms.

time. The saccades predicted by a uniform cost have symmetric velocity profiles (Figure 1.1a) and do not closely resemble the kinematics of real saccades. The principle of minimizing endpoint variance can also be applied to reaching movements (Figure 1.1b, 1.1d), in which case it also leads to reasonable predictions about hand trajectories during point-to-point reaching movements. Exactly as in the case of saccades, a minimum effort cost leads to symmetric velocity profiles. For minimizing endpoint, the effective time-varying cost is greater for motor commands later in the movement resulting in a velocity profile that is slightly skewed towards being faster later in the movement – opposite from the result for saccades. The key difference between oculomotor control and control of the arm is that the arm does not have a natural equilibrium point in the same way that the eye does. Noise does not, therefore, dissipate in the same way that it does for the eye. The effective cost of motor commands in fact decreases towards the end of the movement because noise introduced by motor commands late in the movement does not have time to significantly affect final position. It should also be noted that the dynamics of the arm are less well specified than those of the eye. Different simulations use a wide range of values for inertia and viscosity of the arm – compare e.g. Harris and Wolpert (1998), Liu and Todorov (2007) – which lead to a fairly broad range of behaviors. We used an intermediate set of parameters to illustrate some of the general characteristics of the predictions of these models (1kg mass, 2Nm/s viscosity, 5ms muscle time constant).

Based on these results, one might be tempted to conclude that noise is in fact the thing of primary interest to the motor system. The effort penalty in models which assume deterministic dynamics (or at least non-signal-dependent noise) might be deemed simply as providing a good approximation to the true cost relating to variability. The main difficulty in validating this theory more rigorously is that it is difficult to directly measure the noise associated with a particular motor command at a given instant in time. The best one can do is to examine patterns of variability and show that they are consistent with a particular signal-dependent noise model coupled with an associated optimal controller (van Beers et al., 2004; van Beers, 2007). It is feasible, however, to measure variability in a static force generation task where motor commands are constant. O'Sullivan et al. (2009) measured variability of forces of various magnitude generated by different digits of the left and right hands. They fitted a model in which force and variability were linearly related, and used this model to predict behavior in a second task in which two fingers, one from each hand, cooperated to achieve a desired force goal. Since the relationship between force and variability was different for each finger, this led to a rich set of predictions for how the force should be divided among varying combinations of fingers. Minimizing the variability of the net force did not turn out to predict behavior well. Rather, in order to completely describe the data, it was necessary to invoke an effort cost penalizing the squared force generated by each finger.

The similar mathematical forms of effort and variability make it difficult to disambiguate one from the other. However, it seems unlikely that variability is the sole factor determining how we move. In addition to the results of O'Sullivan et al., it has been suggested that effort may play an important role in deciding how fast to move to acquire rewards of different magnitude (Shadmehr et al., 2010). Perceived effort leads to robust decisions even in a task in which variability plays little role (Körding et al., 2004). Perceived effort also appears to be awry in patients with Parkinson's disease (Mazzoni et al., 2007).

Feedback control

All of the theories presented so far consider only feedforward control policies, in which a pre-determined sequence of commands is rolled out at the time of execution. In reality, however, we are not restricted to being passive spectators to the effects of execution noise or external perturbations. For all but the fastest of movements (such as saccades) we are able to observe deviations from expected behavior and make online adjustments during a movement. A simple feedback control policy can be formed by augmenting a feedforward control policy with a feedback component that tries to cancel out the effects of noise and external perturbations by keeping the movement of the plant close to some planned trajectory. While this kind of strategy will indeed help to negate the impact of noise, it is highly inflexible, attempting to rapidly correct any deviation from desired behavior. These corrections can be quite costly and in many cases largely unnecessary. Better strategies are possible that do not rigidly adhere to a single trajectory, but allow more flexibility in the way errors are corrected.

The key difference between feedforward and feedback control is that a feedback control policy selects motor commands as a function not just of time, but also of the current state:

$$\mathbf{u}_t = \pi(\mathbf{x}_t, t). \tag{1.8}$$

Feedback control thus relies on knowledge of the instantaneous state of the motor apparatus. While this is somewhat unrealistic, given the noisy and delayed nature of available sensory feedback, it serves as a reasonable simplifying approximation to explore the most salient aspects of feedback control. More detailed models replace the exact state of the system $\mathbf{x}_t$ with an estimate of the state $\hat{\mathbf{x}}_t$ that must be generated based on sensory feedback. For now, however, we make the simplifying assumption that if feedback is available at all, the state of the system is known precisely and instantaneously. We focus instead on the question of how this knowledge about the state of the system should be used.

Just as in the feedforward case, the controller coupled with the dynamics of the motor apparatus completely determines the behavior of the system. Optimal feedback control theory addresses the question of what feedback control policy should be used in the same way feedforward models do, by associating a cost with each potential control policy based on the resulting behavior, and identifying the control policy that leads to minimal cost.

Determining the cost associated with each policy and finding the best possible is much harder for feedback control policies, since each policy does not specify motor commands directly, but instead specifies a rule for determining motor commands based on the current state. Fortunately, the general mathematical problem has been extensively studied in control theory (Bertsekas, 1995) and reinforcement learning theory (Sutton & Barto, 1998), and requisite extensions for dealing with features peculiar to the motor system, such as signal-dependent noise, have been developed (Todorov, 2005). The methods used to find the optimal feedback control policies are beyond the scope of this chapter, but a thorough introduction can be found in (Todorov, 2006).

An intuitive way to understand optimal feedback control without delving into technical details is as a feedforward control policy that is continuously re-planned based on the latest state information. For example, if the hand is perturbed from its initial path, the best thing to do is forget about the original intended movement and plan again from scratch given the new, perturbed state as a starting point. This picture provides a reasonable way to intuit the properties of optimal feedback control policies with a few caveats. Firstly, movements are not actually re-planned. The optimal feedback control policy (Equation 1.8) implicitly encodes the optimal course of action starting from any state at any time. Secondly, there is no need to wait for a large perturbation to prompt an adjustment of one's movement. Even small deviations from expected trajectories should prompt a flexible change in motor commands. Thirdly, although in simple cases there is a precise equivalence between optimal feedback control and the idea of continually re-planning an open-loop controller, they do not always lead to the same predicted behavior. Knowing that one will be able to make feedback corrections in the future can influence one's control strategy in the present. For instance, acting under signal-dependent noise, it can be beneficial to introduce noise early in a movement in the knowledge that one will have time to correct it later on. Such 'strategies' emerge naturally within the theoretically optimal policy.

Minimum intervention principle

The critical difference at the behavioral level between optimal control policies and other potential feedback control policies is the way in which the optimal feedback control policy flexibly exploits redundancy in order to minimize costs. This is perhaps best understood through the notion of the minimum-intervention principle: an optimal control policy should only correct perturbations that interfere with the achievement of task goals. If a perturbation is irrelevant to a task, for instance, if your elbow is knocked during a reaching movement without affecting your hand position, then there is no need to correct for it – just maintain the new elbow posture during the rest of the movement. The same applies to deviations occurring because of noise. Not making unnecessary corrections allows one to be more sparing with motor commands, which helps reduce both effort costs and the impact of signal-dependent noise. This gives rise to the more general prediction that movement variability will be greatest far away from task-critical periods of a movement. This is true in the case of point-to-point reaching movements, where variability is highest midway through the move-ment (Liu & Todorov, 2007) and after striking a target object such as when striking a ping-pong ball with a bat (Todorov & Jordan, 2002).

While any motor task naturally contains redundancies, the characteristics of optimal feed-back control policies are most striking in tasks where redundancy is exaggerated. Diedrichsen (2007) demonstrated this in a bimanual task in which subjects controlled a cursor that appeared at the average location of their two hands. Subjects had to move the cursor from an initial starting location (with specific initial positions for each hand) to a goal location. On select trials, one hand experienced a force that perturbed it perpendicular to its movement direction, causing a corresponding perturbation to the cursor. According to a desired trajec-tory hypothesis, subjects should have corrected for this error by returning the right hand back towards its original path. If, however, subjects performed the task according to an optimal control policy, they should exploit the redundancy afforded by the bimanual nature of the task and recruit the other hand to help steer the cursor towards the target. This is precisely what subjects were found to do.

Taking the idea of a redundant task to extreme, Todorov and Jordan (2002) recorded movements of the fingers while subjects scrunched a piece of paper into a ball using one hand. Subjects do not adopt a fixed method, but show huge variability from trial to trial in how they manipulate the object to achieve their goal. Todorov and Jordan interpreted this variability as evidence that subjects were implementing a flexible feedback control policy of the kind predicted by optimal feedback control. In this case, the variability of actions stemmed directly from the variability in the state of the paper in the hand. Although it is difficult to prove that in this case the high variability had anything to do with performing the task well, this experi-ment serves well as an intuitive example to illustrate the fact that acting flexibly in order to be effective – the key principle underlying optimal feedback control – is something that comes very naturally to us.

Scope and limitations of optimal control models

The basic idea behind optimal control theory is actually a fairly simple one – among the many ways one could accomplish a task, pick the one which is the best according to some cost measure. The power of optimal control theory lies in its ability to explain a wide range of behaviors through a small set of widely applicable cost functions. Whereas early motor control research focused on identifying invariants in kinematics, optimal control focuses on identifying invariants at the level of cost functions. Obtaining the 'best' control policy is often the most technically challenging aspect of applying optimal control theory to model motor behavior. In many ways, however, it is the easy part, since the problems one encounters have unambiguous solutions. For a given dynamics model (Equation 1.1) and cost function, various methods exist – albeit in some cases computationally demanding ones – for obtaining the optimal control policy. The real challenges in testing optimal control as a hypothesis about motor control are that it is very difficult to know the dynamics of the motor apparatus precisely and one doesn't know exactly what kind of costs the motor system uses to evaluate one movement relative to another. Different assumptions about the dynamics of the motor apparatus and different kinds of cost functions can lead to very different control policies. If predictions of a model fail to match those of data, there is no way of knowing if one used a poor dynamics model or the wrong cost function, or whether the motor system simply doesn't behave in a way consistent with the premise of optimal control. Conversely, and much more problematically, the optimal control framework is sufficiently flexible that it is often possible to find some combination of dynamics and cost function that will give rise to any given behavior. Suppose that subjects are seen to always accomplish a particular task by moving in a straight line. This phenomenon can be easily 'modeled' by imposing a cost function that penalizes deviations away from a straight line. This does not, of course, constitute a theory about why subjects move in a straight line, nor does it provide any support for the premise of optimality. In general, optimal control theories of the motor system are most compelling when using cost functions decided *a priori* and generate qualitative predictions about behavior, rather than when they are used to provide quantitatively precise descriptions of data.

Optimal control theory has proven a valuable modeling tool that has both explained previously characterized aspects of motor behavior and generated a host of novel predictions that have been demonstrated experimentally. However, not all behavior is consistent with optimal control predictions. For instance, human subjects seem to have a preference for reaching in straight lines, even in the presence of perturbations that make it sub-optimal to do so (Wolpert et al., 1995). It is difficult to explain this tendency in terms of some underlying normative principle. In reality, it is likely that relies heavily on approximations and heuristics in order to achieve behavior that is near-optimal in a variety of contexts. Outside of these contexts, the particular control strategies that are habitually used may no longer be optimal. This appears to be true of the way in which multiple muscles are coordinated across a joint (de Rugy et al., 2012) and a similar idea may explain the tendency for people to want to reach in straight lines.

State estimation: Making sense of sensory feedback

Any movement we make necessarily depends on incoming sensory information. Knowing exactly what action to take will depend on sensory input to specify the goals of the task and the initial state of the body. Once a movement is underway, feedback control requires ongoing monitoring of the state of the body through vision and proprioception. Individual sensory modalities provide imperfect information. Although we may possess high visual acuity, using visual information to specify the location of a point in space additionally requires knowledge of the orientation of the eyes within the head and of the head relative to the body, all of which introduce noise and uncertainty. Proprioceptive estimation of hand location is notoriously imprecise – try aligning your two index fingers above and below an opaque tabletop. Given the noisy nature of sensory feedback, two modalities may often report conflicting estimates of the same thing. How should one estimate the position of one's hand in order to make decisions about movement?

This general problem can be formalized by supposing that our senses yield noisy readouts of the true state of our arm. If the true location of the hand is x, then we can model vision and proprioception as providing independent, unbiased observations of that location, $v \sim N(x, \sigma_v^2)$ and $p \sim N(x, \sigma_p^2)$. A principled way to estimate the hand position x, given these observations, is to ask what hand location is most likely to have led to these observations. This approach is known as Maximum Likelihood Estimation (MLE). The maximum likelihood estimate of hand position can be shown to be equal to a linear combination of the two unimodal observations, with each observation weighted inversely to its variability:

$$\hat{x}_{MLE} = \frac{\sigma_v^{-2}}{\sigma_v^{-2} + \sigma_p^{-2}} v + \frac{\sigma_p^{-2}}{\sigma_v^{-2} + \sigma_p^{-2}} p. \tag{1.9}$$

Thus, the combined estimate should lie somewhere in between the two unimodal estimates. This principle can be applied directly to any pair of modalities, not just vision and proprioception, and can be generalized to incorporate any number of modalities. Maximum Likelihood Estimation provides a normative rationale for how to optimally combine two sensory observations into a single estimate. Individuals' sensory integration strategies can be tested by providing subjects with conflicting cues about a single object and asking them to judge the location of the common cause, provided the cues are not too discrepant (Körding et al., 2007). In order to test whether people use an optimal sensory integration strategy, it is not enough to simply observe that a subject's combined estimate lies somewhere in between the two uni-modal estimates. Maximum likelihood integration provides a quantitatively precise prediction about where the integrated observation should lie. In order to properly test this hypothesis, one needs to measure the reliability (i.e. the variance) associated with each individual sensory modality, yielding a quantitative prediction that can be compared with subject behavior. This approach has been successfully applied in a number of contexts, including integrating visual and haptic estimates of the size of objects (Ernst & Banks, 2002).

In reality, we are rarely interested in one-dimensional quantities. Position, for instance, is three-dimensional. Extending the MLE framework to higher dimensions actually leads to richer predictions about how information should be combined across modalities than in the one-dimensional case. In higher dimensions, each observation is a vector and the variances associated with each modality become covariance matrices: $\mathbf{y}_i \sim N(\mathbf{x}, \Sigma_i)$. The covariance matrix encodes the relative confidence in a given modality along different dimensions. If two different modalities have different covariance matrices, i.e. if they have different relative reliabilities along different dimensions in space, then combining these observations yields a MLE that, perhaps counter-intuitively, does not lie along the straight line between the individual uni-modal estimates (consider performing independent one-dimensional integrations along each dimension to see how this can be the case – see Figure 1.2b). This qualitative behavior is seen when subjects are asked to integrate visual and proprioceptive estimates of hand location in two dimensions (van Beers et al., 1999).

A similar kind of integration also occurs when estimating the state of the limb for feedback control. In this case, rather than combining two sensory modalities, one combines sensory

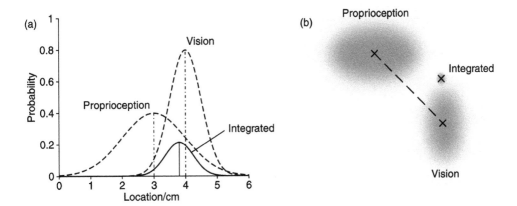

Figure 1.2 Maximum Likelihood Estimation of hand position from visual and proprioceptive estimates. (a) One-dimensional integration of discrepant visual and proprioceptive observations of hand position. Dashed lines indicate single-modality likelihoods—i.e. probability
of hand location given observation. Solid line indicates joint likelihood integrating both
visual and proprioceptive observations. Note that the maximum likelihood estimate is
closer to the visual observation due the visual observation having lower variability.
(b) Integrating two sensory estimates in two dimensions. Anisotropic uncertainties lead to
an integrated estimate that does not lie on a straight line between individual observations.

feedback with an internally-generated predication about the new state of the limb based on
previously issued motor commands. The Kalman filter, a standard approach to atate estimation
in optimal control models, is essentially an iterative version of this idea.

Box 1.1: Forward and inverse models

In addition to using sensory information to estimate the location of our limbs, we can also use
prediction based on known outgoing motor commands. A *forward model* predicts the consequences of outgoing motor commands on the state of our bodies. For example, maintaining an
internal representation of the dynamics of the body given by Equation 1.1 would constitute an
internal forward model. Forward models allow us to overcome the delays inherent in sensory
feedback, enabling early corrections for aberrant outgoing motor commands or for anticipating
unwanted consequences of actions in order to prevent them from occurring.

    A forward model is truly a model in the sense that it is an internal representation of a process
that occurs extrinsically. Historically, motor learning theory has also extensively appealed to the
notion of an inverse model that maps desired changes of state to motor commands. Inverse
models are not really models in the true sense of the word and, in any case, have been superseded
by the more general notion of a control policy.

Motor learning

Optimal control theory, coupled with state estimation, provides a powerful framework for
describing coordination in over-learned tasks in which we are already experts. However,
most movements require extensive practice before we consider ourselves to have reached any

kind of proficiency. The capacity to learn new patterns of movement and improve and adapt existing ones is arguably the most fundamentally important facet of the human motor system. The vast majority of research in motor learning studies this capacity through adaptation paradigms in which a systematic perturbation is introduced to disrupt a well-practiced behavior, such as point-to-point reaching. The imposed perturbation generally causes significant errors at first, however, subject' performance generally returns to near baseline levels within tens of trials. A classic example is adaptation of reaching movements while wearing prism goggles that shift the visual field to the left or right (von Helmholtz, 1962). Other common examples include rotations of visual feedback on a virtual display (Krakauer et al., 2000) or force fields that perturb the dynamics of the arm (Shadmehr & Mussa-Ivaldi, 1994).

In all cases, the course of learning is qualitatively similar. Error measures, such as directional error, tend to decline approximately exponentially across trials. This exponential decline is consistent with the idea that learning between trials is proportional to error size. Formalizing this insight mathematically leads to a so-called 'state-space model' of motor learning (Thoroughman & Shadmehr, 2000; Donchin et al., 2003; Cheng & Sabes, 2006), which we illustrate through the example of adapting reaching movements under a rotation of visual feedback. Suppose that, on a given trial a subject aims their reach in a direction u_i towards a goal located at 0 degrees. In this case, error can be quantified as the directional error at the start of the movement. The error subjects experience on a given trial will be determined by a combination of the subject's chosen action and the perturbation – in this case, a rotation of visual feedback by r degrees:

$$e = u + r. \tag{1.10}$$

The ideal action $u*$ is the one that generates zero error:

$$u* = -r. \tag{1.11}$$

The state-space model assumes that, if an error e_i is experienced on trial i, subjects will adapt their reach angle on the next trial by a proportional amount, i.e.:

$$u_{i+1} = Au_i + \alpha e_i. \tag{1.12}$$

Here α is the learning rate determining the sensitivity to error, while A, which is some number close to but less than 1, captures the natural tendency to return to baseline behavior in the absence of errors (Galea et al., 2011) – often described as 'forgetting'. This summarizes the basic premise of a state-space model. The 'state-space' nomenclature arises from parallels with dynamical systems theory – the trial-to-trial reach errors can be viewed as the output of a simple discrete-time linear dynamical system. This is not, however, a particularly helpful analogy unless one happens to be already familiar with such frameworks. Instead, it is best to think about these models as simply mathematically capturing the idea that changes in behavior are proportional to error size. This example is perhaps the simplest possible such model. Alternative forms of learning, such as adaptation to dynamical perturbations, can be modeled in an identical way with the caveat that one requires an additional parameter to translate between a quantitative measure of the perturbation (typically a force, or viscosity) and a kinematic measure of error (typically a distance or angle) (Donchin et al., 2003; Thoroughman & Shadmehr, 2000).

An important feature of this model is the so-called 'forgetting rate' A in equation 1.12. The existence of this term predicts that, in the absence of observed error, motor commands will gradually return to their original baseline values. For visual perturbations errors can be removed simply by removing visual feedback, in which case behavior does indeed return to baseline. The fact that there is always some forgetting between trials also means that learning can never completely reduce error to zero. Subjects reach an adaptation asymptote at a particular error where there is an equilibrium between trial-to-trial forgetting and the amount of learning due to error. This incomplete adaptation is precisely what is observed experimentally. Mathematically, the asymptotic behavior u_∞ is given by

$$u_\infty = \frac{-\alpha r}{\alpha - (1 - A)}. \tag{1.13}$$

Note that, as A gets closer to 1, u_∞ approaches the 'correct' value of $-r$. The asymptotic error also grows with the size of a learned rotation, suggesting that incomplete adaptation is not simply a reflection of adapting just enough to reach the outside of a target.

Various extensions of this basic model are possible. Most commonly, the basic SSM presented here can be applied in the context of multiple targets. In this case, errors at a given target can also influence learning at neighboring targets (Donchin et al., 2003; Thoroughman & Shadmehr, 2000). Despite the apparently simple assumptions, state-space models describe adaptation behavior remarkably well, even when the imposed perturbation varies randomly from trial to trial (Donchin et al., 2003; Huang et al., 2011).

The success of state-space models in accounting for trial-to-trial behavior in adaptation paradigms tells us little about the underlying mechanisms mediating that behavior. State-space models are thus best viewed as purely phenomenological descriptions of trial-to-trial learning behavior in adaptation paradigms. Nevertheless, state-space models are variously couched in terms of either learning a forward model of the limb (along with any potential external perturbations) or simply directly learning appropriate motor commands. It is difficult to dissociate these interpretations based on trial-to-trial adaptation data, since they lead to identical sets of equations. One important difference, however, is that errors in learning a forward model are prediction errors, whereas errors in learning actions are task errors. In most circumstances, task errors and prediction errors are identical. In certain circumstances they can be dissociated (Mazzoni & Krakauer, 2006; Wong & Shelhamer, 2011), in which case it becomes clear that the motor system is sensitive to *prediction errors*. These findings support the idea that motor learning proceeds by updating an internal forward model that is then subsequently used to guide planning of future movements.

Regardless of the underlying representation, a state-space model offers a principled way to estimate trial-to-trial learning rates. A model that fits the data well can be thought of as a means to compactly summarize the salient features of the data. Comparing parameter fits across conditions then enables one to make inferences about the effect of manipulations on learning rate (Diedrichsen et al., 2005).

Closer inspection of trial-to-trial learning curves reveals that there are typically two phases of learning. An initial, rapid decrease in error is typically by a more prolonged, gradual reduction in error towards asymptote. Consequently, learning curves are far better fit by a sum of two exponentials – one with a fast time constant and one with a slower time constant. This two-rate behavior can naturally be accommodated within the state-space model framework by allowing motor commands to be comprised of two distinct components that learn from the same error, one quickly and one slowly. In addition, the component that learns quickly from error also forgets quickly (i.e. $A \ll 1$), while the slow learning component forgets slowly. Although this two-rate model is motivated by describing data rather than by theory, this model gives rise to an interesting prediction: if subjects are rapidly adapted and then de-adapted, subjects will 'spontaneously recover' previous learning, as the fast component is forgotten to reveal the slow component that still partially reflects the previous learning. This spontaneous recovery has been demonstrated to occur exactly as predicted for both force field learning (Smith et al., 2006) and saccadic gain adaptation (Ethier et al., 2008). Furthermore, components identified through such a model are predictive of long-term retention (Joiner & Smith, 2008) and anterograde interference (Sing & Smith, 2010).

Beyond state-space models: Learning as estimation

Ultimately, working with purely descriptive models is unsatisfying. We really would like unifying concepts to explain why learning should be proportional to error, not just keep a close track of the behavioral consequences of this kind of assumption. The sensitivity to error is also an arbitrary parameter. What determines how fast one learns? Is learning constrained by the mechanics of synaptic plasticity, or are features of the task more important?

Bayes' Rule: Existing beliefs about some variable x in the environment can be expressed as a probability distribution over possible values $p(x)$, with greater probabilities indicating a stronger belief. This distribution is referred to as the prior distribution. A new observation y may prompt us to revise our beliefs about x. If y depends on the true value of x through a probability distribution $p(y|x)$ (the likelihood of x), then Bayes' rule tells us precisely how to revise our existing beliefs in the light of the new evidence:

$$p(x|y) = \frac{p(y|x)p(x)}{p(y)},$$

where $p(x|y)$ (the posterior distribution) represents our revised belief about x in the light of y. $p(y)$ is the overall probability of our observation for any possible underlying x – otherwise known as the marginal probability of y. $p(y)$ is simply a constant if one views $p(x|y)$ as a function of x and serves to normalize the posterior distribution, ensuring that all beliefs sum to 1. In many cases, one will see Bayes' rule written simply as

$$p(x|y) \propto p(y|x)p(x).$$

A critical idea in models of motor learning is that motor learning in adaptation paradigms amounts to estimating the properties of the perturbation. We have seen how a forward model of the dynamics of the motor apparatus can be useful for estimating the state of the limb. In principle, the same forward model can also be used to plan movement. Motor learning research in recent decades has been dominated by this idea.

Korenberg and Ghahramani (2002) pointed out that changing behavior proportionally to the errors one experiences is consistent with the idea of updating one's beliefs about the perturbation from trial to trial in a statistically optimal manner. A basic version of the idea of iteratively revising one's beliefs in this way leads to the commonly employed Kalman filter, which yields updates of one's mean belief that are proportional to prediction errors – precisely as stated by the state-space model framework. The critical difference between this idea and the state-space model formulation is that, in the Kalman filter equations, the learning rate, i.e. the amount one revises one's beliefs in the light of new observations, is not an arbitrary parameter of the model but emerges as a consequence of the statistics of the task. If observations are typically unreliable and the perturbation is expected to remain quite stable over trials, one should be cautious about revising beliefs based on a single large prediction error. On the other hand, if observations are typically reliable and the underlying perturbation is liable to rapid fluctuations, one should trust one's immediate observations more than past experience and employ a high learning rate.

In order to frame this problem mathematically, we require a more precise statement of the variability or confidence associated with observations and of the expected variability in the perturbation over time. Formally, such a model is described as a generative model, since in principle it can be used to generate a realistic sequence of observations. Figure 1.3 illustrates this

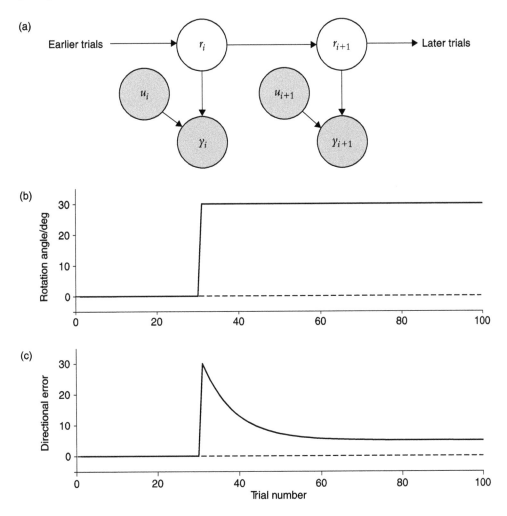

Figure 1.3 Adaptation as Bayesian estimation. (a) Subject's beliefs about the task represented as a graphical model. Shaded circles represent observed variables. Unshaded circles represent unobserved variables to be inferred. Arrows indicate probabilistic dependencies between variables. A rotation r_i is assumed to vary from trial to trial (arrows indicate) following a Gaussian random walk. Each trial yields an observation y that derives from the perturbation and the motor command u. The subject's goal is to infer the rotation on each trial and use this to predict rotations on future trials. (b) and (c) Illustration of the time-course of learning in response to a step-perturbation predicted by the Kalman filter model—in this case also equivalent to a state-space model. Note that asymptotic error is not equal to zero.

model graphically. There are two components that need to be specified. Firstly, there must be a model of how a given perturbation gives rise to observations. In specific cases, this can be as simple as saying the observation is a noisy, unbiased version of the perturbation, i.e. $y \sim N(x, \sigma^2)$, though in general there could be a more complex relationship between the two. Secondly, we require a model of how the perturbation is liable to vary over time. Typically, one assumes that the values of a perturbation follow a random walk over trials, i.e.

$$x_{i+1} \sim N(Ax_i, q^2),$$ (1.14)

where $A < 1$ is some 'forgetting factor', exactly like the forgetting factor in state-space models. In this context, however, A reflects a property of the perturbation itself, rather than the learner's capacity to retain this information. Here, it additionally has the effect of making sure that the perturbation is assumed to not stray too far from zero – the variance remains bounded.

Qualitatively, this model predicts that one should adjust future movements by a fixed proportion of error size – basically recovering the premise of the state-space model from a theoretical grounding. The Bayesian viewpoint is, however, much deeper in its implications than a state-space model. In particular, the Bayesian point of view implies that the learning rate is dictated by particular aspects of the task – the observation noise σ^2 and the perturbation volatility q^2. In practice, estimating these variables is difficult since they reflect implicit beliefs of the subject. In particular, q^2 corresponds to the subject's own estimate of how variable a particular perturbation is from trial to trial. Sensory noise, however, is more amenable to experimental quantification. Burge and colleagues (2008) had subjects adapt to a shift in visual feedback while making point-to-point reaching movements. The cursor representing subjects' hand position was a small circle that was blurred to varying degrees. Consistent with the qualitative Bayesian prediction, the more blurry cursor, which presumably increased observation noise, led to a significant decrease in the rate of learning. This change in adaptation rate was even direction specific – blurring the cursor only along the x-axis led to slower adaptation to shifts in this direction, but not a simultaneously imposed vertical shift. Similar findings have also been reported elsewhere (Wei & Körding, 2010).

The key predictions of the Bayesian interpretation of motor learning are that, 1) as uncertainty in sensory feedback increases, learning should become slower. This is directly analogous to the fact that sensory modalities that are uncertain have less influence over the maximum likelihood estimate of state (Fig. 1.3a). 2) As uncertainty in the perturbation increases, learning should become faster. While reducing certainty in sensory feedback appears to reliably influence learning in the way predicted by the Bayesian framework, manipulating certainty in the perturbation is much harder to achieve. One method that has showed partial success is to simply leave subjects sitting idly in the dark. A period of such inactivity leads to a brief increase in learning rates (Wei & Körding, 2010). A similar increase in learning rates is also observed in monkeys during saccade adaptation paradigms, if the monkey is left in the dark for a period of time (Kojima et al., 2004). In both cases, this faster re-learning has been construed as reflecting an increase in uncertainty of an internal model that guides movement planning (Wei & Körding, 2010; Körding et al., 2007). In the case of manipulating uncertainty in visual feedback, it is possible to empirically determine the extent to which a given manipulation of visual feedback affects confidence in visual observations – through the just-noticeable difference, for instance. Unfortunately, there is no analogy of this approach for measuring a subject's confidence. In their estimate, rather than try to measure confidence in the state mapping, one can try to manipulate it. However, attempts to decrease certainty in the value of a perturbation – usually by having the perturbation itself follow a random walk with varying degrees of volatility – have yielded little success. It is impossible to know whether this is due to a failure on the subject's part to take known variability in the perturbation into account during learning or due to a failure to learn about the variability of the perturbation. In any case, it appears that the motor system is equipped to respond to some forms of variability better than others.

The basic Kalman filter model of adaptation, that we have outlined here, can be extended to include multiple underlying causes for each potential error, such as world versus body (Berniker & Körding, 2008) or deciding whether a given error has been caused by motor execution or sensory miscalibration (Haith et al., 2008). In general, however, such models display a qualitative consistency with patterns of behavior, but are not typically subject to the level of scrutiny that is found in testing models of statistically optimal multi-sensory integration. The Bayesian framework does, however, provide a rich and rigorous framework within which to frame theories of adaptation and generate new hypotheses.

Issues and outlook

We have introduced the main theoretical concepts underlying many recent models of motor control and motor learning. These models, which for the most part adopt a normative approach, have enjoyed considerable success in recent years. Though we have concentrated on theory and have not discussed the underlying neural mechanisms (either at the systems level or neuronal level), bridging the gap between normative and mechanistic models is an important direction for future research. It appears that the cerebellum is of critical importance in motor learning (Bastian, 2006; Tseng et al., 2007; Taylor et al., 2010), likely subserving adaptation through an internal forward model that is updated by sensory prediction errors. The cerebellum also may contribute a forward model that mediates a state estimate used in control (Xu-Wilson et al., 2009). The control policy itself, however, seems likely to reside in primary motor cortex.

While control and learning have been extensively examined individually, surprisingly little work has addressed the intersection of the two. Multiple studies have suggested that control policies are re-optimized after learning novel dynamics (Izawa et al., 2008; Nagengast et al., 2009). What is unclear at present, however, is how knowledge of dynamics stored in a forward model, presumably in the cerebellum, becomes translated into a control policy in motor cortex. An alternative way in which a control policy could be learned or adjusted is through reinforcement of actions that lead to success. Indeed, it appears that such model-free learning is responsible for savings that occur during adaptation paradigms (Huang et al., 2011) and may be responsible for learning in more complex motor tasks (Hosp et al., 2011). Such forms of learning can also in principle be characterized from a normative point of view.

While normative models offer a potentially powerful lens through which to examine the motor system, it is important to bear in mind that the omnipotent brain is only a simplifying approximation. There is unlikely to be any single principle that will account for all behavior. If anything, such models serve to highlight instances where behavior of the motor system deviates from supposed optimal behavior. Ultimately, the true utility of a normative model is not to provide an overarching theory of everything akin to a fundamental law of physics, but to constrain, inform and inspire new ways of thinking about the motor system across multiple levels of analysis.

References

Bastian, A.J. (2006) 'Learning to predict the future: the cerebellum adapts feedforward movement control', *Current Opinion in Neurobiology*, 16(6): 645–9.

Berniker, M. and Kording, K. (2008) 'Estimating the sources of motor errors for adaptation and generalization', *Nature Neuroscience*, 11(12): 1454–61.

Bernstein, N. (1967) *The Co-ordination and Regulation of Movements*, Pergamon Press, New York.

Bertsekas, D.P. (1995) *Dynamic Programming and Optimal Control*, Athena Scientific, Belmont, MA.

Burge, J., Ernst, M.O. and Banks, M.S. (2008) 'The statistical determinants of adaptation rate in human reaching', *Journal of Vision*, 8(4): 20, 1–19.

Cheng, S. and Sabes, P.N. (2006) 'Modeling sensorimotor learning with linear dynamical systems', *Neural Computation*, 18(4): 760–93.

Diedrichsen, J. (2007) 'Optimal task-dependent changes of bimanual feedback control and adaptation', *Current Biology* 17(19): 1675–9.

Diedrichsen, J., Hashambhoy, Y., Rane, T. and Shadmehr, R. (2005) 'Neural correlates of reach errors', *The Journal of Neuroscience*, 25(43): 9919–31.

Donchin, O., Francis, J.T. and Shadmehr, R. (2003) 'Quantifying generalization from trial-by-trial behavior of adaptive systems that learn with basis functions: theory and experiments in human motor control', *The Journal of Neuroscience*, 23(27): 9032–45.

Ernst, M.O. and Banks, M.S. (2002) 'Humans integrate visual and haptic information in a statistically optimal fashion', *Nature*, 415(6870): 429–33.

Ethier, V., Zee, D.S. and Shadmehr, R. (2008) 'Spontaneous recovery of motor memory during saccade adaptation', *Journal of Neurophysiology*, 99(5): 2577–83.

Fagg, A.H., Shah, A. and Barto, A.G. (2002) 'A computational model of muscle recruitment for wrist movements', *Journal of Neurophysiology*, 88(6): 3348–58.

Faisal, A.A., Selen, L.P.J. and Wolpert, D.M. (2008) 'Noise in the nervous system', *Nature Reviews Neuroscience*, 9(4): 292–303.

Flash, T. and Hogan, N. (1985) 'The coordination of arm movements: an experimentally confirmed mathematical model', *The Journal of Neuroscience*, 5(7): 1688–703.

Galea, J.M., Vazquez, A., Pasricha, N., Orban de Xivry, J. and Celnik, P. (2011) 'Dissociating the roles of the cerebellum and motor cortex during adaptive learning: the motor cortex retains what the cerebellum learns', *Cerebral Cortex*, 21(8): 1761–70.

Gribble, P.L., Ostry, D.J., Sanguineti, V. and Laboissière, R. (1998) 'Are complex control signals required for human arm movement?', *Journal of Neurophysiology*, 79(3): 1409–24.

Haith, A.M., Jackson, C.P., Miall, R.C. and Vijayakumar, S. (2008) 'Unifying the sensory and motor components of sensorimotor adaptation', In *Proc. Advances in Neural Information Processing Systems*.

Harris, C.M. and Wolpert, D.M., (1998) 'Signal-dependent noise determines motor planning', *Nature*, 394(6695): 780–4.

Hoffman, D.S. and Strick, P.L. (1999) 'Step-tracking movements of the wrist. IV. Muscle activity associated with movements in different directions', *Journal of Neurophysiology*, 81(1): 319–33.

Hosp, J.A., Pekanovic, A., Rioult-Pedotti, M.S. and Luft, A.R. (2011) 'Dopaminergic projections from midbrain to primary motor cortex mediate motor skill learning', *The Journal of Neuroscience*, 31(7): 2481–7.

Huang, V.S., Haith, A.M., Mazzoni, P. and Krakauer, J.W. (2011) 'Rethinking motor learning and savings in adaptation paradigms: model-free memory for successful actions combines with internal models', *Neuron*, 70(4): 787–801.

Izawa, J., Rane, T., Donchin, O. and Shadmehr, R. (2008) 'Motor adaptation as a process of reoptimization', *The Journal of Neuroscience*, 28(11): 2883–91.

Joiner, W.M. and Smith, M.A. (2008) 'Long-term retention explained by a model of short-term learning in the adaptive control of reaching', *Journal of Neurophysiology*, 100(5): 2948–55.

Jones, K.E., Hamilton, A.F. and Wolpert, D.M. (2002) 'Sources of signal-dependent noise during isometric force production', *Journal of Neurophysiology*, 88(3): 1533–44.

Kojima, Y., Iwamoto, Y. and Yoshida, K. (2004) 'Memory of learning facilitates saccadic adaptation in the monkey', *The Journal of Neuroscience*, 24(34): 7531–9.

Körding, K., Fukunaga, I., Howard, I.S., Ingram, J.N. and Wolpert, D.M. (2004) 'A neuroeconomics approach to inferring utility functions in sensorimotor control', *PLoS Biology*, 2(10): e330.

Körding, K.P., Beierholm, U., Ma, W.J., Quartz, S., Tenenbaum, J.B. and Shams, L. (2007) 'Causal inference in multisensory perception', *PLos One*, 2(9): e943.

Körding, K.P., Tenenbaum, J.B. and Shadmehr, R. (2007) The dynamics of memory as a consequence of optimal adaptation to a changing body', *Nature Neuroscience*, 10(6): 779–86.

Korenberg, A.T. and Ghahramani, Z. (2002) 'A Bayesian view of motor adaptation', *Current Psychology*, 21(4–5): 537–64.

Krakauer, J.W., Pine, Z.M., Ghilardi, M.F. and Ghez, C. (2000) 'Learning of visuomotor transformations for vectorial planning of reaching trajectories', *The Journal of Neuroscience*, 20(23): 8916–24.

Kurtzer, I., Pruszynski, J.A., Herter, T.M. and Scott, S.H. (2006) 'Primate upper limb muscles exhibit activity patterns that differ from their anatomical action during a postural task', *Journal of Neurophysiology*, 95(1): 493–504.

Liu, D. and Todorov, E. (2007) 'Evidence for the flexible sensorimotor strategies predicted by optimal feedback control', *The Journal of Neuroscience*, 27(35): 9354–68.

Marr, D. (1982) *Vision: A Computational Investigation into the Human Representation and Processing of Visual Information*, Freeman, New York.

Mazzoni, P. and Krakauer, J.W. (2006) 'An implicit plan overrides an explicit strategy during visuomotor adaptation', *The Journal of Neuroscience*, 26(14): 3642–5.

Mazzoni, P., Hristova, A. and Krakauer, J.W. (2007) 'Why don't we move faster? Parkinson's disease, movement vigor, and implicit motivation', *The Journal of Neuroscience*, 27(27): 7105–16.

Morasso, P. (1981) 'Spatial control of arm movements', *Experimental Brain Research*, 42(2): 223–7.

Nagengast, A.J., Braun, D.A. and Wolpert, D.M. (2009) 'Optimal control predicts human performance on objects with internal degrees of freedom', *PLoS Computational Biology*, 5(6): e1000419.

O'Sullivan, I., Burdet, E. and Diedrichsen, J. (2009) 'Dissociating variability and effort as determinants of coordination', *PLoS Computational Biology*, 5(4): e1000345.

Robinson, D.A., Gordon, J.L. and Gordon, S.E. (1986) 'A model of the smooth pursuit eye movement system', *Biological Cybernetics*, 55(1): 43–57.

Shadmehr, R. and Mussa-Ivaldi, F.A. (1994) 'Adaptive representation of dynamics during learning of a motor task', *The Journal of Neuroscience*, 14: 3208–24.

Shadmehr, R. and Wise, S.P. (2005) *The Computational Neurobiology of Reaching and Pointing: A Foundation for Motor Learning*, MIT Press, Cambridge, MA.

Shadmehr, R., Orban de Xivry, J., Xu-Wilson, M. and Shih, T. (2010) 'Temporal discounting of reward and the cost of time in motor control', *The Journal of Neuroscience*, 30(31): 10507–16.

Sing, G.C. and Smith, M.A. (2010) 'Reduction in learning rates associated with anterograde interference results from interactions between different timescales in motor adaptation', *PLoS Computational Biology*, 6(8): 1–14.

Smith, M.A., Ghazizadeh, A. and Shadmehr, R. (2006) 'Interacting adaptive processes with different timescales underlie short-term motor learning', *PLoS Biology*, 4(6): e179.

Sutton, R.S. and Barto, A.G. (1998) *Reinforcement Learning: An Introduction*, Cambridge University Press, Cambridge, UK.

Szentesi, P., Zaremba, R., van Mechelen, W. and Stienen, G.J. (2001) 'ATP utilization for calcium uptake and force production in different types of human skeletal muscle fibres', *The Journal of Physiology*, 531: 393–403.

Taylor, J. A., Klemfuss, N.M. and Ivry, R.B. (2010) 'An explicit strategy prevails when the cerebellum fails to compute movement errors', *Cerebellum*, 9: 580–6.

Thoroughman, K.A. and Shadmehr, R. (2000) 'Learning of action through adaptive combination of motor primitives. *Nature*, 407(6805): 742–7.

Todorov, E. (2000) 'Direct cortical control of muscle activation in voluntary arm movements: a model', *Nature Neuroscience*, 3(4): 391–8.

Todorov, E. (2005) 'Stochastic optimal control and estimation methods adapted to the noise characteristics of the sensorimotor system', *Neural Computation*, 17(5): 1084–108.

Todorov, E. (2006) 'Optimal control theory', *Bayesian Brain: Probabilistic Approaches to Neural Coding*, 269–98.

Todorov, E. and Jordan, M.I. (2002) 'Optimal feedback control as a theory of motor coordination', *Nature Neuroscience*, 5(11): 1226–35.

Tseng, Y., Diedrichsen, J., Krakauer, J.W., Shadmehr, R. and Bastian A.J. (2007) 'Sensory prediction errors drive cerebellum-dependent adaptation of reaching', *Journal of Neurophysiology*, 98(1): 54–62.

Uno, Y., Kawato, M. and Suzuki, R. (1989) 'Formation and control of optimal trajectory in human multijoint arm movement. Minimum torque-change model', *Biological Cybernetics*, 61(2): 89–101.

van Beers, R.J. (2007) 'The sources of variability in saccadic eye movements', *The Journal of Neuroscience*, 27(33): 8757–70.

van Beers, R.J., Sittig, A.C. and Gon, J.J. (1999) 'Integration of proprioceptive and visual position-information: An experimentally supported model', *Journal of Neurophysiology*, 81(3): 1355–64.

van Beers, R.J., Haggard, P. and Wolpert, D.M. (2004) 'The role of execution noise in movement variability', *Journal of Neurophysiology*, 91(2): 1050–63.

von Helmholtz, H. (1962). *Treatise on Physiological Optics*, The Optical Society of America.

Wei, K. and Körding, K. (2010) 'Uncertainty of feedback and state estimation determines the speed of motor adaptation', *Frontiers in Computational Neuroscience*, 4: 1–9.

Wolpert, D.M., Ghahramani, Z. and Jordan, M.I. (1995) 'Are arm trajectories planned in kinematic or dynamic coordinates? An adaptation study', *Experimental Brain Research*, 103(3): 460–70.

Wong, A.L. and Shelhamer, M. (2011) 'Sensorimotor adaptation error signals are derived from realistic predictions of movement outcomes', *Journal of Neurophysiology*, 105(3): 1130–40.

Xu-Wilson, M. Chen-Harris, H., Zee, D.S. and Shadmehr, R. (2009) 'Cerebellar contributions to adaptive control of saccades in humans', *The Journal of Neuroscience*, 29(41): 12930–9.

2

WHAT CAN WE LEARN FROM ANIMAL MODELS?

Eric M. Rouiller

DEPARTMENT OF MEDICINE, UNIVERSITY OF FRIBOURG

Conceptualization of paradigms in animal research

In a systemic approach to decipher the subtle mechanisms involved in the control of voluntary movements, a whole organism has to be trained to execute a sensorimotor integration process, in which the experimenter precisely controls the sensory inputs (the cues informing for instance the subject on a choice to be made in a conditional behavioral task), as well as the motor output (by measuring the motor performance), as illustrated in an example presented in this chapter. These behavioral parameters may then be correlated with measurements of neural activity based on different approaches, offering different time and spatial resolutions.

In this context, different animal species come into play as experimental model, depending on specific sensory abilities and/or motor capabilities. Other parameters guiding the choice of an animal model involve experimental constraints (e.g. methods of measurements) and, very importantly, ethical guidelines defining proper use of animals in research protocols. In this context, the final choice results from a compromise, yielding advantages on one hand but inevitably disadvantages on the other hand (see next section). A major issue here concerns the legitimate tendency to choose a specific model for its relative simplicity (for instance some highly valuable invertebrate models such as worms or insects), offering tremendous advantages on the experimental point of view, but often imperfectly reflecting the exquisite complexity of the motor system in human beings. The number of animals involved in an experiment is of course guided by statistical criteria, but also by ethical considerations related to the 'intrinsic value' of the animal: experiments conducted on insects for instance are not limited in terms of the number of animals involved (as such invertebrate experiments usually do not require veterinary authorization or ethical approval), whereas strict guidelines are applied to research conducted on vertebrates. One major consideration is the gravity of the impact of the experiment on an individual animal. Furthermore, the notion of the intrinsic value of the animal species comes into play in order to determine the number of individuals to reasonably include in the study. For this reason, it is generally assumed that larger numbers of rodents can be considered than of non-human primates for instance. As a consequence, it is most often a reasonable choice to elaborate basic principles and mechanisms, as well as therapeutic strategies, in a simple model (in most cases, rodents, with less limitation on the number of animals for instance) and then, in a translation phase, verify and maybe extend these principles to a 'higher' species (e.g. non-human primates), with properties closer to those of humans, including especially the sophisticated attributes of sensorimotor integration processes. This progressive transition from rodents to non-human primates was for instance applied in a long-term (about two decades) research initiative aimed at improving functional recovery from spinal cord lesion (Schnell & Schwab, 1990, 1993; Bregman et al., 1995; Brösamle et al., 2000; Fouad et al., 2004; Schwab, 2004; Schmidlin et al., 2004; Wannier et al., 2005; Liebscher et al., 2005; Gonzenbach & Schwab, 2007; Freund et al., 2006, 2007, 2009).

As far as the control of voluntary movements is concerned, in order to design a relevant experiment in a pertinent animal model, one has to take into account the general organization of the motor system, with its multiple hierarchical levels (effectors, spinal cord, brainstem and cerebral cortex), combined with parallel channels of information processing, affecting more or less directly the motoneurons controlling the skeletal muscles (e.g. corticospinal system, other descending pathways, lateral loops involving the basal ganglia and cerebellum, to name a few). On top of that, it is crucial to recall that motor control cannot be dissociated from sensory processing, either during preparatory and planning phases of a voluntary movement or during execution of the movement itself, in the form of feedback sensory inputs permitting adjustments to reach optimal performance (e.g. sensory responses in the motor cortex: Hatsopoulos & Suminski, 2011; see also Pruszynski et al., 2011; Seki & Fetz, 2012). The mixed hierarchical and parallel organization of the motor system is reflected by the complex motor performance expressed by most mammals, although different levels of complexity can be distinguished. When it comes to the cortical level, it appears that the organization of the motor cortical areas exhibits significant differences between rodents on one hand and primates on the other. As in other cortical systems (visual, auditory, etc.) the motor cortex of primates comprises a huge number of distinct cortical areas, with different functional and anatomical properties whereas rodents display clearly fewer distinct cortical areas (e.g. Neafsey & Sievert, 1982; Neafsey et al., 1986; Rouiller et al., 1993). The main motor cortical areas are illustrated for non-human primates in Figure 2.1, showing the distribution of the corticospinal neurons. The motor cortex in primates comprises four main regions: the primary motor cortex (M1 or area 4 of Brodmann), the premotor cortex (PM or the lateral part of area 6 of Brodmann), the supplementary motor area (SMA or the mesial part of area 6 of Brodmann) and the cingulate motor area (CMA, corresponding to parts of areas 6, 23 and 24 of Brodmann). Further subdivisions of these motor cortical areas are proposed in Figure 2.1. Although M1 appears to be a relatively homogeneous region, there are differences between its rostral and caudal portions (Strick & Preston, 1982; Friel et al., 2007; Rathelot & Strick, 2009). A large panel of criteria (Table 2.1) can be considered to define a cortical region as a separate motor cortical area, criteria which do not need to be all verified simultaneously. Among the main criteria, the connectional properties play an important role. Functional criteria involve essentially the presence of neural activity in a motor cortical area when the subject is engaged in a motor task, with the difficulty to separate sensory attributes from motor attributes, as they are closely interacting in a sensorimotor integration paradigm. The effects of low intensity electrical stimulation in the motor cortex (so-called intracortical microstimulation) as assessed by movements generated in awake monkeys have also been a major criterion to define distinct motor cortical areas, as well as their somatotopic organization (e.g. Asanuma & Rosen, 1972; Muakkassa & Strick, 1979; Macpherson et al., 1982; Mitz & Wise, 1987; Wiesendanger, 1986, 1988; Lemon, 1988; Schieber, 1990; Luppino et al., 1991; Donoghue et al., 1992; Dum & Strick, 1992; Matzusaka et al., 1992; Godschalk et al., 1995; Preuss et al., 1996; Graziano et al., 2002; Boudrias et al.,

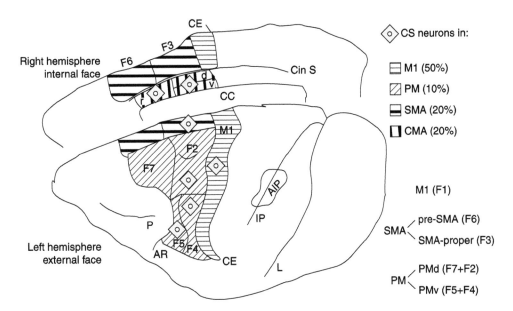

Figure 2.1 Schematic representation of the major motor cortical areas in non-human primates, as seen on a lateral view of the left hemisphere (in front: external surface of the brain) and the right hemisphere (in back: internal surface of the brain). The motor cortical areas are located in the frontal lobe, rostral to the central sulcus (CE). The location and extent of the four principal motor cortical areas (M1, PM, SMA and CMA) are delineated with different shading. The premotor cortex (PM) is divided into two main regions: the dorsal (PMd) and ventral (PMv) premotor cortices, themselves further subdivided in two. PMd comprises the rostral part of PMd (PMd-r or area F7) and the caudal part (PMd-c or area F2). Similarly, in PMv, there are two subdivisions: the rostral part of PMv (PMv-r or area F5) and the caudal part (PMv-c or area F4). The supplementary motor cortex (SMA) was also partitioned in two: rostrally the pre-SMA (or area F6) and caudally SMA-proper (or area F3). The primary motor cortex (M1) corresponds to the area F1, in the nomenclature proposed by Rizzolatti and colleagues (Matelli et al., 1985; Luppino et al., 1993). The cingulate motor area (CMA) comprises three sub-regions, the rostral CMA ('r' in the Figure), the dorsal CMA ('d' in the Figure) and the ventral CMA ('v' in the Figure). The grey diamonds represent the regions of the motor cortical areas where corticospinal neurons are located, in layer V. The percentage distribution of corticospinal neurons in the four different motor cortical areas is indicated (note that some corticospinal neurons present in the parietal lobe have not been considered here). Cin S=cingulate sulcus; CE=central sulcus; AR=arcuate sulcus; P=principal sulcus; IP=intraparietal sulcus; L=lateral sulcus; AIP=anterior intraparietal area (a cortical region in the parietal cortex strongly interconnected with PM, especially area F5, a system specialized for the manipulation and recognition of objects).

2009, 2010). As far as somatotopy is concerned, the classical homunculus representation has been replaced by a more patchy (mosaic) representation of muscles and articulations, more adequate to create synergies of movements involving a highly variable set of muscles (e.g. Schieber, 2001).

The distribution of the corticospinal neurons as shown in Figure 2.1 is consistent with the idea that each motor cortical area has in parallel access to the spinal cord. Tract-tracing experiments have shown that SMA may also establish direct contact with the spinal moto-neurons (Rouiller et al., 1996), as does M1. However, the influence of SMA on spinal motoneurons is less strong than that of M1 (Maier et al., 2002). Along the same line, the

Table 2.1 Summary of criteria to distinguish the various motor cortical areas, as shown in Figure 2.1

	ICMS effect	*(C4-T2) CS neurons**	*Body representation*	*Connection with M1*	*Neuronal activity related to movement*
M1 (F1)	Yes	48%	Full (incl. hand)	Yes	Yes
SMA-proper (F3)	Yes	19%	Full (incl. hand)	Yes	Yes
Pre-SMA (F6)	(Yes)	–	? (eye – hand)	No	Yes
PMd-r (F7)	(Yes)	–	? (eye)	No	Yes
PMd-c (F2)	Yes	7%	? (arm)	Yes	Yes
PMv-r (F5)	Yes	4%	– Hand	Yes	Yes
PMv-c (F4)	Yes	4%	? (arm – face)	Yes	Yes
CMA-r	(Yes)	4%	? (hand)	Yes	Yes
CMA-d	Yes	10%	Full (incl. hand)	Yes	Yes
CMA-v	Yes	4%	Full (incl. hand)	Yes	Yes
AIP	(Yes)	–	– Hand	No	Yes

M1: primary motor cortex (F1).

SMA: supplementary motor area subdivided in SMA-proper (F3) and pre-SMA (F6).

PMd: dorsal premotor area subdivided in rostral (PMd-r) and caudal (PMd-c) parts.

PMv: ventral premotor area subdivided in rostral (PMv-r) and caudal (PMv-c) parts.

CMA: cingulate motor area subdivided in rostral (CMA-r), dorsal (CMA-d) and ventral (CMA-v) parts.

AIP: anterior intraparietal area.

ICMS = intracortical microstimulation.

The distribution of corticospinal (CS) neurons across the multiple motor cortical areas and projecting to the spinal cord (levels C4 to T2) is derived from the data of Dum and Strick (1991).

For body representation (as determined by ICMS), the question mark indicates that it is not yet established whether a full body representation is present: in such a case, body parts observed are indicated between parentheses.

influence of PM on digit muscles is exerted mostly indirectly, via M1 (Schmidlin et al., 2008).

Similarly to the definition of motor cortical areas (see Figure 2.1 and Table 2.1), the influence of motor centers on the effectors, via motoneurons, is mediated via a large set of parallel descending pathways, distinguished by numerous distinct properties or criteria (see Lemon, 2008). Having in mind these organizational properties of the motor system, the conceptualization of an experiment on an animal model is guided mostly by the extent and the precise position of the specific component to be studied in the system. For instance, the rodent model is generally adequate to study mechanisms at the level of the spinal cord (mainly local circuitry), less so however when sophisticated cortical control is involved, especially the so-called corticomotoneuronal system of projection (direct influence of cortical pyramidal neurons on spinal motoneurons), a prerequisite of primates (e.g. Lawrence et al., 1985; Lawrence & Hopkins, 1976; Wiesendanger, 1981; Porter, 1987; Bennett & Lemon, 1996; Lemon, 1993, 2004, 2008). If the goal is to study the specific contribution of a non-primary motor cortical area, such as the premotor cortex (PM), supplementary motor area (SMA) or cingulate motor area (CMA), the recourse to the non-human primate is nearly mandatory due to the difficulty in identifying the counterparts of these distinct motor cortical areas in non-primates.

An important issue is the choice of the methods of investigation. The study of motor control involves at first behavioral observations to assess the performance, ideally with well defined quantitative readouts (e.g. score, reaction time, etc.). The behavioral performance itself represents a strong basis to investigate the fundamental properties of motor control, including motor learning when an individual progressively acquires proficiency for a certain task. The assessment of motor performance implies that the animal used to study voluntary motor control cannot be anesthetized, thus reducing the types of (invasive) interventions that can be conducted in parallel to correlate motor performance with neural activity for instance.

Advantages and disadvantages concerning research on animals

Although the imaging techniques continuously improve in term of resolution, such as structural MRI (magnetic resonance imaging) and approaches like DTI (diffusion tension imaging), these methods cannot (yet) address issues of connectivity at the single cell level in humans. The animal model offers the opportunity to conduct tract-tracing experiments, with subsequent post-mortem histological analysis at high (cellular and sub-cellular) resolution (light microscopy and electron microscopy). In the field of the motor system, based on an elegant intracellular electrophysiological recording coupled to labeling of single corticospinal axons, it was possible to demonstrate the divergent nature of the corticospinal projection: a single corticospinal neuron in the motor cortex innervates, and thus influences, several distinct pools of motoneurons in the spinal cord (Shinoda et al., 1981). Using retrograde neuroanatomical tracers, injected at various levels in the spinal cord, several authors have established the distribution of the corticospinal neurons in the cerebral cortex (e.g. Dum & Strick, 1991; He et al., 1993, 1995; Luppino et al., 1994; Galea & Darian-Smith, 1994; see diamonds in Fig. 2.1), representing a crucial criterion to define a motor cortical area and also indicating that access to the spinal cord from the cerebral cortex is made via parallel descending pathways (originating from M1, SMA, PM and CMA) and not via a single final common pathway arising from M1. Based on retrograde tract tracing (the tracer injected in various muscles of the forelimb), Jenny and Inukai (1983) established the precise distribution of the corresponding motoneurons in the cervical and upper lumbar spinal cord in monkeys. This is highly valuable information in order to design and interpret experiments based on spinal cord lesion (e.g. Galea & Darian-Smith, 1997a,b; Freund et al., 2006, 2007; Nishimura et al., 2007, 2009; Alstermark et al., 2011).

Using tract tracing techniques as well, still in the model of the macaque monkey, the network of connections between the multiple motor cortical areas has been established, both for intrahemispheric (Matelli et al., 1986; Kurata, 1991; Luppino et al., 1993; Ghosh & Gattera, 1995; Tanné-Gariépy et al., 2002) and inter-hemispheric (Jenny & Inukai, 1983; Rouiller et al., 1994a; Liu et al., 2002; Marconi et al., 2003; Boussaoud et al., 2005) connections. Some tracer, such as BDA used as anterograde axonal tracer, yields a cellular resolution and thus makes it possible to follow individual axons and identify axon terminals (boutons) with light microscopy. BDA tracing experiments were used to assess redirection of axon paths following a lesion of the spinal cord (e.g. Freund et al., 2007) or of the motor cortex (Dancause et al., 2005).

Most tract tracing techniques allow the identification of a projection from an area A to an area B, connected by a single synapse. In animal models transynaptic tract-tracing techniques have been developed based mainly on the transport of viruses crossing synapses, an approach which allowed the demonstration of connectivity between remote and well defined brain regions, via a chain of connected neurons. Such transynaptic tract-tracing experiments made it possible to clarify the issue of overlap versus segregation of the projection systems originating respectively from the basal ganglia or the cerebellum to their final targets in the motor cortical areas, via a synapse in the (motor) thalamus (Hoover & Strick, 1993, 1999; Akkal et al., 2007). Before the availability of transynaptic (viral) tracers, the issue of segregated versus separated pathways from basal ganglia and cerebellum to the motor cortex (in the large sense) required difficult combinations of several (retrograde and anterograde) tracers (see e.g. Rouiller et al., 1994b; Sakai et al., 1999, 2002).

In the context of developing animal models for a disease or a trauma of the nervous system, there is a dilemma: when establishing an animal model, in order to study for instance the consequences of a lesion of the central nervous system (spinal cord or brain) on motor control, an important issue is not only the choice of the animal species, but also the type of experimental lesion to perform. From an experimental point of view, with the goal to test the efficacy of a therapeutic strategy, it is generally assumed that it is crucial to reduce the number of variables as much as possible to accurately test the impact of the parameter of interest, namely the treatment. In such context, the experimental lesion is often performed so that it is as reproducible as possible, from one animal to the other. Along this line, for this reason, most spinal cord lesion models are performed with a transection (generally incomplete) of the spinal cord, a procedure more reproducible than for instance a compression lesion. A limitation here is that the gain on the side of reducing variability is detrimental on another side, in the sense that most spinal cord lesions in humans result from traumatism leading to a compression of the spinal cord. The dilemma here is the choice between a lesion model with limited variability in lesion size and/or position, allowing more straightforward interpretation of the results and a lesion model comprising more variability to better reflect the situation in human subjects, in which the lesion properties cover a large range. In an experimental model comprising variability of the lesion properties, this disadvantage in terms of interpretation of the data may be, at least in part, compensated by an appropriate statistical approach which takes the multiplicity of the variables into account, such as the non-parametric multivariate approach (see e.g. Freund et al., 2009 for a model of spinal cord lesion with behavioral assessment of motor control). Recent models of motor cortex lesion in non-human primates included such variability in term of lesion size (e.g. Liu & Rouiller, 1999; Darling et al., 2009; Kaeser et al., 2010, 2011; Bashir et al., 2012) in order to reflect the situation in patients after a stroke for instance. To assess precisely the behavioral consequences of motor cortex lesion and, in a second step, the extent and time course of functional recovery, a large palette of quantitative tests of manual dexterity has been validated over the years (Schmidlin et al., 2011).

Still, in animal models of a trauma affecting the central nervous system in general and the motor system in particular, an exquisite advantage of the animal model is to open the possibility, after functional recovery from the permanent trauma, to either transiently inactivate or permanently lesion a brain region hypothesized to play a major role in the recovery. If this is the case, the transient inactivation or the re-lesion will lead to a loss of the recovered performance, as demonstrated in a few studies dealing with motor cortical lesion (Liu & Rouiller, 1999; McNeal et al., 2010). The latter studies demonstrated that, after a permanent lesion of M1, non-primary motor cortical areas (PM and SMA) are in position to take over, at least partly, the lost function. Non-human primate models were also highly valuable in studies aimed at assessing the changes of cortical motor maps resulting from a lesion of the motor cortex, either at the neonatal or at the adult stage (Rouiller et al., 1998; Liu & Rouiller, 1999; Dancause et al., 2006; Nudo, 2006; Eisner-Janowicz et al., 2008; Dancause & Nudo, 2011) and possible underlying the functional restitution from the cortical lesion. The crucial role played by rehabilitative training to facilitate re-organization of motor maps after motor cortex lesion was demonstrated based on the same animal model (Nudo & Milliken, 1996; Nudo et al., 1996).

The advantage of the animal model regarding the possibility of recording neural activity from individual neurons, while the animal is performing a more or less complex sensorimotor integration task, is however counterbalanced to some extent by the limitation that the animal has to be intensively trained to reach a stable level of performance (e.g. 90 percent of successful trials). The consequence of such overtraining of the experimental subjects preceding the recording period is twofold: First, in most cases, neural activity recording does not take place during the learning phase of the task and therefore it is not possible to decipher the changes of activity associated with the progressive acquisition of the motor task. Second, as the task is often over-trained, the contribution of a given motor cortical area may be different than at less trained stages of the task. For instance, the motor cortex may well, after overtraining, delegate some motor control processing to subcortical regions, a possibility to keep in mind when interpreting data derived from animal models behaviorally engaged in over-trained tasks. Along this line, the animal engaged in a specific task becomes highly specialized for this behavior, whereas human beings are characterized by a much more prominent capability to perform a much larger palette of motor activities at high level and to learn them much more quickly (and efficiently to some extent) than most animal models. Although there are some spectacular experiments in non-human primates, including the use of tools such as a rake extending the reach of the forelimb for reaching objects at remote distance from the body (Hihara et al., 2003; Quallo et al., 2009), the use of tools and devices is obviously a major and sophisticated motor activity of human beings, poorly represented in animal models. In summary, there is a clear discrepancy in terms of motor task complexity practiced by humans on one hand and by experimental animal models on the other hand, although there are examples of fairly complex tasks performed by non-human primates engaged in electrophysiological recording (e.g. Barone & Joseph, 1989; Pellizzer et al., 1995; Sawamura et al., 2002) but once again only after overtraining to reach a sufficiently stable level of performance. The necessity to perfectly train an animal for a motor task, before initiating the physiological recordings, is also a disadvantage with respect to human experiments. In the latter, only short habituation, familiarization with the experimental environment and pre-training is required in most cases whereas, in animals, the training can be rather long – several weeks, if not months, to nearly a year for the most complex tasks in non-human primates. In the design of the experiments on animal models, in particular to investigate motor control in behaving animals, the researchers conceptualize tasks with the aim of targeting specific aspects of the motor function. Furthermore, the nature of the behavioral task is often dictated by technical constraints (e.g. access or not to some expensive technologies) and therefore there is a large variability in the types of motor tasks introduced to study voluntary movements, which on one hand is positive by increasing the diversity of observations and sometimes yielding unexpected results. On the other hand, such a large variability prevents, in the interpretation of the data, more systematic comparisons between studies and there is the difficulty of attributing precisely the reason for divergent observations.

At present, there is also a limitation imposed by the difficulty to apply to behaving animals the functional imaging techniques (such as fMRI) presently used in human subjects, for a direct comparison. Although there are some attempts to conduct fMRI experiments with behaving non-human primates (e.g. Logothetis, 2003; Nelissen & Vanduffel, 2011), this remains a technical challenge accessible to very few research centers. However, the development of such imaging approaches in animal models remains of high priority in order to identify the degree of correspondence between electrophysiological activity and metabolic readouts obtained with fMRI for instance (see e.g. Logothetis, 2002; Logothetis & Wandel, 2004).

Although they are capable of manipulating small objects to some extent (Whishaw & Coles, 1996; Metz & Whishaw, 2000; Whishaw et al., 2003), the rodents do not exhibit comparable performance in term of manual dexterity as non-human primates. Nevertheless, rodents remain a unique model in neurosciences, especially considering the recent development of transgenic animals (e.g. Petersen, 2009). It is now possible to assess the precise impact on the behavior of a well-defined genetic manipulation. Even more promising is the access to optogenetic techniques in rodents (e.g. Mateo et al., 2011; Mao et al., 2011), making it possible to switch on or off individual types of neurons or ionic channels and measure the consequences of such manipulations on the level of behavior, opening the way to better decipher the precise contribution of various cellular components at different locations of the nervous system in the control of motor activities. In this context, optical recording of neural activity using two photons imaging applied to the behaving animal (rodents; e.g. Holtmaat et al., 2009; O'Connor et al., 2010; Lutcke et al., 2010; Komiyama et al., 2010; Kampa et al., 2011) is very attractive in order to visualize the synaptic changes at high resolution (e.g. at the level of dendritic spines), reflecting neuroplasticity associated for instance with (motor) learning, procedural memory, etc.

Major constraints in the transfer of knowledge gained in animal research

The transfer of knowledge from animal models to humans in general, and in the field of motor control in particular, resides in differences in the organization of the motor system between humans and animals. Obviously, the system of control of voluntary movements is far more rudimentary in rodents than in primates, especially if one considers e.g. manual dexterity. Indeed, along a scale reflecting manual dexterity performance ranging from 1 (the least manual dexterity) to 7 (the maximal one), rodents are at level 3 and humans at level 7 (Courtine et al., 2007; Lemon, 2008). This enormous behavioral difference is related to substantial anatomical differences, for instance the link between the motor cortex and spinal motoneurons (the corticospinal projection), indirect in rodents (via interneurons) and predominantly direct for hand control in humans. Not surprisingly, although very useful as a model to elucidate basic principles, the translation of data from rodents to humans may not be straightforward in all situations. For this reason, an intermediate step via non-human primates (mainly macaque monkeys) remains in some cases necessary, as the motor performance of monkeys, including manual dexterity, is fairly comparable to that of humans: along the manual dexterity scale mentioned above, the macaque monkey is placed at level 6.

The motor system is an emblematic example of the inestimable value of the non-human primate model, as put forward by the report published in 2006 by a group of independent experts in the United Kingdom. Based on nearly 400 bibliographic references, these experts reached the conclusions that, for the predictable future (the next 10–15 years), the non-human primate model is indispensable to address biologically relevant issues in the fields of aging, reproduction biology, infectious diseases and neurosciences. In the latter domain, a recent review has emphasized the crucial contribution of the non-human primate model to neuroscience research, including for instance the implementation of therapies in the field of Parkinson's disease (Capitanio & Emborg, 2008). The subtle mechanisms involved in the highly sophisticated control of voluntary movements, such as the hand, can be best elucidated in the non-human primate model, due to its relatively large size, its favorablility for conducting physiological and behavioral investigations, a nd its close proximity to human subjects in terms of performance of motor control. Fairly direct comparisons can be made, even when complex and advanced techniques are engaged, such as MRI (including fMRI) also successfully used in non-human primates.

Translational studies often aim at applying principles elaborated in animal models to the clinics. A major constraint in such transfer resides in the difficulty to confirm proofs of principles obtained in animal models in clinical trials. We have illustrated this difficulty in a recent paper, related to motor control (Kaeser et al., 2010). In the non-human primate model, we have observed that, after unilateral lesion of the motor cortex, there was a clear relationship, in the long term, between the degree of functional recovery of the affected hand (contralesional) and the performance of the unaffected hand: the better the recovery of the contralesional hand, the more dexterous the ipsilesional hand. This result was statistically significant, mainly because the performance of the ipsilesional hand in the animal model can be made pre-lesion versus post-lesion for the same individual. To confirm such a strong relationship in human subjects, as it is impossible to have access to pre-lesion data in patients subjected to a cortical lesion affecting the motor cortex, the only possibility is to make an indirect comparison between a group of intact subjects (reflecting the pre-lesion state) and a separate group of patients, from which post-lesion data are derived. If we apply, however, the same inter-group strategy to our monkeys (Kaeser et al., 2010), namely a comparison of the post-lesion data derived from the lesioned monkeys with data obtained from a separate group of intact monkeys (reflecting pre-lesion data), the relationship between the degree of functional recovery of the affected hand and the long-term performance of the ipsilesional hand cannot be demonstrated any longer. The study mentioned here (Kaeser et al., 2010) belongs to a larger project aimed at translating therapeutic strategies from animal models (rodents first, non-human primates second) to humans, in order to enhance the functional recovery of motor control after a lesion of the spinal cord (e.g. Freund et al., 2009) or of the cerebral cortex (Kaeser et al., 2011). Irrespective of the types of treatment tested with some success in non-human primates (either neutralization of axonal growth inhibitors or transplantation of stem cells), the same difficulty inherent to clinical trials, as illustrated above, will arise when performing inter-group comparisons, due to the large variability across human subjects. One may not exclude that clinical trials yielding negative outcome do not in fact demonstrate the inefficacy of a therapy, but rather the enormous statistical constraint inherent to the design of clinical trials (for instance comparison of a group of patients treated with a group of non-treated patients).

Examples where animal research highlighted the neural mechanisms involved and where this knowledge could be transferred to humans (for therapy, medication, etc.)

In spite of the continuously increasing resolution (either spatial or temporal or both) of several functional methods of investigation in human subjects (e.g. functional magnetic resonance imaging = fMRI; electro-encephalography = EEG; magneto-encephalography = MEG, etc.), recording neural activity at the level of single neurons still remains possible only in animal models, with emphasis on chronic models when the subject performs at the same time a motor task, allowing the correlation between motor performance and pattern of neuronal activity. Studies having reported the pattern of neuronal activities e.g. in the motor cortical areas during a motor performance of the animal are so numerous that an exhaustive survey is simply beyond the scope of the present review. A few examples are, however, presented below.

In a simple reaction time paradigm, we have recently trained macaque monkeys to perform a task of detection of visual or auditory stimuli (Cappe et al., 2010). The animal had to place his hand on a rest platform. When a stimulus was displayed, the animal had to reach out to a target placed in front (to signify detection of the stimulus), allowing measurement of the reaction time. If the monkey responded correctly within a time window locked to the stimulus presentation, the animal was rewarded with solid food pellets. In contrast, when the monkey made an error (no response to a stimulus or response in absence of a stimulus), the trial was interrupted and a 'pause' interval was imposed on the monkey before allowing the monkey to initiate a new trial, possibly giving access to a subsequent reward. The goal of the experiment was to establish a behavioral model suitable for elucidating mechanisms contributing to multisensory facilitation. The hypothesis was that reaction time was shortened when both the acoustic and the visual stimuli were presented simultaneously as compared to unimodal stimulus presentation (acoustic *or* visual). As shown in the top panel of Figure 2.2, the behavioral data confirmed this hypothesis as, in a representative monkey, the reaction time to the bimodal stimulation (visual + acoustic) had an average value of 354 ms, significantly shorter than the reaction time (381 ms) obtained for the most rapid unimodal (acoustic) stimulus. In other words, surprisingly, a unimodal stimulus with a long reaction time (visual one) contributes to shorten the reaction time elicited by a stimulus of another modality (acoustic), when they are presented simultaneously. In line with the principle of 'inverted effectiveness', this effect was more prominent at stimuli intensities close to threshold than at higher (comfortable) intensities. The interpretation of these behavioral data is that there is an early (low level) convergence between the auditory and visual input, which takes place very rapidly. The decrease of the (rapid) reaction time to the acoustic stimulus by simultaneous presentation of a visual stimulus, detected only after a longer reaction time, suggests that the early convergence between the two systems cannot take place, after long unimodal processing, in high level associative cortical areas such as the prefrontal cortex for instance, but rather at low level along the ascending sensory pathways, for instance in the thalamus. This hypothesis of early multisensory integration at the thalamic level is actually consistent with tract-tracing experiments demonstrating the existence of a rapid trans-thalamic route to transfer rapidly information between remote cortical areas underlying the processing of different modalities, associated ultimately with privileged and fast access from the thalamus to motor cortical areas, such as PM, again bypassing long and slow cortico-cortical routes involving high level associative cortical areas (e.g. prefrontal cortex) before access to the premotor areas (Cappe et al., 2009a,b; Shermann & Guillery, 2011).

The task presented above is peculiar in the sense that it requires just a (fast) reaction of the animal to the presentation of the stimulus, in contrast to most studies conducted in non-human primates, based on the paradigm of a conditional task with delay. Typically, in a single trial, after initiation of the trial, the monkey receives a 'cue' stimulus (most often visual) with a random delay to instruct the animal on what is expected in term of motor response. If the

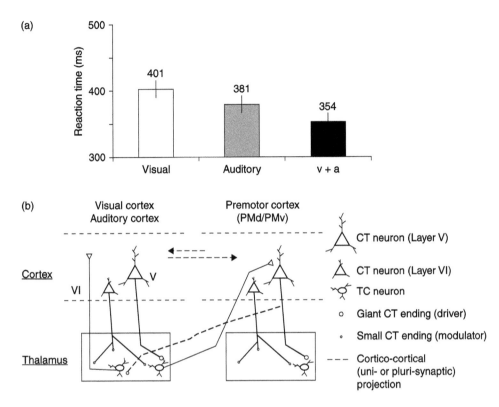

Figure 2.2 (a) Mean reaction times (in ms) obtained from a macaque monkey performing detection of acoustic stimuli (auditory), or of visual stimuli (visual) or to simultaneous presentation of the two sensory modalities (v + a). Note that the mean reaction time is shorter when the two stimuli were presented simultaneously, as compared to unimodal presentation (auditory alone or visual alone). Modified from Cappe et al. (2010). (b) Cartoon illustrating the scenario of a possible fast transmission of information from sensory cortices to premotor cortex, via the thalamus. Such transthalamic transmission of information involves a corticothalamic projection originating from layer V in sensory cortices directed towards thalamic nuclei, which can in turn give fast access to the premotor cortex. The transthalamic transfer of information is likely to be more rapid than cortico-cortical connections. Modified from Cappe et al. (2009b).

correct response is selected (among two or more alternative options), the animal is rewarded. However, the monkey must not give its response immediately after the presentation of the cue, but it has to wait (random delay) for the presentation of a second stimulus (e.g. an acoustic or another visual stimulus) referred to as 'go-signal' to execute the motor response instructed by the 'cue' signal. In parallel to the behavioral performance, neuronal activity was recorded from different brain regions supposed to participate in the control of the task. The advantage of such design is that it is possible to separate the neuronal activity reflecting preparation or planning of the movements (taking place between the 'cue' signal and the 'go-signal', when the animal knows what to do but has to wait until the movement is executed) from the neuronal activity related to the execution of the movement per se (taking place after the 'go-signal'). The behavioral reaction time can be compared to the latency of neuronal activity with respect to the presentation of the 'go-signal'. Most of the studies reported below are based on such a behavioral paradigm, for which monkeys are trained for a very long time before electrophysiological recording can be derived. A difficulty resides in the constraint to fix the head of the animal, permitting control of the movements of the eyes, which may contaminate the recorded activity related to a movement of the hand for instance, and to reduce parasite mechanical artifacts polluting the electrode trace. In this paradigm, the monkey is implanted (under deep anesthesia) with a chronic recording chamber fixed to the skull, allowing electrode penetrations to access different targets in the brain. It is important to note that such electrode penetrations performed in awake and behaving monkeys do not generate pain for the animal, as also reported by awake Parkinson patients in the course of electrode penetrations performed by neurosurgeons to precisely locate the target for deep-brain chronic electrical stimulation, aimed at treating Parkinson symptoms when the patient is or has become resistant to pharmacological treatment (e.g. levodopa).

Based on the above 'conditional task with delay' behavioral paradigm, numerous studies were conducted in order to investigate how single neurons in motor regions (cortex, basal ganglia, cerebellum, brainstem, etc.) encode various attributes of voluntary movements such as force (e.g. Wannier et al., 1991), direction of movements (Georgopoulos et al., 1982; Kalaska et al., 1983; Caminiti et al., 1990, 1991; Kakei et al., 1999: including for some of them the effect of posture), trajectory of a movement (Hocherman & Wise, 1990, 1991), or the temporal sequence of consecutive movements (Barone & Joseph, 1989; Tanji & Shima, 1994). There is evidence that the supplementary motor area (SMA) plays a major role in the control of bimanual action (e.g. Brinkman, 1984; Wiesendanger, 1986; Wiesendanger et al., 1996). This issue was addressed in more detail based on recording of neuronal activity at single cell level in monkeys engaged in bimanual actions, requiring both spatial and temporal coordination of the two hands (e.g. Tanji et al., 1988; Kermadi et al., 1997, 1998, 2000; Kazennikov et al., 1999; De Oliveira et al., 2001; Donchin et al., 1998, 1999, 2001, 2002; Gribova et al., 2002). The results are consistent with a significant role played by the SMA, but other motor cortical areas are involved as well. Most experiments mentioned above and conducted on behaving non-human primates are actually based on such complex paradigms that one cannot imagine them to be performed in rodents, although remaining quite simple for human subjects.

A highly spectacular discovery was made initially by colleagues at the University of Parma in macaque monkeys performing a simple motor task of prehension of small objects using the precision grip (opposition of thumb and index finger). The goal of the experiment was to investigate the discharge properties of single neurons in the premotor cortex (PM), while the monkey grasped small objects (food reward) from a tray presented to the animal by the experimenter. As expected, numerous neurons in the ventral part of PM (PMv), and especially in the rostral part of PM (PMv), exhibited an increase of discharge in relation to the preparation or the execution of hand movement. By accident, another surprising property of some of these neurons emerged: after having recorded the activity of a neuron when the monkey grasped the object presented on the tray, on that particular day one experimenter presented the tray to another experimenter in the laboratory while the monkey watched the scene. The same neuron in rostral PMv, activated when the monkey performed the task, was also similarly activated when the monkey 'just observed' the same sequence of movements performed by another 'hand', although the monkey did not perform any movement itself (e.g. Rizzolatti et al., 1996, 1997; Gallese et al., 1996; Murata et al., 1997; Rozzi et al., 2008; Kraskov et al., 2009; Rochat et al., 2010). Such neurons, exhibiting a comparable activity both when the movement is executed by the monkey itself or by another individual (human or another monkey) observed by the monkey in which the PM neuron is recorded, have been referred to as 'mirror neuron', as they reflect an action either performed or observed by an individual. The mirror neurons are believed to play a crucial role in the recognition of movements performed by others, as well as in the process of learning movements by imitation. The concept of mirror neurons has been extended to a so-called 'mirror neuron system' in human subjects based on fMRI experiments (e.g. Rizzolatti & Craighero, 2004; Rizzolatti & Fabbri-Destro, 2010; Rizzolatti & Sinigaglia, 2010, see also Aglioti et al., 2008 in a context of practice of sport activity). In the studies reporting on mirror neurons as described above, the recognition of the action performed by another subject and observed by the monkey from which neuronal activity is recorded in PMv (or in the parietal cortex), the information is travelling via the visual channel. Further investigations demonstrated that mirror neurons can also be activated when an action is 'observed' via the auditory channel, for an action associated with a well characterized 'noise' (Kohler et al., 2002; Aglioti & Pazzaglia, 2010) (for detailed information please see Chapter 9 below).

A large number of electrophysiological studies describe the mode of representation, for individual neurons in the motor cortex, of various motor attributes (force, direction of movement, etc.), an approach initiated in pioneering works long ago (e.g. Evarts, 1966; Talbot et al., 1968). The coding of the motor parameters established for single neurons in the cerebral cortex can represent a strong basis in order to extract from these signals the intention of an action expressed by the subject, for instance in the premotor cortex or in the parietal cortex (e.g. Andersen & Buneo, 2002; Andersen & Cui, 2009; Scherberger et al., 2005; Baumann et al., 2009; Fluet et al., 2010; Cui & Andersen, 2011; Towsend et al., 2011; Markowitz et al., 2011; O'Doherty et al., 2011). Based on this approach, an outstanding development was the successful implementation in human subjects with lesion of the central nervous system (for instance tetraplegic patients) of a robotic arm guided by the intention of action extracted from the neural signals derived from the subject's brain (e.g. Donoghue et al., 2007; Truccolo et al., 2008; Vargas-Irwin et al., 2010). The same approach allows a human subject, in spite of a paralysis of the forelimbs, to move a cursor on a monitor by extracting his/her intention from the on-going cortical activity recorded from chronic electrodes implanted in the cerebral cortex. Such clinical application is highly illustrative of the long-term effort needed (several decades), involving first fundamental investigations in animal models, before a successful translation to human subjects takes place (for detailed information please see Chapter 4 below).

Although, due to a personal bias, emphasis was placed on the non-human primate model, it remains that the choice of the animal model should be guided by the pertinence of this model for the topic to be studied or the clinical application that is aimed at. A last impressive example is based on the rat model, which proves to be highly suitable to conduct a series of experiments aimed at enhancing the functional recovery from spinal cord lesion in human subjects. Complementary to the approach consisting in enhancing axonal sprouting with neutralization of inhibitors of axonal growth in the central nervous system (see Schwab, 2010), pharmacological and electric stimulation to optimize spinal circuits appears to be highly promising to recover function after spinal cord injury (Courtine et al., 2009; Musienko et al., 2011), an approach generating great hope to successfully treat spinal cord injured patients.

Acknowledgements

The data, originating from personal work and reported here, have been collected with the highly valuable collaboration of several colleagues over the years: M. Wiesendanger, M.E.

Schwab, A. Mir, D. Boussaoud, J. Tanné, A. Morel, F. Liang, V. Moret, Y. Liu, X.Y. Yu, I. Kermadi, A. Tempini, C. Roulin, T. Wannier, E. Schmidlin, A. Belhaj-Saif, P. Freund, A. Wyss, S. Bashir, M. Kaeser, C. Marti, F. Tinguely, O. Kazennikov, A. Babalian, E. Calciati, C. Cappe, M.L. Beaud, F. Lanz, A. Hamdjida, J. Bloch, J.F. Brunet, A.D. Gindrat, J. Savidan, P. Chatagny, S. Badoud.

References

Aglioti, S. M., Cesari, P., Romani, M., and Urgesi, C. (2008), 'Action anticipation and motor resonance in elite basketball players', *Nature Neuroscience*, 11: 1109–1116.

Aglioti, S. M. and Pazzaglia, M. (2010), 'Representing actions through their sound', *Experimental Brain Research*, 206: 141–151.

Akkal, D., Dum, R. P., and Strick, P. L. (2007), 'Supplementary motor area and presupplementary motor area: targets of basal ganglia and cerebellar output', *Journal of Neuroscience*, 27: 10659–10673.

Alstermark, B., Pettersson, L. G., Nishimura, Y., Yoshino-Saito, K., Tsuboi, F., Takahashi, M., and Isa, T. (2011), 'Motor command for precision grip in the macaque monkey can be mediated by spinal interneurons', *Journal of Neurophysiology*, 106: 122–126.

Andersen, R. A. and Buneo, C. A. (2002), 'Intentional maps in posterior parietal cortex', *Annual Review of Neuroscience*, 25: 189–220.

Andersen, R. A. and Cui, H. (2009), 'Intention, action planning, and decision making in parietal-frontal circuits', *Neuron*, 63: 568–583.

Asanuma, H. and Rosén, I. (1972), 'Topographical organization of cortical efferent zones projecting to distal forelimb muscles in the monkey', *Experimental Brain Research*, 14: 243–256.

Barone, P. and Joseph, J. P. (1989), 'Prefrontal cortex and spacial sequencing in macaque monkey', *Experimental Brain Research*, 78: 447–464.

Bashir, S., Kaeser, M., Wyss, A., Hamadjida, A., Liu, Y., Bloch, J., Brunet, J. F., Belhaj-Saif, A., and Rouiller, E. M. (2012), 'Short-term effects of unilateral lesion of the primary motor cortex (M1) on ipsilesional hand dexterity in adult macaque monkeys', *Brain Structure and Function*, 217: 63–79.

Baumann, M. A., Fluet, M. C., and Scherberger, H. (2009), 'Context-specific grasp movement representation in the macaque anterior intraparietal area', *Journal of Neuroscience*, 29: 6436–6448.

Bennett, K. M. B. and Lemon, R. N. (1996), 'Corticomotoneuronal contribution to the fractionation of muscle activity during precision grip in the monkey', *Journal of Neurophysiology*, 75: 1826–1842.

Boudrias, M. H., McPherson, R. L., Frost, S. B., and Cheney, P. D. (2009), 'Output properties and organization of the forelimb representation of motor areas on the lateral aspect of the hemisphere in rhesus macaques', *Cerebral Cortex*, 20(1): 169–186.

Boudrias, M. H., Lee, S. P., Svojanovsky, S., and Cheney, P. D. (2010), 'Forelimb muscle representations and output properties of motor areas in the mesial wall of rhesus macaques', *Cerebral Cortex*, 20: 704–719.

Boussaoud, D., Tanné-Gariépy, J., Wannier, T., and Rouiller, E. M. (2005), 'Callosal connections of dorsal versus ventral premotor areas in the macaque monkey: a multiple retrograde tracing study', *BMC Neuroscience*, 6:67.

Bregman, B. S., Kunkel-Bagden, E., Schnell, L., Dai, H. N., Gao, D., and Schwab, M. E. (1995), 'Recovery from spinal cord injury mediated by antibodies to neurite growth inhibitors', *Nature*, 378: 498–501.

Brinkman, C. (1984), 'Supplementary motor area of the monkey's cerebral cortex: short- and long-term deficits after unilateral ablation and the effects of subsequent callosal section', *Journal of Neuroscience*, 4: 918–929.

Brösamle, C., Huber, A. B., Fiedler, M., Skerra, A., and Schwab, M. E. (2000), 'Regeneration of lesioned corticospinal tract fibers in the adult rat induced by a recombinant, humanized IN-1 antibody fragment', *Journal of Neuroscience*, 20: 8061–8068.

Caminiti, R., Johnson, P. B., Burnod, Y., Galli, C., and Ferraina, S. (1990), 'Shift of preferred directions of premotor cortical cells with arm movements performed across the workspace', *Experimental Brain Research*, 83: 228–232.

Caminiti, R., Johnson, P. B., Galli, C., Ferraina, S., and Burnod, Y. (1991), 'Making arm movements within different parts of space: The premotor and motor cortical representation of a coordinate system for reaching to visual targets', *Journal of Neuroscience*, 11: 1182–1197.

Capitanio, J. P. and Emborg, M. E. (2008), 'Contributions of non-human primates to neuroscience research', *Lancet*, 371: 1126–1135.

Cappe, C., Rouiller, E. M., and Barone, P. (2009a), 'Multisensory anatomical pathways', *Hearing Research*, 256: 28–36.

Cappe, C., Morel, A., Barone, P., and Rouiller, E. M. (2009b), 'The thalamocortical projection systems in primate: an anatomical support for multisensory and sensorimotor interplay', *Cerebral Cortex*, 19: 2025–2037.

Cappe, C., Murray, M. M., Barone, P., and Rouiller, E. M. (2010), 'Multisensory facilitation of behavior in monkeys: effects of stimulus intensity', *Journal of Cognitive Neuroscience*, 22: 2850–2863.

Courtine, G., Bunge, M. B., Fawcett, J. W., Grossman, R. G., Kaas, J. H., Lemon, R., Maier, I., Martin, J., Nudo, R. J., Ramon-Cueto, A., Rouiller, E. M., Schnell, L., Wannier, T., Schwab, M. E., and Edgerton, V. R. (2007), 'Can experiments in nonhuman primates expedite the translation of treatments for spinal cord injury in humans?', *Nature Medicine*, 13: 561–566.

Courtine, G., Gerasimenko, Y., van den Brand, R., Yew, A., Musienko, P., Zhong, H., Song, B., Ao, Y., Ichiyama, R. M., Lavrov, I., Roy, R. R., Sofroniew, M. V., and Edgerton, V. R. (2009), 'Transformation of nonfunctional spinal circuits into functional states after the loss of brain input', *Nature Neuroscience*, 12: 1333–1342.

Cui, H. and Andersen, R. A. (2011), 'Different representations of potential and selected motor plans by distinct parietal areas', *Journal of Neuroscience*, 31: 18130–18136.

Dancause, N., Barbay, S., Frost, S. B., Plautz, E. J., Chen, D. F., Zoubina, E. V., Stowe, A. M., and Nudo, R. J. (2005), 'Extensive cortical rewiring after brain injury', *Journal of Neuroscience*, 25: 10167–10179.

Dancause, N., Barbay, S., Frost, S. B., Zoubina, E. V., Plautz, E. J., Mahnken, J. D., and Nudo, R. J. (2006), 'Effects of small ischemic lesions in the primary motor cortex on neurophysiological organization in ventral premotor cortex', *Journal of Neurophysiology*, 96: 3506–3511.

Dancause, N. and Nudo, R. J. (2011), 'Shaping plasticity to enhance recovery after injury', *Progress in Brain Research*, 192: 273–295.

Darling, W. G., Pizzimenti, M. A., Rotella, D. L., Peterson, C. R., Hynes, S. M., Ge, J., Solon, K., McNeal, D. W., Stilwell-Morecraft, K. S., and Morecraft, R. J. (2009), 'Volumetric effects of motor cortex injury on recovery of dexterous movements', *Experimental Neurology*, 220: 90–108.

De Oliveira, S. C., Gribova, A., Donchin, O., Bergman, H., and Vaadia, E. (2001), 'Neural interactions between motor cortical hemispheres during bimanual and unimanual arm movements', *European Journal of Neuroscience*, 14: 1881–1896.

Donchin, O., Gribova, A., Steinberg, O., Bergman, H., and Vaadia, E. (1998), 'Primary motor cortex is involved in bimanual coordination', *Nature*, 395: 274–278.

Donchin, O., De Oliveira, S. C., and Vaadia, E. (1999), 'Who tells one hand what the other is doing: The neurophysiology of bimanual movements', *Neuron*, 23: 15–18.

Donchin, O., Gribova, A., Steinberg, O., Bergman, H., De Oliveira, S. C., and Vaadia, E. (2001), 'Local field potentials related to bimanual movements in the primary and supplementary motor cortices', *Experimental Brain Research*, 140: 46–55.

Donchin, O., Gribova, A., Steinberg, O., Mitz, A. R., Bergman, H., and Vaadia, E. (2002), 'Single-unit activity related to bimanual arm movements in the primary and supplementary motor cortices', *Journal of Neurophysiology*, 88: 3498–3517.

Donoghue, J. P., Leibovic, S., and Sanes, J. N. (1992), 'Organization of the forelimb area in squirrel monkey motor cortex: Representation of digit, wrist, and elbow muscles', *Experimental Brain Research*, 89: 1–19.

Donoghue, J. P., Nurmikko, A., Black, M., and Hochberg, L. R. (2007), 'Assistive technology and robotic control using motor cortex ensemble-based neural interface systems in humans with tetraplegia', *Journal of Physiology*, 579: 603–611.

Dum, R. P. and Strick, P. L. (1991), 'The origin of corticospinal projections from the premotor areas in the frontal lobe', *Journal of Neuroscience*, 11: 667–689.

Dum, R. P. and Strick, P. L. (1992), 'Medial wall motor areas and skeletomotor control', *Current Opinion in Neurobiology*, 2: 836–839.

Eisner-Janowicz, I., Barbay, S., Hoover, E., Stowe, A. M., Frost, S. B., Plautz, E. J., and Nudo, R. J. (2008), 'Early and late changes in the distal forelimb representation of the supplementary motor area after injury to frontal motor areas in the squirrel monkey', *Journal of Neurophysiology*, 100: 1498–1512.

Evarts, E. V. (1966), 'Pyramidal tract activity associated with a conditioned hand movement in the monkey', *Journal of Neurophysiology*, 29: 1011–1027.

Fluet, M. C., Baumann, M. A., and Scherberger, H. (2010), 'Context-specific grasp movement representation in macaque ventral premotor cortex', *Journal of Neuroscience*, 30: 15175–15184.

Fouad, K., Klusman, I., and Schwab, M. E. (2004), 'Regenerating corticospinal fibers in the Marmoset (*Callitrix jacchus*) after spinal cord lesion and treatment with the anti-Nogo-A antibody IN-1', *European Journal of Neuroscience*, 20: 2479–2482.

Freund, P., Schmidlin, E., Wannier, T., Bloch, J., Mir, A., Schwab, M. E., and Rouiller, E. M. (2006), 'Nogo-A-specific antibody treatment enhances sprouting and functional recovery after cervical lesion in adult primates', *Nature Medicine*, 12: 790–792.

Freund, P., Wannier, T., Schmidlin, E., Bloch, J., Mir, A., Schwab, M. E., and Rouiller, E. M. (2007), 'Anti-Nogo-A antibody treatment enhances sprouting of corticospinal axons rostral to a unilateral cervical spinal cord lesion in adult macaque monkey', *Journal of Comparative Neurology*, 502: 644–659.

Freund, P., Schmidlin, E., Wannier, T., Bloch, J., Mir, A., Schwab, M. E., and Rouiller, E. M. (2009), 'Anti-Nogo-A antibody treatment promotes recovery of manual dexterity after unilateral cervical lesion in adult primates – re-examination and extension of behavioral data', *European Journal of Neuroscience*, 29: 983–996.

Friel, K. M., Barbay, S., Frost, S. B., Plautz, E. J., Stowe, A. M., Dancause, N., Zoubina, E. V., and Nudo, R. J. (2007), 'Effects of a rostral motor cortex lesion on primary motor cortex hand representation topography in primates', *Neurorehabilitation and Neural Repair*, 21: 51–61.

Galea, M. P. and Darian-Smith, I. (1994), 'Multiple corticospinal neuron populations in the macaque monkey are specified by their unique cortical origins, spinal terminations, and connections', *Cerebral Cortex*, 4: 166–194.

Galea, M. P. and Darian-Smith, I. (1997a), 'Corticospinal projection patterns following unilateral section of the cervical spinal cord in the newborn and juvenile macaque monkey', *Journal of Comparative Neurology*, 381: 282–306.

Galea, M. P. and Darian-Smith, I. (1997b), 'Manual dexterity and corticospinal connectivity following unilateral section of the cervical spinal cord in the macaque monkey', *Journal of Comparative Neurology*, 381: 307–319.

Gallese, V., Fadiga, L., Fogassi, L., and Rizzolatti, G. (1996), 'Action recognition in the premotor cortex', *Brain*, 119: 593–609.

Georgopoulos, A. P., Kalaska, J. F., Caminiti, R., and Massey, J. T. (1982), 'On the relations between the direction of two-dimensional arm movements and cell discharge in primate motor cortex', *Journal of Neuroscience*, 2: 1527–1537.

Ghosh, S. and Gattera, R. (1995), 'A comparison of the ipsilateral cortical projections to the dorsal and ventral subdivisions of the macaque premotor cortex', *Somatosensory and Motor Research*, 12: 359–378.

Godschalk, M., Mitz, A. R., Van Duin, B., and Van der Burg, H. (1995), 'Somatotopy of monkey premotor cortex examined with microstimulation', *Neuroscience Research*, 23: 269–279.

Gonzenbach, R. R. and Schwab, M. E. (2007), 'Disinhibition of neurite growth to repair the injured adult CNS: Focusing on Nogo', *Cellular and Molecular Life Sciences*, 65(1): 161–176.

Graziano, M. S. A., Taylor, C. S. R., Moore, T., and Cooke, D. F. (2002), 'The cortical control of movement revisited', *Neuron*, 36: 349–362.

Gribova, A., Donchin, O., Bergman, H., Vaadia, E., and De Oliveira, S. C. (2002), 'Timing of bimanual movements in human and non-human primates in relation to neuronal activity in primary motor cortex and supplementary motor area', *Experimental Brain Research*, 146: 322–335.

Hatsopoulos, N. G. and Suminski, A. J. (2011), 'Sensing with the motor cortex', *Neuron*, 72: 477–487.

He, S.-Q., Dum, R. P., and Strick, P. L. (1993), 'Topographic organization of corticospinal projections from the frontal lobe: Motor areas on the lateral surface of the hemisphere', *Journal of Neuroscience*, 13: 952–980.

He, S.-Q., Dum, R. P., and Strick, P. L. (1995), 'Topographic organization of corticospinal projections from the frontal lobe: Motor areas on the medial surface of the hemisphere', *Journal of Neuroscience*, 15: 3284–3306.

Hihara, S., Obayashi, S., Tanaka, M., and Iriki, A. (2003), 'Rapid learning of sequential tool use by macaque monkeys', *Physiology and Behavior*, 78: 427–434.

Hocherman, S. and Wise, S. P. (1990), 'Trajectory-selective neuronal activity in the motor cortex of rhesus monkeys (*Macaca mulatta*)', *Behavioral Neuroscience*, 104: 495–499.

Hocherman, S. and Wise, S. P. (1991), 'Effects of hand movement path on motor cortical activity in awake, behaving rhesus monkeys', *Experimental Brain Research*, 83: 285–302.

Holtmaat, A., Bonhoeffer, T., Chow, D. K., Chuckowree, J., De, P., V, Hofer, S. B., Hubener, M., Keck, T., Knott, G., Lee, W. C., Mostany, R., Mrsic-Flogel, T. D., Nedivi, E., Portera-Cailliau, C., Svoboda, K., Trachtenberg, J. T., and Wilbrecht, L. (2009), 'Long-term, high-resolution imaging in the mouse neocortex through a chronic cranial window', *Nature Protocols*, 4: 1128–1144.

Hoover, J. E. and Strick, P. L. (1993), 'Multiple output channels in the basal ganglia', *Science*, 259: 819–821.

Hoover, J. E. and Strick, P. L. (1999), 'The organization of cerebellar and basal ganglia outputs to primary motor cortex as revealed by retrograde transneuronal transport of herpes simplex virus type 1', *Journal of Neuroscience*, 19: 1446–1463.

Jenny, A. B. (1979), 'Commissural projections of the cortical hand motor area in monkeys', *Journal of Comparative Neurology*, 188: 137–146.

Jenny, A. B. and Inukai, J. (1983), 'Principles of motor organization of the monkey cervical spinal cord', *Journal of Neuroscience*, 3: 567–575.

Kaeser, M., Wyss, A. F., Bashir, S., Hamadjida, A., Liu, Y., Bloch, J., Brunet, J. F., Belhaj-Saif, A., and Rouiller, E. M. (2010), 'Effects of unilateral motor cortex lesion on ipsilesional hand's reach and grasp performance in monkeys: relationship with recovery in the contralesional hand', *Journal of Neurophysiology*, 103: 1630–1645.

Kaeser, M., Brunet, J. F., Wyss, A., Belhaj-Saif, A., Liu, Y., Hamadjida, A., Rouiller, E. M., and Bloch, J. (2011), 'Autologous adult cortical cell transplantation enhances functional recovery following unilateral lesion of motor cortex in primates: a pilot study', *Neurosurgery*, 68(5): 1405–1416; discussion 1416–1417.

Kakei, S., Hoffman, D. S., and Strick, P. L. (1999), 'Muscle and movement representations in the primary motor cortex', *Science*, 285: 2136–2139.

Kalaska, J. F., Caminiti, R., and Georgopoulos, A. P. (1983), 'Cortical mechanisms related to the direction of two-dimensional arm movements: relations in parietal area 5 and comparison with motor cortex.', *Experimental Brain Research*, 51: 247–260.

Kampa, B. M., Gobel, W., and Helmchen, F. (2011), 'Measuring neuronal population activity using 3D laser scanning', *Cold Spring Harbor Protocols*, 2011: 1340–1349.

Kazennikov, O., Hyland, B., Corboz, M., Babalian, A., Rouiller, E. M., and Wiesendanger, M. (1999), 'Neural activity of supplementary and primary motor areas in monkeys and its relation to bimanual and unimanual movement sequences', *Neuroscience*, 89: 661–674.

Kermadi, I., Liu, Y., Tempini, A., and Rouiller, E. M. (1997), 'Effects of reversible inactivation of the supplementary motor area (SMA) on unimanual grasp and bimanual pull and grasp performance in monkeys', *Somatosensory and Motor Research*, 14: 268–280.

Kermadi, I., Liu, Y., Tempini, A., Calciati, E., and Rouiller, E. M. (1998), 'Neuronal activity in the primate supplementary motor area and the primary motor cortex in relation to spatio-temporal bimanual coordination', *Somatosensory and Motor Research*, 15: 287–308.

Kermadi, I., Liu, Y., and Rouiller, E. M. (2000), 'Do bimanual motor actions involve the dorsal premotor (PMd), cingulate (CMA) and posterior parietal (PPC) cortices? Comparison with primary and supplementary motor cortical areas', *Somatosensory and Motor Research*, 17: 255–271.

Kohler, E., Keysers, C., Umiltà, M. A., Fogassi, L., Gallese, V., and Rizzolatti, G. (2002), 'Hearing sounds, understanding actions: Action representation in mirror neurons', *Science*, 297: 846–848.

Komiyama, T., Sato, T. R., O'Connor, D. H., Zhang, Y. X., Huber, D., Hooks, B. M., Gabitto, M., and Svoboda, K. (2010), 'Learning-related fine-scale specificity imaged in motor cortex circuits of behaving mice', *Nature*, 464: 1182–1186.

Kraskov, A., Dancause, N., Quallo, M. M., Shepherd, S., and Lemon, R. N. (2009), 'Corticospinal neurons in macaque ventral premotor cortex with mirror properties: a potential mechanism for action suppression?', *Neuron*, 64: 922–930.

Kurata, K. (1991), 'Corticocortical inputs to the dorsal and ventral aspects of the premotor cortex of macaque monkeys', *Neuroscience Research*, 12: 263–280.

Lawrence, D. G. and Hopkins, D. A. (1976), 'The development of motor control in the Rhesus monkey: evidence concerning the role of corticomotoneuronal connections', *Brain*, 99: 235–254.

Lawrence, D. G., Porter, R., and Redman, S. J. (1985), 'Corticomotoneuronal synapses in the monkey: light microscopic localization upon motoneurons of intrinsic muscles of the hand', *Journal of Comparative Neurology*, 232: 499–510.

Lemon, R. (1988), 'The output map of the primate motor cortex', *Trends Neurosci.*, 11: 501–506.

Lemon, R. N. (1993), 'The G. L. Brown Prize Lecture. Cortical control of the primate hand', *Experimental Physiology*, 78: 263–301.

Lemon, R. N. (2004), 'Cortico-motoneuronal system and dexterous finger movements', *Journal of Neurophysiology*, 92: 3601.

Lemon, R. N. (2008), 'Descending pathways in motor control', *Annual Review of Neuroscience*, 31: 195–218.

Liebscher, T., Schnell, L., Schnell, D., Scholl, J., Schneider, R., Gullo, M., Fouad, K., Mir, A., Rausch, M., Kindler, D., Hamers, F. P. T., and Schwab, M. E. (2005), 'Nogo-A antibody improves regeneration and locomotion of spinal cord-injured rats', *Annals of Neurology*, 58: 706–719.

Liu, J., Morel, A., Wannier, T., and Rouiller, E. M. (2002), 'Origins of callosal projections to the supplementary motor area (SMA): A direct comparison between pre-SMA and SMA-proper in macaque monkeys', *Journal of Comparative Neurology*, 443: 71–85.

Liu, Y. and Rouiller, E. M. (1999), 'Mechanisms of recovery of dexterity following unilateral lesion of the sensorimotor cortex in adult monkeys', *Experimental Brain Research*, 128: 149–159.

Logothetis, N. K. (2002), 'The neural basis of the blood-oxygen-level-dependent functional magnetic resonance imaging signal', *Philosophical Transactions of the Royal Society of London.B:Biological Sciences*, 357: 1003–1037.

Logothetis, N. K. (2003), 'MR imaging in the non-human primate: studies of function and of dynamic connectivity', *Current Opinion in Neurobiology*, 13: 630–642.

Logothetis, N. K. and Wandell, B. A. (2004), 'Interpreting the BOLD signal', *Annual Review of Physiology*, 66: 735–769.

Luppino, G., Matelli, M., Camarda, R. M., Gallese, V., and Rizzolatti, G. (1991), 'Multiple representations of body movements in mesial area 6 and the adjacent cingulate cortex: An intracortical microstimulation study in the macaque monkey', *Journal of Comparative Neurology*, 311: 463–482.

Luppino, G., Matelli, M., Camarda, R., and Rizzolatti, G. (1993), 'Corticocortical connections of area F3 (SMA-proper) and area F6 (pre-SMA) in the macaque monkey', *Journal of Comparative Neurology*, 338: 114–140.

Luppino, G., Matelli, M., Camarda, R., and Rizzolatti, G. (1994), 'Corticospinal projections from mesial frontal and cingulate areas in the monkey', *NeuroReport*, 5: 2545–2548.

Lutcke, H., Murayama, M., Hahn, T., Margolis, D. J., Astori, S., Zum Alten Borgloh, S. M., Gobel, W., Yang, Y., Tang, W., Kugler, S., Sprengel, R., Nagai, T., Miyawaki, A., Larkum, M. E., Helmchen, F., and Hasan, M. T. (2010), 'Optical recording of neuronal activity with a genetically-encoded calcium indicator in anesthetized and freely moving mice', *Front Neural Circuits*, 4: 9.

Macpherson, J. M., Marangoz, C., Miles, T. S., and Wiesendanger, M. (1982), 'Microstimulation of the supplementary motor area (SMA) in the awake monkey', *Experimental Brain Research*, 45: 410–416.

Maier, M. A., Armand, J., Kirkwood, P. A., Yang, H. W., Davis, J. N., and Lemon, R. N. (2002), 'Differences in the corticospinal projection from primary motor cortex and supplementary motor area to macaque upper limb motoneurons: An anatomical and electrophysiological study', *Cerebral Cortex*, 12: 281–296.

Mao, T., Kusefoglu, D., Hooks, B. M., Huber, D., Petreanu, L., and Svoboda, K. (2011), 'Long-range neuronal circuits underlying the interaction between sensory and motor cortex', *Neuron*, 72: 111–123.

Marconi, B., Genovesio, A., Giannetti, S., Molinari, M., and Caminiti, R. (2003), 'Callosal connections of dorso-lateral premotor cortex', *European Journal of Neuroscience*, 18: 775–788.

Markowitz, D. A., Wong, Y. T., Gray, C. M., and Pesaran, B. (2011), 'Optimizing the decoding of movement goals from local field potentials in macaque cortex', *Journal of Neuroscience*, 31: 18412–18422.

Matelli, M., Luppino, G., and Rizzolatti, G. (1985), 'Patterns of cytochrome oxidase activity in the frontal agranular cortex of the macaque monkey', *Behavioural Brain Research*, 18: 125–136.

Matelli, M., Camarda, R., Glickstein, M., and Rizzolatti, G. (1986), 'Afferent and efferent projections of the inferior area 6 in the macaque monkey', *Journal of Comparative Neurology*, 251: 281–298.

Mateo, C., Avermann, M., Gentet, L. J., Zhang, F., Deisseroth, K., and Petersen, C. C. (2011), 'In vivo optogenetic stimulation of neocortical excitatory neurons drives brain-state-dependent inhibition', *Curr.Biol.*, 21: 1593–1602.

Matsuzaka, Y., Aizawa, H., and Tanji, J. (1992), 'A motor area rostral to the supplementary motor area (presupplementary motor area) in the monkey: Neuronal activity during a learned motor task', *Journal of Neurophysiology*, 68: 653–662.

McNeal, D. W., Darling, W. G., Ge, J., Stilwell-Morecraft, K. S., Solon, K. M., Hynes, S. M., Pizzimenti, M. A., Rotella, D. L., Vanadurongvan, T., and Morecraft, R. J. (2010), 'Selective long-term reorganization of the corticospinal projection from the supplementary motor cortex following recovery from lateral motor cortex injury', *Journal of Comparative Neurology*, 518: 586–621.

Metz, G. A. S. and Whishaw, I. Q. (2000), 'Skilled reaching an action pattern: stability in rat (*Rattus norvegicus*) grasping movements as a function of changing food pellet size', *Behavioural Brain Research*, 116: 111–122.

Mitz, A. R. and Wise, S. P. (1987), 'The somatotopic organization of the supplementary motor area: Intracortical microstimulation mapping', *Journal of Neurophysiology*, 7: 1010–1021.

Muakkassa, K. F. and Strick, P. L. (1979), 'Frontal lobe inputs to primate motor cortex: evidence for four somatotopically organized "premotor" areas', *Brain Research*, 177: 176–182.

Murata, A., Fadiga, L., Fogassi, L., Gallese, V., Raos, V., and Rizzolatti, G. (1997), 'Object representation in the ventral premotor cortex (area F5) of the monkey', *Journal of Neurophysiology*, 78: 2226–2230.

Musienko, P., van den Brand, R., Marzendorfer, O., Roy, R. R., Gerasimenko, Y., Edgerton, V. R., and Courtine, G. (2011), 'Controlling specific locomotor behaviors through multidimensional monoaminergic modulation of spinal circuitries', *Journal of Neuroscience*, 31: 9264–9278.

Neafsey, E. J. and Sievert, C. (1982), 'A second forelimb motor area exists in rat frontal cortex', *Brain Research*, 232: 151–156.

Neafsey, E. J., Bold, E. L., Haas, G., Hurley-Guis, K. M., Quirk, G., Sievert, C. F., and Terreberry, R. R. (1986), 'The organization of the rat motor cortex: a microstimulation mapping study', *Brain Research Reviews*, 11: 77–96.

Nelissen, K. and Vanduffel, W. (2011), 'Grasping-related functional magnetic resonance imaging brain responses in the macaque monkey', *Journal of Neuroscience*, 31: 8220–8229.

Nishimura, Y., Onoe, H., Morichika, Y., Perfiliev, S., Tsukada, H., and Isa, T. (2007), 'Time-dependent central compensatory mechanisms of finger dexterity after spinal cord injury', *Science*, 318: 1150–1155.

Nishimura, Y., Morichika, Y., and Isa, T. (2009) A subcortical oscillatory network contributes to recovery of hand dexterity after spinal cord injury. *Brain*, 132, 709–721.

Nudo, R. J. (2006), 'Mechanisms for recovery of motor function following cortical damage', *Current Opinion in Neurobiology*, 16: 638–644.

Nudo, R. J. and Milliken, G. W. (1996), 'Reorganization of movement representations in primary motor cortex following focal ischemic infarcts in adult squirrel monkeys', *Journal of Neurophysiology*, 75: 2144–2149.

Nudo, R. J., Milliken, G. W., Jenkins, W. M., and Merzenich, M. M. (1996), 'Use-dependent alterations of movement representations in primary motor cortex of adult squirrel monkeys', *Journal of Neuroscience*, 16: 785–807.

O'Connor, D. H., Peron, S. P., Huber, D., and Svoboda, K. (2010), 'Neural activity in barrel cortex underlying vibrissa-based object localization in mice', *Neuron*, 67: 1048–1061.

O'Doherty, J. E., Lebedev, M. A., Ifft, P. J., Zhuang, K. Z., Shokur, S., Bleuler, H., and Nicolelis, M. A. (2011), 'Active tactile exploration using a brain-machine-brain interface', *Nature*, 479: 228–231.

Pellizzer, G., Sargent, P., and Georgopoulos, A. P. (1995), 'Motor cortical activity in a context-recall task', *Science*, 269: 702–705.

Petersen, C. C. (2009), 'Genetic manipulation, whole-cell recordings and functional imaging of the sensorimotor cortex of behaving mice', *Acta Physiologica (Oxford)*, 195: 91–99.

Porter, R. (1987), 'Corticomotoneuronal projections: synaptic events related to skilled movement', *Proceedings of the Royal Society of London B*, 231: 147–168.

Preuss, T. M., Stepniewska, I., and Kaas, J. H. (1996), 'Movement representation in the dorsal and ventral premotor areas of owl monkeys: A microstimulation study', *Journal of Comparative Neurology*, 371: 649–675.

Pruszynski, J. A., Kurtzer, I., Nashed, J. Y., Omrani, M., Brouwer, B., and Scott, S. H. (2011), 'Primary motor cortex underlies multi-joint integration for fast feedback control', *Nature*, 478: 387–390.

Quallo, M. M., Price, C. J., Ueno, K., Asamizuya, T., Cheng, K., Lemon, R. N., and Iriki, A. (2009), 'Gray and white matter changes associated with tool-use learning in macaque monkeys', *Proceedings of the National Academy of Science of the USA*, 106: 18379–18384.

Rathelot, J. A. and Strick, P. L. (2009), 'Subdivisions of primary motor cortex based on cortico-motoneuronal cells', *Proceedings of the National Accidemy of Sciences of the USA*, 106: 918–923.

Rizzolatti, G., Fadiga, L., Gallese, V., and Fogassi, L. (1996), 'Premotor cortex and the recognition of motor actions', *Cognitive Brain Research*, 3: 131–141.

Rizzolatti, G., Fogassi, L., and Gallese, V. (1997), 'Parietal cortex: from sight to action', *Current Opinion in Neurobiology*, 7: 562–567.

Rizzolatti, G. and Craighero, L. (2004), 'The mirror-neuron system', *Annual Review of Neuroscience*, 27: 169–192.

Rizzolatti, G. and Fabbri-Destro, M. (2010), 'Mirror neurons: from discovery to autism', *Experimental Brain Research*, 200: 223–237.

Rizzolatti, G. and Sinigaglia, C. (2010), 'The functional role of the parieto-frontal mirror circuit: interpretations and misinterpretations', *Nat.Rev.Neurosci.*, 11: 264–274.

Rochat, M. J., Caruana, F., Jezzini, A., Escola, L., Intskirveli, I., Grammont, F., Gallese, V., Rizzolatti, G., and Umilta, M. A. (2010), 'Responses of mirror neurons in area F5 to hand and tool grasping observation', *Experimental Brain Research*, 204: 605–616.

Rouiller, E. M., Moret, V., and Liang, F. (1993), 'Comparison of the connectional properties of the two forelimb areas of the rat sensorimotor cortex: Support for the presence of a premotor or supplementary motor cortical area', *Somatosensory and Motor Research*, 10: 269–289.

Rouiller, E. M., Liang, F., Babalian, A., Moret, V., and Wiesendanger, M. (1994a), 'Cerebellothalamocortical and pallidothalamocortical projections to the primary and supplementary motor cortical areas: A multiple tracing study in macaque monkeys', *Journal of Comparative Neurology*, 345: 185–213.

Rouiller, E. M., Babalian, A., Kazennikov, O., Moret, V., Yu, X.-H., and Wiesendanger, M. (1994b), 'Transcallosal connections of the distal forelimb representations of the primary and supplementary motor cortical areas in macaque monkeys', *Experimental Brain Research*, 102: 227–243.

Rouiller, E. M., Moret, V., Tanné, J., and Boussaoud, D. (1996), 'Evidence for direct connections between the hand region of the supplementary motor area and cervical motoneurons in the macaque monkey', *European Journal of Neuroscience*, 8: 1055–1059.

Rouiller, E. M., Yu, X. H., Moret, V., Tempini, A., Wiesendanger, M., and Liang, F. (1998), 'Dexterity in adult monkeys following early lesion of the motor cortical hand area: the role of cortex adjacent to the lesion', *European Journal of Neuroscience*, 10: 729–740.

Rozzi, S., Ferrari, P. F., Bonini, L., Rizzolatti, G., and Fogassi, L. (2008), 'Functional organization of inferior parietal lobule convexity in the macaque monkey: electrophysiological characterization of motor, sensory and mirror responses and their correlation with cytoarchitectonic areas', *European Journal of Neuroscience*, 28(8): 1569–1588.

Sakai, S. T., Inase, M., and Tanji, J. (1999), 'Pallidal and cerebellar inputs to thalamocortical neurons projecting to the supplementary motor area in *Macaca fuscata*: a triple-labeling light microcopic study', *Anatomy and Embryology*, 199: 9–19.

Sakai, S. T., Inase, M., and Tanji, J. (2002), 'The relationship between MI and SMA afferents and cerebellar and pallidal efferents in the macaque monkey', *Somatosensory and Motor Research*, 19: 139–148.

Sawamura, H., Shima, K., and Tanji, J. (2002), 'Numerical representation for action in the parietal cortex of the monkey', *Nature*, 415: 918–922.

Scherberger, H., Jarvis, M. R., and Andersen, R. A. (2005), 'Cortical local field potential encodes movement intentions in the posterior parietal cortex', *Neuron*, 46: 347–354.

Schieber, M. H. (1990), 'How might the motor cortex individuate movements?', *Trends in Neuroscience*, 13: 440–445.

Schieber, M. H. (2001), 'Constraints on somatotopic organization in the primary motor cortex', *Journal of Neurophysiology*, 86: 2125–2143.

Schmidlin, E., Wannier, T., Bloch, J., and Rouiller, E. M. (2004), 'Progressive plastic changes in the hand representation of the primary motor cortex parallel incomplete recovery from a unilateral section of the corticospinal tract at cervical level in monkeys', *Brain Research*, 1017: 172–183.

Schmidlin, E., Brochier, T., Maier, M. A., Kirkwood, P. A., and Lemon, R. N. (2008), 'Pronounced reduction of digit motor responses evoked from macaque ventral premotor cortex after reversible inactivation of the primary motor cortex hand area', *Journal of Neuroscience*, 28: 5772–5783.

Schmidlin, E., Kaeser, M., Gindrat, A. D., Savidan, J., Chatagny, P., Badoud, S., Hamadjida, A., Beaud, M. L., Wannier, T., Belhaj-Saif, A., and Rouiller, E. M. (2011), 'Behavioral assessment of manual dexterity in non-human primates', *J.Vis.Exp.*, 57: 3258.

Schnell, L. and Schwab, M. E. (1990), 'Axonal regeneration in the rat spinal cord produced by an antibody against myelin-associated neurite growth inhibitors', *Nature*, 343 no. 6255: 269–272.

Schnell, L. and Schwab, M. E. (1993), 'Sprouting and regeneration of lesioned corticospinal tract fibres in the adult rat spinal cord', *European Journal of Neuroscience*, 5: 1156–1171.

Schwab, M. E. (2004), 'Nogo and axon regeneration', *Current Opinion in Neurobiology*, 14: 118–124.

Schwab, M. E. (2010), 'Functions of Nogo proteins and their receptors in the nervous system', *Nature Reviews in Neuroscience*, 11: 799–811.

Seki, K. and Fetz, E. E. (2012), 'Gating of sensory input at spinal and cortical levels during preparation and execution of voluntary movement', *Journal of Neuroscience*, 32: 890–902.

Sherman, S. M. and Guillery, R. W. (2011), 'Distinct functions for direct and transthalamic cortico-cortical connections', *Journal of Neurophysiology*, 106: 1068–1077.

Shinoda, Y., Yokota, J. I., and Futami, T. (1981), 'Divergent projection of individual corticospinal axons to motoneurons of multiple muscles in the monkey', *Neuroscience Letters*, 23: 7–12.

Strick, P. L. and Preston, J. B. (1982), 'Two representations of the hand in area 4 of a primate. I. Motor output', *Journal of Neurophysiology*, 48: 139–149.

Talbot, W. H., Darian-Smith, I., Kornhuber, H. H., and Mountcastle, V. B. (1968), 'The sense of flutter-vibration: comparison of the human capacity with response patterns of mechanoreceptive afferents from the monkey hand', *Journal of Neurophysiology*, 31: 301–334.

Tanji, J., Okano, K., and Sato, K. C. (1988), 'Neuronal activity in cortical motor areas related to ipsilateral, contralateral, and bilateral digit movements of the monkey', *Journal of Neurophysiology*, 60: 325–343.

Tanji, J. and Shima, K. (1994), 'Role for supplementary motor area cells in planning several movements ahead', *Nature*, 371: 413–416.

Tanné-Gariépy, J., Rouiller, E. M., and Boussaoud, D. (2002), 'Parietal inputs to dorsal versus ventral premotor areas in the macaque monkey: evidence for largely segregated visuomotor pathways', *Experimental Brain Research*, 145: 91–103.

Townsend, B. R., Subasi, E., and Scherberger, H. (2011), 'Grasp movement decoding from premotor and parietal cortex', *Journal of Neuroscience*, 31: 14386–14398.

Truccolo, W., Friehs, G. M., Donoghue, J. P., and Hochberg, L. R. (2008), 'Primary motor cortex tuning to intended movement kinematics in humans with tetraplegia', *Journal of Neuroscience*, 28: 1163–1178.

Vargas-Irwin, C. E., Shakhnarovich, G., Yadollahpour, P., Mislow, J. M., Black, M. J., and Donoghue, J. P. (2010), 'Decoding complete reach and grasp actions from local primary motor cortex populations', *Journal of Neuroscience*, 30: 9659–9669.

Wannier, T., Schmidlin, E., Bloch, J., and Rouiller, E. M. (2005), 'A unilateral section of the corticospinal tract at cervical level in primate does not lead to measurable cell loss in motor cortex', *Journal of Neurotrauma*, 22: 703–717.

Wannier, T. M. J., Maier, M. A., and Hepp-Reymond, M.-C. (1991), 'Contrasting properties of monkey somatosensory and motor cortex neurons activated during the control of force in precision grip', *Journal of Neurophysiology*, 65: 572–587.

Whishaw, I. Q. and Coles, B. L. K. (1996), 'Varieties of paw and digit movement during spontaneous food handling in rats: Postures, bimanual coordination, preferences, and the effect of forelimb cortex lesions', *Behavioural Brain Research*, 77: 135–148.

Whishaw, I. Q., Gorny, B., Foroud, A., and Kleim, J. A. (2003), 'Long-Evans and Sprague-Dawley rats have similar skilled reaching success and limb representations in motor cortex but different movements: some cautionary insights into the selection of rat strains for neurobiological motor research', *Behavioural Brain Research*, 145: 221–232.

Wiesendanger, M. (1981), 'The pyramidal tract. Its structure and function', *Handbook of Behavioral Neurobiology*, 5: 401–491.

Wiesendanger, M. (1986), 'Recent developments in studies of the supplementary motor area of primates', *Review of Physiology Biochemistry and Pharmacology*, 103: 1–59.

Wiesendanger, M. (1988), 'Output organization of the rolandic cortex as revealed by electrical stimulation,' in *Non-invasive stimulation of brain and spinal cord: fundamentals and clinical applications*, A. R. Liss, ed., pp. 23–35.

Wiesendanger, M., Rouiller, E. M., Kazennikov, O., and Perrig, S. (1996), 'Is the supplementary motor area a bilaterally organized system?', *Advances in Neurology*, 70: 85–93.

3

POSTURAL CONTROL BY DISTURBANCE ESTIMATION AND COMPENSATION THROUGH LONG-LOOP RESPONSES

Thomas Mergner

NEUROLOGY, UNIVERSITY OF FREIBURG, GERMANY

Introduction

This chapter concentrates on the control of human biped stance and balancing as relatively simple prototypes of posture and movement control in general. The simplicity is mainly owing to the fact that the physics of the body sway resembles that of a single or double inverted pendulum. Clinically, postural control is often equated to equilibrium control of stance. However, posture control does not end with body-on-feet balancing. Rather, we balance the head on the trunk, the trunk on the hips, or make the trajectory of an arm during reaching movements resist gravity. Admittedly, there are differences among these tasks, depending on which muscle groups, joints, etc. are concerned. Yet, basic function principles resemble each other when viewed from a 'systems control perspective'. At this level, the muscles are abstracted as actuators. Furthermore the movements of body parts are formalized according to the rules of physics, and the balancing represents the neural control task. Such abstractions help to grasp the essence of the function principles.

The chapter includes a historical perspective on posture control research in the first half of the twentieth century. This perspective may help young scientists to appreciate that natural sciences tend to change over time, and is meant to encourage their own research. Before the twentieth century, scientists were unaware of the fact that humans use a special *'equilibrium sense'* for balancing. In the period 1840–1900, experimental and clinical scientists demonstrated that parts of the inner ears serve this function. Goltz (1870), for example, mechanically irritated these inner ear structures in frogs (nowadays known as the *vestibular organs*) and observed reflexive leg and body movements. These were later called postural reflexes. The first of the three sections of this chapter starts by explaining the reflex concept that then dominated the first half of the twentieth century. The second section covers approximately the second half of the twentieth century when posture control research profited from many important discoveries made in neuroanatomy and neurophysiology. Researchers started to conceptualize how voluntary movements are embedded in posture control. This was also the time when engineering sciences started to boom, and control and signal processing theories were developed. However, only in the last part of this period and the following decade were these theories successfully used for the modeling of sensorimotor control. The modeling will be described in the third part. It defines posture control as a context-dependent, yet mostly automatic disturbance compensation, by which the control of voluntary movements is freed from immense computational burdens. Although building on reflexes, the control principles provide the impressive human flexibility and richness of behavior, including cognitive aspects and voluntary control over the reflexes.

Fundamental levels of postural control research

The reflex concept originated mainly from the work of Sherrington (1900) on muscle function. He established the concept of a sensory feedback circuit, in which an external muscle stretch is sensed through muscle spindle receptors. These receptors then activate motoneurons in the spinal cord, which in turn command the muscle to counteract the stretch. The concept was later applied to postural control in special animal preparations (see Magnus 1924). One of the postural reflexes relates to Goltz's experiment and was called the *vestibulo-spinal reflex* (see Wilson and Melvill Jones 1979). The tonic components of these reflexes originate from activation of gravity-sensitive receptors in the vestibular otolith organs. The brain uses them to sense static body excursion with respect to the gravitational vertical and to adjust leg muscle activity to counteract the excursion. It was later noted that the vestibular canals contribute to the dynamics of the vestibulo-spinal reflexes. Furthermore, it was noted that the vestibulo-spinal reflexes have partners in the cervico(neck)-spinal reflexes, which arise when the head is deflected with respect to the trunk (i.e. they stem from neck muscle proprioception, as we know today). But the functional significance of this partnership remained unclear.

The postural reflexes were thought to cover even complex situations such as standing up following a fall (righting reflexes). This was attributed to a chain of reflexes, where the result of one reflex triggers the next. However, researchers were aware that the reflex concept does not explain the flexibility of human sensorimotor behavior. In this behavior, humans super-impose, for example, voluntary sensorimotor activities on balancing. Furthermore, we automatically adjust the balancing to changing external situations, whether it is a push against the body in a crowd or sequences of tilt and translation of the support surface when standing in a moving bus.

It was recognized that posture and movement should not be considered as two separate activities, even though posture is still often regarded as a static activity and movement as the dynamic counterpart. The control of both rather involves dynamic and static aspects and these two actually belong together functionally. Holmes (1922) conceived that 'the movement itself consists of a series of postures' and, similarly, Denny-Brown (1929) considered movement a modification of posture. In addition, Holmes (1922) extended his notion of posture control, stressing the need of a postural fixation of the moving limb on the body as a prerequisite for an accurate movement. Sherrington (1931) held that posture accompanies movement. This was later called coordination of movement with compensatory and anticipatory postural adjustments (see Gurfinkel 1994, and below).

The research of this period provided a number of 'postural control basics' that still hold today (a–f below). The paragraphs below draw mainly on a review by Hellebrandt and Franseen (1943).

(a) The balancing task

Balancing the inherently unstable body in biped stance usually means maintaining the body's center of mass (COM) over the relatively small supporting base given by the feet. In upright stance, the COM (sometimes also called COG, for center of gravity) is located in the trunk approximately at the 5th lumbar segment (Figure 3.1). With the COM above the feet in quiet stance, balancing is subjectively almost effortless. This applies, although the COM's gravitational vector points not exactly to the rotation axis passing through the two ankle joints, but slightly in front of it.

This COM location is essentially maintained when the body weight distribution is changed by a backpack, trunk bending, etc., in that the body configuration is adjusted accordingly. For example, a heavy backpack is compensated for by a forward body lean.

(b) Spontaneous body sway

In quiet stance, there is always some spontaneous sway. Closing the eyes increases sway amplitude (clinically performed with the feet placed side by side as 'Romberg test'). This suggests that the sway amplitude is related to how accurately our brain senses self-motion. But there exists also the possibility of active interference, for example by making joints stiffer by

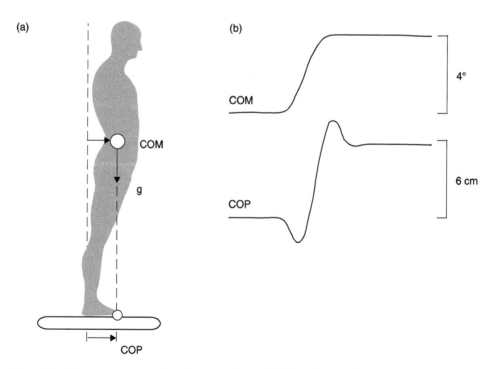

Figure 3.1 Schematic representation of center of mass, COM, and center of pressure, COP. (a) During a static body lean forward, the COM's projection to the support surface determines the COP. (b) During active lean (dynamic condition), the COP also contains the ground reaction force that is generated to accelerate and decelerate the COM.

co-contracting antagonistic muscles (see Nielsen 1998). Thus, sway amplitude should not be the only criterion for clinically diagnosing balancing problems.

(c) Multisensory control

The balancing control depends on co-operations of the muscle stretch reflex with activity arising from vestibular, proprioceptive (joint angle and torque), teleceptive (e.g. vision), and exteroceptive (e.g. touch) neural mechanisms. These mechanisms are usually not experienced consciously, although they involve the cerebral cortex in intact adult individuals, as suggested by lesion studies (Rademaker, 1931; Bard, 1933; Brooks, 1933).

The functional significance of the multisensory co-operations still remained open. The exception was a concept on a co-operation between the vestibulo-spinal and the cervico-spinal reflex (von Holst and Mittelstaedt, 1950; also Roberts, 1978). It suggested that the two reflexes combine in stabilizing the trunk and are not invoked during head rotation on the stationary trunk, in which case they cancel each other. This concept of vestibular–neck reflex interaction together with a modern view of vestibular–neck interaction (Lund and Broberg, 1983; Mergner et al., 1997) is illustrated in Figure 3.2.

(d) Function localization in the central nervous system (CNS)

In human babies, the primitive postural reflexes and their development within the first year of life are routinely tested for diagnostic purposes of brain maturation (Prechtl, 1977). They disappear with progressive maturation of the higher brain centers such as the cerebral cortex, the basal ganglia, and the cerebellum. From this and from comparisons with animal research it was concluded that the proprioceptive reflex mechanisms originate from the spinal cord, while the vestibulo-spinal reflex and its co-operation with the cervico-spinal reflex are processed in the brainstem (in particular the vestibular nuclei and the reticular formation) and the cerebellum. After these higher centers have taken over the control during maturation, the neural substrates of the reflexes remain, which explains why they may reappear upon lesions of these higher centers (Simons, 1923; Walshe, 1923).

(e) Orthostatic effects

Maintaining upright body posture is associated with a number of vegetative functions such as blood pressure control or the excretion of the hormone aldosterone that is related to blood pressure regulation. The need to compensate the hydrostatic effect of gravity, i.e. the return of venous blood to the heart, is quite intuitive. Failure may lead to 'orthostatic collapse'.

(f) Quantification, training and adaptation

Quantitative measurements of body posture and COM motion became possible through photography and film. Using these techniques it was shown that specific posture training tends to improve balancing. It also became evident that adaptation is an important feature of the postural control system, for example when an injury or disease makes it necessary that the neuro-muscular and skeletal systems change in order to regain sensorimotor functions like walking.

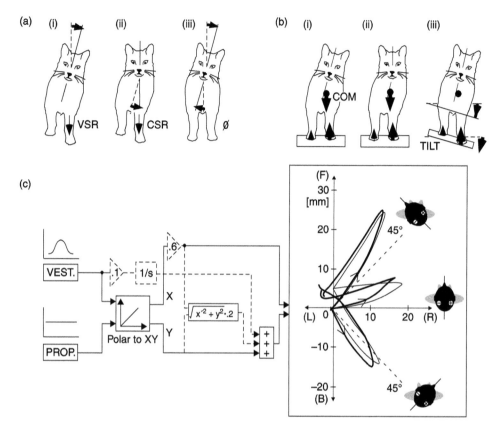

Figure 3.2 Vestibular–neck reflex interaction and re-interpretation in terms of coordinate transformation. (a) Classical view of vestibular–neck interaction in terms of vestibulo-spinal reflex (i, VSR; leg extension on one side and flexion on the other side upon head-in-space rotation) and cervico-spinal reflex (ii, CSR; evoked by trunk rotation with respect to the head). The signs of the two reflexes are such that they cancel each other during head rotation on stationary trunk (iii). (b) Re-interpretation. Tilt of trunk and head together, yielding vestibular head-in-space signal, is associated with COM shift to the same side. The vestibular signal evokes gravity-compensating changes in the leg muscle tone (i). Tilting the trunk with the head remaining vertical yields a similar effect, due to similar COM excursion in space (ii). But here, trunk orientation in space (TS) is perceived from the sum of the vestibular head-in-space signal (HS) and the proprioceptive trunk-to-head signal (TH; TS= HS + TH), which represents a coordinate transformation from body (TH) to space (TS) coordinates (adapted from Mergner 2004 and Mergner et al. 1997). Further down transformation allows estimation of foot-support-in-space tilt with the help of leg-to-trunk proprioceptive signals (iii). C Vestibular–neck coordinate transformation during horizontal head turns. The template shows original body lean response evoked by applying a trans-aural galvanic vestibular stimulus (cathode is at left ear) at different head orientations of a human subject (from above). Body lean results as compensatory reaction to illusory body lean. Direction of illusion turns with the horizontal head rotations due to neck proprioceptive coordinate transformation. (Original data, bold curves, and their model simulations, thin curves, adapted from Hlavacka et al., 1996.)

Emergence of numerous postural control concepts

In the second half of the twentieth century, electrical recordings of muscle and neuron activities and inferences on their relation to sensorimotor functions and behavior became popular in neurophysiology (review articles, Massion, 1992; Horak and Macpherson, 1996). This resulted in novel concepts in posture control research such as the *body schema*. It refers to the internal neural representations of body configuration and dynamic state which the system needs to know in postural control if an external disturbance such as a push against the body has to be compensated. The concepts of *muscle and movement synergies* refer to the fact that several muscles tend to cooperate during certain movements, and that the movements of several body-segments cooperate synergistically during motor behaviors such as walking. Since the movements tend to be structured spatially and in time, they appear to be generated on the basis of complex *central motor programs*. As will be explained below, spatial and temporal sequences of movements and their postural stabilization may as well emerge automatically from hierarchical sensorimotor control principles. Such emergences may also explain certain movement strategies and postural adjustments.

Motor strategies

The concept of motor strategies refers to the fact that a given motor goal may be achieved in different ways. In most situations, standing humans preferentially use the ankle joints for balancing (the 'ankle strategy') while in some special situations they also involve the hip joints ('hip strategy'; Nashner & McCollum, 1985; Nashner & Horak, 1986). Factors that favor the involvement of the hips are restriction of foot support base and fast disturbances (e.g. fast foot support motion; Allum et al., 1989). And involving the hips requires less effort (Kuo and Zajac, 1993b). The ankle strategy is favored at slow disturbance speed (Nashner and Horak, 1986) and sensory restrictions through disease or age (Horak & Macpherson, 1996). Contributions from the knee joints to the balancing are small (Alexandrov et al., 2001). They produce mainly vertical body accelerations, while balancing mostly deals with the compensation of horizontal accelerations of the body's COM (Kuo & Zajac, 1993a).

Postural adjustments

The concept of postural adjustments during voluntary movements was refined by recording electrical muscle activity (electromyogram, EMG) and using biomechanical measurements (overview, Bouisset & Do, 2008). Biomechanics furthermore differentiated the kinetic from the kinematic aspects. *Kinetics* refers to forces and torques. In posture control, it refers for example to adjustments of the ankle joint torque required to prevent an upright standing body from falling over. Superimposed may be a *kinematic* task, for example a righting movement of the trunk to maintain its vertical orientation in space in an attempt to stabilize the workspaces of the eyes and arms.

Quantitative measurements

In many of the above studies, the kinetic result of COM sway was measured using a 'posturographic platform'. It measures the COM-evoked ground reaction forces through a force transducing plate in terms of center of pressure (COP) shifts (Figure 3.1). Also, recordings of COM excursions became possible with the help of optoelectronic devices that measure movements of the body segments. The measurement of the COP provides more information than that of the COM, because it reflects in addition to the ground reaction force (which is holding the COM against gravity) also the joint torques (by which a subject actively moves the COM). Consider an active body lean movement forward (Figure 3.1b). In the beginning of the movement, the COP is transiently shifted backward by a ground reaction force that mirrors the force required to accelerate the COM forward. Then the COP shifts with the COM's projection, before finally an active body deceleration leads to a transient COP overshooting. In other words, in unperturbed stance, the COP measure reflects both the dynamic aspects of balancing, stemming from the ankle torque produced to overcome body inertia, and the static torque from the ankle joint that overcomes the gravity effect resulting from COM excursion. Recording EMG may provide additional information on the spatial distribution and timing of muscle activity during balancing.

Attempts to formalize conceptual frameworks

The recording of COP shifts, body segment and COM movements provides the information that is needed to develop system concepts on how the brain controls posture. The approach requires recording of balancing responses evoked by external disturbances such as translation or rotation of the support surface (e.g. on a motion platform) or well-controlled pushes and pulls acting on the body. In a so-called 'black box' approach of the systems analysis, the input stimuli (often having a sine wave form) are compared to the output response. From the relation between the two, inferences on the information processing in the 'box' can be made. However, this approach is not sufficient to infer the details of signal processing in the CNS and it does not allow a decision between several possible solutions. A variant of the black box approach is the 'grey box' approach. In this case, established knowledge from sensory and motor neurophysiology is included, allowing far-reaching conclusions as will been shown below. Further help comes from a simple scientific rule that demands a search for the simplest solution (Occam's razor rule; Gibbs & Sugihary 1996/97). This meets with nature's success to find simple, parsimonious solutions even for complex mechanisms.

In the further wake of cybernetics and control theory, new tools for control system identification, abstraction and simulation became available and inspired sensorimotor physiologists. Merton hypothesized that voluntary movements are initiated by the gamma motoneurons, setting an 'error' in the muscle spindle to drive the alpha motoneurons ('follow-up servo hypothesis'; Merton, 1953). This inspired others, leading to alternatives to or modifications of this hypothesis (e.g., 'alpha–gamma linkage' of Granit, 1955, and 'servo-assisted motor control' by Mathews, 1972). However, several researchers doubted that, apart from the short-latency primitive reflexes, sensory afferents from the periphery can contribute much to the control of movements. They favored therefore feed forward control and postulated that posture control in the adult individual is mainly centrally controlled (these protagonists were called the 'centralists' as compared to the 'peripheralists' by Mackay, 1980). The reason was that control theory asks for high gains in the control to achieve mechanical stability in the face of external disturbances, which seemed to be incompatible with the biological delay times that are long compared to those in technical systems and may lead to control instability (Rack, 1981). Engineering academia, on the other hand, provided mathematical control methods that required no sensory feedback, at least in their dogmatic forms, or only proprioceptive feedback on the state of the system. Other methods combined predictive fast feed forward control with slow sensory feedback (variants of the so-called Luenberger observer). Disputed observations, originally obtained in monkeys, lent support to this notion (Bizzi et al., 1978, 1984). This development fostered the question as to where the primitive reflexes remain during the development from birth to adulthood.

Developmental aspects

The primitive vestibular and proprioceptive reflexes get integrated into more flexible movement patterns approximately 3–8 months after birth, developing into righting reflexes (reactions) and positive support reactions (Magnus, 1924; Prechtl, 1977). Important in this period of child development is the development of anti-gravity extensor tone. The child starts balancing by learning to raise its head against gravity, before continuing to do so with the upper trunk etc. until it is able to stand upright, balancing by holding with the hands to some support. In this time, it learns to deal with the external field forces (gravity, centrifugal and Coriolis forces). Once learned, their impacts are no longer consciously perceived (for learning new Coriolis force effects in adults, see Lackner & DiZio, 1994).

The first period of child development includes the learning of a 'top-down' strategy (Massion, 1998). Vision helps balancing by fixating the eyes on items in the surroundings, thereby stabilizing gaze. Furthermore, when voluntarily moving in a goal-directed way in space, such a top-down strategy may be adopted so that, during body turns, gaze shifts occur first and are then followed by the body. For balancing with eyes closed or eyes open, however, a 'bottom-up' strategy must also be learned. In this strategy, the lowest joints (ankle joints) balance the body as a whole. This strategy compensates for self produced or externally applied disturbances, which can be support surface motion as well as field and contact forces having impact on the body.

Even though learning plays an important role in development, it was found that the initial responses to rapid perturbations (e.g. support surface translations) still reflect the primitive stereotype short-latency reflexes (SLR). This is in agreement with the observation that these reflexes persist and may re-emerge in adult humans and vertebrate animals after large brain lesions. Therefore, there was *agreement that the balancing responses to external disturbances initially stem from the primitive postural reflexes, after which a transition into voluntary feed forward control evolves* (although some authors were aware of the shortcomings of this notion and of pending new developments from systems analysis and modeling; e.g. Horak & Macpherson, 1996).

Long-latency reflexes

A revival of the reflex concept arose with the discovery of long-latency reflexes (LLR; the long latency originally referred to EMG responses upon electrical or rapid perturbing stimuli). Compared with the SLR, they occur in a context-dependent way and they act through supraspinal pathways to stabilize multiple joints in complex actions (Marsden et al., 1983; Kurtzer et al., 2008). The response and joint stiffness is modified by a subject's prior intentional or volitional set (Hammond, 1956; Marsden et al., 1976). The LLR represent a feedback control process that shares features of voluntary control (Pruszynski et al., 2011). They are affected in basal ganglia diseases, such as Parkinson's disease (PD), in which patients have difficulties in adjusting postural responses to behavioral contexts. The patients produce the responses even when body stability is not endangered, for example in response to a foot rotation stimulus while sitting (Diener et al., 1987).

An instructive example for the context dependency of the LLR in postural control has been reported for transient electrical vestibular stimulation in the adult human (Britton et al., 1993). In free stance, the stimulus evoked leg muscle activation in terms of an initial SLR response and a subsequent LLR response. In contrast, when equilibrium was maintained by the arms holding on to a firm support, the LLR disappeared in the leg muscles and emerged in the arm muscles. It appears that the LLR takes over postural control during development. In experiments on adult animals, for example, lesioning of the fast direct pathways from the vestibular nuclei in the brain stem to target motoneurons in the cervical spinal cord did not produce considerable functional impairment (Wilson & Schor, 1999). Although classical neurophysiology and neuroanatomy primarily focused their work on these direct pathways, their functional role in the adult individual is still open. Possibly, the SLR serves as a kind of 'spare tire' and, because of its short latency, contributes to overall stability of the feedback control loops. Concerning the LLR, they were only recently implemented in a postural control concept (see below).

Recent developments and postural control models

First attempts to describe human quiet stance and balancing with the help of control models reach back about half a century. As mentioned above, engineers' pleas for using high loop gains (which help to resist external disturbances) and their warnings in the face of biological time delays (because they may make the control unstable) prevented the implementation of LLR and favored mainly feed forward ideas of control (Rack, 1981). In a first comprehensive system identification study of human balancing by Fitzpatrick et al. (1996), however, it was shown that the loop gain of human stance control is rather low, which puzzled researchers about how the control might work. Even so, others have provided arguments against the notion that maintaining biped stance is mainly through passive stiffness (e.g., Morasso & Sanguineti, 2002).

Further work offered possible solutions. From a special analysis (stabilogram diffusion analysis) of COP time series, Collins & De Luca (1993) derived two different aspects of control, one called 'short range control' and the other 'long range control'. From this they inferred the existence of two control mechanisms operating during quiet stance, one open-loop and the other closed-loop. This COP behavior, however, can also be explained in terms of noise-related variations in the neural controller and time delay parameters rather than in terms of open-loop versus closed-loop behavior (Peterka, 2000). From observations of muscle length associated with spontaneous body sway during stance, Loram et al. (2005a,b; Lakie & Loram, 2006) hypothesized that stance is maintained by ballistic, catch and throw-like muscular activity with a cyclic pattern favoring a discontinuous control. Others, however, could explain these findings assuming a continuous control (Peterka, 2002; Masani et al., 2006). In the presence of very small activations, nonlinearities such as detection thresholds can, in fact, lead to such cyclic patterns (called limit cycles; compare below, thresholds in the DEC model).

An important aspect in the more recent modeling approaches is the combination of a derivative term with a proportional term in the feedback loop (PD controller). If the loop is used to control the body angle, this mechanism not only tries to minimize deviations of the body angle from the desired value (the P part), but uses in addition body angular velocity information, which precedes position (the D part). In addition, the D part helps to stabilize the system by 'damping' overshoots during rapid movements. Applying it to biological control loops does not necessarily imply that it is implemented in this form in the nervous system. It appears that biological loops tend to feed back both position and velocity sensor signals (as is true for muscle spindle afferents), which for certain purposes can be considered functionally equivalent to PD control. When using PD control and low loop gain, one can now build control loops that include LLR time delays. Interestingly, the low loop gain comes with compliant limbs, which is typical in humans (and is still an aim in robotics where humanoids show rigid limb behavior that may be dangerous to humans).

Another important aspect of human postural control is its multisensory nature, called 'multisensory integration'. This means that, during an external disturbance (e.g. a push having impact on the body), the control system combines proprioceptive joint position, vestibular, and visual signals for the balancing (a 'multisensory integration'). Changes in the relative contribution of the sensors across different disturbance scenarios are often termed 'sensory re-weighting'. In engineering modeling approaches, this was realized using an adaptive 'sensory integration center', in which the sensory signals are combined with additional information ('efference copy'). The approach aimed to find the most accurate sensory representation for a given environmental situation under noise optimization principles using Kalman filters with iterative processing (van der Kooij et al., 1999, 2001; also Kuo, 2005). A much simpler method is realized in the model of Peterka (2002). There, the sensory integration results from combining proprioceptive, vestibular, and visual reflex pathways ('independent channel model'). However, here the modeler performs the sensory re-weighting in relation to changes of the disturbances.

A third solution that combines simplicity in multisensory control and an automatic sensory re-weighting is presented below (DEC model). Before doing so, some building blocks are explained.

Sensor concept

The most relevant information for posture control stems from visual, vestibular, joint angle proprioceptive and joint torque proprioceptive sensors (Horak and Macpherson, 1996). In addition, haptic contact to the support surface has to be known for balance control. Vision improves stance stability, but is dispensable for principle considerations (Mergner et al., 2005). Therefore, the focus here is on the vestibular sensor and the joint angle and torque sensors. Noticeably, these sensors are not identical with the sensory organs (transducers).

For example, the information on self-motion in space and with respect to the gravitational vertical must be derived from combining signals from the vestibular canal and otolith organs, because each of them has flaws. The otolith organs react equally to both the inertial force during head translational acceleration and the gravitational force during head tilt. On the other hand, the canal organs provide information on head angular motion, but this information is not reliable during slow rotations for physical reasons. By way of interactions between canal and otolith signals, the nervous system improves the angular motion information and distinguishes between translation and tilt. The results are not ideal, still allowing self-motion illusions in special situations, but in concert with the other senses they suffice to provide a good estimation of self-orientation (overview in Mergner et al., 2009b).

Another example is the sensor of joint angle and angular velocity. Muscle lengthening during joint rotation leads to firing increase of sensory endings in muscle spindle receptors. These spindles make a major contribution to the joint angle and movement sense ('muscle proprioception', 'kinesthesia'). However, because the muscle is elastic, force signals need to be included in the sense to obtain reliable estimates in both active and passive conditions. The force information appears to stem from transducers in the tendons (Golgi tendon organs, GTOs; see Duysens et al., 2000). With active movements, a contribution from central sources ('effort'; Gandevia, 1987) also seems to play a role. Furthermore, transducers in the skin and in the joint capsules participate (Jones, 1972). Appropriate combinations of the transducer signals then provide estimates of the physical variables. These will from now on be called 'sensors'.

Thus, the sensors are virtual in that their signals arise from interactions within a distributed neural network. This network shows properties similar to artificial neural nets, where information on function typically cannot be drawn from a node within the net because the signal processing is distributed across many nodes. Functions can be inferred, however, from nodes at the input and output sites. An output into perception can be used to learn which information the sensors are providing. This has been done in human psychophysical experiments on conscious perception of self-motion with the help of open loop indication and estimation procedures (overview, Mergner, 2002). They showed that the percepts represent the physical variables such as position and velocity of joint angle or head rotation in space. In view of the mostly valid action–perception congruency in human behavior, it is plausible to assume that the sensors also feed into the sensorimotor systems.

Meta level concept

When subjects give verbal reports in psychophysical self-motion experiments, these typically do not reflect the sensor signals such as the vestibular head angular velocity. Rather, they reflect a reconstruction of the outside world event that was causing the self-motion, such as 'the chair I am sitting on was rotated first leftwards and then rightwards' (in psychological literature this is called 'distal stimulus', in contra-distinction to the 'proximal stimulus' that refers to the fluid pressure on a cupular receptor in the vestibular canals or a photon hitting a receptor cell in the retina). For the internal reconstruction of the outside events, typically several sensor signals are combined. This is performed by a continuous event 'estimation' from noisy sensory inputs, cognition and expectations.

The concept of perceptual reconstruction of external events from multisensory input was applied to the modeling of human balancing experiments, when it became clear that this required a novel concept of sensory processing (Mergner et al., 2003; Mergner, 2004). The question was, which external events in relation to balancing are perceptually reconstructed and from which sensors in which way. It was concluded on theoretical grounds that four external disturbances are reconstructed: support surface rotation and translational acceleration, field forces such as gravity, and contact forces such as a push against the body. Furthermore, it was identified which combinations of the vestibular, joint angle and joint torque sensor signal are required for the estimation of the four external disturbances (overview, Mergner, 2010).

Servo loop concept

Modeling of human stance control requires a concept of the underlying motor mechanism. As mentioned above, sensorimotor researchers were fascinated by engineering servo control (or *servomechanism*) because it shares essential points with the biological *reflex concept* in that it uses *automatic error-sensing negative feedback*. The technical concept is best known from effortlessly moving a steering wheel in a car, which makes the car wheels turn through some mechanical amplification (the car driver is serving as operator who first observes and then, based on certain intentions, etc., controls the servomechanism). In the version of the servo loop described below for stance control, movement performance by the mechanism is virtually without effort in that body inertia is accounted for by corresponding joint torque.

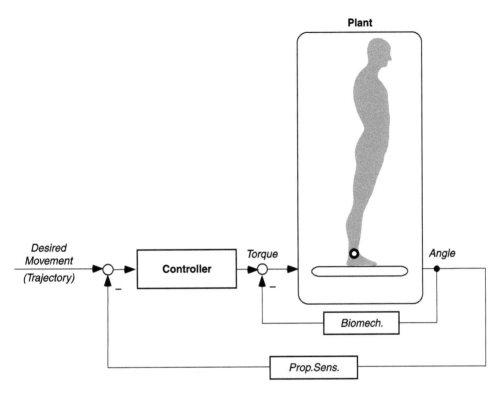

Figure 3.3   Servo control model. The desired movement is driven by error-sensing 'reflexive' negative feedback. Biomech., instrinsic (passive) muscle stiffness and damping. Prop.Sens., proprioceptive sensor of joint angle (here of ankle joint). Given feedback and controller parameters are adjusted to plant dynamics, the loop compensates for body inertia when performing the desired movement trajectory.

In this version, the input is a signal indicating the desired position or movement (displacement trajectory). The difference between the desired trajectory and the actual trajectory, reported by proprioceptive feedback of joint position, drives the movement via a controller (lower loop in Figure 3.3). The controller uses a proportional and a derivative term (PD controller) adjusted to yield a joint torque signal ('motor command') which is adapted to the plant dynamics. The command-to-movement transformation implies a joint actuator with an inner force control loop (not shown in the model). The negative proprioceptive feedback represents a short-latency reflex (SLR) that stabilizes the joint according to the desired position or rotates it according to the desired displacement (McIntyre & Bizzi, 1993, wrote of 'the beauty of movement from posture'). The generated reflexive stiffness combines with passive muscle stiffness (and damping) which provides additional negative feedback with virtually no time delay (box Biomech., for biomechanics; to some extent adjustable through co-contraction of agonist–antagonist muscle pairs).

An advantage of this mechanism is that it does not require an additional feed forward signal of inverse plant dynamics (this argument becomes relevant when additional mechanisms need to be added to the servo; see below). A problem with this mechanism is that the proprioceptive feedback is too weak to account sufficiently for the compensation of external disturbances such as gravity. However, this problem can be eliminated when one uses the estimations of the external disturbances from the meta-level to compensate them.

Disturbance estimation and compensation (DEC) model

Extending the servomechanism by the estimators of the four disturbances and feeding their signals with inversed signs into the input summing junction compensates for the disturbances (strongly simplified DEC model in Figure 3.4). Given full compensation, the servo can function and fulfill its task as if there was essentially no gravity or other external disturbance.

A more detailed picture of the DEC model is given in Figure 3.5 (for anterior–posterior balancing of the body around the ankle joints, biomechanically simplified as if the body was a single inverted pendulum). The fusion of the sensory transducer signals into the sensors is omitted. The figure shows the distributed network of sensor signals feeding into the disturbance estimators. Note that the vestibular sensor, for example, inputs into all four disturbance estimators (shown are actually two vestibular sensors, because the vestibular translation signal feeds into the support translation estimation, while the other vestibular signals deal with angular motion). Furthermore, note that each estimator receives a specific set of inputs, so that changes in the disturbances are automatically associated with changes in the contributions from each sensor – which represents a sensory re-weighting. Another type of the re-weighting comes from detection threshold mechanisms in the estimators. They are causing an amplitude non-linearity in the form that compensation of small disturbances is relatively weak and automatically gets more efficient when the disturbances get larger (in accordance with human behavior; Maurer et al., 2006). Additional effects of the two types of sensory re-weighting concern the reduction of sensor noise in the DEC loops (see Mergner et al., 2009b).

The DEC model was developed in an iterative back and forth between balancing experiments, modeling and model simulations. Eventually, the model described and predicted human balancing in a variety of external disturbance scenarios (Maurer et al., 2006; overview, Mergner et al., 2009b). It also covers the superposition of the different disturbances and of the disturbances and voluntary movements.

Furthermore, it has been extended to also include the hip joints in addition to the ankle joints (double inverted pendulum dynamics). Interconnections of sensors and estimators across the body segments lead to the emergence of human-like movement synergies and postural adjustments. In the example shown in Figure 3.6, trunk bending forward is automatically associated with a backward lean of the legs, because the DEC method tries to maintain the

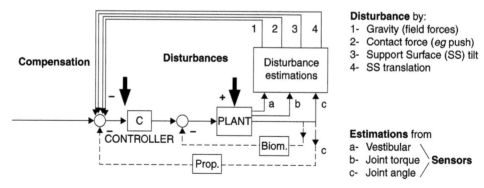

Figure 3.4 Servo control model extended by long latency reflex (LLR) loops for disturbance estimation and compensation ('DEC') model. Estimations of disturbances (1–4) receive multisensory input (a–c). Disturbance compensation occurs through feedback after sign reversal. Given ideal compensation, the servo loop performs the desired movement trajectory as if there were no disturbances.

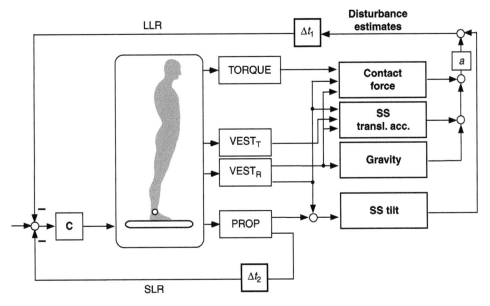

Figure 3.5 More detailed version of DEC model. It shows the network of sensor output signals that feed into the disturbance estimations (VEST$_T$ and VEST$_R$, vestibular sensors of head translation and rotation, respectively). The lower feedback loop represents the servomechanism in terms of short-latency reflex, SLR (lumped time delay is in the order of Δt_2 = 80 ms). The upper loop represents long latency reflexes, LLR (Δt_1 = 120–180 ms or more). The controlled variable is body-in-space angle. The torque signals in the LLR loops (upper three) are transformed into body-in-space angle equivalents (box a; e.g., estimate of COM gravitational torque, box Gravity, relates to body lean angle). Also, the support surface (SS) tilt estimate, a foot-in-space signal, is transformed into a body-in-space signal in that it combines with the body-to-foot signal from the proprioceptive SLR loop (this LLR loop is not shown separtely, for sinplicity).

COM of the body as a whole above the ankle joints. Together with simulations that included 4 DOF, Hettich et al. (2011) demonstrated modularity of the DEC method, meaning that the same control method can be applied as a module for each segment in a stack of segments. Control reconfiguration automatically occurs when a joint becomes fixed (e.g. by a plaster) or may be easily established for use of an additional segment (e.g., by handling a tool).

A basic constituent of the DEC model is the vestibular sensor. But human subjects without vestibular function are able to balance support surface tilts and pull stimuli, and this even with closed eyes (Maurer et al., 2006). When modeling this within the DEC framework, the postural responses of the vestibular loss subjects could be formally described and simulated in software simulations and robot experiments (Schweigart & Mergner, 2008; Mergner et al., 2009b). This implied the assumption that human subjects use force/torque cues to substitute the vestibular signals. However, this substitution endangers control stability and limits the working range of the balancing (Mergner et al., 2009a).

Cognitive aspects of the DEC model

The DEC loops represent LLR in that they perform context-dependent disturbance compensations involving loops through basal ganglia and the cerebral cortex. One aspect of this

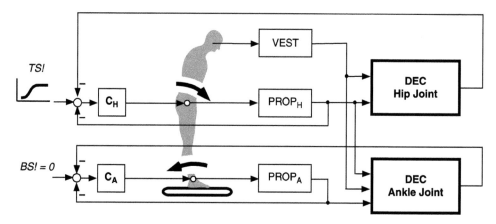

Figure 3.6 Emergence of movement synergy and postural adjustment from hierarchical DEC control (extended to include the hip joints in addition to the ankle joints). Trunk lean forward automatically leads to backward lean of legs. Simplified control model of corresponding double inverted pendulum biomechanics, where the desired trunk-in-space movement signal (*TS!*) is fed into hip control, while maintaining upright position of whole-body COM-in-space is desired for the ankle control (*BS!* = 0°). Ankle control takes into account the change in body geometry by including sensory information of trunk motion. Thus, the leg movement emerges from the DEC principle and does not necessarily require feed forward signals.

context dependency is that compensations are continuously adjusted to external disturbances. For example, when standing in a moving bus and the disturbances are changing from support surface tilt to translational acceleration, the type and strength of the compensations change accordingly. From the outside, the impression might be that an intelligent observer analyzes the scenario and adjusts the compensation accordingly. Actually, it is the DEC mechanism that in real-time automatically extracts the disturbances. The output mechanisms of the estimators then compute the disturbance compensations for the relevant joints (especially ankle, hip, and knee joints). This may involve motor learning and automating in the cerebellum.

Another aspect of the context dependency is that the compensation can voluntarily be influenced. For example, when standing on a tilting platform, an automatic response would be to compensate the tilt in order to remain upright, but it is also possible to voluntarily downscale the compensation and get rotated with the platform.

These two aspects make the DEC processing level an intermediate between the high level of cognition (with attention allocation, planning, reasoning, etc.) and the low levels of basic sensory processing and primitive reflex motor control. There may be even more intermediate level functions, *disturbance expectation and anticipation*, in addition to the context dependency of disturbance compensation.

One idea was that the cognitive level may associate objects and events in the outside world (ride in red bus) with typical disturbances in world or task coordinates. This may be advantageous compared to associations with patterns of muscle activities and joint torques in body coordinates as they change with body configuration. Having learned the associations between outside event and disturbance, a given event (red stop light) may invoke an *expected* disturbance (support surface deceleration) and predictive compensation (anticipatory postural adjustment). The expected disturbance signal would be fed forward from the cerebral cortex into the estimator and fused with sensory feedback estimates. Similarly, such expected disturbance signals can be thought to be derived from desired movement signals during voluntary movements, dealing then with the self-produced external disturbances (e.g., an increasing gravitational ankle torque with active body lean) or internal disturbances (inter-segmental torques).

Both ideas underwent proof-of-principle tests in experiments that used humanoid robots, in which the DEC model was implemented, in a human posture control laboratory for comparison with human subjects (see next section).

Robots and posture control research

Abstracting human experimental data into control models faces the problem that the models of complex mechanisms such as human sensorimotor control tend to be incomplete. For example, they do not usually contain 'real world' noisy and inaccurate sensors or non-linearities such as mechanical dead zones and thresholds. Some features such as foot friction on the support surface may even be very difficult to approximate in such models. Using robots in so-called 'hardware in the loop' simulations in addition to software computer simulation enhances the modeling. Benefits can be expected not only for sensorimotor research, but also for robotics where researchers would like to learn from biology human-like features such as failsafe robust-ness, flexibility, energy and processing efficiency, etc. (Mergner & Tahboub, 2009).

With this in mind, the DEC model was re-embodied in a humanoid robot (Figure 3.7) that was equipped with human-inspired mechatronic sensors and artificial muscles (Mergner et al., 2006). It was tested in the posture control laboratory in direct comparison with humans, demonstrating the validity of the DEC model (Figure 3.8; see Mergner et al., 2009b; Hettich et al., 2011). In a further step, it served the proof-of-principle tests for the predictive (feed

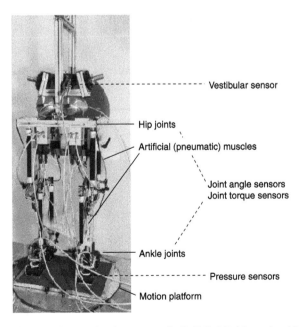

Figure 3.7   Picture of postural control robot Posturob II (2 DOF, hip and ankle joints). It is given human anthropometric parameters, uses artificial muscles and mechatronic artificial sensors and is controlled by the DEC algorithm. It serves as 'real world' model that is tested in the postural control laboratory in the same way as the human subjects.

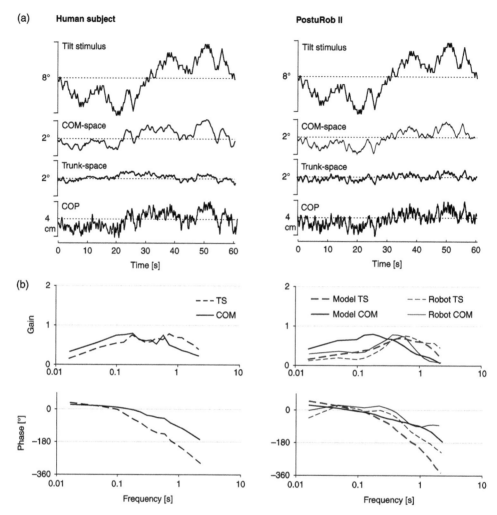

Figure 3.8 Comparison of postural responses between human and robot experiments. (a) Responses to pseudorandom sequence of anterior–posterior support surface tilts. Shown are angular wholebody COM and trunk excursions with respect to the gravitational space vertical and COP shifts (averages over five trials from one subject). (b) Gain and phase of the trunk and COM tilt responses plotted over frequency. Bode plots were calculated using cross-spectral analysis (Fourier transform; human data, averages of six subjects). Zero gain means that COM and trunk remain perfectly vertical, while unity gain means their rotation with the support surface (motion platform). The phase indicates the temporal relation between stimulus and response (adapted from Hettich et al., 2011).

forward) aspects of the DEC model (Mergner, 2010). The results of these robot experiments now provide the basis for future human experiments.

Further benefits obtained by including the robot into research were 'learning by building', making teaching more attractive, and better public reception for this research. Even more important, the robot with its human-like sensorimotor behavior represents a guideline for improving medical assistive devices such as prostheses and exoskeletons.

Summary

When we reach out with our arm or maintain a body posture, we compensate for mechanical disturbing factors such as gravity usually without being aware of it. The need to stabilize is more intuitive when we consider our activities during biped stance since we need to balance almost continuously. The balancing may be challenged not only by external factors such as gravity or compliant or rough terrain, but also by our own actions such as trunk bending or foot lifting. It has been known for a century that sensory feedback control is a key to understanding such stabilizations. Originally, it was thought to occur in the form of primitive postural reflexes. These reflexes, however, do not explain humans' rich and flexible sensorimotor behavior. Only recently have researchers discovered how the primitive reflexes become superseded by context-dependent sensory feedback in automated stabilizing reactions, which develop during sensorimotor maturation in childhood. These automated reactions provide the necessary behavioral flexibility. The underlying mechanisms have been abstracted and formalized for use in computer simulations and humanoid robots. This research helps to improve therapies and assistive devices for sensorimotor-impaired patients.

Acknowledgements

The author would like to thank his colleagues L. Assländer and G. Hettich for valuable discussions and their help in editing the chapter.

References

Alexandrov, A.V., Frolov, A.A. and Massion, J. (2001) 'Biomechanical analysis of movement strategies in human forward trunk bending. I. Modeling', *Biological Cybernetics*, 84: 425–34.

Allum, J.H.J., Honegger F. and Pfaltz, C.R. (1989) 'The role of stretch and vestibulo-spinal reflexes in the generation of human equilibrating reactions', *Progress in Brain Research*, 80: 399–409.

Bard, P. (1933) 'Studies on the cerebral cortex I. Localized control of placing and hopping reactions in the cat and their normal management by small cortical remnants', *Archives of Neurology and Psychiatry*, 30: 40–74.

Bizzi, E., Dev, P., Morasso, P. and Polit, A. (1978) 'The effect of load disturbances during centrally initiated movements', *Journal of Neurophysiology*, 41: 542–55.

Bizzi, E., Accornero, N., Chapple, W. and Hogan, N. (1984) 'Posture control and trajectory formation during arm movement', *Journal of Neuroscience*, 4: 2738–44.

Bouisset, S. and Do, M.C. (2008) 'Posture, dynamic stability, and voluntary movement', *Neurophysiologie Clinique*, 38: 345–62.

Britton, T.C., Day, B.L., Brown, P., Rothwell, J.C., Thompson P.D. and Marsden, C.D. (1993) 'Postural electromyographic responses in the arm and leg following galvanic vestibular stimulation in man', *Experimental Brain Research*, 94: 143–51.

Brooks, C.M. (1933) 'Studies on the cerebral cortex II. Localized representation of hopping and placing reactions in the rat', *Archives of Neurology and Psychiatry*, 30: 162–71.

Collins, J.J. and De Luca, C.J. (1993) 'Open-loop and closed-loop control of posture: a random-walk analysis of center-of-pressure trajectories', *Experimental Brain Research*, 95: 308–18.

Denny-Brown, D. (1929) 'On the nature of postural reflexes', *Proceedings of the Royal Society*, 104: 252–301.

Diener, C., Scholz, E., Guschlbauer, B. and Dichgans, J. (1987) 'Increased shortening reaction in Parkinson's disease reflects a difficulty in modulating long loop reflexes', *Movement Disorders* 2: 31–6.

Duysens, J., Clarac, F. and Cruse, H. (2000) 'Load-regulating mechanisms in gait and posture: comparative aspects', *Physiological Reviews*, 80: 83–132.

Fitzpatrick, R., Burke, D. and Gandevia, S.C. (1996) 'Loop gain of reflexes controlling human standing measured with the use of postural and vestibular disturbances', *Journal of Neurophysiology*, 76: 3994–4008.

Gandevia, S.G. (1987) 'Roles for perceived voluntary motor commands in motor control', *Trends in Neurosciences*, 10: 81–5.

Gibbs, P., and Sugihara, H. (1996/97) What is Occam's Razor? *Usenet Physics FAQ*, http://math.ucr. edu/home/baez/physics/General/occam.html.

Goltz, F. (1870) 'Ueber die physiologische Bedeutung der Bogengänge des Ohrlabyrinths', *Archiv der Physiologie*, 3: 172–92.

Granit, R. (1955) *Receptors and Sensory Perception*, London: Yale University Press.

Gurfinkel, V.S. (1994) 'The mechanisms of postural regulation in man', *Soviet Science Review for Physics, Genetics and Biology* 7: 59–89.

Hammond, P.H. (1956) 'The influence of prior instruction to the subject on an apparently involuntary neuro-muscular response', *Journal of Physiology*, 132: 17–18P.

Hellebrandt, F.A. and Franseen, E.B. (1943) 'Physiological study of the vertical stance of man', *Physiological Reviews*, 33: 202–55.

Hettich, G., Fennel, L. and Mergner, T. (2011) 'Double inverted pendulum model of reactive human stance control', Multibody Dynamics 2011, ECCOMAS Thematic Conference, Brussels.

Hlavacka, F., Krizkova, M., Bodor, O. and Mergner, T. (1996) 'Adjustment of equilibrium point in human stance following galvanic vestibular stimulation: Influence of trunk and head position', in Stuart D. (ed.) *Motor Control VII* (pp. 289–293), Tucson, AZ: Motor Control Press.

Holmes G. (1922) 'The Croonian lectures on the clinical symptoms of cerebellar disease and their interpretation', in C.G. Phillips (ed.) *Selected Papers of Gordon Holmes* (pp. 186–203), New York: Oxford University Press.

Horak, F.B. and Macpherson, J.M. (1996) 'Postural orientation and equilibrium', in L. Rowell and J. Shepherd (eds) *Handbook of Physiology* (pp. 255–92), New York: Oxford University Press.

Jones, E.G. (1972) 'The development of the 'muscular sense' concept during the nineteenth century and the work of H. Charlton Bastian', *Journal of the Medicine and Allied Sciences*, 27: 298–311.

Kuo, A.D. (2005) 'An optimal state estimation model of sensory integration in human postural balance', *Journal of Neural Engineering*, 2: 235–49.

Kuo, A.D. and Zajac, F.E. (1993a) 'Human standing posture: Multi-joint movement strategies based on biomechanical constraints', *Progress in Brain Research*, 97: 349–58.

Kuo, A.D. and Zajac F.E. (1993b) 'A biomechanical analysis of muscle strength as a limiting factor in standing posture', *Journal of Biomechanics*, 26: 137–50.

Kurtzer, I.L., Pruszynski, J.A. and Scott, S.H. (2008) 'Long-latency reflexes of the human arm reflect an internal model of limb dynamics', *Current Biology*, 18: 449–53.

Lackner, J.R. and DiZio, P. (1994) 'Rapid adaptation to Coriolis force perturbations of arm trajectories', *Journal of Neurophysiology*, 72:299–313.

Lakie, M. and Loram, I.D. (2006) 'Manually controlled human balancing using visual, vestibular and proprioceptive senses involves a common, low frequency neural process', *Journal of Physiology*, 577: 403–16.

Loram, I.D., Maganaris, C.N. and Lakie, M. (2005a) 'Active, non-spring-like muscle movements in human postural sway: how might paradoxical changes in muscle length be produced?', *Journal of Physiology*, 564: 281–93.

Loram, I.D., Maganaris, C.N. and Lakie, M. (2005b) 'Human postural sway results from frequent, ballistic bias impulses by soleus and gastrocnemius', *Journal of Physiology*, 564: 295–311.

Lund, S. and Broberg, C. (1983) 'Effects of different head positions on postural sway in man induced by a reproducible vestibular error signal', *Acta Physiologica Scandinavica*, 117: 307–9.

Mackay, W.A. (1980) 'The motor program: back to the computer', *Trends in Neuroscience*, 3: 97–100.

Magnus, R. (1924) *Körperstellung*, Berlin: Springer.

Marsden, C.D., Merton, P.A. and Morton, H.B. (1976) 'Stretch reflex and servo action in a variety of human muscles', *Journal of Physiology*, 259: 531–60.

Marsden, C.D., Rothwell, J.C. and Day, B.L. (1983) 'Long-latency automatic responses to muscle stretch in man: origin and function' *Advances in Neurology* 39: 509–39.

Masani, K., Vette, A.H., and Popovic, M.R. (2006) 'Controlling balance during quiet stance: proportional and derivative controller generates preceding motor command to body sway position observed in experiments', *Gait & Posture*, 23, 164–72.

Massion, J. (1992) 'Movement, posture and equilibrium: Interaction and coordination', *Progress in Neurobiology*, 38:35–56.

Massion, J. (1998) 'Postural control systems in developmental perspective', *Neuroscience and Biobehavioral Reviews*, 22:465–72.

Mathews, P.B.C. (1972) *Mammalian Muscle Receptors and Their Central Actions*, Baltimore: Williams & Wilkins.

Maurer, C., Mergner, T. and Peterka, R.J. (2006) 'Multisensory control of human upright stance', *Experimental Brain Research*, 171: 231–50.

McIntyre, J. and Bizzi, E. (1993) 'Servo hypotheses for the biological control of movement', *Journal of Motor Behavior*, 25: 193–202.

Mergner, T. (2002) 'The Matryoshka Dolls principle in human dynamic behavior in space – A theory of linked references for multisensory perception and control of action', *Current Psychology of Cognition*, 21: 129–212.

Mergner, T. (2004) 'Meta level concept versus classic reflex concept for the control of posture and movement', *Archives Italiennes de Biologie*, 142: 175–98.

Mergner, T. (2010) 'A neurological view on reactive stance control', *Annual Reviews in Control*, 34: 177–98.

Mergner, T. and Tahboub, K.A., (2009) 'Neurorobotics approaches to human and humanoid sensori-motor control', *Journal of Physiology Paris*, 103: 115–18.

Mergner, T., Huber, W. and Becker, W. (1997) 'Vestibular-neck interaction and transformations of sensory coordinates', *Journal of Vestibular Research*, 7: 119–35.

Mergner, T., Huethe, F., Maurer, C. and Ament, C. (2006) 'Human equilibrium control principles implemented into a biped robot', in T. Zielinska and C. Zielinski (eds.) *Robot Design, Dynamics, and Control* (Romansy 16), CISM Courses and Lectures, 487: 271–9, Vienna: Springer.

Mergner, T., Maurer, C. and Peterka, R.J. (2003) 'A multisensory posture control model of human upright stance', *Progress in Brain Research*, 142: 189–201.

Mergner, T., Schweigart, G. and Fennell, L. (2009b) 'Vestibular humanoid postural control', *Journal of Physiology Paris*, 103: 178–94.

Mergner, T., Schweigart, G., Fennell, L. and Maurer, C. (2009a) 'Posture control in vestibular loss patients', *Annals of the New York Academy of Sciences*, 1164: 206–15.

Mergner, T., Schweigart, G., Maurer, C. and Bluemle, A. (2005) 'Human postural responses to motion of real and virtual visual environments under different support base conditions', *Experimental Brain Research*, 167: 535–56.

Merton, P.A. (1953) 'Speculations on the servo control of movement', in G.E.W. Wolstenholme (ed) *The Spinal Cord* (pp. 247–255), Boston, MA: Little Brown.

Morasso, P.G. and Sanguineti, V. (2002) 'Ankle muscle stiffness alone cannot stabilize balance during quiet standing', *Journal of Neurophysiology*, 88: 2157–62.

Nashner, L.M. and Horak, F. B. (1986) 'Central programming of postural movements: adaptation to altered support-surface configurations', *Journal of Neurophysiology*, 55: 1369–81.

Nashner, L.M. and McCollum, G. (1985) 'The organization of human postural movements: a formal basis and experimental synthesis', *Behavioral and Brain Sciences*, 8: 135–72.

Nielsen, J.B. (1998) 'Co-constraction of antagonistic muscles in man', *Danish Medical Bulletin*, 45: 423–35.

Peterka, R.J. (2000) 'Postural control model interpretation of stabilogram diffusion analysis', *Biological Cybernetics*, 82: 335–43.

Peterka, R.J. (2002) 'Sensorimotor integration in human postural control', *Journal of Neurophysiology*, 88:1097–118.

Prechtl, H.F.R. (1977) 'The neurological examination of the full-term newborn infant' In *Clinics in Developmental Medicine*, No. 63 (pp. 1–68). 2nd edn, London: Heinemann.

Pruszynski, J.A., Kurtzer, I., Nashed, J., Omrani, M., Brouwer, B. and Scott, S.H. (2011) 'Primary motor cortex underlies multi-joint integration for fast feedback control', *Nature*, 478: 387–90.

Rack, P.M.H. (1981) 'Limitations of somatosensory feedback in control of posture and movement' in *Handbook of Physiology: The Nervous System II* (ch. 7), Baltimore, MD: Williams and Willkins.

Rademaker, G.G.J. (1931) *Das Stehen*, Berlin: Springer.

Roberts, T.D.M. (1978). *Neurophysiology of Postural Mechanisms*, Part I, ch. 10 (Reflexes of balance), London, Boston: Butterworths.

Schweigart, G. and Mergner, T. (2008) 'Human stance control beyond steady state response and inverted pendulum simplification', *Experimental Brain Research*, 185: 635–53.

Sherrington, C.S. (1900) 'The muscular sense', in E.A., Schafer & J. Young (eds) *Textbook of Physiology* (pp. 1002–25), Edinburgh, London: Pentland.

Sherrington, C.S. (1931) 'Quantitative management of contraction in lowest level coordination' (Hughlings Jackson Lecture), *Brain*, 54: 1–28.

Simons, A. (1923) 'Kopfhaltung und Ruhetonus', *Zeitschrift für die gesamte Neurologie und Psychiatrie*, 80: 499–549.

van der Kooij, H., Jacobs, R., Koopman, B. and Grootenboer, H. (1999) 'A multisensory integration model of human stance control', *Biological Cybernetics*, 80: 299–308.

van der Kooij, H., Jacobs, R., Koopman, B. and van der Helm, F. (2001) 'An adaptive model of sensory integration in a dynamic environment applied to human stance control', *Biological Cybernetics*, 84: 103–15.

von Holst, E. and Mittelstaedt, H. (1950) 'Das Reafferenzprinzip (Wechselwirkung zwischen Zentralnervensystem und Peripherie)', *Naturwissenschaften*, 37: 464–76.

Walshe, F.M.R. (1923) 'On certain tonic or postural reflexes in hemiplegia, with special reference to the so-called "associated movements"', *Brain*, 46: 1–37.

Wilson, V. and Melvill Jones, G. (1979) *Mammalian Vestibular Physiology.* New York, NY: Plenum Press.

Wilson, V.J. and Schor, R.H. (1999) 'The neural substrate of the vestibulocollic reflex. What needs to be learned', *Experimental Brain Research*, 129: 483–93.

4

MOTOR LEARNING EXPLORED WITH MYOELECTRIC AND NEURAL INTERFACES

Andrew Jackson and Kianoush Nazarpour

INSTITUTE OF NEUROSCIENCE, NEWCASTLE UNIVERSITY, UNITED KINGDOM

Introduction

Most of the time we take for granted the effortless way in which our thoughts are transformed into actions; upon feeling thirsty we reach for a glass of water without pausing to reflect on the ease with which multiple muscles were co-ordinated to bring the hand to the glass, grasp it between fingers and thumb and lift it smoothly to the mouth without spilling a drop. Yet this task is a significant challenge for even the most sophisticated robots, and the loss of motor function can be one of the most devastating consequences of disease or injury to the nervous system.

In recent years, it has become possible to record electrical signals directly from the nervous system to enable communication or control over external devices such as computers and artificial prostheses. Myoelectric interfaces (controlled by signals recorded from muscles) and neural interfaces (controlled directly from the brain) effectively provide new motor pathways for patients who have lost the use of a limb through amputation or paralysis. Unfortunately the performance of these artificial interfaces rarely matches the speed and accuracy of natural limb movements. Their practical utility therefore depends critically on the extent to which patients can learn new motor strategies appropriate for these devices.

Studying how the brain acquires control over a new motor pathway provides a unique window onto the mechanisms of motor learning. Experimental myoelectric and neural interfaces can create simplified sensorimotor worlds in which the map from motor commands to effectors can be precisely controlled. These abstract sensorimotor paradigms offer an opportunity to explore further fundamental motor learning questions that would otherwise be obscured by the anatomical and biomechanical complexity of the limbs. By introducing highly artificial and unusual sensorimotor mappings, we may ask whether the human motor system is constrained to naturalistic behaviors or can adapt to circumstances outside the normal ethological repertoire. Ultimately, it is hoped that a better understanding of these processes will inform the development of improved interfaces designed to exploit our remarkable capacity for motor learning to restore or even enhance function.

In this chapter we will see how operation of myoelectric interfaces reveals the flexibility inherent in the motor system to optimize behavior for abstract sensorimotor environments.

Along the way, we will highlight a number of concepts with relevance also to the control of natural movements, for example the influence of neuromotor noise on movement accuracy, hierarchical levels of redundancy and the flexible use of muscle synergies. We will find that three broad themes emerge from these studies:

1   Initially people find it easiest to operate interfaces that resemble natural limb function, which we will refer to as *biomimetic interfaces*. However, with sufficient training, they can learn to use interfaces that bear no relationship to the limb (*abstract interfaces*). In fact, the ultimate performance levels obtained with both biomimetic and abstract interfaces are often equivalent.
2   A key problem for generating accurate movements using a myoelectric interface is that the control signals recorded from the muscles are corrupted by *noise*. When operating abstract interfaces, the brain must learn new strategies to minimize the influence of this noise on task success. This is achieved by exploiting *redundancy* in the mapping from control signals to effector movements, and through flexible use of divergent pathways to form new, task-specific *muscle synergies*.
3   The *distal muscles* acting on the hand and fingers exhibit greater flexibility compared with shoulder and arm muscles, and are therefore more suited to forming unnatural synergies appropriate for abstract interfaces. This corresponds to an increased contribution to control of these muscles in primates by direct connections from the cortex via the corticospinal pathway.

We will then introduce the principles underlying the operation of direct neural interfaces and discuss their conceptual similarity with myoelectric interfaces. We will end with some speculation as to how the themes we have highlighted might inform the future direction of neural interface research.

Clinical and experimental applications of myoelectric interfaces

The electrical signal generated by muscles, known as the *electromyogram* (EMG), is readily recorded by surface electrodes placed over the skin. The EMG signal comprises the summation of motor unit action potentials generated within the targeted muscles, increasing in amplitude to tens of millivolts as motor units are recruited and the muscle contracts. There is an approximately linear relationship between the rectified EMG and force under conditions of isometric muscle contraction. Myoelectric interfaces process EMG signals from one or more muscles to allow real-time control of a device. For example, an amputee who has lost the hand at the wrist may open and close a myoelectric hand prosthesis using extensor and flexor muscles located in the forearm. After amputation at a higher level, signals from shoulder muscles may to some extent substitute for more distal musculature, especially if these can be surgically reinnervated by severed peripheral nerves.

The extent to which the brain can incorporate a myoelectric prosthesis as an extension of the body can be studied in the laboratory using a simple myoelectric cursor control task in healthy subjects. The typical set-up is summarized in Figure 4.1a (Radhakrishnan et al., 2008). Subjects control the interface by making isometric contractions of arm and hand muscles with the upper limb restrained. The task is to move a cursor to targets that are presented on a screen. The EMG signals from several muscles are first rectified and then smoothed to produce control signals. Each muscle is associated with a different direction of action (DoA) in the screen space. Cursor position is determined by the sum of vectors aligned

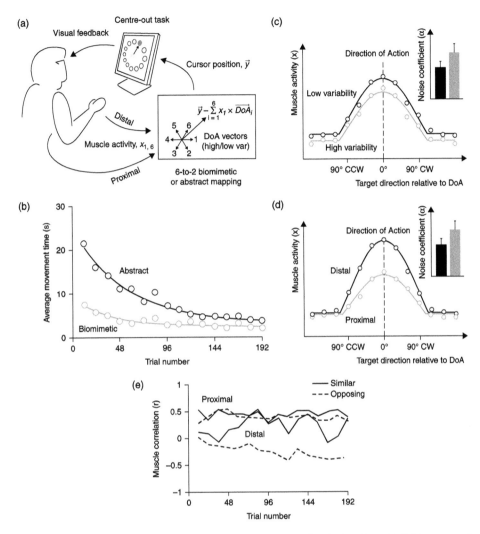

Figure 4.1 Exploring abstract sensorimotor learning with a myoelectric interface. (a) Subjects control a myoelectric cursor using smoothed EMG activity recorded from proximal (shoulder/elbow) and distal (hand) muscles of the upper limb. The summation of DoA vectors scaled by muscle activity is used to calculate cursor position according to a biomimetic or abstract scheme. The variability of specific muscles is altered by changing the smoothing window. The resultant cursor position is displayed to the subject on a monitor, forming a closed sensorimotor loop. (b) Learning curves for control of the myoelectric interface with biomimetic or abstract mappings. Plotted on the vertical axis is the average time taken to move the cursor to target and successfully complete a trial. Performance is initially poor using an abstract mapping, but quickly converges to the same level as under biomimetic control. (c) Tuning functions for muscles with high and low variability. The activity of each muscle is plotted for different target directions, relative to the DoA for that muscle. Tuning curves are fitted by truncated cosine functions. Muscles with a larger coefficient of variability (inset) contribute less to cursor control. (d) Tuning functions for distal and proximal muscles reveal differential contribution to cursor control despite comparable coefficients of variability (inset). (e) Correlation between pairs of muscles that have similar (solid lines) and opposing (dashed lines) actions on the cursor. For pairs of distal muscles, the sign of correlation diverges with training to reflect appropriate muscle use. For pairs of proximal muscles, the correlation remains positive, revealing subjects' difficulty in dissociating proximal muscle synergies. Adapted from Radhakrishnan et al. (2008).

to the DoAs, scaled by the control signals derived from corresponding muscles. The subject receives visual feedback of the resultant cursor position in real-time on a computer screen. As such, the user myoelectric interface and visual feedback effectively form a new closed sensorimotor loop, and acquiring accurate control of the device provides a simple model of *sensorimotor learning.*

Learning new sensorimotor relationships

Studies of sensorimotor learning date back to Hermann von Helmholtz in the late nineteenth century. Using prism lenses to displace laterally the perceived position of objects within the field of view, he found that movement errors due to the visuomotor misalignment diminish quickly with practice. Once the lenses are removed, errors in the opposite direction (*after-effects*) demonstrate that the adaptation is due to a recalibration of the transformation from visual to motor co-ordinate frames in the brain. More recently, Shadmehr and Mussa-Ivaldi (1994) have used a robotic manipulandum to exert perturbing forces to the hands of subjects while they made reaching movements. Initially this perturbation causes subjects to make curved movements as their hand is pushed off course, but with training the trajectories become straight and accurate. When the perturbation is removed, after-effects are observed in the form of trajectories that curve in the opposite direction. Again, this demonstrates specific adaptation of an internal representation of the dynamics of hand movement (rather than a general strategy to cope with uncertainty such as stiffening the arm). Prism and forcefield studies have become paradigmatic examples of sensorimotor adaptation, demonstrating the brain's apparent flexibility to recalibrate control strategies in altered sensorimotor environments. But while such tasks require the brain to learn novel sensory consequences of motor commands, these commands nevertheless act through forces and torques generated by the limb's normal musculature and mechanics. In fact, these experimental manipulations resemble situations we might expect to encounter in our normal lives, for example as we attempt to open the front door whilst carrying a heavy shopping bag. It is perhaps not surprising to find that the brain, optimized over evolutionary and developmental time-scales, is capable of adapting quickly to such circumstances.

By contrast, operation of a device using electrical activity recorded directly from the nervous system, bypassing movements of the limbs, has only become possible due to the technological advances of recent years. There is no guarantee that we should necessarily come pre-equipped with innate neural architectures capable of controlling behavior that is entirely dissociated from the body. Therefore, the myoelectric interface provides a powerful paradigm to explore the extent to which our motor system is limited to naturalistic limb movements or can transcend bodily constraints to produce behavior that is optimized for abstract sensorimotor environments.

To address this question, we can compare people's ability to learn a *biomimetic interface* (designed to mimic the natural action of muscles on the limb) with one that is *abstract* (unrelated to the normal muscle configuration of the limb) (Radhakrishnan et al, 2008). During biomimetic control, muscles act along DoAs that are consistent with their action in the posture in which the limb is restrained (for example, with the palm of the right-hand down, *first dorsal interosseous* acts to abduct the index finger to the left, hence this muscle moves the cursor to the left on the screen). During abstract control, the DoA of each muscle is chosen randomly at the beginning of the session, meaning that subjects must learn an unnatural mapping between muscle activity and movement.

Subjects are instructed to make 'centre-out' cursor movements to targets that appear around the edge of the screen. Initial performance with the abstract interface is considerably worse than with a biomimetic algorithm. However, after several hundred trials of practice (about half an hour), subjects can move the cursor quickly and accurately to all targets. The rate at which subjects successfully complete trials converges to the same level under both control algorithms (Figure 4.1b). Thus it seems that although the biomimetic interface is more intuitive, with sufficient training subjects can learn to control abstract myoelectric interfaces with similar speed and accuracy.

Optimal control of an abstract myoelectric interface

Acquiring mastery of an abstract myoelectric interface involves two important components. First, the brain must explore the relationship between motor signals and resultant sensory feedback (the *forward problem*). Second, the brain must find the best movement to achieve a desired target state (the *inverse problem*) (Wolpert & Ghahramani, 2000). For myoelectric interfaces as well as most natural movements the inverse problem is ill-posed due to *redundancy* in the mapping from motor commands to sensory states. In other words there is more than one way to achieve any particular goal. For natural limb movements, the abundance of choices faced by the motor system is sometimes referred to as the 'degrees of freedom' problem (see Box 4.1). The myoelectric interface shown in Figure 4.1a provides a simple model of a redundant mapping because six effectors (muscles) control the cursor in only two dimensions (screen co-ordinates) so there are multiple ways to acquire any target. For example, a target presented in a direction close to one muscle's DoA could be reached by activating only that single muscle. On the other hand, subjects could use a combination of multiple muscles acting in other directions on the cursor. Since this choice is not explicitly determined by the structure of the task, it may reveal other constraints on behavior that are imposed by the organization of the motor system.

Box 4.1 Redundancy and optimal control theory

The ill-posed nature of the inverse problem can be illustrated by considering the computations that must be performed by the participants of a game of cricket. The bowler can choose from a near infinite set of elevations and speeds that will propel the ball towards the wicket. Likewise, the batsman has a choice of shots involving a variety of postures (e.g. orientations of the shoulder, elbow and wrist) to bring the bat to the ball. In both cases, even if the full time-course of every joint in the cricketer's body is fully specified, different patterns of muscle activity and stiffness can generate the same overall behavior because more than one muscle acts at each joint. In other words, natural movements are under-determined by their specific goals. This is sometimes known as the 'degrees of freedom problem' and its formulation is credited to the Russian neuroscientist Nikolai Bernstein who wrote: 'It is clear that the basic difficulties for co-ordination consist precisely in the extreme abundance of degrees of freedom' (Bernstein, 1967: p. 107)

Despite this apparent abundance of solutions, a large body of experimental data shows that human goal-directed movements are remarkably stereotyped. This is true not only for the highly-skilled batsman who executes a text-book cover drive; during everyday movements such as reaching for an object the hand follows a straight-line trajectory with a consistent bell-shaped velocity profile. The reasons why we choose one particular strategy over the many others that could achieve the same goal have been the subject of much speculation. For example, it has been proposed that intrinsic hard-wired constraints within the nervous system limit the number of degrees of freedom that can actually be controlled by the motor system.

By contrast, Optimal Control Theory assumes that while people are free to choose any movement, they try to find the solution that incurs a minimal *cost*. A number of intrinsic cost factors have been proposed, for example energy expenditure, fatigue or jerkiness of movement, in order to explain the smooth efficient trajectories characteristic of biological motion. Alternatively, Harris and Wolpert (1998) suggest that an extrinsic cost to movement is imposed by noise in the motor system. Consider again the problem facing the batsman in cricket; while there are many ways to bring bat to ball on average, uncertainty in the activity of muscles caused by motor noise means that each choice of shot can be associated with a probability distribution over the space of movement endpoints. The task of the batsman is to choose the shot which maximizes the likelihood of hitting the ball or, in other words, minimizing the effect of neuromotor noise on the movement goal. In general, while on average any solution to the inverse problem will achieve the desired goal, only one solution is optimal when the control signals are corrupted by noise.

As an example of how minimizing the effect of signal-dependent noise resolves muscular redundancy, consider two agonist muscles which move the limb (or myoelectric interface) in the same direction. In this simple 2-to-1 redundant system, we can imagine that the total force (or cursor displacement), x, is the sum of the intended contribution from each muscle, x_1 and x_2, plus some zero-mean neuromuscular noise, ε_1, ε_2:

$$x = x_1 + \varepsilon_1 + x_2 + \varepsilon_2. \tag{1}$$

Let us suppose the goal of the motor system is to produce an average total force equal to x_T which can be achieved by any choice of x_1 and x_2 that satisfy:

$$E(x) = x_T = x_1 + x_2. \tag{2}$$

However, each of these solutions is associated with different overall variability. Following Harris and Wolpert (1998) we assume the noise is signal-dependent and scales with the square of the muscle activity, with constants of proportionality α_1 and α_2. Therefore the overall task variability is:

$$Var(x) = Var(\varepsilon_1) + Var(\varepsilon_2) = \alpha_1 . x_1^2 + \alpha_2 . x_2^2. \tag{3}$$

From here it is straightforward to show that the particular solution to muscle activity x_1 and x_2 in (2) that minimizes the overall variance (3) is:

$$x_1 = \frac{\alpha_2}{\alpha_1 + \alpha_2} . x_T, \quad x_2 = \frac{\alpha_1}{\alpha_1 + \alpha_2} . x_T \tag{4}$$

In other words, the optimal strategy is one in which the effort of movement is divided among agonist muscles, according to their relative accuracy. Note, however, that a similar result could be achieved by minimizing an intrinsic cost which scales with the square of the muscle activity (effort). In practice for natural movements it can be difficult to distinguish the predictions of minimization of intrinsic effort costs from the minimization of inaccuracies produced by signal-dependent noise.

The behavior of subjects while performing a myoelectric interface task can be represented by 'tuning functions' that describe how each muscle is used for targets presented in different directions on the screen. If a muscle is used only for targets in the direction it acts, the tuning function would be narrow with a sharp peak at the DoA. In fact, the observed tuning curves shown in Figure 4.1c are broad, with individual muscles active over a range of target directions up to 90° either side of the DoA. This pattern is well described by the truncated cosine functions shown as solid lines fitted to the experimental data.

Interestingly, while the muscles controlling an abstract myoelectric interface are acting in a very different way to how they would normally move the limb, this pattern of *cosine-tuning* bears a remarkable similarity to that seen under natural conditions. For instance, when people make forces in different directions with the hand, multiple muscles in the forearm act cooperatively on the wrist and their activity follows a cosine-shaped dependence on force direction. This suggests we should look for a 'high-level' explanation for this behavior that is common to both myoelectric and natural movements but unrelated to the specific action of muscles. It can be shown that cosine-tuning minimizes a quantity known as *effort*, defined as the sum of the squared muscle activation (Fagg et al., 2002). In brief, to minimize effort, activity should be distributed among muscles according to their efficiency at contributing to the desired goal. But why should the motor system want to minimize effort?

One answer is suggested by the observation that noise in the motor system is *signal-dependent*; muscles become more variable as they increase their activity (Harris & Wolpert, 1998). More specifically, it has been proposed that the variance of this neuromotor noise scales with the square of the activation. Minimizing overall effort will therefore be the best strategy for minimizing the uncertainty about cursor position that results in the presence of such noise.

We can test this hypothesis by changing the smoothing window used to derive control signals for some muscles. Smoothing the EMG over a shorter time window increases the natural variability of these control signals so if subjects are attempting to minimize cursor variability, there should be a corresponding decrease in the contribution of these muscles to movements. Figure 4.1c shows experimental data for muscles with low and high variability that were co-operating in a myoelectric interface task. Both tuning functions are cosine-shaped but low variability (long smoothing window) muscles are active to a greater extent than high variability (short smoothing window) muscles, consistent with the predictions of a strategy to minimize the influence of noise on cursor position. However, it should be noted that the reduction in overall activity between low and high variability muscles (by a ratio of 1.3 to 1) is less than would be predicted by their relative variability coefficients (1 to 1.7; inset). It may be that over the period of this experiment, subjects did not have enough exposure time to learn the noise properties of each muscle in order to fully optimize the accuracy of their movements. Alternatively it may be that despite the importance of accuracy for successful myoelectric control, there are also intrinsic 'low-level' constraints on the extent to which performance on abstract interfaces can be optimized.

Co-ordination patterns of proximal muscle are less flexible than distal muscles

While many features of the tuning functions seen during myoelectric control can be explained by a strategy to cope with signal-dependent motor noise, people are not entirely free to use their muscles in ways that are optimal according to 'high-level' criteria. The most striking deviation from theoretical predictions is seen when tuning functions for proximal muscles (acting on the shoulder or elbow, e.g. *biceps*) are compared with distal muscles (intrinsic to the hand acting on the fingers and thumb, e.g. *first dorsal interosseous*). As shown in Figure 4.1d, distal muscles are used to a considerably greater extent despite comparable noise coefficients for each group (inset). Since the best strategy to maximize accuracy would be to distribute effort evenly between muscles, this bias towards the distal muscles appears to be suboptimal.

The unequal recruitment of proximal and distal muscles arises because people find it diffi-cult to dissociate the activity of proximal muscles, even when they act on the cursor in oppo-site directions. As shown in Figure 4.1e, during training pairs of proximal muscles remain positively correlated irrespective of whether they have similar or opposing DoAs. This suggests that proximal muscle groups may be constrained to fixed, naturalistic patterns of co-contraction, or *synergies,* even when these are inappropriate for an abstract interface. By contrast, with practice the correlation between distal muscles becomes positive or negative according to their relative effect on the cursor. Therefore distal muscles are not limited to a small number of fixed synergies, but can be flexibly dissociated according to the high-level requirements of an abstract myoelectric interface. The inflexibility of proximal synergies is a low-level constraint that limits the degree to which muscles can best be used for abstract control. However, as we will see, activating muscles together in coordinated synergies is not always suboptimal for myoelectric interface tasks because in some circumstances this is a good strategy for minimizing the effects of motor noise on movement accuracy.

Using muscle synergies for myoelectric control

The experiments described in the preceding section used circular targets that penalize subjects equally for errors in all directions. However, for many real world movements particular dimensions of a task may be more or less important for success. For example, holding an object steadily in the hand requires that no net force is exerted on the object (Figure 4.2). The absolute force exerted by the finger is not particularly important so long as it is balanced by

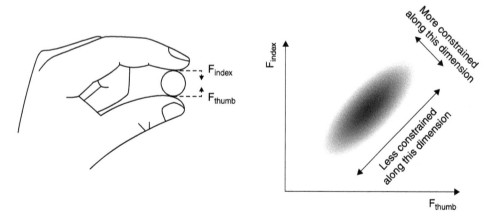

Figure 4.2   When holding an object steady, the net force on the object is more important than the absolute force exerted by either finger or thumb muscles. The shaded area of the left plot signifies the region of the effector space compatible with this criterion (the uncontrolled manifold). The mechanics of the task therefore impose different constraints along dimen-sions of positive and negative correlation between finger and thumb. Activation of a synergy comprising finger and thumb muscles would be an appropriate strategy given these constraints.

the force of the thumb. The task of balancing these forces is made difficult because of noise in the activity of the individual finger and thumb muscles. Some of this noise arises from unpredictable variations in the response of muscles to descending commands, and will have unavoidable consequences for task performance. However, another component of motor noise arises from central sources of variability within the brain and its effect can be mitigated if it is transmitted to *all* muscles equally. Then any increase in the force exerted by the finger will be associated with an equal and opposite increase in thumb force with no impact on object stability. A possible strategy to couple the variability of muscles in this way is to co-activate them as a *synergy* (see Box 4.2).

Box 4.2 The Uncontrolled Manifold and Optimal Feedback Control

Consider reaching for a wine bottle; since it is taller than it is wide, more inaccuracy can be accommodated in the vertical position of the hand than along the horizontal axis without affecting our ability to grasp the bottle successfully. Experiments using natural pointing and aiming tasks have shown that the structure of movement variability can be shaped to reflect such high-level task constraints. This suggests a potential resolution to Bernstein's degrees of freedom problem; the motor system identifies and controls only those dimensions within the effector space that are important for task success. Improved accuracy in these task-relevant dimensions may then be achieved by buffering variability along task-irrelevant dimensions, known as the *uncontrolled manifold* (Scholz & Schöner, 1999).

There are two general ways in which the movement variability can become structured in the effector space. One is through the divergence that exists in the feed-forward structure of the motor system. For example, many muscles act around more than one joint; noise in the activity of these muscles will lead to correlated variability in joint space. Similarly, many of the neural pathways that drive muscles are divergent. By using pathways that couple muscles into functional synergies, noise in the neural drive will result in structured variability in muscle space. If the motor system can exploit divergence such that correlation structure is aligned to the uncontrolled manifold, then noise will effectively be 'buffered' into task-irrelevant dimensions.

A second way in which structured variability can emerge is through the intelligent use of feedback. The principle of *minimum intervention* suggests that only task-relevant errors should be corrected, leaving variability to accumulate in the uncontrolled manifold (Todorov & Jordan, 2002). This makes intuitive sense because the correction of errors in task-irrelevant dimensions is itself subject to noise and risks introducing other, task-relevant errors. Such a strategy may be incorporated within the framework of Optimal Control Theory by assuming there is an effort cost associated with making corrections. Optimal Feedback Control formalizes the minimum intervention principle in terms of feedback control policies that are optimized for particular task goals. To overcome the time delays and noise associated with sensory feedback, an optimal feedback controller must estimate the current state of the limb by combining sensory evidence with an internal prediction.

Optimal Feedback Control is one of our most complete descriptions of the computations that must be performed by the motor system. But it says little about how minimum intervention policies are learned and implemented by neural elements. It remains to be seen whether its predictions can be reconciled with the neuroanatomical evidence for divergent pathways underlying muscle synergies.

This raises an interesting question: is the range of available synergies limited to a small number of common behaviors such as grasping, or do new synergies emerge as appropriate for more abstract requirements? It is difficult to address this question by studying natural movements because the constraints imposed by the biomechanics of the limbs are hard to manipulate experimentally. However, with an abstract myoelectric interface, we can impose high-level task requirements that are unrelated to any physical properties of the body. We have already seen that hand muscles are dissociated when they move a myoelectric cursor in opposite directions. But can they be recombined into new synergies that cooperate to restrict errors along particular dimensions of the task space?

In recent experiments, we trained subjects to move a myoelectric cursor to targets that are circular or elliptical (Nazarpour et al., 2012). The task shown in Figure 4.3a uses a well-posed two (muscle) to two (screen co-ordinates) mapping with target ellipses oriented along axes of positive or negative correlation in muscle space. The dots in Figure 4.3b show the distribution of movement end-points over repeated trials for a well-trained subject, with different shading indicating movements to different target shapes. The effect of motor noise can be seen in the variability of movement end-point from one trial to the next, described by covariance ellipses which are overlaid on the data. The distribution of errors closely matches the respective target orientation, with greatest trial-to-trial variability along the less constrained, long axis of

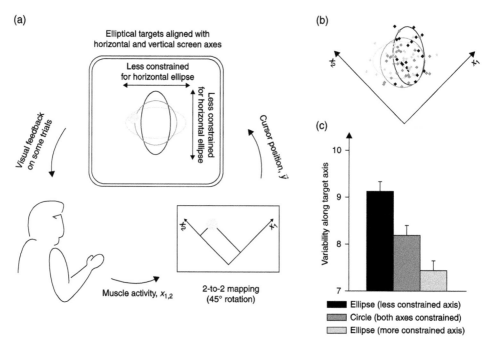

Figure 4.3 Exploring flexible control of synergies using a myoelectric interface. (a) Subjects control a myoelectric cursor with two muscles that act along orthogonal axes. Elliptical targets impose different accuracy constraints on dimensions of positive and negative muscle correlation. (b) The distribution of movement end-points across repeated trials mirrors the target constraints. (c) Variability is greater along the less constrained, long axis of elliptical targets than along the more constrained, short axis. Note that accuracy along the more constrained axis of elliptical targets is also smaller than when the target is circular, suggesting subjects can exploit the leeway afforded by an unconstrained dimension to reduce task-relevant errors. Adapted from Nazarpour et al. (2012).

elliptical targets. We find that with training, comparable results can be obtained irrespective of which distal muscles are used to control the task. Moreover, average movement errors along the constrained, short axis of elliptical targets are smaller than errors made when moving to circular targets (Figure 4.3c). This suggests that the motor system can learn to exploit the leeway granted by an unconstrained dimension to improve task-relevant accuracy.

Several mechanisms contribute to the emergence of task-specific correlation structure in motor variability. For example, subjects can use visual feedback to correct errors in task-relevant dimensions, but may choose not to correct errors in task-irrelevant dimensions (a strategy known as *minimum intervention* control; Box 4.2). However, while target-dependent corrections do occur during myoelectric control, target-dependent structure of cursor variability persists even in the absence visual feedback. To understand how subjects achieve this, we need to examine the anatomy of the descending pathway used to operate myoelectric interfaces.

Neural substrates for skilled myoelectric interface control

Volitional control of skilled actions like those required for control of a myoelectric interface is mediated by the *corticospinal* pathway. In primates including man, many corticospinal neurons in the primary motor cortex make monosynaptic excitatory connections to moto-neurons, as well as disynaptic inhibitory projections via spinal interneurons. These direct connections bypass synergies hard-wired into subcortical and spinal circuitry, allowing the motor cortex considerable fine control over specific muscles. The distal bias of cortico-motoneuronal connections likely explains why these muscles are better suited to control of abstract myoelectric interfaces than proximal muscles.

Lesions of the corticospinal pathway particularly impair independent movements of the fingers. However, while direct cortico-motoneuronal inputs are undoubtedly important for *fractionated* movements, corticospinal axons branch extensively at the spinal level to target motoneurons of more than one muscle. The extent of excitatory and inhibitory connections from a single corticomotoneuronal cell to muscles can be revealed in monkeys using the technique of *spike-triggered averaging*. Sections of rectified EMG aligned to the occurrence of the action potentials (spikes) of a single cortical neuron are averaged (Figure 4.4a). When sufficient spikes (usually several thousand) are sampled such that background fluctuations are averaged out, a small post-spike facilitation or suppression can sometimes be seen at latencies consistent with mono- and disynaptic corticospinal transmission. A typical cell may exhibit post-spike effects in several forearm and hand muscles, which defines the *muscle field* for that cell (Fetz & Cheney, 1978; Jackson et al., 2003). Across the population a wide variety of different muscle fields are seen for individual corticospinal neurons (Figure 4.4b).

It is worth noting that the relative importance of direct corticospinal control in higher primates parallels a phylogenetic transition in forelimb use from predominantly locomotor function to skilful manipulation of objects. The diversity of muscle fields observed for individual corticomotoneuronal cells presumably reflects the variety of high-level, abstract goals involved in grasping and manipulating objects, and ultimately making and using tools. We speculate that it is this diversity that provides a rich substrate for finding optimal synergies of distal muscles for control of abstract myoelectric interfaces.

Exploiting redundancy at multiple levels of the motor hierarchy

Figure 4.5a shows a simple model to illustrate how the convergence of neural drive from cortical neurons with different muscle fields might allow the corticospinal system to optimize

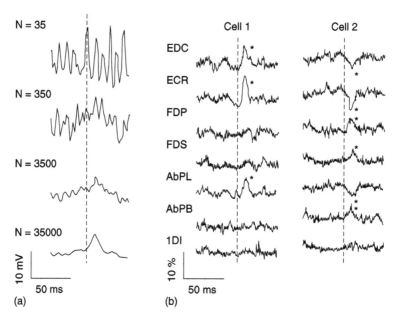

Figure 4.4 Spike-triggered averaging reveals monosynaptic facilitation of muscles via the corticospinal pathway. (a) A spike-triggered average is compiled from sections of EMG aligned to the time of action potentials from a single cortical neuron (indicated by dashed line). As more sections of data are included in the average, a small post-spike facilitation is observed, consistent with monosynaptic conduction latencies. (b) Typical corticospinal neurons exhibit post-spike facilitation of multiple hand and forearm muscles. Cell 2 also shows post-spike suppression of some muscles (EDC, ECR and AbPL), mediated by inhibitory spinal interneurons.

patterns of motor variability (Nazarpour et al., 2012). The model includes pools of input neurons that activate each muscle individually (u_2 and u_6). Further populations of neurons have fields that span both muscles via divergent pathways ($u_{3,4,5}$) and are responsible for synergistic activation of the muscles (grey shading). For completeness we also include neurons which activate one muscle while suppressing the other (u_1 and u_7). Each input neuron pool is subject to independent signal-dependent noise, as are the output motoneurons. For any pattern of activity across units $u_{1..7}$ we can calculate the distribution of cursor locations that would be expected once the effects of this signal-dependent noise have been taken into account (Figure 4.5b). Then, for each target we can calculate the particular pattern of cortical activity that maximizes the overlap between this distribution and the target shape as a prediction of the optimal neural drive for successful task performance (Figure 4.5a, upper plot). For the circular target, what emerges is the now familiar cosine tuning function. This is to be expected, since the radial arrangement of effects in the muscle space (Figure 4.5b) mirrors the radial arrangement of the DoAs of muscles in the cursor space used in the previous experiment (Figure 4.1a). However, the solution for the vertical ellipse target (which constrains errors in the direction of negative correlation in the muscle space) predicts increased activity of the divergent input units which activate the muscles synergistically. By contrast, the solution for the horizontal ellipse (which constrains positively correlated muscle fluctuations) involves reducing the contribution of these divergent pathways. Note that all three solutions activate the muscles to the same extent (and hence bring the cursor to the

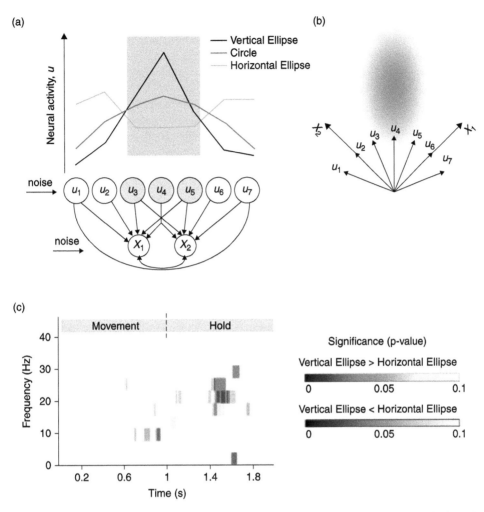

Figure 4.5 Divergence and convergence in the corticospinal system allows flexible control of muscle synergies. (a) A simple model of control of two muscles (x_1 and x_2) by populations of cortical neurons ($u_{1..7}$) with different patterns of facilitation and suppression. The optimal cortical activity for reaching different target shapes in the presence of signal-dependent noise is shown above. The model predicts target-dependent modulation of neurons which facilitate both muscles through divergent connections (shaded). (b) Direction of action of each cortical population in the muscle space. The shaded region represents the probability density of cursor end-points when signal-dependent noise is taken into account. (c) Inter-muscular coherence provides evidence for task-modulation of divergent drive. Coherence in the beta band around 20 Hz (encircled) is elevated when subjects hold the myoelectric cursor inside a vertical target ellipse compared with a horizontal ellipse. This is consistent with increased divergent corticospinal drive predicted by the model. Adapted from Nazarpour et al. (2012).

same location on the screen). What differs between them is the pattern of trial-to-trial variability resulting from the effect of signal-dependent noise, which fits well with the experimental data.

While this model can explain the behavior of subjects, is there any direct evidence for a modulation of the divergent drive to muscles according to target shape? This is revealed by analysis of the coherence between the EMG signals from the two muscles (Nazarpour et al., 2012). Coherence is a frequency-domain measure of correlation that reveals oscillatory components common to each signal, so by applying coherence analysis to the rectified EMG signals we can identify oscillatory inputs common to the two muscles. Corticospinal neurons are synchronized to oscillations in the beta frequency around 20 Hz, and therefore divergent corticospinal drive leads to inter-muscular coherence at this frequency. As shown in Figure 4.5c, movements to the vertical target are associated with an increased coherence at around 20 Hz, which fits the prediction that synergistic drive is appropriate for this target.

We have already seen how redundancy in the mapping from muscles to movement can be exploited by distributing effort optimally across multiple *muscles*. We now find that redundancy in the mapping from cortical neurons with divergent muscle fields provides a further opportunity to distribute effort among multiple *muscle synergies*. The picture that emerges is one of hierarchical levels of redundancy within the motor system (Figure 4.6). At each level, *convergence* onto fewer dimensions in the level below means a multitude of redundant activity patterns are consistent with the desired goal. At the same time, *divergence* in the descending pathways allows behavior to be optimized for high-level task goals like accuracy. However, a high-dimensional control space is a double-edged sword; the flexibility to optimize the motor pattern for any abstract high-level goals comes at the cost of a large search space within which this pattern must be sought. The learning mechanisms that allow the motor system to achieve optimality are the subject of much current research which is outside the scope of this

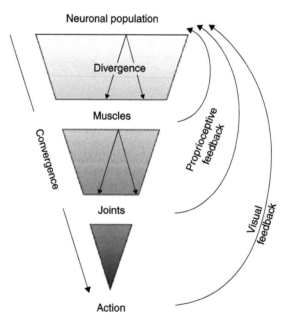

Figure 4.6 Schematic of hierarchal redundancy within the motor system. There is an overall convergence from a large number of neurons onto fewer muscles, from muscles onto fewer joints etc. At the same time there is divergence in the descending pathway since a single cortical neuron can facilitate multiple muscles, and a single muscle can act around multiple joints. Also shown is the feedback from each level of the hierarchy mediated by proprioceptive and other sensory feedback. Proprioceptive feedback will be profoundly altered for myoelectric and neural interfaces compared with natural movement.

chapter. It is worth noting that extensive sensory feedback returns from all levels of the motor system during natural movements, for example through proprioceptive receptors in the muscles and tendons, and undoubtedly plays an important role in shaping cortical activity patterns. Nevertheless, the fact that learning can occur even under the relatively impoverished sensory conditions imposed by myoelectric interfaces suggests that, to a significant extent, actions can be optimized solely through feedback of their effect in an abstract task space. This will become particularly important as we consider learning in the context of neural interfaces, when the muscles and their proprioceptive sensors are bypassed altogether.

From myoelectric to neural interfaces

Myoelectric interfaces are advantageous for experimental studies and prosthetics applications because EMG can be easily recorded using surface electrodes. Unfortunately they are unsuitable for patients in whom degeneration of motoneurons or injury to the spinal cord has rendered the muscles paralyzed. For these patients, volitional control of communication devices and prostheses could, however, be provided by electrical signals obtained from motor areas of the brain. This is the concept of a direct neural interface, also known as a *Brain–Machine Interface* (BMI) or *Brain–Computer Interface* (BCI). While surface electrodes can detect electrical fields generated by brain activity, the *electroencephalogram* (EEG) is typically smaller than EMG by around three orders of magnitude and difficult to interpret since multiple, often independent, neural sources contribute to the potential recorded at the scalp. The best spatial and temporal resolution is obtained by placing electrodes inside the cortex to record extracellular action potentials from individual neurons. The invasive nature of these approaches means that to date most such experiments have been carried out in animals, usually monkeys, although several trials in human patients are now underway (Hochberg et al., 2006).

The typical set-up for a BMI experiment is shown in Figure 4.7. The action potential discharge of tens to hundreds of motor cortex neurons is converted into control signals to drive an effector, in this case a robotic arm. Typically, this conversion is achieved via a biomimetic decoding scheme based on an observed relationship between neural rates firing and the kinematics of natural movements. For instance, studies of center-out reaching movements by Apostolos Georgopoulos and colleagues in the 1980s found that the activity of individual primary cortex neurons is modulated by the direction of arm movements (Georgopoulos et al., 1986). Tuning functions are again cosine-shaped, with maximal firing rate for movements in one particular direction known as the preferred direction (PD) of that cell. Different neurons have different PDs which are distributed approximately equally in all directions. By adding up vectors aligned to the PD of each cell, scaled by the firing rate of that cell, a population vector can be constructed that closely matches the observed direction of movement. For BMI applications, the PDs of cells are usually calculated off-line from a training dataset of real reaching movements. However, once these directions are known, the population vector can be calculated on-line, independently of any actual movement, and used to drive a closed-loop neural interface. In recent years, a number of landmark studies have shown that monkeys can use neural interfaces to control computer cursors in two and three dimensions, and even take pieces of food to eat with a robotic arm (Taylor et al., 2002; Carmena et al., 2003; Velliste et al., 2008). Note the conceptual similarity with the myoelectric decoding algorithm shown in Figure 4.1, although here velocity rather than position is decoded. In particular, since a large number of cortical neurons are used to control only a small number of degrees of freedom, the mapping is ill-posed and there is a high degree of redundancy at the neural level.

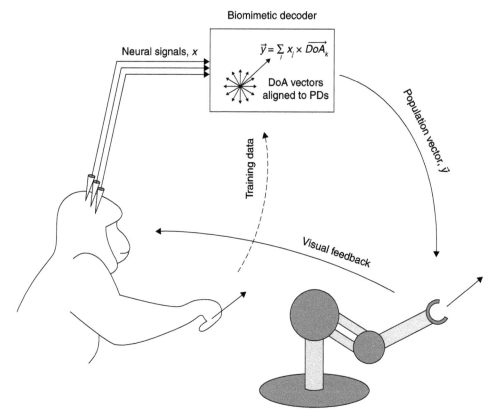

Figure 4.7 Schematic of a Brain–Machine Interface experiment. The firing rate of multiple cortical neurons is converted into a population vector and used to control a robotic prosthesis. A biomimetic decoder is constructed from training data obtained during natural limb movements. During 'brain control' each neuron moves the arm in a Direction of Action (DoA) that is aligned to the neurons' Preferred Direction (PD) determined from the training data. Visual feedback of robot movements closes the sensorimotor loop.

Although biomimetic decoders are trained to predict natural reaching movements made under 'open-loop' conditions, once feedback is provided by the interface a number of changes occur during 'closed-loop' operation. The movements made by the actual limb may change or drop out completely while the artificial effector continues to reach for targets. Furthermore, the relationship between neuron firing and effector movement may change: for example, some studies report that the PD of neurons shifts away from their decoded DoA (Taylor et al., 2002) while others find that correlations among neurons increase (Carmena et al., 2003). These changes in neural tuning are not particularly surprising once we realize that however well the biomimetic decoder performs, operation of a neural interface is very different to natural limb control. For example, only a small fraction of the neurons in the motor system are sampled and afferent feedback from the limb is profoundly altered. Therefore, as with an abstract myoelectric interface, accurate control of BMIs requires a new relationship between motor commands and sensory feedback to be learned. Indeed, given the results from myoelectric interface studies described earlier, one might question whether biomimetic decoding is required at all so long as the processes of sensorimotor adaptation are able to optimize the tuning of cortical neurons to be appropriate for an abstract mapping. It has been known since the operant conditioning studies of Eberhard Fetz in the 1960s that, given appropriate feedback, monkeys can modulate the activity of neurons according to arbitrary reward rules (Fetz 1969). Indeed, a recent study by Ganguly and Carmena (2009) suggests that with sufficient practice, monkeys can learn to control an abstract interface in which the DoAs of neurons are randomized. Unlike the abstract myoelectric interface which is learned quickly within a single session, training over several days is required for the appropriate pattern of neural activity to emerge. This may reflect the higher dimensionality of the neural space in which the optimal control strategy must be sought (Figure 4.6).

Improving neural interface performance is an active area of current research with many questions to be addressed before these devices can achieve widespread clinical application. The conceptual similarity between neural and myoelectric interfaces leads us to speculate that insights gained from the former may usefully inform the development of Brain–Machine Interfaces. To this end, we will conclude this chapter with three unresolved questions which should be the focus of future research:

Are changes in neural activity seen during 'brain control' consistent with the minimization of neural effort and/or inaccuracy? We have described how the minimization of effort during myoelectric control leads to the distribution of workload across multiple muscles and that this reduces the influence of neuromotor noise on task-relevant control dimensions. Can similar principles explain the distributed cortical activity during neural interface experiments? The ubiquity of cosine-tuning in the activity of muscles and neurons during natural movements suggests similar optimality criteria may apply at different levels of the motor hierarchy. Certainly coping with the inaccuracies caused by noise in individual neurons is critical for Brain–Machine Interface control, and accuracy improves as the number of neurons used for decoding is increased (Carmena et al., 2003). At the same time, neurons not involved in control become less modulated as learning progresses, consistent with the minimization of an overall 'neural effort'. This may explain why overt movements of the limb change or drop out completely over time. On the other hand, while signal-dependent noise in muscles has been well characterized, the properties of neural noise are less understood. In particular, dense reciprocal cortical connectivity means that neural noise is not statistically independent across the population. Therefore it is hard to predict from high-level optimality principles how effort should best be distributed at the neural level.

Can the brain use 'neural synergies' to buffer variability into task-irrelevant dimensions during control of a redundant neural interface? By analogy with muscle synergies, we may wonder whether correlated noise in neuronal activity during normal activity may reflect a strategy to exploit redundancy in the descending corticospinal pathways and buffer noise into task-irrelevant control dimensions. Since the redundant mapping from neuronal ensembles to population vector parallels the dimensional reduction of natural motor control, it is interesting to speculate whether new patterns of neural correlation emerging in BMI experiments reflect structured neuronal variability appropriate for the new high-level constraints imposed by 'brain control'.

Do intrinsic, hard-wired constraints limit the optimality of neural interface control? The divergent connections between cortical neurons may provide a rich substrate to construct 'neural synergies' appropriate for abstract control. However, the true extent of cortical flexibility during brain control is as yet unknown, particularly in the absence of proprioceptive feedback. Connectivity is strongest between cells that drive the same muscles (Jackson et al., 2003);

these cells may be most difficult to dissociate even if they have opposing effects on a neural interface. Such hard-wired circuits configured for naturalistic limb movements would represent an intrinsic constraint on the brain's ability to optimize control of a direct neural interface. At the muscle level, intrinsic constraints seem particularly to limit the flexibility of the proximal musculature, while direct cortical control of distal muscles liberates the hand for a wide repertoire of tasks such as object manipulation and tool use. However, cortico-motoneuronal neurons are located in a part of primary motor cortex found in the bank of the central sulcus that is hard to target with electrode arrays. Thus far most studies have focused on more accessible motor areas on the convexity. If the greater flexibility of distal muscles is reflected in the architecture of their corresponding cortical areas, then these neural circuits may be ideally suited to operate abstract tools through neural interfaces that are dissociated entirely from the constraints of the body.

Summary

In this chapter we have seen how myoelectric interface experiments reveal the inherent flexibility of the motor system to optimize behavior under conditions of abstract sensorimotor relationships. The activity patterns of upper-limb muscles, in particular those controlling distal wrist and hand joints, can be rapidly dissociated and recombined into new synergies appropriate for minimizing the impact of neuromotor noise on task performance. We have speculated that a similar optimization process may allow accurate operation of devices controlled directly from brain activity, especially if they interface with the neurons that control the distal musculature. There is still much to be understood about how learning takes place: for example, which brain structures are responsible and what role is played by various modalities of feedback, reinforcement and error signals from the periphery. Many of these issues are discussed in the other chapters of this book in relation to natural movements of the limbs. It is to be hoped that a better understanding of these mechanisms will aid the development of new clinical devices based on myoelectric and neural interfaces designed to harness the remarkable adaptive abilities of the motor system to restore function following motor injury.

References

1. Bernstein, N.A. (1967) *The co-ordination and regulation of movement*, Oxford: Pergamon Press.
2. Carmena, J.M., Lebedev, M.A., Crist, R.E., O'Doherty, J.E., Santucci, D.M., Dimitrov, D.F., Patil, P.G., Henriquez, C.S. and Nicolelis, M.A. (2003) 'Learning to control a brain–machine interface for reaching and grasping by primates', *PLoS Biology*, 1: E42.
3. Fagg, A.H., Shah, A. and Barto, A.G. (2002) 'A computational model of muscle recruitment for wrist movements', *Journal of Neurophysiology*, 88: 3348–3358.
4. Fetz, E.E. and Cheney, P.D. (1978) 'Muscle fields of primate corticomotoneuronal cells', *Journal of Physiology (Paris)*, 74: 239–245.
5. Ganguly, K. and Carmena, J.M. (2009) 'Emergence of a stable cortical map for neuroprosthetic control', *PLoS Biology*, 7(7): e1000153.
6. Georgopoulos, A.P., Schwartz, A.B. and Kettner R.E. (1986) 'Neuronal population coding of movement direction', *Science*, 233: 1416–1419.
7. Harris, C.M. and Wolpert, D.M. (1998) 'Signal-dependent noise determines motor planning', *Nature* 394: 780–784.
8. Hochberg, L.R., Serruya, M.D., Friehs, G.M., Mukand, J.A., Saleh, M., Caplan, A.H., Branner, A., Chen, D., Penn, R.D. and Donoghue, J.P. (2006) 'Neuronal ensemble control of prosthetic devices by a human with tetraplegia', *Nature*, 442: 164–171.
9. Jackson, A., Gee, V.J., Baker, S.N. and Lemon R.N. (2003) 'Synchrony between neurons with similar muscle fields in monkey motor cortex', *Neuron*, 38: 115–125.

10. Nazarpour, K., Barnard, A. and Jackson, A. (2012) 'Flexible cortical control of task-specific muscle synergies', *J Neuroscience*, 32(36): 12349–12360.

11. Radhakrishnan, S.M., Baker, S.N. and Jackson, A.J. (2008) 'Learning a novel myoelectric-controlled interface task', *Journal of Neurophysiology*, 100: 2397–2408.

12. Scholz, J.P. and Schöner, G. (1999) 'The uncontrolled manifold concept: identifying control variables for a functional task', *Experimental Brain Research*, 126: 289–306.

13. Shadmehr, R. and Mussa-Ivaldi, F.A. (1994) 'Adaptive representation of dynamics during learning of a motor task', *Journal of Neuroscience*, 14: 3208–3224.

14. Taylor, D.M., Tillery, S.I. and Schwartz, A.B. (2002) 'Direct cortical control of 3D neuroprosthetic devices', *Science*, 296: 1829–1832.

15. Todorov, E. and Jordan, M.I. (2002) 'Optimal feedback control as a theory of motor coordination', *Nature Neuroscience*, 5: 1226–1235.

16. Velliste, M., Perel, S., Spalding, M.C., Whitford, A.S. and Schwartz, A.B. (2008) 'Cortical control of a prosthetic arm for self-feeding', *Nature*, 453: 1098–1101.

17. Wolpert, D.M. and Ghahramani, Z. (2000) 'Computational principles of movement neuroscience', *Nature Neuroscience*, 3: 1212–1217.

5

BIOMECHANICAL AND NEUROMECHANICAL CONCEPTS FOR LEGGED LOCOMOTION

Computer models and robot validation

Andre Seyfarth,[1] *Sten Grimmer,*[1] *Daniel Häufle,*[2] *Horst-Moritz Maus,*[1] *Frank Peuker*[1] *and Karl-Theodor Kalveram*[1,3]

[1] LAUFLABOR LOCOMOTION LAB, INSTITUTE OF SPORT SCIENCE, TECHNISCHE UNIVERSITÄT DARMSTADT
[2] INSTITUTE OF SPORT AND MOVEMENT SCIENCE, UNIVERSITY OF STUTTGART
[3] CYBERNETICAL PSYCHOLOGY, UNIVERSITY OF DÜSSELDORF

Complex motor tasks such as human or animal locomotion require a concerted action of body mechanics, muscle dynamics, and sensor function. For successful movements a careful combination of fast mechanical pre-flexes (Brown & Loeb, 2000) with feedforward and feedback control is required (Dickinson et al., 2000). Surprisingly, during fast and dynamic legged movements like running or hopping, the body function resembles simple movement templates (Full & Koditscheck, 1999), which can be characterized as an overall spring-like response of the body during ground contact. This unifying behavior can be found in human and animal locomotion and has motivated a class of simple gait models based on the spring-mass model (Blickhan 1989; McMahon & Cheng, 1990), also called the *spring-loaded-inverted-pendulum* (SLIP) model (Schwind, 1998).

In this chapter we describe how biomechanical models of different levels of complexity can help to analyze and better understand legged locomotion at different gaits and speeds. In particular we will address how spring-like leg operation during fast locomotion can help to maintain gait stability at different speeds and how the segmented body contributes to the overall movement dynamics. Furthermore, we ask how this knowledge can be transferred to robotics in order to demonstrate our level of understanding. With this transfer to robotics we aim at validating and enhancing our level of understanding, which provides the basis for applying this knowledge to engineering, e.g. for rehabilitation devices or legged robots.

Compliant leg function in human walking and running

During human walking up to 71 percent of the joint moments and up to 74 percent of the joint power could be realized by a physical network of elastic structures spanning the joints of the segmented leg (van den Bogert, 2003). By introducing a network of switchable spring-damper elements mimicking the muscle function of the human leg and by applying hip actuation, the joint function during walking can be well represented (Endo et al., 2006). This indicates that spring-like leg function as found in human running and hopping (Blickhan, 1989) can also be found at joint and inter-joint level during walking. Spring-like leg function means that leg force increases proportional to the amount of leg compression during stance phase (Figure 5.1a).

Based on a distribution of elastic structures within a segmented leg, transition from walking to running can be introduced in a simulation model by changing the actuation of the hip (Seyfarth et al., 2009). Similar gait transitions were also observed in the humanoid robot test by Jena Walker (Iida et al., 2008; Seyfarth et al. 2009). The co-existence of walking and running gaits based on compliant leg operation was previously predicted based on a unifying gait model, the bipedal SLIP (B-SLIP) model (Figure 5.1b, Geyer et al., 2006). In this simulation model, different regions of stable gait patterns could be identified (Figure 5.1c) depending on the combination of leg stiffness k, angle of attack α_0 and system energy E (i.e. speed).

It is important to note that these movements emerge by the system dynamics and are not prescribed or controlled based on kinematic trajectories. This emergent behavior of gait based on spring-like leg behavior is called *self*-stability (Blickhan et al., 2003), as no explicit control law is required to achieve steady-state locomotion. At high speeds the model predicted a large running domain, whereas at lower speeds a walking domain was found (Figure 5.1a).

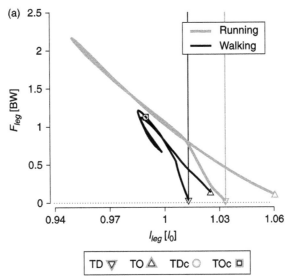

Figure 5.1 (a) Experimental evidence for spring-like leg behavior in human walking and running provided by mean leg force vs. leg length traces of 21 subjects (adapted from Lipfert, 2010). TD = touchdown, TO = take-off, index 'c' represents contra-lateral leg, BW = body weight, l_0 = CoM height at upright standing. (b) The bipedal SLIP model (Geyer et al., 2006) predicts walking and running patterns based on the assumption of spring-like leg behavior. (c) Stable gait patterns can be characterized by three parameters (leg stiffness k, angle of attack α_0, system energy E_S). At faster speeds (above 3m/s) a large domain of stable running characterized by single-humped force patterns is found. At moderate and lower speeds (below 1.6m/s) the model predicts stable gait patterns with two (walking) or more humps. Figure adapted from Geyer et al., 2006.

(Continued)

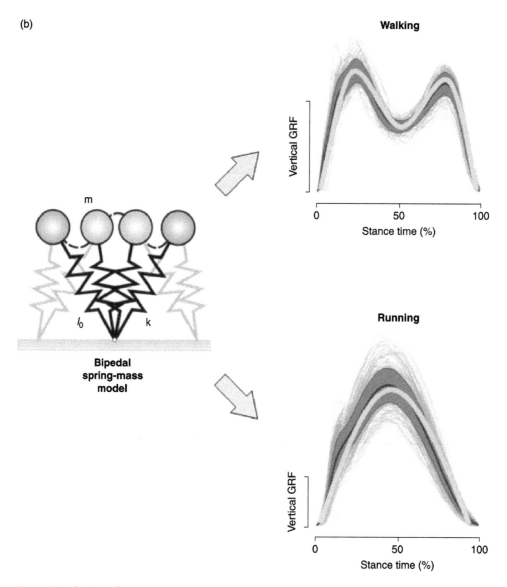

Figure 5.1 Continued

(c)

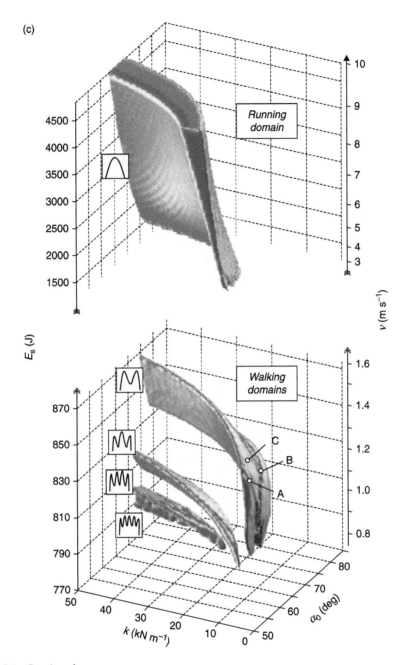

Figure 5.1 Continued

Running gaits were characterized by single-humped patterns of the ground reaction force and walking typically exhibited double-humped force profiles, similar to findings in human walking and running. These different gait-specific force patterns can be explained based on the same spring-like leg behavior, which can be found in running and preferred walking speeds (Figure 5.1a; Lipfert, 2010).

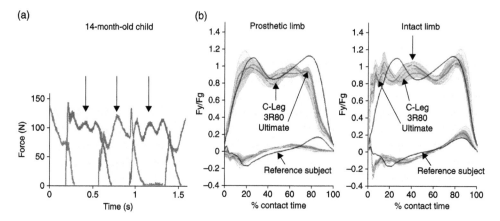

Figure 5.2 (a) Ground reaction forces of a 14-month-old child walking several steps on an instrumented treadmill. (b) Ground reaction forces of a person with long-term unilateral leg amputation using three different kinds of knee joints (C-Leg and 3R80, both Otto Bock; Ultimate knee, Ohio Willowood). In both situations (a, b) three-humped force patterns are found experimentally (a on both sides, b on one side: the intact leg). Data provided by Margrit Schaarschmidt (unpublished data) and Schaarschmidt (2007).

At even lower speeds (Figure 5.1b) additional regions of stable gaits were predicted in the B-SLIP model. Here, the ground reaction forces show profiles with multiple (at least three) peaks. Such force patterns can be found in human locomotion, for instance in early childhood when walking slowly (Figure 5.2a). It is important to note that these gait patterns are less steady, which is reflected in a high step-to-step variability and a high sensitivity to perturbations.

A similar observation can be found in walking of people with unilateral transfemoral leg amputation (Schaarschmidt et al., 2012). In this study one subject with unilateral leg amputation showed a gait pattern with three-humped force patterns on the intact side and double-humped ground reaction forces on the prosthetic limb (Figure 5.2b). These results indicate that three-humped force patterns are indeed found in human locomotion, even though they are less recognized than the double-humped patterns typically found in human walking.

Varying leg properties during locomotion: Step-to-step changes

The previous two examples demonstrate not only that SLIP-based models describe body dynamics found during locomotion, but also that previously less recognized gait patterns can be predicted by these models. This includes also step-to-step changes in leg properties, which may occur during locomotion and will affect the gait pattern and gait stability. For instance, when running on a step upward, leg parameters change to lower leg stiffness and flatter angle of attack (Grimmer et al., 2008; Müller & Blickhan, 2010). But even on level ground, variations in leg operation may occur, e.g. systematic deviations in leg operation between left and right leg, as is typically found in amputee walking (Figure 5.2b). However, asymmetric leg function is found even in natural human walking (Sadeghi et al., 2000) and is reflected by an adapted gait pattern (Figure 5.3a–c).

The effects of these asymmetries can be investigated in a bipedal SLIP model (Geyer et al., 2006) by introducing different leg parameters between the two limbs. A systematic increase

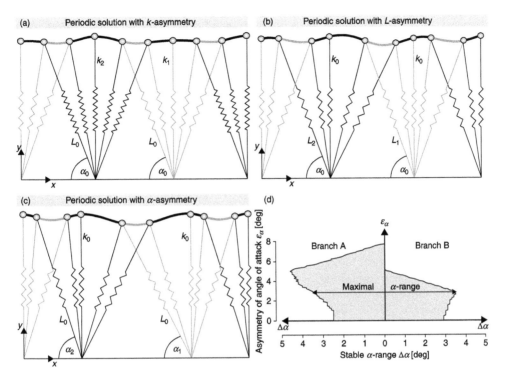

Figure 5.3   Asymmetric walking described in the bipedal SLIP model (B-SLIP, see Fig. 5.1b). Three different kinds of left–right asymmetries can be distinguished: (a) k-asymmetry, (b) L-asymmetry and (c) α-asymmetry. (d) In contrast to k- and L-asymmetry with moderate differences $\varepsilon\alpha$ in the angle of attack between both legs (α-asymmetry: $\alpha_1 = \alpha_0 - \varepsilon_\alpha$, $\alpha_2 = \alpha_0 + \varepsilon_\alpha$) an increased range of perturbations in the angle of attack α_0 (α-range $\Delta\alpha$) can be tolerated. Figure adapted from Merker et al., 2011.

in leg stiffness of one leg (k-asymmetry) compared to the other leg results in mostly symmetric contact phase, but causes different center of mass heights during mid-stance (Figure 5.3a). In contrast, changed leg angles at touchdown between the two limbs (α-asymmetry) result in an asymmetry in each single stance phase, but symmetric double support phases (Figure 5.3c). If the leg length of the two limbs is different (L-asymmetry), both single and double support phases are asymmetric (Figure 5.3b; Merker et al., 2011). One would expect that leg asymmetries challenge walking gait stability. This was indeed found for stiffness asymmetry and leg length asymmetry. In contrast, differences in the angle of attack between the two limbs (α-asymmetry) were less critical for walking stability. For moderate asymmetries in angle of attack even a larger single disturbance in leg angle could be tolerated by the system (Figure 5.3d). This indicates that the effect of asymmetric leg operation on gait stability is specific to the kind and amount of asymmetry and may even provide advantages as demonstrated for the case of α-asymmetry. Hence, asymmetries, as observed in natural or in impaired gait patterns, do not necessarily compromise gait stability and may be compensated by the mechanics of the locomotor system.

Thus, legged robots or artificial limbs (like leg prostheses and orthoses) could take advantage of the stabilizing mechanisms provided by asymmetric leg function. With this, new system designs could be developed, which do not require high precision but benefit from passive stability provided by asymmetrically operating compliant legs. As shown, with unilateral leg amputation, even more pronounced leg asymmetries can be achieved by entering a hybrid gait with alternating double and three-humped force patterns.

Varying leg properties during locomotion: Changes within one step

Leg spring properties may change not only from step to step, but even within one step. For instance, during leg compression leg stiffness may be higher than during leg lengthening (Figure 5.1a). This is reflected by the take-off asymmetry during landing, which is commonly found in human and animal locomotion (Cavagna, 2010). Here, the breaking phase of the leg is shorter than the push phase. One reason for this asymmetry could be eccentric muscle function (McMahon, 1984) of the leg extensors muscles during leg compression, as suggested by Cavagna (2010). Another origin could be the segmental arrangement of the leg. For instance, in human heel-toe running leg stiffness could be higher during leg loading than during leg extension as the heel of the foot is in contact with the ground until mid-stance. After mid-stance the heel is lifting, only the forefoot remains in contact with the ground, and hence the foot segment (ankle joint) adds a serial spring to the leg spring. This may reduce the resultant leg stiffness in the second half of the contact (Maykranz et al., 2009). However, these potential sources for landing–take-off asymmetry could be easily compensated by tuning the muscle function (neural control) or by changing the foot placement strategy (e.g. forefoot running).

Obviously, different mechanisms could equally cause landing–take-off asymmetry depending on body structure and control. Here, a more detailed analysis based on neuromuscular models would be of help (e.g. Geyer & Herr, 2010). Another, even more fundamental question, however, would be to understand *why* this asymmetry is commonly found during human and animal locomotion? For this, let us assume that leg properties (leg stiffness k, rest length L_0) would continuously change with time during ground contact (Riese & Seyfarth, 2012) with the rates R_K and R_L:

$$k(t) = k_{TD} + R_K \star (t - t_{TD}),$$

$$L_0(t) = L_{TD} + R_L \star (t - t_{TD}).$$

With this assumption one can calculate the combinations of R_K and R_L, which lead to periodic bouncing (hopping) for a certain initial height $y_0 = y_0\star$ of the center of mass (Figure 5.4b). Periodic bouncing means that after each step the same apex height $y_0\star$ is reached. Such periodic solutions can only be found for the following two cases: (case 1) $R_K > 0$ and $R_L < 0$ or (case 2) $R_K < 0$ and $R_L > 0$. The stability of these periodic solutions can be easily quantified: If the initial height y_0 slightly deviates from the steady-state condition $y_0\star$ (i.e. $y_0 = y_0\star + \Delta y_0$), this deviation Δy_0 will be amplified after one cycle in case 1 ($\Delta y_1 > \Delta y_0$) or reduced in case 2 ($\Delta y_1 < \Delta y_0$). Hence, only in the second case ($R_K < 0$ and $R_L > 0$), stable periodic hopping cycles are predicted (Riese & Seyfarth, 2012).

This indicates that fluctuations in ground level, as in running over uneven ground, could be automatically compensated if small variations are added to the original spring-mass model (SLIP model). The predicted deviations nicely fit the experimentally observed landing–take-off asymmetry during human and animal locomotion (Cavagna, 2010). Similar deviations from spring-like leg function were also predicted by a neuromuscular hopping model (Geyer et al., 2003). With changes in ground level, the landing–take-off asymmetry is even

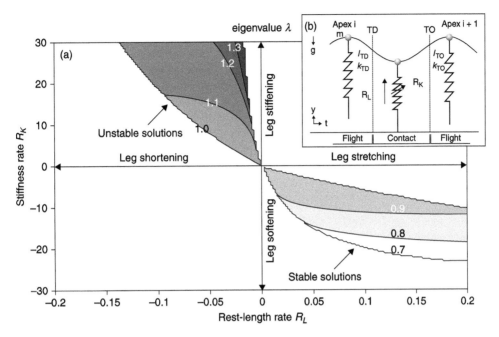

Figure 5.4　Stability in spring-mass hopping with variable leg parameters $k(t)$ and $L_0(t)$ during contact. (a) Shaded regions represent periodic hopping solutions. Only for $R_K < 0$ and $R_L > 0$ stable hopping solutions exist as indicated by eigenvalues λ (see values at contour lines) smaller than 1. (b) Variations in leg parameters (R_K, R_L) of the leg spring are introduced during contact phase. Figure adapted from Riese & Seyfarth, 2012.

enhanced in order to compensate for the energy fluctuation experienced (e.g. due to changed ground level).

These findings indicate that asymmetric contact phases may reflect the ability of compliant systems to manage system energy in the case of unpredictable ground conditions. These conceptual models (Riese & Seyfarth, 2012; Geyer et al., 2003) were developed for hopping on the spot. So far, it is not well understood how these mechanisms are affected by forward speed as present in walking and running gaits. Also the detailed neuromuscular (Geyer et al. 2003; Häufle et al., 2012) or muscle–mechanical interactions (Müller & Blickhan, 2010) within the segmented leg responsible for the landing–take-off asymmetry need to be understood in much more detail. Some of these mechanisms will be described in the following sections.

Human locomotion: Keeping the body upright

Spring-like leg operation during stance phase as described by SLIP-type models is not sufficient for stable locomotion. One additional issue is to guarantee that the body orientation stays within certain limits, e.g. not falling over. A simple approach to achieve postural stability during locomotion is to measure the angular orientation of the trunk and to apply restoring torques (e.g. at the hip joint; Geyer & Herr, 2010) in the case of deviations from a desired configuration. Trunk orientation with respect to gravity can be estimated by the vestibular system or approximated by visual information.

There are technical systems, like a spinning top, a roly-poly toy or a bicycle, which can stabilize an upright configuration even without the need of sensory feedback. In the case of a vestibular defect, it would be very useful to have such an alternative mechanism to achieve postural stability in locomotion without the need of information on the current trunk orientation. The idea of a roly-poly toy can easily be transferred to swimming: In fish and many animals the location of the center of mass (CoM) is below the center of the body volume (Figure 5.5a). Simply by gravity the body will rotate until the CoM is finally located directly below the center of the body volume.

During locomotion on land the situation is different, as the body is not surrounded by water and thus the body does not passively align its orientation. A simple solution mimicking the situation in water would be to design a virtual pivot point (VPP), which is – similar to the roly-poly toy – located above the CoM. This VPP concept can be realized by deviating

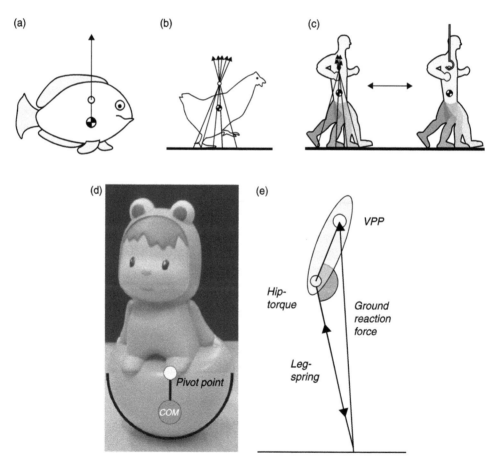

Figure 5.5   The VPP concept: upright body orientation can be achieved by creating a center of rotation above the center of mass (CoM). This can be realized by morphological means (a) fish, (d) poly-doly doll) or by neuromuscular control (b) chicken, (c) humans) such that the ground reaction forces intersect in a virtual pivot point, VPP, above CoM. (e) Based on an extended SLIP model with upright trunk, hip torques can be calculated to redirect the leg forces from the leg spring axis to point to the VPP (adapted from Maus et al., 2010).

the ground reaction forces from the leg axis (from center of pressure, CoP, to CoM) to point to a virtual intersection point *above* the CoM.

To implement and test this idea, we replaced the point mass of the SLIP model (Geyer et al., 2006) by a rigid body. With the help of such a model, hip torques can be calculated to redirect the leg forces to the VPP. If the VPP is located on the body axis above the CoM, the model exhibits stable walking and running patterns, which can cope with changes in ground level. In walking, the predicted hip torque patterns are surprisingly similar to those measured in human walking (Maus et al., 2008). As assumed by the VPP concept, the experimentally observed ground reaction forces (Figure 5.5c) intersect in one point above the CoM. Moreover, the VPP concept not only appears to hold for human locomotion, but can also be found in other animals like birds (Figure 5.5b) and dogs (Maus et al., 2010).

By shifting the VPP backward with respect to the body axis, the trunk tilts forward and locomotion speed starts to increase. In contrast, if the VPP lies in front of the body axis, the trunk leans backward and speed decreases. Hence, by controlling the VPP position two objectives can be realized: (1) an upright trunk posture in walking and running and (2) a change in forward velocity by shifting the VPP horizontally. The corresponding adaptations in trunk orientation (forward lean during acceleration, backward lean during deceleration) are similar to those observed during speed changes in human locomotion.

Running in 3D

All the above-mentioned simulation studies investigated 2D sagittal movements. This approach represents in most cases a valid reduction for human walking and running as the center of mass trajectory remains on an almost straight line with little lateral excursions. Even the leg placement largely remains within the sagittal plane. In animals, however, leg operation during locomotion can be quite different to that in humans. Here, the leg may be placed laterally with respect to the sagittal plane when touching the ground (Figure 5.6a). This is similar to the situation in early childhood when infants start to walk with sprawled leg

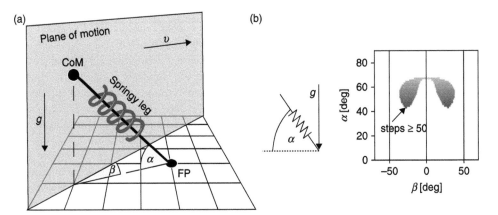

Figure 5.6 (a) Extension of the SLIP model to 3D running. Leg placement is described by the angle of attack α and the lateral leg angle β. (b) Predicted region of 3D stable running with more than 50 steps for leg stiffness $k = 20$ kN/m and forward speed $v_0 = 5$ m/s. With lateral leg placement (e.g. $\beta > 0$) the region of stable running with respect to the angle of attack α is significantly enlarged (Peuker & Seyfarth, 2010).

configurations (Adolph et al., 2003). The step-width continuously decreases with age, finally leading to the mainly sagittal gait pattern found in adults (Donelan et al., 2001).

In the 3D-SLIP model (Figure 5.6a) gait patterns can be described by the same parameters as in the SLIP model (Figure 5.1c) plus one additional leg parameter, the lateral leg angle β with respect to the plane of motion (Figure 5.6a). In contrast to a previous study (Seipel & Holmes, 2005) the lateral leg angle β is not defined with respect to a desired direction, but to the current motion direction of the CoM.

With this assumption stable 3D running patterns can be identified (Figure 5.6b), which extend the region of stable sagittal plane running ($\beta = 0$) of the original spring-mass model (Seyfarth et al., 2002) to non-sagittal patterns ($\beta \neq 0$). Even larger lateral leg angles ($\beta > 35$) can be tolerated by the system. For moderate lateral leg angles ($\beta \approx 25$) a large range of angles of attack α can be tolerated for stable running. This indicates that running with sprawled leg posture, as found in many animals like lizards or salamanders, may benefit from mechanical advantages provided by larger lateral leg angles (Figure 5.6b) requiring a more relaxed tuning of the leg angle of attack α. During evolution leg placement has subsequently changed to the parasagittal leg placements of therian mammals (Fischer & Blickhan, 2006). For such a development, the 3D running model predicts the need for a more careful tuning of leg parameters for stable locomotion in the sagittal plane.

A second extension of the SLIP model towards 3D running is to use the velocity vector of the CoM (spanning the plane of motion, Figure 5.6a) not only as a reference for the lateral leg placement, but also as a reference for the angle of attack within the motion plane (Peuker et al., 2012). During flight phase the velocity vector first points upwards and after passing the apex it starts to point downwards. This change in the direction of the CoM velocity vector caused by gravity results in a rotation of the leg angle in preparation for touchdown, which can be observed in human running over uneven ground (Müller & Blickhan, 2010) and is consistent with the concept of swing leg retraction (Seyfarth et al., 2003).

Role of muscle properties for movement stability

As a consequence of the muscles' intrinsic properties their ability to produce force is limited depending on muscle fiber length and shortening velocity (McMahon, 1984). These muscle properties have been shown to increase the robustness of movements against variations in control commands (e.g. changes in muscle activation) in jumping (Van Soest & Bobbert, 1993) and periodic movements like walking (Gerritsen et al., 1998), swinging (Wagner & Blickhan, 1999) or hopping on the spot (Geyer et al., 2003; Häufle et al., 2010). Intrinsic mechanical properties of muscles are able to adapt muscle force (e.g. in the case of an unexpected change in ground properties, Van der Krogt et al., 2009) faster than feedback loops based on time-delayed sensory signals. Therefore, the stabilizing mechanical responses of muscles are also termed 'pre-flex' (Brown & Loeb, 2000).

Most importantly the force–velocity relationship (Van Leeuwen, 1992) of the muscle fibers supports stability in hopping (Geyer et al., 2003; Häufle et al., 2010). Here the required stimulation signal to appropriately activate the muscle can be equally provided as a prescribed (learned) feed-forward pattern (Häufle et al., 2010) or by reflex pathways using direct sensory feedbacks (Geyer et al., 2003; Häufle et al., 2012). For stable hopping feedback pathways based on fiber length and muscle force (provided by muscle spindles or Golgi tendon organs, respectively) turned out to be most useful for achieving effective hopping heights (Geyer et al., 2003). Interestingly, a positive feedback scheme of the extensor muscle force (positive force feedback, PFF) was predicted to be most suited as a reflex pathway for hopping.

As muscle function in hopping is similar to a rebounding spring, these stabilizing muscle properties may support not only hopping and swinging (Wagner & Blickhan, 1999), but also other gaits such as running or walking with stretch-shortening cycles of the leg extensor muscles. For instance Geyer and Herr (2010) could successfully extend the feedback schemes for hopping to a sagittal plane seven-segment human walking model. In parallel the neuromuscular model was used to control a powered ankle–foot prosthesis (Eilenberg et al., 2010) and successfully demonstrated in human walking. The energy stabilizing properties of the implemented muscle-reflex system resulted in stable walking patterns including the ability to walk on stairs. More recently, this approach was extended to a transtibial prosthesis, which could adapt the ankle net power output depending on walking speed based on the dynamics of the neuromuscular model (Markowitz et al., 2011).

Combination of feedforward and feedback control

It is often assumed that feedforward control is not able to sufficiently adapt to changes in the environment such as locomotion on uneven terrain. With the help of muscle properties as described above, a different observation can be found. Also with a simple playback of an appropriate feedforward command, stable hopping cycles can be achieved even in face of substantial ground level changes (up to 20 percent of leg lengths as predicted by Häufle et al., 2010). Also changes in ground stiffness can be compensated by a corresponding counteracting stiffness change of the muscle without the need to adapt the muscle activation patterns (Van der Krogt et al., 2009).

Such an 'exploitive' actuation protocol – taking advantage of the mechanical properties of the muscle-skeletal system – could not only be present in humans, but is also supported by studies on insect locomotion (Koditschek et al., 2004). Whereas feedforward actuation might be crucial in insect locomotion, it could well be complemented by feedback schemes based on reflex pathways[1] as found in mammals (Pearson et al., 2006). The ability of feedforward and feedback schemes to perform a given motor task largely depends on the way in which sensors and actuators mediate between neural control and the body mechanics. Zuur et al. (2010) suggested that both control schemes might coexist. This control redundancy requires distinct actuator properties, like the non-linear Hill-type force–velocity dependency found in muscles (Häufle et al., 2012). Surprisingly, the two control schemes nicely complement each other and can provide additional stability benefits if mixed appropriately (Häufle et al., 2012). This opens a wide field for exploration, e.g. in the face of motor learning and automation of motor skills. The muscle properties might have evolved during evolution so that motor learning is facilitated by taking advantage of attractive feedforward and feedback control schemes.

---

1 In this context, feedforward and feedback do not refer to feedforward control and (negative) feedback control as used in engineering. The present approach argues that organisms perform in low level limb actuation without any recourse to control in the sense of control theory. So, feedforward simply means that an internal signal is linked with a limb actuation, and feedback means that a sensory signal has influenced such a feedforward signal. Mittelstaedt (1957, 1990) denoted such processing as 'feedforward by proprioception'.

Generalization to multi-segment systems

The results of the reduced models discussed earlier show on a fundamental level the ability of the neuromuscular system to provide efficient and robust control schemes for stable locomotion. This ability largely relies on the properties of the muscles and a proper integration of sensory feedback and feedforward programs for muscle activation. With this, spring-like function of extensor muscles can be achieved during contact phases. How could these results translate to more complex models and finally help to better understand human movements (Figure 5.7)?

With muscle fibers being arranged in series with elastic tendons, the operation can be energy efficient and stable with respect to energetic perturbations (e.g. changed ground levels). As muscle forces directly contribute to joint torques, compliant muscle function also translates to compliant joint function. However, this compliant joint function is not necessarily sufficient to achieve stable leg operation (Seyfarth et al., 2001). For instance, when hitting the ground with the forefoot (toes) during walking, ankle flexion may lead to a sudden (over-) extension of the knee joint. If the foot strikes first with the heel (as typically found in human walking) this unfortunate situation can be avoided. This illustrates that effective strategies are required to achieve an appropriate integration of the muscle and joint function in a segmented leg. Possible strategies include (1) the selection of joint configuration and tuning of joint stiffness in preparation for ground contact, (2) the generation of nonlinear

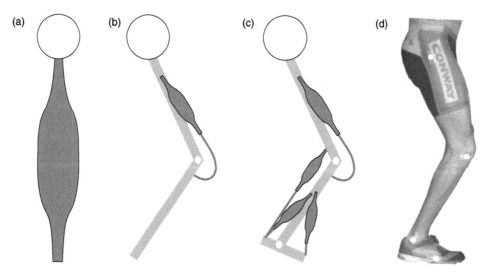

Figure 5.7 Model evolution: the path from a reductive model to more elaborated representative models for human hopping. (a) The one-dimensional neuromechanical template model incorporates muscle characteristics (Häufle et al., 2010). This model predicts the center of mass trajectory, ground reaction forces and reactions to perturbations in landing height. (b) Model with two segments and one muscle–tendon complex. This model can accurately predict elastic leg behavior and vertical hopping stability (Geyer et al., 2003). (c) More elaborate neuromechanical models incorporate several muscles and segments (van Soest & Bobbert, 1993; van der Krogt et al., 2009; Geyer & Herr, 2010). (d) The goal is to understand the functional characteristics of human and animal locomotion. Therefore, the more advanced anchor models implement the concepts found in the template models towards a specific human/animal and task (figure from Häufle et al., 2012)

spring characteristics at joint level (e.g. due to nonlinear tendon properties), (3) the activation of muscles spanning two joints (biarticular muscles) and (4) the inclusion of joint constraints (e.g. heel contact). So far it is not well understood to what extent the coordination of joint function requires neural control, or whether the mechanical interplay of muscles distributed across the different leg joints is sufficient. The model of Geyer and Herr (2010) suggests that only minimal neural information processing across leg joints and muscles could be sufficient to achieve stable and human-like walking patterns. Spring-like body dynamics as described by the SLIP model does not necessarily require the leg joints to operate elastically, nor does an elastic leg joint operation necessarily contribute to spring-like function of the contacting leg (e.g. when the joint is close to fully extended). Hence, the remaining body (trunk, upper limbs, swing leg) may equally contribute to the overall spring-like body dynamics as found in human gaits (Lipfert, 2010).

Robots as a tool for validation of concepts and their application

In the previous sections different aspects of legged locomotion were considered and described with the help of conceptual biomechanical and neuromuscular models. These models were simplified to an extent that may provide a better understanding on how the motion dynamics and control interact to achieve stable locomotion. However, this under-standing is often based on key assumptions made in the model for the sake of simplicity. Therefore, it is important to prove whether the assumptions made are justified and practical. The transfer of conceptual models of legged locomotion to robots can be considered an appropriate research tool to prove the real world validity of theoretical models (Kalveram & Seyfarth, 2009). By comparing the behavior of the robot that relies on previously identified principles of locomotion with experimental findings, the value of the concept can be estimated.

With this approach important features of the real system can be identified, which are missing in the model. This can guide the development of future model generations with the ultimate goal to bridge the 'reality-gap' between model-based concepts and the biological counterpart. It is not sufficient to copy the naturally observed behavior to the model and consequently to a robot, even though this is a challenge on its own. More importantly, it is required that the responses of the locomotor system to unexpected changes in the body (e.g. changed morphology) or in the environment (e.g. changed ground level) of the robot are comparable to the model and to the biological system (Kalveram & Seyfarth, 2009). This needs to be tested by a 'test trilogy' comparing motor behaviors in the face of perturbations at all three levels (biology – concept – robot) as different strategies (or combinations of them) may be used to generate the same steady-state movement.

The ultimate goal is to identify the interplay of different levels of the neuro-mechanical system including central pattern generators (CPGs), sensory feedback, mechanical pre-flexes and the environment (Dickinson et al., 2000). For instance, the integration of CPGs and sensory feedback for leg angle control in the SLIP model may greatly enhance robustness of locomotion in the face of parameter changes (Pouya et al., 2011).

For legged locomotion three key features must be realized in order to achieve stable gait patterns (Figure 5.8a):

(1) Axial leg function: The leg must be able to rebound elastically in order to withstand the effect of gravity. This results in phases of leg compression and leg extensions during stance and phases where the leg leaves the ground (e.g. swing phase). Energy changes, e.g.

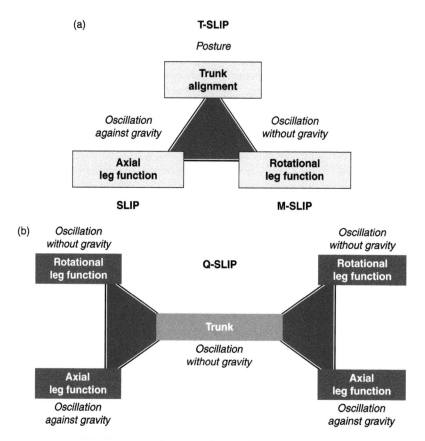

Figure 5.8 (a) Schematic approach to legged locomotion based on three functional levels: axial leg function, rotational leg function, and trunk alignment. (b) Extension of this approach to quadrupedal locomotion.

by changed ground levels or non-elastic ground properties, must be compensated by the neuromechanical system which requires to deviate from a spring-like leg function.

(2) Rotational leg function: The leg must be able to swing forward and backward resembling the function of a mechanical pendulum. This results in leg oscillations perpendicular to the direction of gravity required to propel the body forward during stance phase (leg retraction) and to propel the leg forward during swing phase (leg protraction).

(3) Trunk alignment: The forces generated during ground contact must be such that the supported upper body (trunk, arms, head) keeps in a desired posture (e.g. upright in human walking and running).

The first feature required for a legged locomotion, the *axial leg function*, can be tested using robots performing hopping experiments (Figure 5.9). This was done in several approaches with prismatic legs (e.g. Marco Hopper, Kalveram et al., 2012) or segmented legs (e.g. Hosoda et al., 2010; BioBiped, Maufroy et al., 2011). In the case of the Marco Hopper (Figure 5.8a) a first direct comparison of leg function between human hopping, simulated hopping and robot hopping with unexpected changes in ground level was performed

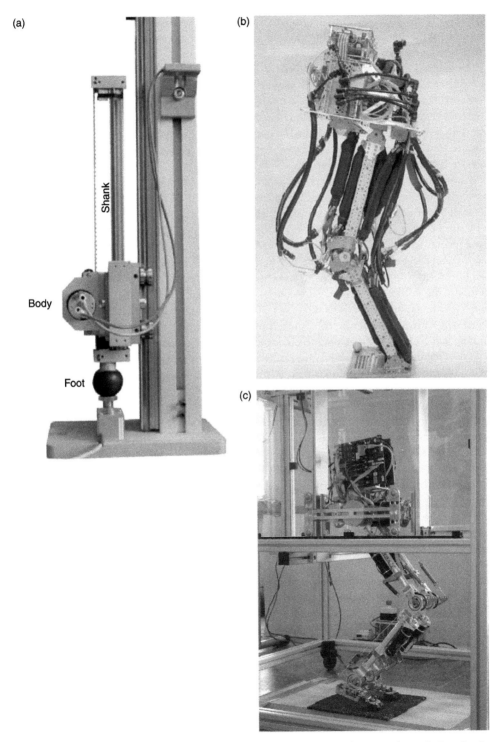

Figure 5.9 Conceptual robots to investigate the interplay of body mechanics and muscle function in hopping: (a) Marco Hopper (Kalveram et al., 2012), (b) segmented hopper with pneumatic muscle (Hosoda et al., 2010) and (c) segmented BioBiped robot with serial elastic actuators (Maufroy et al., 2011).

(Kalveram et al., 2012). A similar comparison between robot and human leg function was performed with the BioBiped robot (Figure 5.9c) in order to analyze its ability to generate spring-like leg behavior.

The analysis showed that these robotic systems are capable of stable hopping, but exhibit clear differences in the axial leg function as compared to humans. For instance, the analyzed robots exhibited high impact forces occurring during touchdown. Also during contact different energy losses (e.g. joint friction, motor properties) significantly shaped the leg function. This is different to human or animal locomotion where impacts and energy losses during contacts are carefully managed, e.g. by landing strategies or by taking advantage of muscle properties and neuromuscular control.

The second requirement for locomotor systems is the *rotational leg function*, meaning an appropriate timing and amplitude of leg protraction and leg retraction. This oscillation can be tuned by introducing a rotational spring between the trunk and the leg. This was realized in the first Passive Dynamic Runner (Owaki et al., 2011) consisting of two legs with prismatic springs and knee joints that are able to bend during the swing phase. This is the first purely passive legged robot being able to run down a slope without any actuation or control for more than 30 subsequent steps. Depending on the selection of leg spring stiffness and rotational hip stiffness, different gaits could be generated by the robot. Based on a corresponding simulation model (Owaki et al., 2008), it was predicted that for walking gaits higher leg stiffness and lower rotational hip stiffness values were required compared to running gaits. This general gait-specific dependency of the stiffness distribution along the leg axis (axial leg spring) and at the hip joint (rotational leg spring) was indeed reproduced by the robot. This is in line with the idea that passive properties of the hip joint may support the generation of gait patterns (Riener & Edrich, 1999) and that neuromuscular control may help to tune these mechanical properties depending on gait and speed. Future research needs to clarify how this rotational leg function is indeed controlled and how this control level is interconnected to the axial leg function found in hopping motions.

The third body function required for legged locomotion is to *keep the trunk aligned* in a certain posture that is mostly an upright posture in human walking and running. Simulation studies based on the VPP concept (see above; Maus et al., 2010) showed that a stable trunk posture can be guaranteed based on a simple control scheme using leg force and leg orientation with respect to the trunk as inputs to control the hip torques. These hip torques not only affect the orientation of the trunk, but also contribute to the rotational leg function. Here two different objectives (trunk alignment, rotational leg function) could be realized based on rotational leg control if the requirements of the VPP concept are fulfilled. Currently, this idea is used to design the controller for the upcoming generations of the BioBiped robot (Figure 5.9c).

The three functional elements of the control of legged systems (Figure 5.8a) not only apply to bipedal systems, but can also be extended to quadrupeds (Figure 5.8b) or to animals or robots with even more legs (e.g. hexapods). The trunk orientation is then usually horizontal, and postural control becomes less difficult as more (e.g. four or six) limbs are supporting the body. This is reflected also in the VPP concept, which is represented to a lesser degree in dogs as compared to birds or humans (Maus et al., 2010). Consequently simple control schemes can be used for highly dynamic quadrupedal and hexapod robots such as Scout II (Papadopoulos & Buehler, 2000), iSprawl (Kim et al., 2006) or RHex (Saranli et al., 2001). All these robots truly rely on their built-in compliant leg function with actuated rotational leg function (Figure 5.8b) powered by electric motors. In iSprawl and RHex, a mechanical coupling between rotational and axial leg function was achieved by the careful mechanical designs of the legs. In all three robots (Scout II, iSprawl and RHex) the mechanical design was inspired by the compliant leg function commonly found in animal locomotion. A direct comparison of axial or rotational leg function with respect to biological gait data could help to identify the design and control differences with respect to the three functional levels of legged locomotion (Figure 5.8a).

During the past decade substantial progress has been made in investigating different aspects of human and animal locomotion based on conceptual biomechanical and neuromechanical models. These models are promising as they provide valuable insights on the organization of legged systems. In future these models need to be further integrated into more complex *anchor* models that inherit key properties of the underlying conceptual models (Full & Koditscheck, 1999). Only such more realistic models will be able to bridge the existing gap between today's conceptual models and real biological and robotic systems. Already many of the conceptual models can be combined (e.g. VPP concept and SLIP model) and predict reasonable design and control strategies of stable locomotion. This methodology needs to be further developed defining the rules for successful combinations of conceptual models. Complementing the mechanical layout of legged systems based on template models, a matching layout on the control side needs to be developed. Similar to the mechanical system, the control system should equally be able to be extendable from the template level to more complex systems with a high level of inheritance of primitive behaviors. Robots can be used to transfer these mechanical and control concepts in order to identify logical or practical shortcomings of the concepts. In comparison to experimental data, required changes of concepts can then be localized (Kalveram & Seyfarth, 2009).

It is the authors' opinion that this close interaction of biological observations, conceptual models and concept-driven robots will become a powerful tool to enhance our understanding in movement organization and motor control.

Acknowledgements

During the preparation of this chapter the authors were supported by DFG grant SE1042/6 and by the European Commission under the FP7, grant # 231688 (Locomorph project).

References

Adolph, K. E., Vereijken, B. & Shrout, P. E. 2003. What changes in infant walking and why, *Child Development* 74, 475–497.

Blickhan, R. 1989. The spring-mass model for running and hopping. *Journal of Biomechanics* 11/12, 1217–1227.

Blickhan, R., Wagner, H. & Seyfarth, A. 2003. Brain or muscles? In: Pandalai, S.G., editor, *Recent research developments in biomechanics*, Trivandrum, India, Vol. 1. pp. 215–245.

Brown, I. E. & Loeb, G. E. 2000. A reductionist approach to creating and using neuromusculoskeletal models. In *Biomechanics and neural control of posture and movement* (eds J. M. Winters & P. Crago), pp. 148–163. New York, NY: Springer.

Cavagna, G. A. 2010. Symmetry and asymmetry in bouncing gaits. *Symmetry* 2, 1270–1321 (doi:10.3390/sym2031270).

Dickinson, M. H., Farley, C. T., Full, R. J., Koehl, M. A. R., Kram, R. & Lehman, S. 2000. How animals move: an integrative view. *Science* 288, 100–106. (doi:10.1126/science. 288.5463.100).

Donelan, J. M., Kram, R. & Kuo, A. D. (2001). Mechanical and metabolic determinants of the preferred step width in human walking. *Proceedings of the Royal Society of London B* 268, 1985–1992.

Eilenberg, M. F., Geyer, H., Herr. H. 2010. Control of a powered ankle–foot prosthesis based on a neuromuscular model. *IEEE Transactions on Neural Systems and Rehabilitation Engineering* 18, 164–173.

Endo, K., Paluska, D. & Herr, H. 2006. A quasi-passive model of human leg function in level-ground walking. IEEE/RSJ International conference on intelligent robots and systems (IROS). Beijing, China.

Farley, C. T., Blickhan, R., Saito, J. & Taylor, C. R. 1991. Hopping frequency in humans: a test of how springs set stride frequency in bouncing gaits. *Journal of Applied Physiology* 71, 2127–2132.

Fischer, M. S. & Blickhan, R. 2006. The tri-segmented limbs of therian mammals: Kinematics, dynamics, and self-stabilization – a review. *Journal of Experimental Zoology* 305A: 935–952.

Full, R. F., Koditschek, D. E. (1999) D. Template and anchors: neuromechanical hypothesis of legged locomotion on land *Journal of Experimental Biology* 202, 1–8.

Gerritsen, K. G. M., Van Den Bogert, A. J., Hullinger, M. & Zernicke, R. F. 1998. Intrinsic muscle properties facilitate locomotor control – a computer simulation study. *Motor Control* 2, 206–220. See http://www.ncbi.nlm.nih.gov/pubmed/9644290.

Geyer, H., Seyfarth, A. & Blickhan, R. 2003. Positive force feedback in bouncing gaits? *Proceedings of the Royal Society of London B* 270, 2173–2183. (doi:10.1098/rspb.2003.2454)

Geyer, H., Seyfarth, A. & Blickhan, R., 2006. Compliant leg behaviour explains basic dynamics of walking and running. *Proceedings of the Royal Society B: Biological Sciences* 273 (1603), 2861.

Geyer, H. & Herr, H. 2010. A muscle-reflex model that encodes principles of legged mechanics produces human walking dynamics and muscle activities, *IEEE Transactions on Neural Systems and Rehabilitation Engineering* 18: 263–273.

Grimmer, S., Ernst, M., Günther, M., & Blickhan, R. 2008. Running on uneven ground: Leg adjustment to vertical steps and self-stability. *Journal of Experimental Biology*, 211, 2989–3000.

Häufle, D. F. B., Grimmer, S. & Seyfarth, A. 2010. The role of intrinsic muscle properties for stable hopping – stability is achieved by the force–velocity relation. *Bioinspir. Biomim.* 5, 016004. (doi:10.1088/1748-3182/5/1/016004)

Häufle, D. F. B., Grimmer, S., Kalveram, K. T. & Seyfarth, A. 2012. Integration of intrinsic muscle properties, feed-forward and feedback signals for generating and stabilizing hopping. *Journal of the Royal Society* Interface. Published online. (doi:10.1098/rsif.2011.0694)

Hosoda, K., Sakaguchi, Y., Takyama, H., Takuma, T. 2010. Pneumatic-driven jumping robot with anthropomorphic muscular skeleton structure. *Autonomous Robot* 28: 306–316.

Iida, F., Rummel, J. & Seyfarth, A., 2008. Bipedal walking and running with spring-like biarticular muscles. *Journal of Biomechanics* 41 (2008) 656–667.

Kalveram, K. T. & Seyfarth, A. (2009). Inverse biomimetics: How robots can help to verify concepts concerning sensorimotor control of human arm and leg movements. *Journal of Physiology – Paris*, 103, 232–243. DOI: 10.1016/j.jphysparis.2009.08.006

Kalveram, K. T., Häufle, D., Seyfarth, A. and Grimmer, S. (2012) Energy management generating terrain following versus apex preserving hopping. Comparison of simulated, robotic, and human bouncing. *Biological Cybernetics* 106 (1), 1–13, DOI: 10.1007/s00422-012-0476-8

Kim, S., Clark, J. E. & Cutkosky, M. R. 2006. iSprawl: Design and tuning for high-speed autonomous open-loop running. *International Journal of Robotics Research* Vol. 25, No. 9: 903–912.

Koditschek, D. E., Full, R. J. & Buehler, M. 2004. Mechanical aspects of legged locomotion control. *Arthropod. Struct. Dev.* 33, 251–272. (doi:10.1016/j.asd.2004. 06.003)

Lipfert, S. W. 2010. *Kinematic and dynamic similarities between walking and running.* Hamburg: Verlag Dr. Kovac.

Markowitz, J., Krishnaswamy, P., Eilenberg, M. F., Endo, K., Barnhart, C. & Herr, H. 2011. Speed adaptation in a powered transtibial prosthesis controlled with a neuromuscular model. *Philosophical Transactions of the the Royal Society B* 2011 366, 1621–1631 (doi: 10.1098/rstb.2010.0347)

Maufroy, C, Maus, H.-M., Radkhah, K., Scholz, D., von Stryk, O., Seyfarth, A. 2011. Dynamic leg function of the BioBiped humanoid robot. In: *Proc. 5th Intl. Symposium on Adaptive Motion of Animals and Machines (AMAM)*, Oct. 11–14, 2011.

Maus, H. M., Rummel, J., Seyfarth, A. (2008) Stable Upright Walking and Running using a simple Pendulum based Control Scheme, Advances in Mobile Robotics: *Proc. of 11th CLAWAR*, Marques L., Almeida A., Tokhi M. O., Virk G. S. (Eds), World Scientific: 623–629. Coimbra, Portugal.

Maus, H. M., Lipfert, S. W., Gross, M., Rummel J. & Seyfarth, A. (2010) Upright human gait did not provide a major mechanical challenge for our ancestors. *Nature Communications.* DOI: 10.1038/ncomms1073.

Maykranz, D., Grimmer, S., Lipfert, S. W. & Seyfarth, A. 2009. Foot function in spring mass running. *Autonome Mobile Systeme*, Berlin, Springer, 81–88.

McMahon, T. A. 1984. *Muscles, reflexes and locomotion*, Princeton, NJ: Princeton University Press.

McMahon, T. A., Cheng, G. C. 1990. The mechanics of running: How does stiffness couple with speed? *Journal of Biomechanics* 23 (Supplement 1):65–78.

Merker, A., Rummel, J. & Seyfarth, A. 2011. Stable walking with asymmetric legs. *Bioinspir. Biomim.* 6 (doi:10.1088/1748-3182/6/4/045004)

Mittelstaedt, H. 1957. *Recent advantages in invertebrate physiology–a symposium*. Chap: Prey capture in Mantida. University of Oregon Publications, Eugene, pp 51–71

Mittelstaedt, H. 1990. *The perception and control of self-motion*. Chap: Basic solutions to the problem of head-centric visual localization. Erlbaum, Hillsdale, NJ, pp. 267–286.

Müller, R. & Blickhan, R. 2010. Running on uneven ground: Leg adjustments by muscle pre-activation. *Human Movement Science* 29, 299–310.

Owaki, D., Koyama, M., Yamaguchi, S., Kubo, S. & Ishiguro, A. 2011. A 2-D passive-dynamic-running biped with elastic elements. *IEEE Transactions on Robotics*, Vol. 27: 156–162.

Owaki, D., Osuka, K., & Ishiguro, A. 2008. On the Embodiment That Enables Passive Dynamic Bipedal Running. 2008 IEEE International Conference on Robotics and Automation (ICRA), Pasadena, CA, USA.

Papadopoulos, D. & Buehler, M. 2000. Stable Running in a Quadruped Robot with Compliant Legs. IEEE International Conference on Robotics and Automation (ICRA), San Francisco, USA.

Pearson, K. G., Ekeberg, O. & Büschges, A. 2006. Assessing sensory function in locomotor systems using neuro-mechanical simulations. *Trends in Neuroscience* 29, 625–631. (doi:10.1016/j.tins.2006.08.007)

Peuker, F. & Seyfarth, A. 2010. Adjusting Legs for Stable Running in Three Dimensions. 6th World Congress of Biomechanics (Singapore), IFMBE Proceedings vol 31, Berlin: Springer.

Peuker, F., Renjewski, D., Gross, M., Grimmer, S. & Seyfarth, A. 2011. Involuntary morphosis modular simulation of limb damage in quadrupeds. Dynamic Walking conference proceedings. Jena, Germany.

Peuker, F., Maufroy, C. & Seyfarth, A., 2012. Leg adjustment strategies for stable running in three dimensions. *Bioinspir. Biomim.*, 7, 036002.

Pouya, S., Moeckel, R., Peuker, F., Seyfarth, A. & Ijspeert, A. J. 2011. Stability augmentation of SLIP-like legged locomotion. Presented at Climbing and walking robots (CLAWAR). Paris, France, August 2012.

Riener, R. & Edrich, T. 1999. Identification of passive elastic joint moments in the lower extremities. *Journal of Biomechanics* 32, 539–544.

Riese, S. & Seyfarth, A. 2012. Stance leg control: variation of leg parameters supports stable hopping. *Bioinspir. Biomim.* 7 (doi:10.1088/1748-3182/7/1/016006)

Sadeghi, H., Allard, P., Prince, F. & Labelle, H. 2000. Symmetry and limb dominance in able-bodied gait: a review. *Gait and Posture* 12, 34–45.

Saranli, U., Buehler, M. & Koditschek, D. E. 2001. RHex – a simple and highly mobile hexapod robot. *International Journal of Robotics Research* 20: 616–631.

Schaarschmidt, M., Lipefert, S. W., Meier-Gratz, C., Scholle, H. C., Seyfarth, A. 2012. Functional gait asymmetry of unilateral transfemoral amputees. *Human Movement Science*. (doi: 10.1016/j.humov.2011.09.004)

Schaarschmidt, M. 2007. Biomechanik des Prothesenlaufs – Von der Analyse zur Intervention. Diplomarbeit. Friedrich-Schiller-Universität, Jena, Germany.

Schwind, W. J. 1998. Spring Loaded Inverted Pendulum Running: A Plant Model. Doctoral Dissertation. University of Michigan, Ann Arbor, MI, USA.

Seipel, J. E. & Holmes, P. 2005. Running in three dimensions: analysis of a point-mass sprung-leg model. *International Journal of Robotics Research*, 24(8), 657–674.

Seyfarth, A., Günther, M. & Blickhan, R. 2001. Stable operation of an elastic three-segment leg. *Biological Cybernetics*, 85: 365–382.

Seyfarth, A., Geyer, H., Günther, M., & Blickhan, R. 2002. A movement criterion for running. *Journal of Biomechanics*, 35, 649–655.

Seyfarth, A., Geyer, H. & Herr, H. 2003. Swing-leg retraction: A simple control model for stable running. *Journal of Experimental Biology*, 206, 2547–2555.

Seyfarth, A., Iida, F., Tausch, R., Stelzer, M., Von Styk, O., Karguth, A. 2009. Towards bipedal jogging as a natural result of optimizing walking speed for passively compliant three segmented legs. *International Journal of Robotics Research*, 28, 257–265.

Van Den Bogert, A. J. 2003. Exotendons for assistance of human locomotion. *Biomedical Engineering Online*, 2–17.

Van der Krogt, M., De Graaf, W. W., Farley, C. T., Moritz, C. T., Casius, L. J. R & Bobbert, M. F. 2009. Robust passive dynamics of the musculoskeletal system compensate for unexpected surface changes during human hopping. *Journal of Applied Physiology* 107:801–808. (doi:10.1152/japplphysiol.91189.2008)

Van Leeuwen, J. L. 1992. Muscle function in locomotion. In *Advances in comparative and environmental physiology 11: Mechanics of animal locomotion* (ed. R. McN. Alexander), pp. 191–242. Heidelberg, Berlin: Springer-Verlag.

Van Soest, A.J. & Bobbert, M.F. 1993. The contribution of muscle properties in the control of explosive movements. *Biological Cybernetics* 69, 195–204. (doi:10.1007/ BF00198959).

Wagner, H. & Blickhan, R. 1999 Stabilizing function of skeletal muscles: an analytical investigation. *J. Theor. Biol.* 199, 163–179. (doi:10.1006/jtbi.1999.0949)

Zuur, A. T., Lundbye-Jensen, J., Leukel, C., Taube, W., Grey, M. J., Gollhofer, A., Nielsen, J. B. & Gruber, M. 2010. Contribution of afferent feedback and descending drive to human hopping. *Journal of Physiology* 588, 799–807. (doi:10.1113/jphysiol.2009.182709).

PART II

Basic aspects of motor control and learning

6

VISUAL ACTIVATION OF SHORT LATENCY REINFORCEMENT MECHANISMS IN THE BASAL GANGLIA

Nicolas Vautrelle, Mariana Leriche and Peter Redgrave

NEUROSCIENCE RESEARCH UNIT, DEPARTMENT OF PSYCHOLOGY, UNIVERSITY OF SHEFFIELD, UNITED KINGDOM

Introduction

What is an 'action' and where does it come from? This is a fundamental question in motor control that will be addressed in the current chapter. As infants, it is likely that we start with sensorimotor 'babbling' (Thelen, 1979) that, by chance, causes something to happen in the external world. The 'baby babble' is then gradually refined until the causal components embedded within the initial multidimensional movements are identified. At this point, we can say that a new action has been acquired. The defining characteristic of an action developed in this manner is that it refers to a movement or sequence of movements that causes a predicted outcome in the external world. Thorndike reported an early example of an experiment that demonstrated the process of acquiring a novel action, the results of which led him to formulate his famous Law of Effect (Thorndike, 1911). He placed a hungry cat in a cage with a door that was held closed by a pin. A pedal in the cage was connected to the pin, so that, if the cat pressed the pedal, the pin was released and the door fell open. Outside the cage was a piece of fish. As the animal explored the cage to find a way to the fish, the pedal was initially pressed 'by accident', i.e. with no knowledge of the outcome. At this stage, door opening was not predicted, and therefore, when it happened, was a surprise to the animal. However, through trial and error, the cat gradually discovered that the critical component of its initial exploratory behavior was to place its foot on the pedal and press. This form of learning effectively equipped the animal with a novel act (pedal press), which produced a predicted outcome (door opening giving access to the fish). Formally, Thorndike proposed that 'any act which in a given situation produces satisfaction becomes associated with that situation so that when the situation recurs the act is more likely than before to recur also' – the Law of Effect (Thorndike, 1911).

Had single unit electrophysiological recording been available to Thorndike, he could have recorded the activity of ventral midbrain dopaminergic (DA) neurones during his experiment. What we now know about the activity of DA neurones suggests that the unpredicted movement or sound of the pin being released would have caused a short latency short duration burst of DA activity, which is referred to as the 'phasic' dopamine response (Schultz, 1998). Evidence is now emerging to suggest that this neural response would have occurred as soon as the pin was released, even before the cat had turned to see what was happening, long before the door had fallen open, and even longer before the cat had the 'satisfaction' of eating the fish (Redgrave et al., 1999b). We have argued (Redgrave and Gurney, 2006, Redgrave et al., 2008), that the phasic response of DA neurones provides the learning signal in circuitry that would allow the cat to discover *exactly* what movements it had to make, when and where to make them, to release the pin; in other words, to reinforce the development of an entirely novel action. This suggestion will be contrasted with the view that phasic DA responses reinforce the selection of actions that maximize the long term acquisition of reward (Montague et al., 1996, Montague et al., 2004, Schultz, 1998, Schultz, 2006, Niv, 2009). We will suggest that another reinforcement mechanism may be at work to ensure this happens. However, to appreciate these arguments it will be necessary first to introduce some background information about the circuitry, within which the ascending dopaminergic projections are a vital component for action discovery.

The basal ganglia

The basal ganglia are one of the vertebrate brain's fundamental information processing units (Figure 6.1) (Gerfen and Wilson, 1996, Redgrave, 2007). As such, they receive excitatory inputs, directly from cerebral cortex, limbic structures, and the thalamus, or indirectly from midbrain and brainstem regions, via the thalamus. Thus, they receive inputs from most regions of the brain engaged in motivation, emotion, cognition and sensorimotor control of behavior (DeLong and Wichmann, 2010). The brain structures that provide input to the basal ganglia receive, in return, direct (thalamus, brainstem structures) or indirect inhibitory signals (cortex, limbic structures) from the basal ganglia output nuclei (Kimura et al., 2004, Smith et al., 2004,

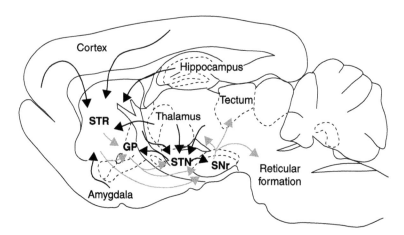

Figure 6.1 Principal components of the mammalian basal ganglia. The main input nuclei are the striatum (STR) and the subthalamic nucleus (STN). Direct connections to both input nuclei are from the thalamus, cerebral cortex and limbic structures (amygdala and hippocampus). The main output nuclei are the substantia nigra pars reticulata (SNr) and the internal globus pallidus (not shown). The external globus pallidus (GP) is an intrinsic nucleus as most of its connections are with the input and output nuclei of the basal ganglia. Black arrows denote excitatory connections while grey arrows represent inhibitory connections.

Redgrave et al., 1992, Takakusaki et al., 2004). The arrangement of these connections between the basal ganglia and external structures can be viewed as a series of partially segregated parallel projecting re-entrant loops (Alexander et al., 1986, McHaffie et al., 2005) (Figure 6.2). A critical aspect of this arrangement is that the spatial segregation between the loops is maintained throughout the intrinsic nuclei of the basal ganglia (Figure 6.2a). This

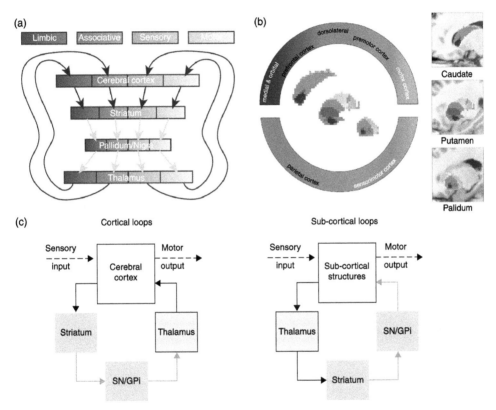

Figure 6.2   Cortico-basal ganglia-cortical loops in animals and humans: (a) An important component of the architecture connecting the cerebral cortex with the basal ganglia is the series of parallel projecting loops or channels (Alexander et al., 1986). Functional territories represented in the cerebral cortex are maintained throughout the basal ganglia nuclei and thalamic relays. Relay points in the cortex, basal ganglia, and thalamus, offer opportunities for activity inside the loop to be modified/modulated by signals from outside the loop. (b) Spatially segregated 'rostral caudal gradient' of human frontal cortical connectivity in caudate, putamen, and pallidum. The greyscale-coded ring denotes limbic, associative and sensorimotor regions of cerebral cortex in the sagittal plane (after Haber (Haber, 2003). Using probabilistic tractography on MR-diffusion weighted imaging data Draganski and colleagues (Draganski et al., 2008) identified regions of the striatum receiving the strongest input from the identified cortical regions (sagittally oriented images contained within the ring). To aid orientation, segmented basal ganglia nuclei are superimposed on T1-weighted structural images in the far right of the diagram (rostral-left, caudal-right). (c) Cortical and sub-cortical sensorimotor loops through the basal ganglia. The position of the thalamic relay for cortico-basal ganglia loops is on the return link of the loop (Alexander et al., 1986), while for all sub-cortical loops the position of the thalamic relay is on the input side (McHaffie et al., 2005). Abbreviations: SN/GPi, substantia nigra/globus pallidus. Part (b) is reproduced by permission from Draganski et al (2008) © Society for Neuroscience.

imposes a territorial segregation of function, which is determined by the functional status of inputs from the external structures. For example, in the case of afferents to the basal ganglia's principal input nucleus, the striatum, ventromedial striatal territories receive signals from brain structures primarily concerned with motivation and emotion, dorsomedial territories from brain regions engaged in cognition, and dorsolateral territories from brain areas involved in sensorimotor control (Alexander et al., 1986) (Figure 6.2a and 6.2b). However, an intriguing characteristic of the basal ganglia is that although different regional territories are processing inputs linked to widely differing functions, the intrinsic microcircuits that transform inputs of the basal ganglia to outputs are qualitatively similar across the widely varying functional territories (Voorn et al., 2004). Within this architecture, a vital component is the dopaminergic projections that ascend from cell bodies located in the ventral midbrain and broadcast widely throughout the different functional territories of intrinsic basal ganglia nuclei (Lindvall and Bjorklund, 1974, Bjorklund and Dunnett, 2007). Insofar as computational functions are emergent properties of the ways in which neurons are connected, it is likely that similar input–output computations are being applied to the widely differing motivational/affective, cognitive and sensorimotor signals that are received by the basal ganglia (Redgrave et al., 1999a). In other words, whatever it is that the basal ganglia contribute to motivational, cognitive and sensorimotor processing, it is likely that the contribution is the same or similar for all. A second important functional constraint imposed by the anatomical architecture of the basal ganglia is their early appearance in vertebrate brain evolution. The basal ganglia appeared at least 400 million years ago and their cell types, intrinsic connectivity and their patterns of projections with external structures have been largely conserved throughout vertebrate evolution (Reiner, 2010). This suggests that the computational issues addressed by the basal ganglia are likely to be common, not only to different functional systems within organisms, but also among organisms across the vertebrate phyla. We will now consider what candidate problems might have the required generality.

Selection and reinforcement functions of basal ganglia architecture

Despite suggestions that the basal ganglia are involved in a wide range of functions including perception, learning, memory, attention, many aspects of motor function, even analgesia and seizure suppression, accumulating evidence points to an underlying role in generic selection processes (Grillner et al., 2005, Mink, 1996, Redgrave et al., 1999a, Gurney et al., 2001a, Prescott et al., 2006, Humphries et al., 2006, Hikosaka, 2007). The problem of having to make selections is common to all multifunctional systems. For example, most vertebrates including humans have multiple neural circuits, each specialized for expressing different classes of behavior, including homeostatic regulation of energy, fluid and temperature, looking for opportunities to mate, and escaping from, or avoiding harm (Prescott et al., 1999). While these neural systems normally process in parallel, their behavioral outputs cannot usually be expressed simultaneously. Therefore, most animals are continuously faced with the problem of having to decide, at any time, which functional system or sensory event should be allowed to control the motor resources which are shared by all – i.e. the final common motor path (Redgrave et al., 1999a).

Selection by the basal ganglia

Selection is a property that would emerge naturally from the re-entrant looped architecture that connects the basal ganglia with external structures (Alexander et al., 1986, McHaffie et al., 2005) (Figures 6.2 and 6.3). Remember, a critical feature of this architecture is the last link of the loop that connects the basal ganglia output nuclei with external structures (thalamus and other subcortical targets) (Figure 6.2c). These links are inhibitory (i.e. they use the inhibitory transmitter GABA) and are tonically active, thereby keeping their target structures in a default state of inhibition (Chevalier and Deniau, 1990). Selective removal of this tonic inhibitory influence from some loops while maintaining or increasing it on others, in effect, allows the disinhibited functional system(s) access to motor mechanisms capable of

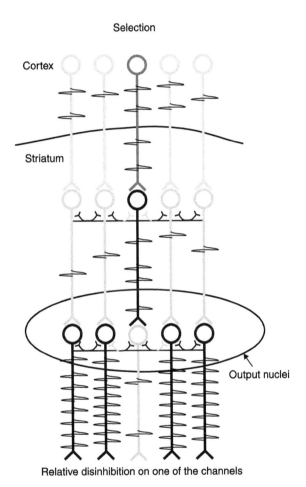

Figure 6.3 Selective disinhibition within the re-entrant parallel looped architecture of the basal ganglia represents a mechanism for selection (Gurney et al., 2001a, Gurney et al., 2001b, Humphries et al., 2006, Prescott et al., 2006). The model assumes that the relative input saliencies to the striatum (denoted by the relative numbers of spikes) in competing loops are resolved by intrinsic basal ganglia micro-circuitry so that the channel(s) with the greatest net inhibitory activity impinging on the output nuclei effectively disinhibit their target structure(s) (Chevalier and Deniau, 1990), while tonic inhibitory output is maintained elsewhere. Disinhibitory output from the basal ganglia to 'selected' external targets allows them to guide movement. To simplify the diagram, return relays via the thalamus have been omitted, as are the connections to the subthalamic nucleus and external globus pallidus. Cortico–striatal projections are excitatory while striatal and output nuclei projections are inhibitory.

generating behavioral output (Redgrave et al., 1999a, Prescott et al., 2006). Operating in this manner, the basal ganglia may be seen as a generic neural mechanism for selecting between competing options, be they motivational, cognitive or sensorimotor (Redgrave et al., 1999a). Independent of any biological considerations, a similar 'central-selection' control architecture was devised by control engineers to select the actions of an autonomous mobile robot (Snaith and Holland, 1990). We have since confirmed that a biologically constrained model of basal ganglia architecture can do likewise (Prescott et al., 2006). Proposals that the basal ganglia represent a vertebrate solution to generic selection problems (Redgrave et al., 1999a, Mink, 1996, Grillner et al., 2005, Hikosaka, 2007) are consistent both with the ancient evolutionary origins of the basal ganglia and the conserved nature of their anatomical connections and neurotransmitters systems (Reiner, 2010). Indeed, it is important to recognize that the need for multifunctional vertebrates to select between competing motivations and stimuli competing for attention has not changed materially over the course of evolution. Thus, an ancient species such as the crocodile deciding which wildebeest to bite as they cross the Mara River in Africa is faced with essentially the same computational problem as a human driver deciding which vehicle to look at on a crowded highway. Thus, despite great changes in the range, power and sophistication of the neural systems competing for expression, it appears that the neural mechanism(s) responsible for deciding between options have been retained in more or less their original form.

Reinforcement in the basal ganglia

Another consistent theme in the basal ganglia literature is their association with reinforcement learning, in particular, instrumental conditioning (Niv, 2009, Berridge, 2007, Wise, 2004, Doya, 2007, Graybiel, 2008, Schultz, 2006). This should not be surprising. Since reinforcement-based learning operates to bias future behavioral selections (increasing or decreasing the probability that previously reinforced selections will be re-selected – Thorndike's Law of Effect (Thorndike, 1911)), there may be a necessary association between selective and reinforcing functions. Therefore, it is likely that reinforcement will be a process operating within a neural network whose primary function is selection. However, while there is general agreement that the basal ganglia play a central role in instrumental reinforcement learning, the details of that role remain to be determined.

The most widely accepted view is that the basal ganglia represent a mechanism where sensory evoked dopamine (DA) responses act as 'teaching signals' in the form of 'reward prediction errors' (Montague et al., 1996, Schultz, 2006, Montague et al., 2004). A reward prediction error occurs when the value of a behavioral outcome is better or worse than expected. Thus, when an unexpected reward evokes phasic DA responses, the actions that are associated with those rewards are chosen more frequently in the long term. The correspondences between the phasic activity of DA neurones and the 'reward prediction error' term in computational models of reinforcement learning continue to provide each other with mutual support (Montague et al., 1996, Schultz, 2006, Montague et al., 2004, Niv, 2009). Behavioral and computational analyses of the reinforcing function of phasic DA signaling are further supported by studies of neural plasticity within the basal ganglia.

The mechanisms by which phasic DA signals might act to bias selection in the basal ganglia have been widely investigated. Long term potentiation (LTP) of corticostriatal transmission, which is thought to be associated with increasing selection probabilities, and long term depression (LTD), which correspondingly is associated with reducing selection probabilities, both appear critically dependent on the presence/absence of DA (Calabresi et al., 2007, Surmeier et al., 2007, Reynolds and Wickens, 2002, Wickens et al., 2003, Wickens, 2009). In such models of the learning process, it is assumed that inputs from the cerebral cortex, which represent competing actions, induce patterns of activity in striatal neurones that are reinforced by phasic DA signaling when they happen to be associated with unpredicted reward. Required selectivity within this system is achieved by restricting the effects of reinforcement (LTP/LTD) to the specific sub-sets of corticostriatal inputs which recently preceded or are concurrently active with the reinforcing DA input (Arbuthnott and Wickens, 2007).

In summary, the conventional view of instrumental conditioning is that actions are selected, at least in part, by the basal ganglia. Those that are associated with unpredicted reward are reinforced by the action of phasic DA signaling, which is mediated via the cellular mechanisms of LTP and LTD. Consequent changes in the relative efficiency of transmission associated with reinforced inputs to the basal ganglia bias future selections in favor of reward-related actions. Despite an impressive amount of support for this model there are aspects of this theory that have been questioned by recent evidence.

Reinforcement in the basal ganglia – unresolved questions

Until recently (see below) we have expressed concern about the 'reward prediction error' hypothesis of phasic DA signaling (Redgrave et al., 1999b, Redgrave and Gurney, 2006, Redgrave et al., 2008). Our concern has been on the grounds that, in a natural environment, typical latencies of the phasic DA response (generally ~100ms) (Schultz, 1998) would require early sensory and perceptual processing to assess and signal the value of an event, before the animal had an opportunity to turn and see what the event was (normal saccadic latencies 150–200ms) (Jay and Sparks, 1987, Hikosaka and Wurtz, 1983). However, recent evidence (Matsumoto and Takada, 2011) suggests that phasic DA responses can be evoked by the discovery of a reward predictor on the basis of visual search, that is, after stimulus identification. In the first part of this section, we will therefore re-consider the circumstances where visual stimuli can elicit phasic DA activity. A second issue arises around the growing body of literature demonstrating that rewards can have profound effects on the processing of sensory stimuli in neural structures that provide inputs to the basal ganglia. Thirdly, we will note that the experimental paradigms showing that cortico-striatal plasticity is critically dependent on the presence/absence of DA, normally do so without due regard to behaviorally relevant timing (Dickinson, 2001). We will now discuss these issues in more detail as a means of introducing an expanded view of the mechanisms of reinforcement in the re-entrant loop architecture of the basal ganglia.

Phasic DA signaling

To generate a reward prediction error signal, the value of what has just happened or has been discovered must be compared with what has been predicted (Montague et al., 2004, Niv, 2009). There are now several lines of evidence that suggest this is not always the computation that is being performed by DA neurons – they can be itemized as follows:

1   For some time now it has been appreciated that neutral stimuli that are novel and elicit orienting, but are unrelated to reward and have no reinforcement consequences, can evoke strong phasic DA responses provided they are unanticipated (Horvitz, 2000, Bromberg-Martin et al., 2010). Counterclaims that such stimuli must therefore be rewarding are circular.

2   There are frequent reports that conditioned stimuli, indicating a new trial is to commence, can also elicit robust and reliable DA neuronal responses (Bromberg-Martin et al., 2010). There is also no direct association between these stimuli and reward (significant numbers of trials are not rewarded and for those that are, reward can be unpredictably delayed) (Fiorillo et al., 2008). They are, however, behaviorally significant as they require the animal to switch its attention and behavioral resources to start a new experimental trial.

3   Recent investigations by ourselves (Coizet et al., 2006, Coizet et al., 2010) and others (Brischoux et al., 2009, Matsumoto and Hikosaka, 2009, Joshua et al., 2008) have reported sub-populations of DA neurons, where some are inhibited and others are excited by aversive stimuli and their predictors. Further evidence suggests that the DA neurones that respond differently to aversive stimuli have different spatial distributions in the ventral midbrain (Brischoux et al., 2009, Matsumoto and Hikosaka, 2009). These findings cast doubt on the popular concept of population-based phasic DA responses broadcasting reward prediction errors widely to targets in the basal ganglia and elsewhere (Montague et al., 2004, Niv, 2009, Schultz, 2006). This idea is further undermined by the accumulating reports that use fast-scan cyclic voltammetry to detect the local release of DA (Clark et al., 2010, Wightman and Robinson, 2002). In these studies, a spatially heterogeneous pattern of DA release in targeted structures is frequently observed.

4   In mammals, an unexpected sensory stimulus typically elicits an orienting gaze-shift that brings it onto the fovea for analysis by cortical visual systems (Rousselet et al., 2004, Thorpe and Fabre-Thorpe, 2001). The latency of these gaze-shifts is typically in the range of 150–200ms (Hikosaka and Wurtz, 1983, Jay and Sparks, 1987). In many of the early studies reporting phasic DA responses to sensory stimuli (Schultz, 1998), the latencies of phasic DA neuronal responses were normally in the range 70–100ms after stimulus onset. In these experiments, therefore, DA responses must have been initiated on the basis of pre-attentive, pre-gaze-shift sensory processing (Redgrave et al., 1999b). However, recent investigations by ourselves (Hudgins et al., 2009) and others (Joshua et al., 2009, Nomoto et al., 2010, Bromberg-Martin et al., 2010), have found that under certain conditions, complex stimuli can evoke biphasic DA responses comprising an initial fast component (latency <100ms) followed by a second slower component (latency ~150–200ms). In all cases so far, the initial component of the biphasic response appears to be value independent (i.e. the same for high and low value stimuli), while the second component reflects stimulus value. It has been suggested (Bromberg-Martin et al., 2010, Joshua et al., 2009) that the initial component of the biphasic response plays more of an alerting function designed to facilitate immediate responding, while the second value-based component represents a reward prediction error-based reinforcement signal acting to maximize future reward acquisition. However, this interpretation requires a clear temporal separation of the two phases, which is not always the case (Nomoto et al., 2010). Also, whether the temporal smearing of DA receptor activation associated with 'volume transmission' in targeted structures (Arbuthnott and Wickens, 2007, Fuxe and Agnati, 1991) would allow them to discriminate and respond differently to the two signals, remains to be seen.

5   An important experiment is currently being conducted in Japan by Matsumoto and Takada, the initial findings of which were recently reported in abstract form (Matsumoto and Takada, 2011). In this study a monkey was shown a target stimulus (the reward predictor), which subsequently it had to search for among a fixed number of physically similar distracters. When the record of DA neurone activity was aligned to the end of

the saccade that brought the target stimulus on to the fovea, a short duration (~150ms) phasic DA response was observed with a delay of approximately +150–200ms. Alignment of the DA record to distracter fixation revealed no significant change in DA activity. The phasic DA responses in this experiment may not, however, signal reward prediction errors. During the training for this study the monkey learned that on each trial the target stimulus was always hidden somewhere among the distracters. All it had to do is search the visual display and show that the target stimulus had been found by fixating it. With the target always present, the greatest reward prediction error would occur when it was the first stimulus inspected and smallest when it was the last. In other words, the greater the number of items in the display the monkey inspected without finding the target, the higher the probability that the next item would be the target. For example, if all but one of the items had been inspected without finding the target, the monkey could predict with certainty that the last item would be the target. However, the experimental finding was that the magnitude of the DA responses evoked by the target increased with the number of items inspected before finding the target, rather than the other way round, as would be expected if they were signaling reward prediction errors. These findings need to be confirmed. However, this experiment is also important because it demonstrates for the first time, that, in addition to biologically significant stimuli with phasic onsets, phasic DA responses can also be evoked by the discovery of static sensory stimuli. Despite this extension of the range of events that can evoke DA responses, it is important to note that they all appear to precede any behavior that is evoked by the sensory events themselves. From a computational perspective, this is significant because having the reinforcing 'teaching signal' (the phasic DA response) precede evoked behavior simplifies the credit-assignment problem (Izhikevich, 2007) (see below).

6   In the case of visual stimuli, and consistent with previous points, we have published a series of studies showing that the short-latency component of visual input to DA neurones derives largely, if not exclusively, from subcortical visual processing in the midbrain superior colliculus (Comoli et al., 2003, Dommett et al., 2005, McHaffie et al., 2006, May et al., 2009). While collicular neurones are exquisitely sensitive to the location of luminance changes, they are largely insensitive to color, static contrast (geometric configuration), and high spatial frequency (Boehnke and Munoz, 2008). Consequently, at pre-gaze-shift latencies, the superior colliculus alone would be unable to determine the identity of unexpected stimuli. Recent evidence indicates, however, that color signals can be recorded in the primate superior colliculus with latencies only ~30ms longer than the initial response to luminance onset (Kim and Basso, 2008, White et al., 2009). This suggests that pre-attentive cortical processing of color can also be made available to the superior colliculus at short (pre-saccadic) latencies. Recently, experiments in our laboratory have discovered that stimulation of visual cortex in rats can drive phasic responses in DA neurones via a relay in the superior colliculus (Vautrelle et al., November 2011). It is now clear, therefore, that early sensory processing by the cerebral cortex can also initiate phasic DA activity, possibly by engaging the evolutionary prior tectonigral projection (Comoli et al., 2003, McHaffie et al., 2006, May et al., 2009). In the case of phasic DA responses being evoked by search-based discovery of biologically significant stimuli (Matsumoto and Takada, 2011), the afferent source of sensory control of DA will be an important research question.

From this analysis we can conclude that stimulus value can modulate both the initial (Schultz, 1998, Schultz, 2006) and late phase of pre-attentive phasic DA responses (Hudgins et al.,

2009, Joshua et al., 2009, Nomoto et al., 2010, Bromberg-Martin et al., 2010). Value can also influence phasic DA responses to stimuli discovered by visual search (Matsumoto and Takada, 2011). However, in all cases so far, the observed phasic DA responses occur prior to behavior elicited by the evoking stimuli. At this point an important unresolved question is, how DA neurones come to be modulated by the value of unpredicted sensory events at such short latencies following stimulus onset, or foveation.

Training sensory systems to discriminate value

Sensory systems have been optimized to detect change. From a formal perspective, visual processing of an unpredicted sensory event (e.g. by retina, superior colliculus, lateral geniculate nucleus, or cerebral cortex) will generate a temporal difference sensory prediction error. What this means is that when there is no sensory event, the prediction for the next time-step will continue to be 'no-sensory-event'. If, however, 'no-event' is predicted, and there is an event, a sensory prediction error will be registered in all relevant sensory structures. So, when for example, the superior colliculus detects a sensory stimulus, it will pass this information directly to ventral midbrain DA neurones (Comoli et al., 2003, McHaffie et al., 2006, May et al., 2009), in the form of a sensory prediction error. The question is, by what mechanism this sensory prediction error can be transformed into the frequently reported reward prediction error signal (Montague et al., 2004, Niv, 2009, Schultz, 2006).

With few exceptions (Pilz et al., 2004), there has been little recent work on the mechanisms of sensory habituation in mammals. However, early investigations (Harris, 1943) show that sensory responses in primary sensory areas habituate rapidly if they are not, in some way, associated with reward (Grantyn, 1988, Horn and Hill, 1966, Sprague et al., 1968). An important property of reward is, therefore, to block the default process of sensory habituation in primary sensory structures (El-Amamy and Holland, 2006). Moreover, recent evidence indicates that early sensory processing of neutral stimuli can be potentiated when the neutral stimulus is associated with reward (Ikeda and Hikosaka, 2003, Bunzeck et al., 2009, Franko et al., 2010, Hickey et al., 2010, Hui et al., 2009, Igelstrom et al., 2010, Mogami and Tanaka, 2006, Serences and Saproo, 2010, Weil et al., 2010). This suggests that reward-related modulation of processing in primary sensory structures offers a mechanism whereby the magnitude/salience of afferent sensory signals that are relayed to DA neurones can reflect stimulus value.

Could this be the mechanism that allows DA neurones seemingly to discriminate the complex characteristics of visual stimuli used to signal different reward magnitudes and probabilities, prior to the gaze-shift that brings the event onto the fovea (Bromberg-Martin et al., 2010, Fiorillo et al., 2008, Fiorillo et al., 2003, Joshua et al., 2009, Nomoto et al., 2010, Morris et al., 2004, Schultz, 2006, Tobler et al., 2003, Tobler et al., 2005)? With few exceptions (Morris et al., 2004, Fiorillo et al., 2008), many of the experiments reporting that phasic DA activity signals reward prediction errors confound the location of stimulus presentation and the stimulus value. Thus, prior to the recording of DA neurones, subjects are typically given many training trials where stimuli at specific locations are associated with the delivery of different value rewards. In cases where value discriminations can be made on the basis of spatial location, it is therefore likely that retinotopic maps in the superior colliculus have been conditioned, prior to DA recording, to respond differentially to stimuli of different value based on their locus of presentation (Ikeda and Hikosaka, 2003). If these different magnitude responses were relayed to the DA neurones (Comoli et al., 2003, Dommett et al., 2005, McHaffie et al., 2006, May et al., 2009), the sensory prediction error would then be a signal conditioned by value – a 'reward prediction error' by any other name. Pre-conditioning of afferent sensory structures would therefore explain how the initial component of the phasic DA response could be made to be sensitive to the value of sensory events at such short latencies (Schultz, 1998, Schultz, 2006). However, in cases of complex stimuli that cannot be discriminated by sub-cortical visual processing, value-based conditioning of more sophisticated cortical visual processing (e.g. color, high spatial-frequency information or random dot motion detection), would be required (Franko et al., 2010, Hickey et al., 2010, Hui et al., 2009, Mogami and Tanaka, 2006, Serences and Saproo, 2010, Weil et al., 2010). If value conditioned cortically based sensory processing was made available to DA neurones, albeit at slightly longer latencies, this could offer an explanation for the longer latency value-based component of biphasic DA responses (Joshua et al., 2009, Hudgins et al., 2009, Nomoto et al., 2010, Bromberg-Martin et al., 2010). The concept of reward-modulated sensory processing accords well with the framework proposed by Mackintosh (Mackintosh, 2009) for understanding perceptual learning based on well-established principles of associative learning theory.

Value-based conditioning of early sensory processing: implications

In the context of reinforcement mechanisms associated with the re-entrant loop architecture of the basal ganglia (Figure 6.2), the modulation of cortical and sub-cortical sensory processing by reward has two important implications. First, if the detection of an unpredicted sensory event can be considered formally as a sensory prediction error, and prior associations with reward have conditioned initial sensory responses, formal reward prediction errors will be immediately available in the primary sensory networks following event onset, without any further calculations required. If these signals are relayed directly to the ventral midbrain (Comoli et al., 2003, Dommett et al., 2005, McHaffie et al., 2006, May et al., 2009) the responses of DA neurones evoked by predictive sensory cues will inevitably equate with the temporal-difference reward prediction errors in the formal reinforcement learning models. The subsequent response of DA neurons to the delivery, or withholding, of reward will depend on the accuracy of the prediction associated with the preceding sensory cue. Thus, the magnitude of DA responses to non-habituated afferent sensory signals associated with reward delivery depends on the extent to which a timed inhibitory cancelling signal has developed (Schultz, 1998, Pan et al., 2005). However, the question of whether the sensory-evoked phasic DA responses should be interpreted according to the conventional theory, as a teaching signal for maximizing the acquisition of future reward (Schultz, 1998, Schultz, 2006, Montague et al., 2004, Niv, 2009), as a means of incentivizing subsequent behavior (Berridge, 2007), and/or as signals to reinforce an appreciation of agency and discovery of novel actions (Redgrave and Gurney, 2006, Redgrave et al., 2008), remains open. We will address these questions in the next sections. However, part of the answer we will propose depends on a second, perhaps more important implication of having sensory signal processing modulated by reward. Independent of any direct involvement of short-latency phasic DA responses, reward-related modulation of early sensory processing may itself represent an alternative mechanism for biasing selection processes within the re-entrant loop architecture of the basal ganglia.

Selection bias in the basal ganglia re-entrant loop architecture

In the early sections of this chapter we concluded that selective disinhibition within the re-entrant loops through the basal ganglia could provide the brain with a generic mechanism for selecting between parallel processing motivational, cognitive and sensorimotor systems seeking access to limited motor resources. Biologically constrained computational investigations of this architecture (Gurney et al., 2001a, Gurney et al., 2001b, Humphries et al., 2006, Prescott et al., 2006) show that the relative magnitude or salience of signals in competing channels within the looped architecture represents a 'common-currency' according to which selections are made. In other words, stronger signals get selected over weaker ones. We also proposed that reinforcement learning can be thought of as processes that cause future selections to be biased by the value of previous outcomes – i.e. selections that are followed by good outcomes are more likely to be selected in future (Redgrave and Gurney, 2006, Redgrave et al., 2008). If this analysis is correct, then the overall network property of selection in the basal ganglia re-entrant loop architecture could potentially be biased at each relay point within the parallel loops – in the brain regions where the loops originate, at synapses within the intrinsic nuclei of the basal ganglia, and in the thalamic relay nuclei. We will now argue that selective biasing by DA reinforcement at relay points within the striatum serves one function in instrumental conditioning, while reward-related modulation of processing in structures that project to the striatum serves another.

Agency and the development of novel actions

A fundamental requirement, which must be satisfied by all mobile agents that interact with the world, is an appreciation of which sensory events they cause to happen and which ones are the result of an intervention from an external source. This is what we call an appreciation of 'agency'. Following an analysis of functional architecture and signal timing within the basal ganglia, we have proposed (Redgrave and Gurney, 2006, Redgrave et al., 2008) that the reinforcement associated with short-latency phasic DA signaling is ideally suited to determine agency and to reinforce the development of novel actions. To understand why these short-latency responses are particularly suitable to reinforce the discovery of agency and action development, it is necessary to consider the other inputs to the basal ganglia likely to be present at the time of phasic DA release. This is important because it is with these signals that sensory-evoked release of DA will most likely interact. Relevant literature suggests there will be at least three additional afferent signals present in the striatum at the time of phasic DA release (Redgrave and Gurney, 2006, Redgrave et al., 2008) (Figure 6.4).

Sensory: Insofar as significant early visual input is relayed to DA neurons via the superior colliculus (Comoli et al., 2003, Dommett et al., 2005, May et al., 2009, McHaffie et al., 2006), we have shown that many tectonigral axons also have co-lateral projections that innervate particularly those regions of the thalamus providing input to the striatum (Coizet et al., 2007). This branched architecture would ensure that an unexpected sensory event that triggered a phasic DA signal would also evoke a widely broadcast glutamatergic input to the striatum relayed from the intralaminar thalamic nuclei. The timing of these two signals (Schulz et al., 2009, Dommett et al., 2005) suggests that they have the potential to converge (Figure 6.4a).

Contextual: Striatal neurones are influenced by variables related to the general sensory, metabolic and cognitive state of the animal (Nakahara et al., 2004, Apicella et al., 1997). Such inputs are likely to originate in afferent cortical, limbic and subcortical (thalamic) structures (Gerfen and Wilson, 1996, Alexander et al., 1986).

Motor-copy: Anatomical and physiological data (Bickford and Hall, 1989, Crutcher and DeLong, 1984, Levesque et al., 1996, McHaffie et al., 2005, Reiner et al., 2003) indicate that

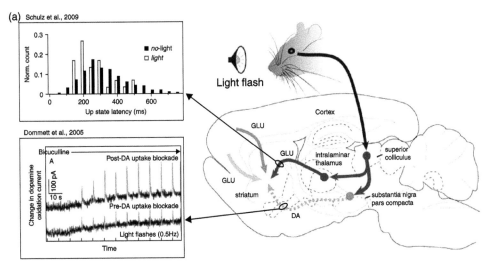

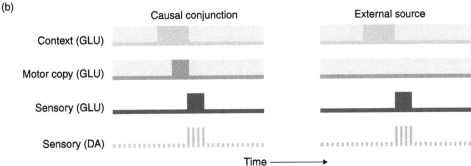

Figure 6.4 A neural network configured to determine agency and the discovery of novel actions. Knowledge of relevant neuroanatomy and signal timing points to the likely convergence of afferent signals in the striatum when the agent's behavior causes an unpredicted, biologically salient event. (a) Relevant signals include: (i) a running copy of motor output delivered from collaterals of projections from motor structures to the brainstem (Reiner et al., 2003) (grey); (ii) a short latency (~120ms) sensory input from subcortical structures such as the superior colliculus via relays in the intralaminar thalamus (Schulz et al., 2009) (grey); and (iii) a sensory-evoked short-latency (~120ms) phasic input of dopamine from substantia nigra pars compacta (Dommett et al., 2005) (light grey). (b) Causal conjunction: whenever the subject causes an unpredicted sensory event, relevant components of the contextual and motor efference copy inputs will directly precede the near simultaneous short-latency glutamateric sensory input from the thalamus and the phasic dopaminergic input from substantia nigra. External source: when no aspect of motor copy inputs reliably precedes the phasic sensory inputs (glutamateric and dopaminergic), the event will have been caused by an external source. The panels in (a) are reproduced with permission from Schulz et al. (2006 © Society for Neuroscience, and Dommett et al. (2005) © American Association for the Advancement of Science.

copies of motor commands, originating in both cortical and subcortical sensorimotor structures and sent to the brainstem structures that determine motor output, are also relayed to the striatum via branching collaterals. These efference-copy signals of action decisions and motor commands provide the striatum with a running record of the agent's behavioral output (Figure 6.4a).

We have proposed that these potentially converging signals could be used to determine whether any aspect of the agent's behavioral output was the likely cause of the onset of a biologically significant event (Redgrave and Gurney, 2006, Redgrave et al., 2008). When the agent is responsible, the critical movement(s) that caused the event will somewhere be embedded in the preceding record of behavioral output (motor copy in Figure 6.4b). Alternatively, if the event was caused by something else (i.e. not the agent), then no consistent component in the record of motor copy would reliably predict event onset (Figure 6.4b). Our suggestion is, that the reinforcing role of phasic DA release is to increase the probability of the agent to immediately re-select what it has just done to see if the unpredicted event can be caused again (Figure 6.5a). We term this particular reinforcing effect 'repetition bias'. However, this simple story hides a major computational problem. Typically, the contextual and motor record of behavioral output that precedes an unexpected salient event will be exceedingly complicated (multiple contextual stimuli, different movements of multiple muscles etc.). Among this mass of information, possibly only a few components within the multiple dimensions of behavioral output may be causally relevant. The problem for the agent is to discover which aspects of what it is doing are causing the salient event. However, if the repetition bias reinforced by phasic DA release evokes an imperfect repetition of prior behavioral selections then by trial and error – that is, on each iteration promoting activity in loops when the event occurs (LTP), and suppressing activity in loops when it fails (LTD) (Calabresi et al., 2007, Surmeier et al., 2007, Reynolds and Wickens, 2002, Wickens et al., 2003, Wickens, 2009) – the system should converge on the components of behavioral output that are causing the event. At this point, it would be recognized that a specific sequence of movements now perfectly predicts the caused sensory event and a new action–outcome routine would have been established. The question is now, to find out where such action–outcome routines are acquired and stored.

We have suggested that 'repetition bias' would cause representations of the sensory event that is being caused to appear more frequently in regions of the brain external to the basal ganglia. It is in these structures that models of action–outcome associations are probably generated. Such systems are likely to include the amygdala (Paton et al., 2006, Balleine and Killcross, 2006), hippocampus (Lisman and Grace, 2005) and cerebral cortex (Schultz et al., 2000, Schoenbaum et al., 2003, Corbit and Balleine, 2003). It is in the circuitry of these structures that the long-term action–outcome associations are probably established and stored. Consequently, the repetition biasing mechanism for identifying critical components of agency by the basal ganglia would facilitate action–outcome learning by providing relevant training data to the external neural systems with the capacity to generate actions and recognize outcomes.

This agency hypothesis also provides an insight into why the short-latency phasic DA signaling is so short. Further, we established that a critical feature of sensory-evoked DA responses is that they typically occur before any behavior is triggered in response to the sensory event. This arrangement would avoid having the record of potentially causal motor-copy contaminated by irrelevant behavior associated with responses subsequently directed to the event. Such confusing contamination would make the credit assignment problem more difficult. The credit assignment problem is how to project reinforcement effects back in time, so that only the components of behavioral output that caused the event are reinforced (Izhikevich, 2007).

In summary, we are proposing that sensory-evoked DA reinforcement operates, in part, at relay points within the basal ganglia (Figure 6.5a) to identify those external events that are caused by the agent and to discover exactly what is being done to cause them, i.e. to develop

(a)   Intrinsic reinforcement   (b)   Extrinsic reinforcement

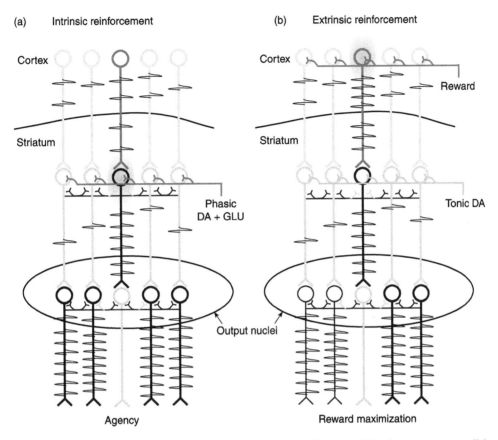

Figure 6.5 Separate mechanisms of reinforcement could bias selection within the re-entrant parallel
looped architecture of the basal ganglia. (a) Intrinsic reinforcement sensitises the striatal
response to cortico-striatal inputs (indicated by relative numbers of spikes of striatal projection neurones in different channels). Thus, transmission in recently active (selected)
channel(s) is reinforced by the phasic release of dopamine evoked by an unpredicted salient
sensory event. This reinforcement occurs prior to any subsequent behavior elicited by the
dopamine-triggering event. Non-active channels do not have the 'eligibility' traces (evoked
by recent activity) required for DA-reinforced plasticity (Izhikevich, 2007, Pan et al., 2005).
The 'repetition bias' induced by the DA-reinforced plasticity causes versions of recent
behavioral output to be re-selected enabling the system to converge on the causal components of behavior (Redgrave and Gurney, 2006, Redgrave et al., 2008). (b) Association
with reward can potentiate sensory processing in structures providing afferent signals to
the striatum. In the absence of such association, early sensory processing is suppressed
(habituation). Insofar as the model of selection by the re-entrant loop architecture is influenced by the relative magnitude of input saliences (Gurney et al., 2001a, Gurney et al.,
2001b, Humphries et al., 2006, Prescott et al., 2006), reward-related modulation of afferent
signals would effectively bias selection to favour reward-related inputs. This extrinsic mechanism of reinforcement would have the potential to maximize future reward acquisition.

novel actions. The products of this mechanism would be an expanding library of action–
outcome associations (behavioral options), stored in external structures that would be
available for re-use whenever appropriate (Singh et al., 2005). Now, the problem for
the agent is how to exploit this knowledge so that, in the future, the items it selects from the
library will maximize its chances of survival and reproduction. We will now argue that

the reward-related modulation of processing in the external structures that provide afferent signals to the striatum could ensure this happens.

Maximization of future reward acquisition

A basic requirement for agents that interact with a sometimes hostile world is to appreciate which stimuli are most urgent, and how best to respond to them (Jovancevic-Misic and Hayhoe, 2009). Assuming that reward-related modulation of early sensory processing can influence the magnitude/salience of afferent signals competing in the basal ganglia for behavioral resources, we would expect that reward-boosted inputs would have a higher probability of being selected and allowed to guide behavior. This suggestion is supported by evidence indicating that the primary sensory signals that are relayed to DA neurones in the ventral midbrain are also relayed to the other major input nuclei of the basal ganglia (Middleton and Strick, 1996, Harting and Updyke, 2006, Coizet et al., 2009) (Figure 6.4a). In our computational analyses of the basal ganglia's re-entrant loop architecture (Gurney et al., 2001a, Gurney et al., 2001b, Humphries et al., 2006, Prescott et al., 2006), we show that the relative magnitude or salience of competing input signals importantly influences overall selection outcome. The discovery that an association with the presence or absence of reward can adjust processing in afferent sensory structures opens the possibility of an additional mechanism for biasing overall selection decisions within the re-entrant loop network (Figure 6.5b). If so, this mechanism would be important because it could satisfy the requirements of formal reinforcement learning theory (Sutton and Barto, 1998, Montague et al., 2004, Schultz, 2006, Niv, 2009), and Thorndike's Law of Effect (Thorndike, 1911), that behavioral selections associated with reward should, in the future, be selected with high priority. However, much additional work will be required to test this suggestion. While it is known that reward-related modulation of primary sensory processing can both bias and speed-up perceptual decisions (Lee et al., 2002, Hickey et al., 2010, Seitz et al., 2009, Kristjansson et al., 2010, Pleger et al., 2008, Weil et al., 2010, Wunderlich et al., 2010), it is unclear whether these effects result from reward-related adjustments of relative input saliences to the basal ganglia's re-entrant loops. It is also not currently known by what mechanisms the presence/absence of reward is able to modulate early sensory processing. Thus, the means by which modulations are achieved and the time course of modulation all remain to be determined.

Summary and conclusions

The theoretical positions proposed in this chapter evolved from our systems-level analyses of the functional anatomy and signal timing within the re-entrant looped architecture of the basal ganglia. We sought to address two important issues in motor control – how actions are developed, and how they are subsequently exploited to maximize the effect. We first proposed that selective disinhibition within the re-entrant loop architecture at the level of the basal ganglia output nuclei can act as a generic selection mechanism (Redgrave et al., 1999a). This architecture is likely to have appeared early in vertebrate brain evolution to resolve the problems of selection arising from parallel multifunctional processing (Redgrave et al., 1999a, Gurney et al., 2001a, Humphries et al., 2006, Prescott et al., 2006). Secondly, insofar as reinforcement learning acts to bias future selections, reinforcement-driven adaptations should operate within the mechanism(s) of selection, i.e. at points within the re-entrant looped architecture. Therefore, we identify two complementary mechanisms that together could discover and exploit the actions that would maximize reward acquisition (Figure 6.5): (i) The first is phasic DA-associated reinforcement that adjusts the sensitivity of internal basal ganglia circuitry to specific inputs so that immediately preceding behavioral output is repeated (Figure 6.5a). This mechanism would facilitate the discovery of what external events are caused by the agent and the development of novel actions (Redgrave and Gurney, 2006, Redgrave et al., 2008) (Figure 6.4). (ii) Reward-related modulation within the external source structures of the re-entrant loops (those providing afferents to the basal ganglia) could provide a potent mechanism to promote the selection of reward-associated behaviors (Fig 6.5b). Clearly, much experimental and computational work will be required to test these informed speculations. Strategies highlighted by our analysis would test whether suitable manipulations of afferent circuitry would produce the predicted cortico-striatal plasticity. Experiments are also required to analyze the behavioral parameters of 'repetition bias' and predicted association with phasic DA neurotransmission. The current analysis also provides a blueprint of biologically inspired, system-level architectures for action acquisition and exploitation. Computational modeling of the proposed basal ganglia architectures and their deployment to control behavioral selections and adaptation in autonomous artificial agents (robots) will provide ultimate tests of functionality.

Acknowledgements

During the preparation of this chapter the authors were supported by the Wellcome Trust (080943 and 091409), and by the European Commission under the FP7 Cognitive Systems, Interaction, and Robotics Initiative, grant # 231722 ImClever project.

References

Alexander, G.E., DeLong, M.R. and Strick, P.L. (1986) 'Parallel organization of functionally segregated circuits linking basal ganglia and cortex.', *Ann. Rev. Neurosci.*, 9: 357–81.

Apicella, P., Legallet, E. and Trouche, E. (1997) 'Responses of tonically discharging neurons in the monkey striatum to primary rewards delivered during different behavioral states', *Exp. Brain Res.*, 116: 456–66.

Arbuthnott, G.W. and Wickens, J. (2007) 'Space, time and dopamine', *Trends Neurosci.*, 30: 62–9.

Balleine, B.W. and Killcross, S. (2006) 'Parallel incentive processing: an integrated view of amygdala function', *Trends Neurosci.*, 29: 272–9.

Berridge, K.C. (2007) 'The debate over dopamine's role in reward: the case for incentive salience', *Psychopharmacol. (Berl.)*, 191: 391–431.

Bickford, M.E. and Hall, W.C. (1989) 'Collateral projections of predorsal bundle cells of the superior colliculus in the rat.', *J. Comp. Neurol.*, 283: 86–106.

Bjorklund, A. and Dunnett, S.B. (2007) 'Dopamine neuron systems in the brain: an update', *Trends Neurosci.*, 30: 194–202.

Boehnke, S.E. and Munoz, D.P. (2008) 'On the importance of the transient visual response in the superior colliculus', *Curr. Opin. Neurobiol.* 18: 544–51.

Brischoux, F., Chakraborty, S., Brierley, D.I. and Ungless, M.A. (2009) 'Phasic excitation of dopamine neurons in ventral VTA by noxious stimuli', *Proc. Natl. Acad. Sci. USA*, 106: 4894–9.

Bromberg-Martin, E.S., Matsumoto, M. and Hikosaka, O. (2010) 'Dopamine in motivational control: rewarding, aversive, and alerting', *Neuron*, 68: 815–34.

Bunzeck, N., Doeller, C.F., Fuentemilla, L., Dolan, R.J. and Duzel, E. (2009) 'Reward Motivation accelerates the onset of neural novelty signals in humans to 85 milliseconds', *Curr. Biol.*, 19: 1294–1300.

Calabresi, P., Picconi, B., Tozzi, A. and Di Filippo, M. (2007) 'Dopamine-mediated regulation of corticostriatal synaptic plasticity', *Trends Neurosci.*, 30: 211–19.

Chevalier, G. and Deniau, J.M. (1990) 'Disinhibition as a basic process in the expression of striatal functions.', *Trends Neurosci.*, 13: 277–81.

Clark, J.J., Sandberg, S.G., Wanat, M.J., Gan, J.O., Horne, E.A., Hart, A.S., Akers, C.A., Parker, J.G., Willuhn, I., Martinez, V., Evans, S.B., Stella, N. and Phillips, P.E.M. (2010) 'Chronic microsensors for longitudinal, subsecond dopamine detection in behaving animals', *Nat. Meth.*, 7: 126–U58.

Coizet, V., Dommett, E.J., Klop, E.M., Redgrave, P. and Overton, P.G. (2010) 'The parabrachial nucleus is a critical link in the transmission of short-latency nociceptive information to midbrain dopaminergic neurons', *Neuroscience*, 168: 263–72.

Coizet, V., Dommett, E.J., Redgrave, P. and Overton, P.G. (2006) 'Nociceptive responses of midbrain dopaminergic neurones are modulated by the superior colliculus in the rat', *Neuroscience*, 139: 1479–93.

Coizet, V., Graham, J.H., Moss, J., Bolam, J.P., Savasta, M., McHaffie, J.G., Redgrave, P. and Overton, P.G. (2009) 'Short-latency visual input to the subthalamic nucleus is provided by the midbrain superior colliculus', *J. Neurosci.*, 29: 5701–9.

Coizet, V., Overton, P.G. and Redgrave, P. (2007) 'Collateralization of the tectonigral projection with other major output pathways of superior colliculus in the rat', *J. Comp. Neurol.*, 500: 1034–49.

Comoli, E., Coizet, V., Boyes, J., Bolam, J.P., Canteras, N.S., Quirk, R.H., Overton, P.G. and Redgrave, P. (2003) 'A direct projection from superior colliculus to substantia nigra for detecting salient visual events', *Nat. Neurosci.*, 6: 974–80.

Corbit, L.H. and Balleine, B.W. (2003) 'The role of prelimbic cortex in instrumental conditioning', *Behav. Brain Res.*, 146: 145–57.

Crutcher, M.D. and DeLong, M.R. (1984) 'Single cell studies of the primate putamen. II. Relations to direction of movement and pattern of muscular activity', *Exp. Brain Res.*, 53: 244–58.

DeLong, M. and Wichmann, T. (2010) 'Changing views of basal ganglia circuits and circuit disorders', *Clinical EEG Neurosci.*, 41: 61–7.

Dickinson, A. (2001) 'The 28th Bartlett Memorial Lecture. Causal learning: An associative analysis', *Quart. J. Exp. Psych. B Com. Phy. P*, 54: 3–25.

Dommett, E., Coizet, V., Blaha, C.D., Martindale, J., Lefebvre, V., Walton, N., Mayhew, J.E., Overton, P.G. and Redgrave, P. (2005) 'How visual stimuli activate dopaminergic neurons at short latency', *Science*, 307: 1476–9.

Doya, K. (2007) 'Reinforcement learning: Computational theory and biological mechanisms', *Hfsp J.*, 1: 30–40.

Draganski, B., Kherif, F., Kloeppel, S., Cook, P.A., Alexander, D.C., Parker, G.J.M., Deichmann, R., Ashburner, J. and Frackowiak, R.S.J. (2008) 'Evidence for segregated and integrative connectivity patterns in the human basal ganglia', *J. Neurosci.*, 28: 7143–52.

El-Amamy, H. and Holland, P.C. (2006) 'Substantia nigra pars compacta is critical to both the acquisition and expression of learned orienting of rats', *Eur J. Neurosci.*, 24: 270–6.

Fiorillo, C.D., Newsome, W.T. and Schultz, W. (2008) 'The temporal precision of reward prediction in dopamine neurons', *Nature Neuroscience*, 11: 966–73.

Fiorillo, C.D., Tobler, P.N. and Schultz, W. (2003) 'Discrete coding of reward probability and uncertainty by dopamine neurons', *Science*, 299: 1898–1902.

Franko, E., Seitz, A.R. and Vogels, R. (2010) 'Dissociable neural effects of long-term stimulus-reward pairing in macaque visual cortex', *J. Cog. Neurosci.*, 22: 1425–39.

Fuxe, K. and Agnati, L.F. (eds.) (1991) *Volume transmission in the brain: Novel mechanisms for neural transmission*, New York: Raven Press.

Gerfen, C.R. and Wilson, C.J. (1996) 'The basal ganglia', in L.W. Swanson, A. Bjorklund and T. Hokfelt (eds.) *Handbook of chemical neuroanatomy, Vol 12: Integrated systems of the CNS, Part III*, Amsterdam: Elsevier.

Grantyn, R. (1988) 'Gaze control through superior colliculus: structure and function.', in J.A. Buttner-Ennever (ed.) *Neuroanatomy of the oculomotor system*, Amsterdam: Elsevier.

Graybiel, A.M. (2008) 'Habits, rituals, and the evaluative brain', *Ann. Rev. Neurosci.*, 31: 359–87.

Grillner, S., Helligren, J., Menard, A., Saitoh, K. and Wikstrom, M.A. (2005) 'Mechanisms for selection of basic motor programs – roles for the striatum and pallidum', *Trends Neurosci.*, 28: 364–70.

Gurney, K., Prescott, T.J. and Redgrave, P. (2001a) 'A computational model of action selection in the basal ganglia. I. A new functional anatomy', *Biol. Cybern.*, 84: 401–10.

Gurney, K., Prescott, T.J. and Redgrave, P. (2001b) 'A computational model of action selection in the basal ganglia. II. Analysis and simulation of behaviour', *Biol. Cybern.*, 84: 411–23.

Haber, S.N. (2003) 'The primate basal ganglia: parallel and integrative networks', *J. Chem. Neuroanat.*, 26: 317–30.

Harris, J.D. (1943) 'Habituatory response decrement in the intact organism.', *Psych. Bulletin*, 40: 385–422.

Harting, J.K. and Updyke, B.V. (2006) 'Oculomotor-related pathways of the basal ganglia', *Prog. Brain Res.*, 151: 441–60.

Hickey, C., Chelazzi, L. and Theeuwes, J. (2010) 'Reward changes salience in human vision via the anterior cingulate', *J. Neurosci.*, 30: 11096–103.

Hikosaka, O. (2007) 'GABAergic output of the basal ganglia', *Prog. Brain Res.*, 160: 209–26.

Hikosaka, O. and Wurtz, R.H. (1983) 'Visual and oculomotor function of monkey substantia nigra pars reticulata. I. Relation of visual and auditory responses to saccades.', *J. Neurophysiol.*, 49: 1230–53.

Horn, G. and Hill, R.M. (1966) 'Effect of removing the neocortex on the response to repeated sensory stimulation of neurones in the mid-brain', *Nature*, 211: 754–5.

Horvitz, J.C. (2000) 'Mesolimbocortical and nigrostriatal dopamine responses to salient non-reward events', *Neuroscience*, 96: 651–6.

Hudgins, E.D., McHaffie, J.G., Redgrave, P., Salinas, E. and Stanford, T.R. (2009) 'Putative midbrain dopamine neurons encode sensory salience and reward prediction at different latencies', poster presented at Society for Neuroscience, Chicago, Illinois, October.

Hui, G.K., Wong, K.L., Chavez, C.M., Leon, M.I., Robin, K.M. and Weinberger, N.M. (2009) 'Conditioned tone control of brain reward behavior produces highly specific representational gain in the primary auditory cortex', *Neurobiol. Learn. Mem.*, 92: 27–34.

Humphries, M.D., Stewart, R.D. and Gurney, K.N. (2006) 'A physiologically plausible model of action selection and oscillatory activity in the basal ganglia', *J. Neurosci.*, 26: 12921–42.

Igelstrom, K.M., Herbison, A.E. and Hyland, B.I. (2010) 'Enhanced c-Fos expression in superior colliculus, paraventricular thalamus and septum during dearning of cue-reward association.', *Neuroscience*, 168: 706–14.

Ikeda, T. and Hikosaka, O. (2003) 'Reward-dependent gain and bias of visual responses in primate superior colliculus', *Neuron*, 39: 693–700.

Izhikevich, E.M. (2007) 'Solving the distal reward problem through linkage of STDP and dopamine signaling', *Cereb. Cortex*, 17: 2443–52.

Jay, M.F. and Sparks, D.L. (1987) 'Sensorimotor integration in the primate superior colliculus. I. Motor convergence.', *J. Neurophysiol.*, 57: 22–34.

Joshua, M., Adler, A. and Bergman, H. (2009) 'The dynamics of dopamine in control of motor behavior', *Curr. Opin. Neurobiol.*, 19: 615–20.

Joshua, M., Adler, A., Mitelman, R., Vaadia, E. and Bergman, H. (2008) 'Midbrain dopaminergic neurons and striatal cholinergic interneurons encode the difference between reward and aversive events at different epochs of probabilistic classical conditioning trials', *J Neurosci.*, 28: 11673–84.

Jovancevic-Misic, J. and Hayhoe, M. (2009) 'Adaptive gaze control in natural environments', *J. Neurosci.*, 29: 6234–8.

Kim, B. and Basso, M.A. (2008) 'Saccade target selection in the superior colliculus: A signal detection theory approach', *J. Neurosci.*, 28: 2991–3007.

Kimura, M., Minamimoto, T., Matsumoto, N. and Hori, Y. (2004) 'Monitoring and switching of cortico-basal ganglia loop functions by the thalamo-striatal system', *Neurosci. Res*, 48: 355–60.

Kristjansson, A., Sigurjonsdottir, O. and Driver, J. (2010) 'Fortune and reversals of fortune in visual search: Reward contingencies for pop-out targets affect search efficiency and target repetition effects', *Atten. Percept. Psychophys.*, 72: 1229–36.

Lee, T.S., Yang, C.F., Romero, R.D. and Mumford, D. (2002) 'Neural activity in early visual cortex reflects behavioral experience and higher-order perceptual saliency', *Nat. Neurosci.*, 5: 589–97.

Levesque, M., Charara, A., Gagnon, S., Parent, A. and Deschenes, M. (1996) 'Corticostriatal projections from layer V cells in rat are collaterals of long-range corticofugal axons', *Brain Res.*, 709: 311–15.

Lindvall, O. and Bjorklund, A. (1974) 'The organization of the ascending catcholamine neuron systems in the rat brain as revealed by the glyoxylic acid fluoresence method.', *Acta Physiol. Scand.*, Suppl. 412: 1–48.

Lisman, J.E. and Grace, A.A. (2005) 'The hippocampal-VTA loop: Controlling the entry of information into long-term memory', *Neuron*, 46: 703–13.

Mackintosh, N.J. (2009) 'Varieties of perceptual learning', *Learn. Behav.*, 37: 119–25.

Matsumoto, M. and Hikosaka, O. (2009) 'Two types of dopamine neuron distinctly convey positive and negative motivational signals', *Nature*, 459: 837–U4.

Matsumoto, M. and Takada, M. (2011) 'Midbrain dopamine neurons represent behavioral relevance in a working memory task', poster presented at Society for Neuroscience, Washington, DC, November.

May, P.J., McHaffie, J.G., Stanford, T.R., Jiang, H., Costello, M.G., Coizet, V., Hayes, L.M., Haber, S.N. and Redgrave, P. (2009) 'Tectonigral projections in the primate: a pathway for pre-attentive sensory input to midbrain dopaminergic neurons', *Eur. J. Neurosci.*, 29: 575–87.

McHaffie, J.G., Jiang, H., May, P.J., Coizet, V., Overton, P.G., Stein, B.E. and Redgrave, P. (2006) 'A direct projection from superior colliculus to substantia nigra pars compacta in the cat', *Neuroscience*, 138: 221–34.

McHaffie, J.G., Stanford, T.R., Stein, B.E., Coizet, V. and Redgrave, P. (2005) 'Subcortical loops through the basal ganglia', *Trends Neurosci.*, 28: 401–7.

Middleton, F.A. and Strick, P.L. (1996) 'The temporal lobe is a target of output from the basal ganglia', *Proc Natl Acad Sci USA*, 93: 8683–7.

Mink, J.W. (1996) 'The basal ganglia: Focused selection and inhibition of competing motor programs', *Prog. Neurobiol.*, 50: 381–425.

Mogami, T. and Tanaka, K. (2006) 'Reward association affects neuronal responses to visual stimuli in macaque TE and perirhinal cortices', *J. Neurosci.*, 26: 6761–70.

Montague, P.R., Dayan, P. and Sejnowski, T.J. (1996) 'A framework for mesencephalic dopamine systems based on predictive Hebbian learning', *J. Neurosci.*, 16: 1936–47.

Montague, P.R., Hyman, S.E. and Cohen, J.D. (2004) 'Computational roles for dopamine in behavioural control', *Nature*, 431: 760–7.

Morris, G., Arkadir, D., Nevet, A., Vaadia, E. and Bergman, H. (2004) 'Coincident but distinct messages of midbrain dopamine and striatal tonically active neurons', *Neuron*, 43: 133–43.

Nakahara, H., Itoh, H., Kawagoe, R., Takikawa, Y. and Hikosaka, O. (2004) 'Dopamine neurons can represent context-dependent prediction error', *Neuron*, 41: 269–80.

Niv, Y. (2009) 'Reinforcement learning in the brain', *J. Math. Psychol.*, 53: 139–54.

Nomoto, K., Schultz, W., Watanabe, T. and Sakagami, M. (2010) 'Temporally extended dopamine responses to perceptually demanding reward-predictive stimuli', *J. Neurosci.*, 30: 10692–702.

Pan, W.X., Schmidt, R., Wickens, J.R. and Hyland, B.I. (2005) 'Dopamine cells respond to predicted events during classical conditioning: Evidence for eligibility traces in the reward-learning network', *J. Neurosci.*, 25: 6235–42.

Paton, J.J., Belova, M.A., Morrison, S.E. and Salzman, C.D. (2006) 'The primate amygdala represents the positive and negative value of visual stimuli during learning.', *Nature*, 439: 865–70.

Pilz, P.K.D., Carl, T.D. and Plappert, C.F. (2004) 'Habituation of the acoustic and the tactile startle responses in mice: Two independent sensory processes', *Behav. Neurosci.*, 118: 975–83.

Pleger, B., Blankenburg, F., Ruff, C.C., Driver, J. and Dolan, R.J. (2008) 'Reward facilitates tactile judgments and modulates hemodynamic responses in human primary somatosensory cortex', *J. Neurosci.*, 28: 8161–8.

Prescott, T.J., Gonzalez, F.M.M., Gurney, K., Humphries, M.D. and Redgrave, P. (2006) 'A robot model of the basal ganglia: Behavior and intrinsic processing', *Neural Netw.*, 19: 31–61.

Prescott, T.J., Redgrave, P. and Gurney, K.N. (1999) 'Layered control architectures in robots and vertebrates', *Adapt. Behav.*, 7: 99–127.

Redgrave, P. (2007) *Basal ganglia*, Scholarpedia 2(6):1825. Available: http://www.scholarpedia.org/article/Basal_ganglia.

Redgrave, P. and Gurney, K. (2006) 'The short-latency dopamine signal: a role in discovering novel actions ?', *Nat. Rev. Neurosci.*, 7: 967–75.

Redgrave, P., Gurney, K. and Reynolds, J. (2008) 'What is reinforced by phasic dopamine signals?', *Brain Res. Rev.*, 58: 322–39.

Redgrave, P., Marrow, L.P. and Dean, P. (1992) 'Topographical organization of the nigrotectal projection in rat: Evidence for segregated channels', *Neuroscience*, 50: 571–95.

Redgrave, P., Prescott, T. and Gurney, K.N. (1999a) 'The basal ganglia: A vertebrate solution to the selection problem ?', *Neuroscience*, 89: 1009–23.

Redgrave, P., Prescott, T.J. and Gurney, K. (1999b) 'Is the short latency dopamine response too short to signal reward error ?', *Trends Neurosci.*, 22: 146–51.

Reiner, A., Jiao, Y., DelMar, N., Laverghetta, A.V. and Lei, W.L. (2003) 'Differential morphology of pyramidal tract-type and intratelencephalically projecting-type corticostriatal neurons and their intrastriatal terminals in rats', *J. Comp. Neurol.*, 457: 420–40.

Reiner, A.J. (2010) 'The conservative evolution of the vertebrate basal ganglia', in H. Steiner and K.Y. Tseng (eds.) *Handbook of basal ganglia structure and function*, Burlington, MA: Academic Press.

Reynolds, J.N. and Wickens, J.R. (2002) 'Dopamine-dependent plasticity of corticostriatal synapses', *Neural Netw.*, 15: 507–21.

Rousselet, G.A., Thorpe, S.J. and Fabre-Thorpe, M. (2004) 'How parallel is visual processing in the ventral pathway?', *Trends Cogn. Sci.*, 8: 363–70.

Schoenbaum, G., Setlow, B., Saddoris, M.P. and Gallagher, M. (2003) 'Encoding predicted outcome and acquired value in orbitofrontal cortex during cue sampling depends upon input from basolateral amygdala', *Neuron*, 39: 855–67.

Schultz, W. (1998) 'Predictive reward signal of dopamine neurons', *J. Neurophysiol.*, 80: 1–27.

Schultz, W. (2006) 'Behavioral theories and the neurophysiology of reward', *Annu. Rev. Psychol.*, 57: 87–115.

Schultz, W., Tremblay, L. and Hollerman, J.R. (2000) 'Reward processing in primate orbitofrontal cortex and basal ganglia', *Cereb. Cortex*, 10: 272–83.

Schulz, J.M., Redgrave, P., Mehring, C., Aertsen, A., Clements, K.M., Wickens, J.R. and Reynolds, J.N.J. (2009) 'Short-latency activation of striatal spiny neurons via subcortical visual pathways', *J. Neurosci.*, 29: 6336–47.

Seitz, A.R., Kim, D. and Watanabe, T. (2009) 'Rewards evoke learning of unconsciously processed visual stimuli in adult humans', *Neuron*, 61: 700–7.

Serences, J.T. and Saproo, S. (2010) 'Population response profiles in early visual cortex are biased in favor of more valuable stimuli', *J. Neurophysiol.*, 104: 76–87.

Singh, S.P., Barto, A.G. and Chentanez, N. (2005) 'Intrinsically motivated reinforcement learning', in L.K. Saul, Y. Weiss and L. Bottou (eds.) *Advances in Neural Information Processing Systems*, 17: 2181–8.

Smith, Y., Raju, D.V., Pare, J.F. and Sidibe, M. (2004) 'The thalamostriatal system: a highly specific network of the basal ganglia circuitry', *Trends Neurosci.*, 27: 520–7.

Snaith, S. and Holland, O. (1990) 'An investigation of two mediation strategies suitable for behavioural control in animals and animats', in J.A. Meyer and S. Wilson (eds.) *From Animals to Animats: Proceedings of the First International Conference on the Simulation of Adaptive Behaviour*, Cambridge, MA: MIT Press.

Sprague, J.M., Marchiafava, P.L. and Rixxolatti, G. (1968) 'Unit responses to visual stimuli in the superior colliculus of the unanesthetized, mid-pontine cat', *Arch. Ital. Biol.*, 106: 169–93.

Surmeier, D.J., Ding, J., Day, M., Wang, Z.F. and Shen, W.X. (2007) 'D1 and D2 dopamine-receptor modulation of striatal glutamatergic signaling in striatal medium spiny neurons', *Trends in Neurosciences*, 30: 228–35.

Sutton, R.S. and Barto, A.G. (eds.) (1998) *Reinforcement Learning – An Introduction*, Cambridge, MA, MIT Press.

Takakusaki, K., Saitoh, K., Harada, H. and Kashiwayanagi, M. (2004) 'Role of basal ganglia-brainstem pathways in the control of motor behaviors', *Neurosci. Res.*, 50: 137–51.

Thelen, E. (1979) 'Rhythmical stereotypies in normal human infants', *Anim. Behav.*, 27: 699–715.

Thorndike, E.L. (ed.) (1911) *Animal Intelligence*, New York, Macmillan.

Thorpe, S.J. and Fabre-Thorpe, M. (2001) 'Seeking categories in the brain', *Science*, 291: 260–3.

Tobler, P.N., Dickinson, A. and Schultz, W. (2003) 'Coding of predicted reward omission by dopamine neurons in a conditioned inhibition paradigm', *J. Neurosci.*, 23: 10402–10.

Tobler, P.N., Fiorillo, C.D. and Schultz, W. (2005) 'Adaptive coding of reward value by dopamine neurons', *Science*, 307: 1642–5.

Vautrelle, N., Leriche, M., Dahan, L., Redgrave, P. and Overton, P.G. (2011) 'Cortical control of dopaminergic neurons', poster presented at Society for Neuroscience, Washington, DC, November.

Voorn, P., Vanderschuren, L.J., Groenewegen, H.J., Robbins, T.W. and Pennartz, C.M. (2004) 'Putting a spin on the dorsal-ventral divide of the striatum', *Trends Neurosci.*, 27: 468–74.

Weil, R.S., Furl, N., Ruff, C.C., Symmonds, M., Flandin, G., Dolan, R.J., Driver, J. and Rees, G. (2010) 'Rewarding feedback after correct visual discriminations has both general and specific influences on visual cortex', *J. Neurophysiol.*, 104: 1746–57.

White, B.J., Boehnke, S.E., Marino, R.A., Itti, L. and Munoz, D.P. (2009) 'Color-related signals in the primate superior colliculus', *J. Neurosci.*, 29: 12159–66.

Wickens, J.R. (2009) 'Synaptic plasticity in the basal ganglia', *Behav. Brain Res.*, 199: 119–28.

Wickens, J.R., Reynolds, J.N.J. and Hyland, B.I. (2003) 'Neural mechanisms of reward-related motor learning', *Curr. Opin. Neurobiol.*, 13: 685–90.

Wightman, R.M. and Robinson, D.L. (2002) 'Transient changes in mesolimbic dopamine and their association with 'reward'', *J. Neurochem.*, 82: 721–35.

Wise, R.A. (2004) 'Dopamine, learning and motivation', *Nat. Rev. Neurosci.*, 5: 483–494.

Wunderlich, K., Rangel, A. and O'Doherty, J.P. (2010) 'Economic choices can be made using only stimulus values', *Proc. Nat. Acad. Sci. USA*, 107: 15005–10.

7

THE ROLE OF AUGMENTED FEEDBACK IN HUMAN MOTOR LEARNING

Christian Leukel[1,2] and Jesper Lundbye-Jensen[3,4]

[1] DEPARTMENT OF SPORT SCIENCE, UNIVERSITY OF FREIBURG, GERMANY
[2] DEPARTMENT OF MEDICINE, MOVEMENT AND SPORT SCIENCE, UNIVERSITY OF FRIBOURG, SWITZERLAND
[3] DEPARTMENT OF EXERCISE AND SPORT SCIENCES, UNIVERSITY OF COPENHAGEN, DENMARK
[4] DEPARTMENT OF NEUROSCIENCE AND PHARMACOLOGY, UNIVERSITY OF COPENHAGEN, DENMARK

Introduction

As kids, we were playing a game called 'Hit the pot' where a pot was placed somewhere on the floor in a room by one child. The challenge for another child was to search for the pot, while crawling on the floor, making knocking movements using a wooden spoon. A tinny sound indicated when the pot was hit. Not a very challenging game, you may argue, because one instantly sees where the pot was placed. However, you miss the important information, that the kid who was searching for the pot was blindfolded.

The word 'blindfolded' could technically be described as 'the temporary loss of visual (sensory) feedback'. As in our example, sensory feedback helps us to move successfully, and also to acquire new motor skills and refine existing skills. This chapter focuses on sensory feedback, more specifically a particular kind of sensory feedback called augmented feedback, and discusses its influence on motor learning. Note that although different types of learning have different substrates, in this chapter we do not distinguish between 'motor adaptation', often associated with the refinement of existing movements, and 'motor learning', the acquisition of new movements (i.e. skill learning), but instead use the term 'motor learning' for all processes.

Sensory feedback arises from many sources in our human body, including e.g. audition, vision, proprioception, or tactile sensation. The sensory feedback serves to inform us about ourselves, and our environment. In this case, 'ourselves' specifically means the state of our own body, and the 'environment' specifically reflects the state of the external world in which we are moving. These are the two important divisions of interest (Wolpert & Ghahramani 2000). For instance, with respect to the state of our body, sensory feedback informs us whether our arm is flexed or extended while crawling over the floor. With respect to the state of the external world, sensory feedback informs us whether there is a couch or a chair in between us and the pot.

Internal intrinsic, internal extrinsic, and augmented feedback

Given the many types of sensors with many different origins, how can sensory feedback be classified? First, sensory feedback can be subdivided into the modality or source of the information, e.g. whether the information originates from vision, audition, or proprioception. Besides this categorization, sensory feedback has traditionally been divided with respect to the division of interest (Schmidt & Lee 1999). There are basically three subdivisions, called (a) internal intrinsic, (b) internal extrinsic, and (c) external (augmented) feedback (see Figure 7.1).

The first subdivision (a) is the sensory feedback informing us about the state of our own body. This was exemplified with the estimation of the state (simplified as an extention or flexion) of our own arm. This estimation can be performed with several sources like proprioception, vision, either alone or in combination. Combining more sources may improve the estimate. The reason for this was argued to be the noisiness of sensory signals processed in the human body (Faisal et al. 2008) and the reduction of noise when using multiple signals with different qualities and therefore noise profiles (Ernst & Banks 2002).

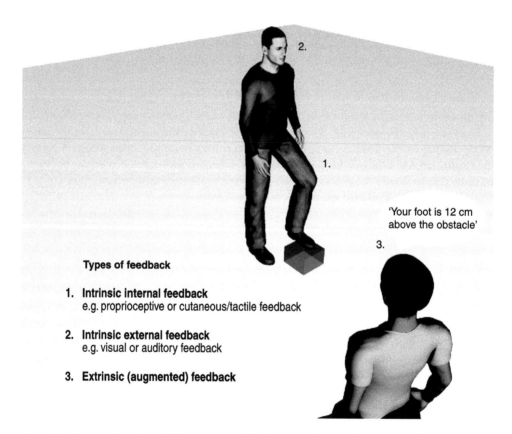

Types of feedback

1. **Intrinsic internal feedback**
   e.g. proprioceptive or cutaneous/tactile feedback

2. **Intrinsic external feedback**
   e.g. visual or auditory feedback

3. **Extrinsic (augmented) feedback**

'Your foot is 12 cm above the obstacle'

Figure 7.1 Types of feedback. Different types of feedback may be an inherent or intrinsic part of the motor task. The intrinsic feedback may originate from internal inputs e.g. from proprioception and/or external inputs e.g. from visual or auditive feedback. In addition to the intrinsic feedback, augmented feedback may be provided.

The second type of sensory feedback is intrinsic external feedback (b), informing us about the state of the external world. Again, as with intrinsic internal feedback, the estimation of the nature of our environment can be done with a single source, e.g. vision, but is often performed with multiple sources of information. The reason for it is, again, to improve the estimate. As an example, seeing and hearing an approaching train is better with regards to its location and speed than just seeing or hearing it.

In addition to these two types of feedback through which we continuously monitor ourselves and the surrounding world there is a specific type of feedback, called extrinsic feedback or augmented feedback (c) (see Figure 7.2). This type of feedback informs us how we interacted with the external world and it is therefore particularly relevant in all motor learning

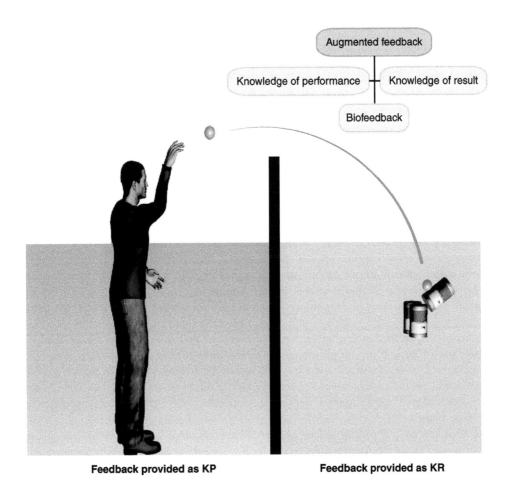

Feedback provided as KP   **Feedback provided as KR**

Figure 7.2 Augmented feedback. Augmented feedback may be provided to the learner based on different parameters during motor learning. Knowledge of Result (KR) is information given to the learner after completion of a movement, which describes the outcome of the movement in terms of the movement goal. Knowledge of Performance (KP) is information that describes the quality of the movement pattern that led to the performance outcome. KP differs from KR in terms of what aspects of performance the information refers to. In addition to performance and outcome, feedback may be given based movement kinematics, EMG etc. as biofeedback.

settings (Adams 1987). In the following, we will just use the word 'augmented' feedback, which equals the expression 'external'. The adjective 'augmented' refers to adding or enhancing feedback with an external source, providing an explicit (quantified) knowledge of the result(s) of the motor performance. In other words, augmented feedback informs the person about how well (in quantitative terms) the task was performed (Winstein 1991). Importantly, this information is adjusted to explicitly guide the acting person in improving the motor outcome. As an example, informing a basketball player whether the basket was hit or not provides a simple form of augmented feedback. This simplicity can be arbitrarily increased: for instance, by providing the basketball player with distance errors (30 cm left, 15 cm below) relative to the basket.

Why augmented feedback?

Many experimental, especially motor learning, studies concentrated on augmented feedback because it is better controllable in quantitative terms and can be modified in an experimental setting. In contrast to intrinsic internal and intrinsic external feedback, the intensity of augmented feedback is better controlled on an inter-individual basis. The perception of the arm while shooting a basketball may differ between persons. Consequently, the same angularity of the elbow joint may lead to differences in the sensation of the angularity. On the contrary, augmented feedback about the distance error between the ball and the basket is definite information.

Furthermore, this controllability allows modification(s) of the feedback. This characteristic of augmented feedback was used in many studies aiming to investigate how the nervous system learns movements. For instance, a recent, frequently used paradigm has been the so-called 'visuomotor rotation', where subjects start a drawing or reaching movement from a centre point of a circle to target locations equally placed on the outer bound of a circle (Krakauer 2009). The subjects typically cannot see their moving hand drawing on a touchpad or interacting with an arm of a robot but instead see the projected position of their hand in a virtual setup on a computer screen. While aiming to a target, subjects receive visual feedback about trajectory and/or the endpoint of their movement. In particular, the error between their movement endpoint and the target provides explicit and quantitative information about the performance guiding future movements and can therefore be ascribed to augmented feedback. Now, the position of the target can be rotated artificially so that straight movements of the hand will result in a line deviated by 45° to the left on the computer screen. Subjects will use the explicit visual information on the computer screen to adapt their motor behaviour and will end by making deviated movements of 45° rightwards (compensate) to reach the target.

A final argument for using augmented feedback is that it is very powerful to facilitate performance. There are even some learning paradigms where performance enhancements are difficult with just intrinsic internal and intrinsic external feedback. Learning to improve fast ballistic contractions with the foot requires a quantified feedback of the acceleration of the contraction. This information is provided to the learner. Previous studies investigated the effect of few (30–50) ballistic contractions (called short term learning) on performance enhancements (Lundbye-Jensen et al. 2011). Considerable improvements in performance were only seen when augmented feedback was provided. This feedback consisted of a visual presentation of the mathematical calculation of the acceleration of the movement after each contraction. We will come back to the importance of augmented feedback for performance improvements in more detail later in this chapter.

Augmented feedback and different processes of learning

So far, we have used the term 'motor learning' to refer to the process of performance improvement due to repetitive motor practice. However, there have been different processes of motor learning described in the literature (Wolpert et al. 2011). One process is called reinforcement learning. It refers to a formation of directed action selection based on rewards, and was extensively discussed in the previous chapter. In reinforcement learning, augmented feedback may be provided to facilitate the change in behaviour. Specifically, additional explicit information about behaviour (e.g. whether or not a playground swing went higher (Wolpert et al. 2011)) may help to bias future behaviour (e.g. that the playground swing actually will go higher). The information provided to the acting person in reinforcement learning is, however, limited. Basically, the content of the information is success or failure. In contrast, augmented feedback can contain richer information in an error-based learning process (Wolpert et al. 2011). A classical error-based paradigm has been discussed above. In visuomotor rotation, the learner receives an explicit error signal, indicated by the distance between the target and the endpoint of the movement to the target. This distance can be mathematically described, as, for example, 15 mm to the left and 33 mm below. Thus, in error-based learning, augmented feedback can be easily and extensively used to minimize the error. In our example, minimizing the error in a visuomotor rotation paradigm would mean to directly hit the target.

A third learning process is use-dependent learning. Use-dependent learning means that behaviour can change by pure repetition of (a) specific movement(s) (Wolpert et al. 2011). The person who performs the movement(s) does not receive any information about the outcome. Consequently, augmented feedback is – by definition – not part of the process of use-dependent learning.

Knowledge of performance versus knowledge of result

When mentioning the visuomotor rotation paradigm, we discussed exactly two forms of augmented feedback, which are differently treated in the literature. First, there is the error *while moving* towards a target, referring to the deviation between the trajectory made by the learner and the desired trajectory (e.g. a straight line between the starting point and the target). Second, there is an error *when the movement is finished*, meaning the deviation between the movement endpoint and the target. Traditionally, the first error has been called 'knowledge of performance' and the second has been called 'knowledge of result' (KR) (see Figure 7.2).

The term knowledge of performance refers to feedback on the sequence of the movement. A synonym of knowledge of performance is 'kinematic feedback' (Schmidt & Lee 1999) as the kinematics (path and time) of the moving body are the basis for this form of augmented feedback. For instance, a tennis player aiming to serve a ball could be informed about the paths relative to time of the moving arm and hand. Now, one may argue that the terms paths and time are very imprecise. Specifically, the question is on which parameter one should focus while serving: Should it be the elbow angle while the arm swings up to hit the ball? Or should it be something else? Especially in practical settings like a tennis serve, the determination of parameters is not easy. One problem of the determination relates to the fact that there exist multiple degrees of freedom to perform an action with the same outcome (Bernstein 1967). In other words, in a movement like a tennis service, the trajectory of a hand can look different in-between players but the endpoint (and even the motor outcome, e.g. the speed of the ball or the final destination) could be the same.

In contrast to the difficulties with knowledge of performance as augmented feedback, the variable KR is much clearer. KR provides explicit information about the result(s) of the accomplished movement. In a tennis serve, this explicit information could be whether the ball hit the ground within the opposite service box. Here, the word 'explicit' has to be emphasized. Otherwise, it could be just intrinsic external feedback as the tennis player immediately sees whether the ball hits within the service box after the serve. Explicit information in this example could mean that the coach tells whether the ball was in or out of the service box. As mentioned above, more precise information after the tennis serve could mean to provide the exact location of the touchdown within the service box. A synonym for KR is 'information feedback' or 'reinforcement' (Schmidt & Lee 1999). The term reinforcement is basically correct as the augmented feedback indeed serves to reinforce future behaviour (e.g. minimizing the error). However, it contrasts with the previously mentioned 'reinforcement learning' and thereby may lead to misunderstandings. Precisely, we refer to the fact that the content of the information in reinforcement learning is limited (success versus failure) whereas the content of the information in the KR is extendible.

Dimensions of the augmented feedback

When applying augmented feedback to a learner, several aspects of the transfer of the information need to be considered. One of these aspects was discussed in the previous paragraph. Namely, it is of relevance if the augmented feedback is provided with respect to characteristics of the movement execution (knowledge of performance) or with respect to the final aim of the movement (KR). Besides that, there are several more aspects, which can be divided, according to their nature, in three groups. The first group (i) incorporates the aspect of the content of the information. The second group (ii) contains the aspect of the frequency of the augmented feedback, and the third group (iii) comprises the timing of the feedback relative to the movement execution. We will discuss these three in the following (see Figure 7.3).

Content of the information

The group content of the information considers the variable(s) 'extracted' from the movement and presented to the learner. For instance, it contains the question: which characteristics of the movement and/or of the final result of the movement are presented? Therefore, the distinction between knowledge of performance and KR belongs to this group.

A second aspect is the richness of the information. This aspect has been discussed above. A basketball player may be informed about whether or not the basket was hit. Alternatively, one may provide detailed information about the shot, e.g. that the distance error between ball and basket was 30 cm to the right and 15 cm up. In the literature, this difference has been termed qualitative versus quantitative feedback (Schmidt & Lee 1999). In the former, it is simply differentiated between success and failure. In the latter, additional information, e.g. about the size or the direction of the movement error, is provided.

Furthermore, the augmented feedback may comprise true or false information. For instance, when mentioning the visuomotor rotation paradigm, we discussed the possibility of altering the actual trajectory of the movement. Thus the content of the information is wrong. You may question why it could be reasonable to feed wrong information? One reason leads to basic research, specifically to the question how the nervous system acquires movements based on the integration of sensory information. Another reason is also related to basic science, namely how intrinsic feedback conflicts with augmented feedback in the process of learning.

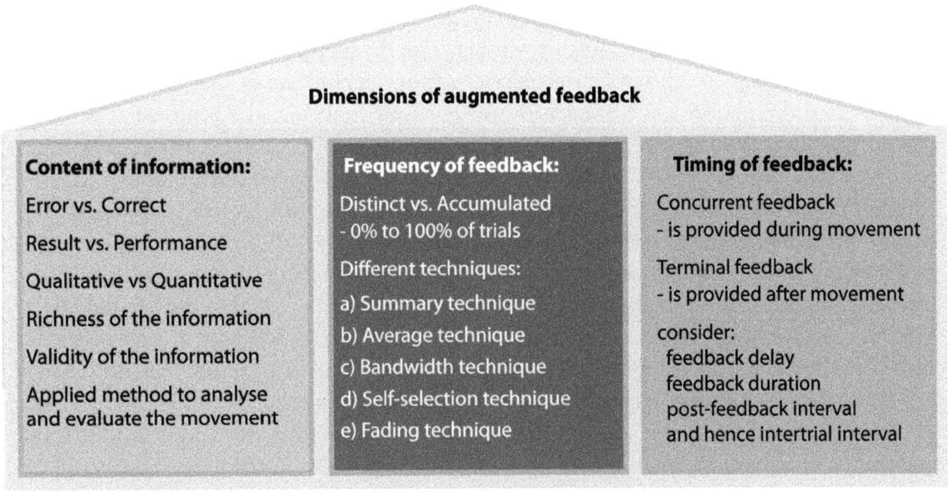

Dimensions of augmented feedback

Content of information:	Frequency of feedback:	Timing of feedback:
Error vs. Correct	Distinct vs. Accumulated - 0% to 100% of trials	Concurrent feedback - is provided during movement
Result vs. Performance		
Qualitative vs Quantitative	Different techniques:	Terminal feedback - is provided after movement
Richness of the information	a) Summary technique	
Validity of the information	b) Average technique	consider: feedback delay
Applied method to analyse and evaluate the movement	c) Bandwidth technique d) Self-selection technique e) Fading technique	feedback duration post-feedback interval and hence intertrial interval

Figure 7.3 Dimensions of augmented feedback. When augmented feedback is conveyed to the learner, several aspects need to be considered. These aspects can be divided into three categories according to the nature of the information, namely (i) the content of the information, (ii) the frequency, and (iii) the timing of the feedback.

Apart from questions in basic research, applying wrong augmented feedback in a practical setting may be considered with caution. Previous experiments have indicated that augmented feedback may be superior to intrinsic feedback (see Magill 2001). A learner may therefore become hooked on augmented information and ignore the intrinsic feedback. A consequence would be the acquisition of wrong behaviour.

A last aspect we want to emphasize is the applied method to analyse the movement and convey the information. The information may be based on kinematic or kinetic analysis of the movement. For instance, one can present the elbow angle or the recorded counterforce when making a punching movement. Furthermore, physiological parameters could be used, such as the heart rate, blood pressure, muscle or brain activity measured with surface electromyography and electroencephalography. These physiological parameters have often been called 'biofeedback' in the literature. Finally, it is of relevance how the information is transported, e.g. by video, audio, or tactile sensation.

Frequency of the augmented feedback

The group 'frequency' is concerned with the question about how often a learner receives augmented feedback. This could be expressed as a percentage: 100 percent could thereby indicate that augmented feedback was provided in each of the executed movements and 0 percent would consequently mean that no feedback was provided at all. Besides the possibilities of providing feedback in each (or part) of the corresponding movements, several other techniques exist in the literature. Four of these techniques will be discussed: (a) the summary technique, (b) the average technique, (c) the bandwidth technique, and (d) the self-selection technique. For the (a) summary technique, the learner receives feedback after a defined number of trials for all of the executed movements performed so far. For instance, a tennis player may serve five times in a row and may then get the maximum velocity of the ball from each of the five services. A summary technique is especially suitable if there is a series of movements rapidly succeeding each other. The (b) average technique is similar to the summary technique. The difference with the average technique is that feedback is not presented from the performance of each of the previous trials but as a representative of all previous trials. In our example, the tennis player may receive the average of the maximum speed of the past five services. Consequently, this would be one value instead of five.

In the (c) bandwidth technique, a range is predefined where a movement is successful. For instance, in a tennis serve, it could be determined that the serve was successful when the ball hit somewhere within the opposite outer box. It is only if the tennis player misses the box that a distance error in the form of augmented feedback is provided (e.g. 1 m to the left and 50 cm behind the service box). You may argue that the learner receives feedback in any case – no matter if the error was in or out of the predefined range. This is certainly true, as subjects know that, when they do not receive feedback, the movement was correct and this is indeed a kind of feedback. The reason to use the bandwidth technique was argued to be the impossibility of correcting small errors of movement, especially in the early learning phase (see Magill 2001). When using the bandwidth technique only gross movement errors may be reported.

The last technique we will discuss is the (d) self-selection technique. Here, the learner decides himself/herself when he/she wants to receive feedback. The advantage of this technique is that the learner may pay heightened attention to the execution of the movement in order to judge when the movement was wrong and augmented feedback is required.

In particular, this last point may solve a potential problem also known as 'guidance hypothesis' (Schmidt & Lee 1999). As mentioned, augmented feedback is powerful and can therefore influentially guide the formation of new behaviour. The problem arises when the learner begins to depend solely on augmented feedback and loses confidence in the intrinsic perception. As a consequence, progress in learning can be hindered when augmented feedback is stopped.

Timing of the augmented feedback

The timing of the augmented feedback refers to the question of when, meaning in which phase of the movement, feedback is provided. There are two main ways to apply feedback. The first possibility is to provide feedback during movement execution. This is also called 'concurrent feedback'. The second possibility is to provide feedback after the movement is finished. This is often called 'terminal feedback'. A classical example of concurrent feedback is the monitoring of the heart rate while running. For the terminal feedback, two different aspects can be considered. The first aspect refers to the timing of the augmented feedback in relation to the end of the respective movement. In other words: it is concerned with the question how long one should wait to provide feedback after the movement is finished? The second aspect relates to the timing of the feedback with respect to the consecutive movement, essentially the point of how long one should wait for the execution of the subsequent movement after augmented feedback is provided.

With the discussion about the different dimensions of augmented feedback, we would like to finish the fundamental part of this chapter. So far, we mainly described basic facets of augmented feedback, starting with its definition, its use in the different processes of motor learning, and the different dimensions inherent in this subdivision of sensory feedback. In the following, we would like to apply this knowledge. Consequently, we will discuss the effect of augmented feedback in different areas, laboratory and field studies.

Augmented feedback can facilitate motor performance and learning

The acquisition of motor skills is fundamental to human life. The ability of a person to acquire with practice the proficiency to execute coordinated motor actions enables that person to have a wide range of human experiences. The experiences of motor skill learning may range from tying shoelaces, or relearning to walk after a stroke, to coping with a demanding surgical operation or acquiring a complex sport skill. It is widely known that task performance appears to improve in a consistent manner with practice (i.e. Fitts 1964, Adams 1987, Schmidt & Lee 1999, Magill 2001). It is, however, not just the amount of training, but also the training conditions and quality of practice that can have a significant effect on the rate of learning and final performance (e.g. Schmidt & Lee 1999).

One of the most critical variables affecting motor skill learning, aside from practice itself, is augmented feedback (Newell 1991). Augmented task-related feedback supplements the response-produced intrinsic feedback obtained from vision, audition, and proprioception. The learner can achieve a certain skill level with task-intrinsic feedback, but in order to attain a faster learning or a higher level of expertise, augmented feedback may indeed be beneficial (Magill 1994). In the tasks where the information from intrinsic sources does not provide the feedback needed to determine the appropriateness of the performance, or when the learner cannot adequately access the information critical to learning the skill, augmented feedback can play an essential role in effective skill acquisition (see Magill 1994).

Although it is a challenging task to optimize skill acquisition, it is also exceedingly important both theoretically in terms of understanding human learning and practically, for those seeking to devise training and rehabilitation paradigms, to enhance performance and/or improve quality of life (Green & Bavalier 2008). The use of augmented feedback (together with consideration of other learning determinants such as task difficulty, practice structure and motivation) does offer the potential to promote motor performance and learning.

Further evidence from lab and field studies on the role of augmented feedback

Ever since the beginning of the twentieth century and the seminal experiments of Thorndike (1927), augmented feedback has been considered critically important for learning motor skills, and research in motor skill learning has focused both on elucidating the mechanisms underlying reinforcement and feedback-mediated learning and on identifying key factors for optimization of learning. A variety of tasks and experimental paradigms have been used for studying motor skill learning and the effects of augmented feedback from basic lab experiments in animals and humans to learning in more ecological settings as in sports and clinical practice. Each of these approaches to motor learning offers different perspectives. Whereas studies of learning in sports, music and rehabilitation often yield knowledge relating to optimization of practice, the basic lab studies seek insights in the fundamental principles and mechanisms underlying learning and effects of augmented feedback.

With practice, there are improvements in performance that characterize skilled behaviour and these changes are accompanied by learning-dependent changes in the functional networks of the brain (for review see e.g. Seidler, 2010). Motor learning leads to behavioural changes that are implemented by processes that occur during practice as well as processes that evolve after practice ends. These processes and related network changes are thought to represent the formation of motor memory, which relates to encoding, consolidation and retrieval (Fuster 1995). The study of motor learning focuses on understanding the motor memory processes as well as practice-related factors that influence these memory processes. Learning mechanisms surely vary in their specific implementation across different domains, but some mechanisms and determinants of learning appear to be shared across domains. One of the most prominent practice-related determinants in motor learning is augmented feedback.

Augmented feedback can play different roles in the motor skill learning process (Schmidt & Lee 1999). First, augmented feedback may provide information about response errors and utility of reward via task-related information about the skill being performed or just performed. This information can be descriptive in whether the performance was successful or not, or more prescriptive in informing the learner about the errors made and/or what the learner should do to correct those errors. Whereas augmented feedback has long been believed to function primarily as reward, it can also be used to determine the nature of errors. This is critical in planning and correcting subsequent movements. Augmented feedback may in this sense also contribute to the development of accurate error detection and correction mechanisms through comparison of intrinsic internal and intrinsic external feedback (Young et al. 2001).

Second, augmented information feedback may motivate the learner by making the task seem more interesting and enjoyable, keep the learner alert and thereby lead learners to increase their efforts to achieve their learning goals (see Adams 1987, Little & McGullagh 1989, Annesi 1998, Silverman et al. 1998). As a result of motivation induced by augmented feedback, learners are inclined to strive longer, more frequently and with more effort (Schmidt & Lee 1999). Whereas motivation and arousal have been largely overlooked in the field of skill learning (see, however, Ackerman et al. 1995 and Ackerman & Cianciolo 2000) these factors are considered to be critically important components of most major theories of learning in other areas, e.g. social psychology and education (Green & Bavalier 2008). It is indeed logical that deliberate practice is an important variable for learning (Ericsson et al. 1993) and if augmented feedback increases it, it may certainly enhance learning. Recent studies do, however, also suggest a more direct motivational effect of augmented feedback on learning (Schmidt & Lee, 1999, Chiviacowski & Wulf 2007, Lewthwaite & Wulf 2010).

The heading of the present section states that augmented feedback can facilitate motor performance and learning. Although augmented feedback has repeatedly been demonstrated to facilitate changes in motor performance, it is important to emphasize the 'learning–performance distinction' (Salmoni et al. 1984, Schmidt & Lee 1999). Although performance and learning are inter-related, and we expect motor skill learning to be accompanied by an improved capability for motor performance, the essence of learning lies in its permanence over time (Kantak & Winstein 2012). The learning–performance distinction discriminates between the motor performance observed during practice and the resilience of this perform-ance that develops during practice and is sustained over time (Cahill et al. 2001, Schmidt & Bjork 1992). In order to assess learning effects, motor performance should be measured not only during practice but also in retention tests at different time points following practice since this will reflect the efficiency of the memory processes evolved at that time (Kantak & Winstein 2012). Although augmented feedback may indeed facilitate motor performance during practice, retention tests are consequently necessary in order to assess the potential beneficial effects of augmented feedback on learning.

Although performance and acquisition are two distinct phenomena, they are intricately linked. It has been demonstrated that subjects who practise a motor task in a reduced feedback condition perform better in a delayed retention test compared to subjects who practise with augmented feedback during or following every practice trial (Schmidt et al. 1989, Sherwood 1998, Winstein & Schmidt 1990, Guay et al. 1999, Anderson et al. 2005). Although a poor performance in certain practice setting may ultimately lead to better retention, this may not be appreciated very well by the learner, the trainer or the therapist who wish to experience excellence in performance during practice (Lee & Wishart 2005). Consequently, it is a major challenge to structure practice sessions that may motivate the learner, enhance performance as well as maximize retention in delayed tests.

While augmented feedback may indeed enhance motor performance and learning (Lee et al. 1995, Swinnen et al. 1997, Swinnen 2002, Puttemans et al. 2005) the optimal implementation of augmented feedback is not straightforward and there may be potential pitfalls depending on the specific task and learner characteristics (Magill 1994). First, performance changes do occur in the absence of augmented feedback indicating a role of error detection and correction processes for the learning. It is critical that augmented feedback aids these processes and does not replace them. Second, when augmented feedback does facilitate learning, the learner does not require it on every trial. Third, in some instances augmented feedback may even be detrimental to learning effects observed in delayed tests (Swinnen 1996). In recent decades extensive evidence from behavioural studies has suggested that providing augmented feedback during training improves performance, whereas its removal during subsequent retention tests or conditions may result in performance deterioration. This has come to be known as the 'feedback-guidance hypothesis' (see above), suggesting that availability of augmented feedback during training guides the learner towards proper motor output, but its subsequent removal may lead to performance decrements and/or suboptimal retention (Salmoni et al. 1983, Schmidt et al. 1989, 1990, Winstein and Schmidt 1990, Swinnen 1996). This is presumably a consequence of the learner becoming too dependent on augmented feedback, possibly at the expense of relying on the intrinsic feedback to support performance under nonaugmented test conditions (Salmoni et al. 1984, Swinnen 1996, Schmidt and Lee 1999, Magill 2007). Recently, Ronsse et al. (2011) provided evidence on the neural basis of the feedback-guidance hypothesis by the use of fMRI. Whereas frequent feedback may guide the learner to a correct response during practice and interfere with or replace active problem-solving processes, reduced feedback practice conditions are hypothesized to increase information-processing demands during practice that are advantageous to the relatively permanent motor learning effects observed in delayed retention tests (Sullivan et al. 2008). In order to implement augmented feedback in an optimal way and diminish feedback-dependency effects, it is consequently necessary to carefully consider the specific learner and task characteristics when designing skill learning paradigms (Magill 1994, Swinnen 1996).

The exact role that augmented feedback plays in learning is still a subject of much debate within the field. Numerous examples have demonstrated that feedback is necessary for learning, whereas other counterexamples have not been able to demonstrate a necessity of augmented feedback for learning. A complication is that learning may relate to reward-prediction errors and even when experimenter-generated augmented feedback is not provided, participants will nevertheless have varying degrees of confidence that their performance was correct, which could act as a de facto feedback signal (Mollon & Danilova 1996, Green & Bavalier 2008). Interestingly, learners often have an accurate impression of how they perform. Recent studies have shown beneficial learning and retention effects of providing augmented feedback following 'good' trials (throwing to a target) compared to 'poor' trials (Chiviacowski & Wulf 2007), and further studies have demonstrated enhanced motor learning by providing (false) positive relative to negative normative feedback (Wulf et al. 2010a, Wulf & Lewthwaite 2010, Lewthwaite & Wulf 2010). In one study, two groups of subjects practised a balance task. In addition to receiving veridical feedback on their performance after each trial, subjects also received normative feedback if they were better or worse than average. The 'better' group were led to believe that their performance was better than average, whereas the opposite was the case for the 'worse' group. In transfer tests, the 'better' group demonstrated more effective learning than the 'worse' group (Wulf & Lewthwaite 2010) and this was also the case in another study where the 'better' group was compared to controls who received only veridical feedback (Lewthwaite & Wulf 2010). Thus, the mere conviction of being 'good' enhances learning and feedback cannot merely be viewed as information, which is processed without affective connotation. On the contrary, the valence of feedback can indeed enhance learning (Wulf et al. 2010b).

Essentially, the principal question is not whether augmented feedback is necessary, but if augmented feedback may be beneficial for learning and if so, how is it best implemented in order to promote learning. In all likelihood, the importance of augmented feedback for motor skill learning depends heavily on the characteristics of the specific skill or task and the learner, and this may indeed explain the seemingly inconsistent findings of different studies.

Considering the characteristics of the task and the learner, Newell (1974) demonstrated how important augmented feedback is in a situation in which the intrinsic feedback needed to perform a skill is not available or the learner is not yet capable of using it. Subjects had to make a specific lever movement in 150 ms. Although they could see their arm, the lever and the target, their success in learning the task depended on how many times they had received KR about their performance. The results indicated that in the early learning phase the learner did not have a good internal model for the movement. They needed KR for more than 50 trials to establish this, and by then the movement could be performed without augmented feedback (Newell 1974). Different tasks also have different requirements and hence also different implications for augmented feedback. The role of augmented feedback in motor learning has been investigated for a wide range of tasks and skills from simple tasks with single degree of freedom movements in isolated lab experiments to complicated coordination tasks in more complex ecological settings.

Fowler and Turvey (1978) suggested that the required information content of the provided feedback must contain as many degrees of constraint as there are degrees of freedom in the action to be coordinated. In simple motor skills with single degree of freedom, or in practice of complex tasks where the focus is specifically on single parameters or the scaling of a given coordination pattern, KR usually specifies all the information that is needed for learning, e.g. in a ballistic task (see Salmoni et al. 1984, Newell 1974; see Swinnen 1996, for a review). Lavery (1962) found beneficial effects of KR in a simple ballistic task, where a ball had to be propelled to a target, and this finding has beem replicated by many studies. In a recent study we also found ballistic motor learning in a single-joint task to be highly enhanced by KR (Lundbye-Jensen et al. 2012). Relating to ballistic learning but in a more functional setting, Moran et al. (2011) investigated practice of service speed in tennis and found that KR is needed in order for players to improve with practice. This finding does indeed make sense since the learners do not acquire a new coordination pattern, but focus on improving a single parameter in a pre-existing skill. KR may, in other words, have a sufficiently prescriptive function in simple tasks. In addition to a beneficial role of KR, Lavery (1962) also demonstrated beneficial effects of summary KR relative to single-trial KR and this effect was also found and extended in more recent studies (e.g. Guadagnoli et al. 1996; Schmidt et al. 1989, 1990; Yao et al. 1994). The optimal frequency of augmented feedback does indeed interact with task complexity in affecting motor skill learning, and there is some evidence that the learning of simple motor tasks may benefit from a reduction in augmented feedback (Winstein & Schmidt 1990, Lai & Shea 1999). One explanation for this could be that when the task and the performance measure are isomorphic, as often happens in simple tasks, learners do not need augmented feedback on every trial (Magill 1994). Learning of a simple task might also be enhanced by a reduced feedback frequency, because withholding feedback makes practice more difficult or challenging, and it forces the learner to develop his or her own internal error detection and correction mechanisms (Wulf & Shea 2002).

While learning of simple skills may benefit from KR and reducing or delaying the augmented feedback, the usefulness of KR to a learner in acquiring wholebody actions or complex skills where the learner needs to establish new coordination modes has been questioned for a long time (Newell 1991, Gentile 1972). There is indeed evidence to suggest that different and more frequent feedback can be beneficial for the learning of more complex skills (Schmidt et al. 1990, Yao et al. 1994, Guadagnoli et al. 1996, Wulf & Shea 2002). Wulf and Shea (2002) proposed that because complex tasks require different components to be coordinated in order to produce skilled performance, the learner also has to rely on intrinsic feedback, and augmented feedback is generally not as prescriptive as in simple tasks. In complex tasks, the learner presumably benefits from information about the dynamics of the recent task performance in addition to the KR (Newell 1991), and KP describing movement kinematics or kinetics has been demonstrated to be more important than KR alone in complex skill learning (e.g. Newell & Walter 1981, Newell & Carlton 1987, Schmidt & Lee 1999).

In complex skill learning, augmented feedback may be given based on multiple different performance characteristics and learning inherently involves higher demands on control, error-detection and problem-solving. Based on this, the learning processes may not benefit from increasing the demands imposed on the learner further as it has been seen for simple tasks, e.g. by reducing feedback frequency (see Wulf & Shea 2002). On the contrary, learning may be enhanced by more frequent feedback (Magill 1994, Wulf & Shea 2004). Guadagnoli et al. (1996) directly demonstrated that task complexity and task-related experience interacted with the optimal number of trials summarized in the augmented feedback. Whereas reduced feedback frequencies benefited the learning of a simple striking task for novice and experienced participants, frequent (single-trial) feedback was more effective than longer feedback summaries in the learning of a more complex double-striking task, particularly for novices.

Several studies have replicated the finding that frequent (100 percent) feedback can be beneficial in complex skill learning, e.g. involving bimanual coordination (Swinnen et al. 1997, Wulf & Shea 2002) and, although frequent visual feedback typically promotes strong feedback dependency (at least for simple skills), this was not found to the same extent for these more complex tasks. It may be that this reduced susceptibility to feedback dependency in complex learning relates to a different role of intrinsic feedback (Wulf & Shea 2004). With regard to complex motor tasks, a number of studies have found that the performance and learning of a skill, such as dance routines (Clarkson et al. 1986), cycling (Sanderson & Cavanagh 1990, Broker et al. 1993), and swimming (Chollet et al. 1988), is enhanced when concurrent feedback is provided. These studies indicate that augmented feedback in real time can have a powerful effect on performance in certain sport tasks. Additionally, the added use of kinematic and kinetic information feedback facilitates motor learning beyond the level reached by presentation of KR alone (Newell 1991). Kinematic feedback has been demonstrated to be beneficial for learning a golf shot and increasing power output in a leg press exercise (Hopper et al. 2003). The effectiveness of kinematic feedback does, however, depend on the relevance of the feedback to the success of the movement or task goal. Furthermore, many studies of kinematic feedback have not obtained results in delayed retention or transfer tests.

Augmented feedback, properly employed, may also have practical implications for rehabilitation therapy since the re-acquisition of motor skills is an important part of functional motor recovery (Winstein 1991). An important question for physiotherapists working in rehabilitation is whether the research findings for healthy subjects apply to patients, e.g.

persons with stroke. This depends on whether people with stroke learn in the same way as people with an intact nervous system and whether the tasks which are practised in rehabilitation are comparable to the tasks investigated in research studies (van Vliet & Wulf 2006). Concerning the second question, a rehabilitation professional may find it challenging to implement principles derived from studies of laboratory-based tasks since these findings may not generalize into a clinical setting.

Concerning the first question, the principles of learning and the factors influencing learning are to a large extent identical between groups, but there may indeed be differences in learning ability, strategy etc. between individuals and groups of patients. Some patients may have a compromised ability to process intrinsic feedback due to neurological sensory impairments and some patients with cognitive and perceptual impairments may not be able to use intrinsic feedback to guide their performance. These factors may indeed influence the role of augmented feedback in learning and consequently also have implications for how augmented feedback should be provided.

There are indications that addition of augmented feedback to exercises can stimulate the learning process in rehabilitation therapy by making patients more aware of their performance (Holden 2005, Winstein & Stewart 2006) and systematic reviews indicate that augmented feedback in general has an added value for e.g. stroke rehabilitation. Molier et al. (2010) found trends in favour of providing augmented KP, augmented auditory feedback and combined sensory and visual feedback. There were no consistent effects on motor relearning for summary or faded, terminal or concurrent, solely visual or solely somatosensory augmented feedback. Although augmented feedback may be beneficial in rehabilitation, it is difficult to identify patients or groups of patients who might be more likely to benefit from a specific type of intervention due to heterogeneity of patients, groups and conducted trials, and Van Dijk et al. (2005) found no general differences in effectiveness between different therapeutic interventions using augmented feedback, i.e. electromyographic biofeedback, kinetic feedback, kinematic feedback, and KR.

Augmented feedback may indeed be beneficial for the reacquisition of motor skills in rehabilitation. Winstein (1991) suggested that it is appropriate to use the principles obtained through laboratory experimentation as guidelines rather than as exact recommendations when applying basic research findings to clinical practice. It is, however, not yet possible to formulate to what extent principles of augmented feedback are properly employed (van Dijk et al. 2005). Future studies should focus more on the content, form, and timing of augmented feedback in the therapeutic interventions and importantly distinguish performance and learning effects by assessing learning in delayed retention tests. Additionally, it is important to incorporate considerations about the role of the specific characteristics of the individual learner or patient (Magill 1994).

Individual differences in the benefits of feedback for learning

As discussed previously, different skills and tasks have different characteristics and learners are also different. From a motor skill acquisition point of view learners have individual differences in motor experience and baseline task performance, which will most likely influence skill acquisition (Magill 1994). Additionally, learning requires the use of cognitive resources and effort during practice. While advantageous for some people, the demands of a task may exceed the optimal capability for other individuals, especially those with reduced or impaired information-processing abilities (Sullivan et al. 2008, Kelley & McLaughlin 2012).

The amount of cognitive resources that learners possess is positively related to learning. Learners with more cognitive resources and higher abilities related to the task, learn faster than those lower in those abilities (Craik & Salthouse, 2000; Engle & Kane, 2004). The advantage for learners with higher resources may be explained through cognitive load theory (Kelley & McLaughlin 2012). Guadagnoli and Lee (2004) have proposed a framework which suggests that motor learning depends on the level of challenge emerging from an interaction of the information-processing capability of the learner, task demands, and practice condition. According to this framework, there is a point of optimal challenge that yields maximum practice benefits when optimal cognitive effort is invoked. A level of challenge below or above this optimal challenge point may attenuate learning. That is, conditions that demand too much cognitive effort may interfere with learning effects (Sullivan et al. 2008). One way to ensure that the demands of the task are optimally aligned to the resources at hand is to modify the task demands and/or to provide adjusted augmented feedback. In a recent study, Kelley & McLaughlin (2012) found positive interactions between individual differences in cognitive resources and adjusted augmented feedback during the process of learning. Those with higher abilities for the type of demands imposed by the task were more likely to benefit from reduced feedback.

Based on the discussion in the preceeding paragraphs, we suggest the following considerations for feedback design: Incorporate learner characteristics and task demands when designing learning support via augmented feedback (Magill 1994). Optimal feedback characteristics depend on individual differences in the learners' ability levels for the demands of the task being learned (Kelley and McLaughlin 2012). This is of outmost importance not only in skill learning relating to sports and daily activities but even more so in clinical practice.

Subliminal augmented feedback and motor learning

Augmented feedback may not only have an effect when it is consciously perceived by the learner. During daily life the central nervous system constantly receives and processes sensory input providing information on the state of our body and the surrounding world. Although we do not consciously perceive all sensory inputs, these may nevertheless have consequences for our future behaviour. Although evidence of subconscious processing has been debated for several decades, and the literature is not free of controversy, it is a well-described phenomenon that we may respond to features of our surroundings without being aware of them (e.g. Goodale et al. 1991, Goodale & Milner 1992, Pessiglione et al. 2007, Goodale 2008).

Subliminal stimuli may be presented to subjects as auditory stimuli, e.g. as tones with frequencies outside the perceptible range, at extremely low intensities or in some scrambled form. They may also be presented visually as low-contrast images presented for very brief durations or masked by other figures. In addition to these modalities, subliminal cues may also be presented as e.g. augmented sensory feedback, e.g. induced by tactile stimulation, vibration etc.

Several studies have demonstrated correct reactions in spite of conscious visual perception being disrupted through temporary knock-out of the visual cortex (Amassian et al. 1989, Christensen et al. 2008), and basic research studies manipulating only the augmented feedback characteristics have also provided evidence for processing of subliminal cues leading to behavioural changes. Eimer & Schlaghecken (2003) observed changes in reaction times and Taylor & McCloskey (1996) observed behavioural changes in a choice reaction task based on subliminal visual information. In recent studies it has also been observed that subliminal vibrotactile stimulation can lead to increased postural stability and balance in young and elderly subjects and in diabetes and stroke patients. Furthermore this subliminal stimulation can also reduce gait variability in elderly fallers (Galica et al. 2009, Priplata et al. 2002, Priplata et al. 2003).

Since learning may be reinforced by augmented feedback, and subliminal stimuli, which are not consciously perceived, may affect behavior, it may indeed be hypothesized that learning could also be facilitated by subliminal augmented feedback on motor performance. Pessiglione et al. (2006) demonstrated that modulation of dopamine-dependent striatal activity during learning can account for how the human brain uses reward prediction errors to improve future decisions, and Pessiglione et al. (2007) also used functional magnetic resonance imaging to investigate the neural basis of processing subliminal stimuli and the translation to behaviour. The study focused on an incentive force task and used money as reward, which was presented either subliminally or for a longer time leading to a conscious perception of the stimulus. Even when subjects could not report how much money was at stake, they nevertheless deployed more force for higher amounts. The findings imply that expected rewards may indeed energize behaviour, without the need for the subjects' awareness. This indicates that motivational processes involved in boosting behaviour are qualitatively similar, despite whether subjects are conscious or not of the reward at stake. Consistently, the same basal forebrain region underpinned subliminal and conscious motivation.

While there have been some positive results (e.g. Masters et al. 2009), motor skill learning based on subliminal augmented feedback has not yet been consistently demonstrated. It is, however, a very interesting area. Pessiglione et al. (2007) focused on motivation, rather than motor learning, but the results are consistent with the notion (and our recent observation) that subliminal augmented feedback can also facilitate motor learning (Lundbye-Jensen et al. 2012). As it may be beneficial to combine augmented feedback modalities in motor learning, it may also potentially be beneficial for learning to combine supraliminal and subliminal augmented feedback.

Conclusions and perspectives

As we have seen in this chapter, augmented feedback may be very powerful to facilitate performance. This type of feedback can be applied in a variety of motor learning processes and in a variety of ways. Regarding the possible ways, we mentioned and discussed three dimensions. The 'content of the information' considers the variable(s) 'extracted' from the movement and presented to the learner. The 'frequency of the augmented feedback' comprises the aspect of the number of instances of feedback in relation to the number of executed movements. Finally, the group 'timing of the feedback' is concerned with the timing of the feedback relative to the phase of the movement.

From our point of view, the power of the augmented feedback to facilitate performance in learning is of great interest for both, scientific research as well as for the praxis. Regarding its power, we have already mentioned the fact (and the potential risk) that learners may become hooked on augmented feedback and even lose trust in their internal sensation(s). Consequently, augmented feedback may be optimally applied in a practical setting such as motor rehabilitation or in the early phase of learning where external and explicit guiding is significant. This guiding may be necessary because of uncertainties in the internal sensation after injuries or diseases and/or because of non-existing experiences (referring to either the movement per se or the external world with which the person has to interact).

However, we argue that, to reasonably apply augmented feedback in these settings, important future work needs to be done. We specifically relate this future work to the efficiency of the outcome when learning with augmented feedback. For instance, consider a sensorimotor training in the rehabilitation after a stroke. What would you recommend as the content of the information provided in the form of augmented feedback? What should be the frequency of the information relative to the executed movement and what should be the timing of the feedback to maximize its effect? You do not know? Neither do we. In our opinion, these questions require substantial research work to clarify in which setting augmented feedback should be applied in which form, how often, when, and furthermore how individual characteristics should be accounted for. Practically, these evaluations would need to be done in two steps. In the first step, laboratory work is required. In the second step, the acquired knowledge would need to be tested in practical (real world) settings. This is important as the implementation finally takes place in praxis and not in the laboratory.

A last thing we want to focus on is the combination of supraliminal and subliminal augmented feedback. We have already mentioned that, in addition to augmented feedback consciously perceived by the learner, subconscious (subliminal) information is processed in the central nervous system. Not only is it processed but it may also facilitate performance in motor learning. Therefore, it would be interesting to know whether the combination of supraliminal augmented feedback and subliminal augmented feedback can facilitate performance to a greater extent than each of these alone. A rationale for an enhanced facilitation may be the postulated processing of the information, specifically the notion that supraliminal and subliminal information is differently processed (presumably relating to the involved pathways) in the central nervous system.

At this point we want to end the chapter. We hope that we have made clear the definition of augmented feedback and its influence in motor learning. Furthermore, we hope that we have made clear its important capability, when appropriately adjusted, to facilitate motor performance and promote learning processes.

References

Ackerman, P.L., Kanfer, R. & Goff, M. (1995) Cognitive and noncognitive determinants and consequences of complex skill acquisition. J. Exp. Psychol: Applied 1: 270–304

Ackerman, P.L. & Cianciolo, A.T. (2000) Cognitive perceptual-speed, and psychomotor determinants of individual differences during skill acquisition. J. Exp. Psychol: Applied 6: 259–290. PMID: 11218338.

Adams, J.A. (1987) Historical review and appraisal of research on the learning, retention, and transfer of human motor skills. Psychol. Bull. 101:41–74

Amassian, V.E., Cracco, R.Q., Maccabee, P.J., Cracco, J.B., Rudell, A., Eberle, L. (1989) Suppression of visual perception by magnetic coil stimulation of human occipital cortex. Electroencephalogr. Clin.Neurophysiol. 74:458–462

Anderson, D.I., Magill, R.A., Sekiya, H., Ryan, G. (2005) Support for an explanation of the guidance effect in motor skill learning. J. Mot. Behav. 2005;37:231–238

Annesi, J. J. (1998) Effects of computer feedback on adherence to exercise. Percept. Motor Skills 87: 723–730

Bernstein, N.A. (1967) The co-ordination and regulation of movements. Pergamon Press, Oxford

Broker, J.P., Gregor, R.J. & Schmidt, R.A. (1993) Extrinsic feedback and the learning of kinetic patterns in cycling. J. Appl. Biomech. 9: 111–123

Cahill, L., McGaugh, J.L., Weinberger, N.M. (2001) The neurobiology of learning and memory: some reminders to remember. Trends Neurosci. 24:578–581

Christensen, M.S., Kristiansen, L., Rowe, J.B., Nielsen, J.B. (2008) Action-blindsight in healthy subjects after transcranial magnetic stimulation. Proc. Natl. Acad. Sci. USA 105:1353–1357

Chiviacowski, S. & Wulf, G. (2007) Feedback after good trials enhances learning. Res. Qu. Exerc. Sport, 78: 40–47

Chollet, D., Micallef, J.P. & Rabischong, P. (1988) Biomechanical signals for external biofeedback to improve swimming techniques. In B. E. Ungerechts, K. Wilke & K. Reischle (eds.) Swimming V. Champaign, IL: Human Kinetics, 389–396

Clarkson, P.M., James, R., Watkins, A. & Foley, P. (1986) The effect of augmented feedback on foot pronation during barre exercise in dance. Res. Qu. Exerc. Sport 57: 33–40

Craik, F. & Salthouse, T.A. (eds) (2000) The Handbook of Aging and Cognition. Mahwah, NJ: Lawrence Erlbaum.

Eimer, M. & Schlaghecken, F. (2003) Response facilitation and inhibition in subliminal priming. Biol. Psychol. 64:7–26

Engle, R.W. & Kane, M.J. (2004) Executive attention, Worning memory capacity, and a two-factor theory of cognitive control. Psychol. Learn. Motiv., 44, 145–198.

Ernst, M.O. & Banks, M.S. (2002) Humans integrate visual and haptic information in a statistically optimal fashion. Nature 415:429–433

Ericsson, K.A., Krampe, R.Th. & Tesch-Römer, C. (1993) The role of deliberate practice in the acquisition of expert performance. Psychol. Rev., 100(3): 363–406

Faisal, A.A., Selen, L.P., Wolpert, D.M. (2008) Noise in the nervous system. Nat. Rev. Neurosci. 9:292–303

Fitts, P.M. (1964) Perceptual-motor skill learning. In A.W. Melton (ed.) Categories of human learning. New York: Academic Press, 243–285

Fowler, C.A. & Turvey, M.T. (1978) Skill acquisition: an event approach with special reference to searching for the optimum of a function of several variables. In G.E. Stelmach (ed.) Information processing in motor control and learning. New York: Academic Press, 1–40

Fuster, J.M. (1995) Memory in the cerebral cortex: an empirical approach to neural networks in the human and nonhuman primate. Cambridge, MA: MIT Press

Galica, A.M., Kang, H.G., Priplata, A.A., D'Andrea, S.E., Starobinets, O.V., Sorond, F.A., Cupples, L.A., Lipsitz, L.A. (2009) Subsensory vibrations to the feet reduce gait variability in elderly fallers. Gait Posture 30(3):383–387

Gentile, A.M. (1972) A working model of skill acquisition with application to teaching. Quest 17: 3–23

Goodale, M.A., Milner, A.D., Jakobson, L.S., Carey, D.P. (1991) A neurological dissociation between perceiving objects and grasping them. Nature 349:154–156

Goodale, M.A. & Milner, A.D. (1992) Separate visual pathways for perception and action. Trends Neurosci. 15: 20–25

Goodale, M.A. (2008) Action without perception in human vision, Cogn. Neuropsychol. 25:7:891–919

Green, C.S. & Bavalier, D. (2008) Exercising Your Brain: A review of human brain plasticity and training-induced learning. Psychol Aging. December; 23(4): 692–701. doi:10.1037/a0014345

Guadagnoli, M.A., Dornier, L.A. & Tandy, R. (1996) Optimal length for summary of results: the influence of task related experience and complexity. Res. Qu. Exerc. Sport 67: 239–348

Guadagnoli, M.A. & Lee, T.D. (2004) Challenge point: a framework for conceptualizing the effects of various practice conditions in mo- tor learning. J. Mot. Behav. 36: 212–224

Guay, M., Salmoni, A., Lajoie, Y. (1999) The effects of different knowledge of results spacing and summarizing techniques on the acquisition of a ballistic movement. Res. Q. Exerc. Sport. 70:24–32

Holden, M.K. (2005) Virtual environments for motor rehabilitation: review, Cyberpsychol. Behav. June:8(3):187–211; discussion 212–19

Hopper, D., Berg, M., Andersen, H. and Madan, R. (2003) The influence of visual feedback on power during leg press on elite women field hockey players, Phys. Ther. Sport 4: 182–186

Kantak, S.S., Winstein, C.J. (2012) Learning–performance distinction and memory processes for motor skills: A focused review and perspective. Behav. Brain Res. March 1; 228 (1): 219–231, doi: 10.1016/ j.bbr.2011.11.028

Kelley, C.M. & McLaughlin, A.C. (2012) Individual differences in the benefits of feedback for learning. In Human Factors: J. Human Factors and Ergonomics Society 54: 26.

Krakauer, J.W. (2009) Motor learning and consolidation: the case of visuomotor rotation. Adv. Exp. Med. & Biol. 629:405–421. doi: 10.1007/978-0-387-77064-2_21

Lai, Q. & Shea, C.H. (1999) The role of reduced frequency of knowledge of results during practice. Res. Qu. Exerc. Sport 70: 33–40

Lavery, J.J. (1962) Retention of simple motor skills as a function of type of knowledge of results. Can. J. Psychol. 16: 300–311

Lee, T., Swinnen, S., Verschueren, S. (1995) Relative phase alterations. during bimanual skill acquisition. J. Mot. Behav. 27:263–274

Lee, T.D. & Wishart, L.R. (2005) Motor learning conundrums (and possible solutions). Quest 57:67–78

Lewthwaite, R. & Wulf, G. (2010) Social-comparative feedback affects motor skill learning. Qu. J. Exp. Psychol., 63: 738–749

Little, W.S. & McGullagh, P. (1989) Motivation orientation and modelled instruction strategies: The effects of form and accuracy. J. Sport Exerc. Psychol. 11: 41–53

Lundbye-Jensen, J., Petersen, T.H., Rothwell, J.C., Nielsen, J.B. (2011) Interference in ballistic motor learning: Specificity and role of sensory error signals. PloS One March 9; 6(3): e17451, pp. 1–15, doi: 10.1371/journal.pone.0017451

Lundbye-Jensen, J., Leukel, C. & Nielsen, J.B. (2012) Learning without knowing: subliminal visual feedback facilitates motor learning. Manuscript submitted for publication

Magill, R.A. (1994) The influence of augmented feedback on skill learning depends on characteristics of the skill and the learner. Quest 46: 314–327

Magill, R.A. (2001) Augmented feedback in motor skill acquisition. In R.N. Singer, H.A. Hausenblas and C.M. Janelle (eds) Handbook of sport psychology. New York: John Wiley & Sons

Magill, R.A. (2007) Motor learning and control: concepts and applications. 8th edn. New York: McGraw-Hill

Marchese, R., Diverio, M., Zucchi, F., Lentino, C. & Abbruzzese, G. (2000) The role of sensory cues in the rehabilitation of Parkinsonian patients: a comparison of two physical therapy protocols. Mov. Disord., 15(5), 879–883

Masters, R.S.W., Maxwell, J.P. & Eves, F.F. (2009) Marginally perceptible outcome feedback, motor learning and implicit processes. Conscious. Cogn. 18: 639–645

Mollon, J.D. & Danilova, M.V. (1996) Three remarks on perceptual learning. Spat. Vis. 10(1): 51–58

Molier, B.I., van Asseldonk, E.H.F., Hermens, H.J., Jannink, M.J.A. (2010) Nature, timing, frequency, and type of augmented feedback; does it influence motor relearning of the hemiparetic arm after stroke? A systematic review. Disabil. Rehab.: 32(22): 1799–1809

Moran, K., Kieran, A., Murphy, C. & Marshall, B. (2011) The need and benefit of augmented feedback on service speed in tennis. Med. Sci. Sports Exerc.: April: 44(4): 754–760

Newell, K.M. (1974) Knowledge of result and motor learning. J. Mot. Behav. 6: 235–244

Newell, K.M. & Walter, C.B. (1981) Kinematic and kinetic parameters as information feedback in motor skill acquisition. J. Hum. Mov. Stud. 7, 235–254

Newell, K.M. & Carlton, M.J. (1987) Augmented information feedback and the acquisition of isometric tasks. J. Mot. Behav. 19: 4–12

Newell, K.M. (1991) Motor skill acquisition. Ann. Rev. Psychol. 42: 213–237

Pessiglione, M., Schmidt, L., Draganski, B., Kalisch, R., Lau, H., Dolan, R.J. and Frith, C.D. (2007) How the brain translates money into force: A neuroimaging study of subliminal motivation. Science. May 11: 316(5826): 904–906

Pessiglione, M., Seymour, B., Flandin, G., Dolan, R.J. & Frith, C.D. (2006) Dopamine-dependent prediction errors underpin reward-seeking behaviour in humans. Nature 442: 1042–1045

Priplata, A., Niemi, J., Salen, M., Harry, J. & Lipsitz, L. (2002) Noise-enhanced human balance control. Phys. Rev. Lett. 3–6

Priplata, A., Niemi, J., Harry, J. & Lipsitz, L. (2003) Vibrating insoles and balance control in elderly people. The Lancet 362: 1123–1124

Puttemans, V., Wenderoth, N., Swinnen, S.P. (2005) Changes in brain activation during the acquisition of a multifrequency bimanual coordination task: from the cognitive stage to advanced levels of automaticity. J. Neurosci. 25:4270–4278

Ronsse, R., Puttemans, V., Coxon, J.P., Goble, D.J., Wagemans, J., Wenderoth, N. & Swinnen, S.P. (2011) Motor learning with augmented feedback: modality-dependent behavioral and neural consequences. Cereb. Cortex June: 21:1283–1294

Salmoni, A.W., Schmidt, R.A., Walter, C.B. (1984) Knowledge of results and motor learning: a review and critical reappraisal. Psychol. Bull. 1984;95:355–86

Salmoni, A.W., Ross, D., Dillb, S., Zoellerb, M. (1983) Knowledge of results and perceptual-motor learning. Hum. Mov. Sci. 2:77–89

Sanderson, D.J. & Cavanagh, P.R. (1990) Use of augmented feedback for the modification of the pedaling mechanics of cyclists. Can. J. Sport Sci. 15: 38–42

Schmidt, R.A., Lange, C., Young, D.E. (1990) Optimizing summary knowledge of results for skill learning. Hum. Mov. Sci. 9:325–348

Schmidt, R.A., Young, D.E., Swinnen, S., Shapiro, D.C. (1989) Summary knowledge of results for skill acquisition: support for the guidance hypothesis. J. Exp. Psychol. Learn. Mem. Cogn. 15:352–359

Schmidt, R.A. & Bjork, R.A. (1992) New conceptualization of practice: common principles in three paradigms suggest new concepts for training. Psychol. Sci. 1992;3:207–217

Schmidt, R.A. & Lee, T.D. (1999) Human control and learning: A behavioral emphasis. Champaign, IL: Human Kinetics

Seidler, R.D. (2010) Neural correlates of motor learning, transfer of learning, and learning to learn. Exerc. Sport Sci. Rev., 38(1): 3–9

Sherwood, D.E. (1988) Effect of bandwidth knowledge of results on movement consistency. Percept. Mot. Skills. 66:535–542

Silverman, S. S., Woods, A. M. & Subramaniam, P. R. (1998) Task structures, feedback to individual students, and student skill level in physical education. Res. Qu. Exerc. Sport 69, 420–424

Sullivan, K.J., Kantak, S.S., Burtner, P.A. (2008) Motor learning in children: Feedback effects on skill acquisition. Phys. Ther. 88:720–732

Swinnen, S. P. (1996) Information feedback for motor skill learning: A review. In N. Zelaznik (ed.) Advances in motor learning and control. Champaign, IL: Human Kinetics, 37–66

Swinnen, S.P., Lee, T.D., Verschueren, S., Serrien, D.J., Bogaerds, H. (1997) Interlimb coordination: learning and transfer under different feedback conditions. Hum. Mov. Sci. 16:749–785

Swinnen, S.P. (2002) Intermanual coordination: from behavioural principles to neural-network interactions. Nat. Rev. Neurosci. 3:348–359

Taylor, J.L. & McCloskey, D.I. (1996) Selection of motor responses on the basis of unperceived stimuli. Exp. Brain Res. 110:62–66

Thorndike, E.L. (1927) The law of effect. Am. J. Psychol., 39, 212–22

van Dijk, H., Jannink, M.J.A. & Hermens, H.J. (2005) Effect of augmented feedback on motor function of the affected upper extremity in rehabilitation patients: a systematic review of randomized controlled trials. J. Rehab. Med. 2005, 37(4), 202–211

van Vliet P.M. & Wulf G. (2006) Extrinsic feedback for motor learning after stroke: what is the evidence? Disabil. Rehabil. July: 28(13–14):831–40

Winstein, C.J. & Schmidt, R.A. (1990) Reduced frequency of knowledge of results enhances motor skill learning. J. Exp. Psychol. Learn. Mem. Cogn. 1990;16:677–691

Winstein, C.J. (1991) Knowledge of results and motor learning–implications for physical therapy. Phys. Ther. 71:140–149

Winstein, C.J. & Stewart, J.C. (2006) Textbook of neural repair and rehabilitation. Volume 2, Medical neurorehabilitation. Cambridge: Cambridge University Press, 89–102

Wolpert, D.M., Diedrichsen, J., Flanagan, J.R. (2011) Principles of sensorimotor learning. Nature reviews. Neuroscience 12:739–751 doi: 10.1038/nrn3112

Wolpert, D.M. & Ghahramani, Z. (2000) Computational principles of movement neuroscience. Nat. Neurosci. 3 Suppl:1212–1217

Wulf, G. & Shea, C. (2002) Principles derived from the study of simple skills do not generalize to complex skill learning. Psychonom. Bull. Rev. 9, 185–211

Wulf, G. & Shea, C.H. (2004) Understanding the role of augmented feedback: The good, the bad, and the ugly. In A.M. Williams, & N.J. Hodges (eds.), Skill acquisition in sport: Research, theory and practice (pp. 121–44). London: Routledge

Wulf, G., Chiviacowsky, S., Lewthwaite, R. (2010a) Normative feedback effects on the learning of a timing task. Res. Q. Exerc. Sport Dec.:81(4):425–31

Wulf, G. & Lewthwaite, R. (2010) Social-comparison feedback and conceptions of ability: effects on motor learning. Q. J. Exp. Psychol. (Hove). Apr.:63(4):738–749

Wulf, G., Shea, C. & Lewthwaite, R. (2010b) Motor skill learning and performance: a review of influential factors. Med. Educ. 2010: 44: 75–84

Yao, W., Fishman, M.G. & Wang, Y.T. (1994) Motor skill acquisition and retention as a function of average feedback, summary feedback, and performance variability. J. Mot. Behav. 26, 273–82

Young, D. E., Schmidt, R. A., & Lee, T.D. (2001) Skill learning: Augmented feedback In W. Karwowski (Ed.) International encyclopedia of ergonomics and human factors (pp. 558–561), London: Taylor and Francis

8

NEUROSCIENTIFIC ASPECTS OF IMPLICIT MOTOR LEARNING IN SPORT

Frank Zhu,[1,2] *Jamie Poolton*[1] *and Rich Masters*[1]

[1] INSTITUTE OF HUMAN PERFORMANCE, THE UNIVERSITY OF HONG KONG
[2] DEPARTMENT OF SURGERY, THE UNIVERSITY OF HONG KONG

Motor learning has been defined as 'a set of [internal] processes associated with practice or experience leading to relatively permanent changes in the capability for responding' (Schmidt, 1988: 346). When the motor task is sufficiently complex that it requires the coordination of multiple degrees of freedom, as in the skills required for proficient performance in most sports activities, the learner tends to take a proactive role in aspects of the learning process that can be consciously monitored or controlled. At the heart of this chapter is the role of verbal-analytical processes in motor control and learning, taking as a primary distinction the contrast between processing of explicit, declarative or implicit, procedural knowledge during motor performance (e.g., Anderson, 1983; Anderson & Lebiere, 1998; Schneider & Shiffrin, 1977). We will first describe the nature of the knowledge that is involved in verbal-analytical processes and explore how accretion, storage and use of the knowledge are mediated by working memory. We will discuss a range of studies that provide insight into cortical aspects of working memory processes during learning and performance and we will introduce an overview of implicit motor learning, which has been developed as an approach to suppress verbal-analytical involvement during motor performance by controlling working memory input during learning. We will present neurophysiological evidence suggesting that implicit motor learning promotes neural efficiency by suppressing verbal-analytical involvement in motor performance and will try to show how this relates to individual differences in the propensity for conscious motor processing (reinvestment). Finally, we will briefly discuss studies that provide insight to neurophysiological aspects of implicit motor learning in rehabilitation, before trying to summarize the current state of our understanding of neuroscientific aspects of implicit motor learning in sport.

Two fundamental types of knowledge

Declarative knowledge refers to verbalizable rules, techniques or methods that are applied by verbal-analytical processes in an effort to advance learning and to achieve optimal performance, whereas procedural knowledge refers to generalized motor programs and schemas that drive actions automatically (e.g., Keele, 1968; Schmidt, 1977). Theoretical models of motor learning commonly propose that motor control progresses from an initial effortful, explicit and cognitive stage to an effortless, implicit and autonomous stage (e.g., Adams, 1971; Fitts & Posner, 1967). During early motor learning, the learner tends to analyze the different components of the skill and test hypotheses about how to improve motor performance. From trial to trial, successful experiences are repeated and failed attempts are disregarded, resulting in accumulation of task-relevant declarative knowledge. The process is often augmented by additional instructions from coaches or therapists. Much later in learning, the declarative knowledge that initially supported performance consolidates as procedural knowledge (Salmoni, 1989) allowing verbal-analytical processes to be withdrawn and motor performances to run automatically, with little conscious input.

The verbal-analytical processes that are prevalent during the initial stage of motor learning are likely mediated by working memory (Baddeley, 1986). Working memory, as it was first described by Baddeley and Hitch (1974), is a short term, limited capacity system that holds information, via rehearsal mechanisms, in temporary storage so that it can be acted upon. As such, working memory is implicated in the acquisition, manipulation and application of declarative knowledge during motor control and learning (e.g., MacMahon & Masters, 2002; Masters & Maxwell, 2004; Masters & Poolton 2012; Maxwell, Masters & Eves, 2003). Current views hold that working memory is a four-component model (see for a review Baddeley, 2007), which consists of an attention controller and three subsidiary systems (i.e., the phonological loop, the visuospatial sketchpad, and the episodic buffer). The phonological loop is believed to hold verbal information by sub-vocal repetition (Baddeley, 1999). The visuospatial sketchpad is thought to maintain visual and spatial information from sensory input and to process temporal images (Logie, 1995). The episodic buffer is proposed to integrate and store information from the phonological and visuospatial loops, and multi-dimensional information from long-term memory (Baddeley, 2000). The attentional controller acts as a central executive (Baddeley, 1996) that combines information from the three subsidiary systems to generate coherent declarative output (Baddeley, 2000).

Working memory, therefore, provides a temporary storage space for holding and processing incoming information during motor learning, such as verbal instructions from a coach or a therapist, visual feedback representing performance outcome, as well as proprioceptive and tactile sensory feedback that results from the motor performance (Maxwell et al., 2003). Once processed and temporarily stored, the information can be used to form and test hypotheses about how to generate more effective movements, and significant declarative knowledge associated with the outcome is stored in long-term memory. This declarative knowledge can later be retrieved to support motor performance. In short, working memory is used to manage verbal-analytical aspects of motor control and learning, such as utilization of verbal instructions, monitoring and control of performance, formation and testing of hypotheses, correction of errors, and accumulation, retrieval and application of declarative knowledge (Masters & Maxwell, 2004, 2008; Masters & Poolton 2012;).

Working memory's neural network

Although working memory has been argued to play an important role in managing verbal-analytical aspects of motor control and learning (e.g., Masters & Maxwell, 2004; Masters & Poolton 2012; Maxwell et al., 2003), the neural network underlying working memory activity is less clear. Landau and co-authors (2004) used fMRI to examine the effect of practice on brain activity during a working memory dependent face recognition task. The task required participants to encode a series of faces so that they could be retrieved after an eight second delay. The fMRI data showed that neural activity decreased with practice in all the encoding-related brain regions except for left middle temporal gyrus, left postcentral gyrus and left insula, indicating that activation of these three regions is likely to be the necessary component of encoding processes associated with working memory. Other work has argued that the medial temporal lobe (MTL) plays an important role in the retrieval of task-relevant declarative facts and events from long-term memory (e.g., Clark & Squire, 1998; Schacter, 1997; Squire, 1992). Such regions, therefore, may be active in early stages of motor learning, given that working memory is essentially responsible for cognitive encoding of visual and proprioceptive feedback for error detection and correction (Maxwell et al., 2003) and retrieval and application of declarative knowledge to support motor performance (Masters & Maxwell, 2004, 2008). Outside the motor domain, neuroimaging studies suggest that the anterior cingulate cortex (ACC) may play an executive role in the human attentional system, monitoring and responding to conflicts in the processing of information by initiating specific changes in the amount of cognitive control used during subsequent performance (see Botvinick, Carter, Braver, Barch & Cohen, 2001, for a review). While the conflict monitoring hypothesis has not been applied specifically to motor control and learning, cognitive control, presumably mediated through working memory in some way, is the mechanism by which performance is managed, errors are detected and even hypotheses formed during motor learning (e.g., Carter, Braver, Barch, Botvinick, Noll & Cohen, 1998; Lungu, Liu, Waechter, Willingham & Ashe, 2007; MacDonald, Cohen, Stenger & Carter, 2000).

Insight into the cortical activity associated with working memory's involvement in motor control and learning of complex psychomotor tasks, such as baseball batting (e.g., Radlo, Janelle, Barba & Frehlich, 2001), rifle shooting (e.g., Hatfield, Landers & Ray, 1984; Konttinen & Lyytinen, 1993), archery (Salazar, Landers, Petruzzello, Han, Crews & Kubitz, 1990) and golf putting (Baumeister, Reinecke, Liesen & Weiss, 2008a; Crews & Landers, 1993), has been provided by electroencephalography (EEG) methodology. This work has contended that verbal-analytical processing, of the kind conducted by working memory, can be inferred by EEG alpha power at the left temporal region (T3, e.g., Hatfield et al., 1984; Haufler, Spalding, Santa Maria & Hatfield, 2000; Kerick, McDowell, Hung, Santa Maria, Spalding & Hatfield, 2001), EEG alpha power at the parietal region (P3, Pz, P4, e.g., Baumeister et al., 2008a; Smith, McEvoy & Gevins, 1999), EEG theta power at the frontal region (F3, Fz, F4, e.g., Baumeister et al., 2008a; Zhu, Maxwell, Hu et al., 2010), and EEG coherence between the left temporal region and the frontal region (T3–Fz coherence, e.g., Deeny, Hillman, Janelle & Hatfield, 2003; Zhu, Poolton, Wilson, Hu, Maxwell & Masters, 2011a; Zhu, Poolton, Wilson, Maxwell & Masters, 2011b). Figure 8.1 shows the location of the brain regions of interest mentioned above.

In their classic study of rifle shooting, Hatfield et al. (1984) conducted an EEG spectral power analysis of the left temporal region (T3) and right temporal region (T4) of skilled marksmen during a 7.5 s preparatory period prior to a shot. T3 is associated with verbal-analytical processes (Cohen, 1993), whereas T4 is associated with visual-spatial processing. Hatfield et al. (1984) found that while alpha power (8–12 Hz) at T4 remained relatively constant through the preparation period, alpha power at T3 continued to increase as the pull on the rifle's trigger neared. EEG alpha power is believed to inversely relate to cortical activation at specific regions of the cerebral cortex (Pfurtscheller, 1992), implying that the increased T3 alpha power shown by Hatfield et al. (1984) related to the suppression of verbal-analytical processes as shot initiation approached. This interpretation of Hatfield et al.'s (1984) findings is supported by empirical evidence from a longitudinal learning study conducted by Landers, Han, Salazar, Petruzzello, Kubitz and Gannon (1994). EEG activity from the T3 and T4 regions

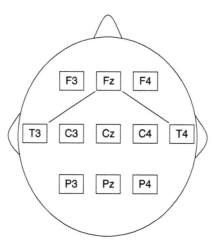

Figure 8.1   The location of the brain regions of interest for studies using EEG methodology to investigate cortical activity during motor performance.

during the aiming period of an archery task was measured before and after 14 weeks of practice. An increase in T3 alpha power was shown only after practice, consistent with the notion that verbal-analytical processing by working memory is withdrawn as learning advances (e.g., Fitts & Posner, 1967). Kerick, Douglass and Hatfield (2004) later replicated Landers et al.'s (1994) findings in a group of novice rifle shooters who practiced over a 12-14 week learning period. An analysis of T3 alpha power during the aiming period of a rifle-shooting task was also shown to differentiate expert marksmen from novice shooters (Haufler et al., 2000).[1]

Frontal theta (4–7 Hz) power is argued to reflect ACC activity (Gevins, Smith, McEvoy & Yu, 1997) associated with executive control (Karni, Meyer, Rey-Hipoli et al., 1998; Luks, Simpson, Feiwell & Miller, 2002). High frontal theta power is interpreted to indicate focused attention (e.g., Haufler et al., 2000; Smith et al., 1999; Zhu et al., 2010) and suppressed working memory activity (Baumeister et al., 2008a). Haufler et al. (2000) also carried out EEG spectral power analysis that included theta power at the frontal region and found that power was higher during the preparation period of a rifle-shooting task in expert marksmen compared to novice shooters. Smith et al. (1999) showed that frontal theta power increased after a short period of practice of a video game motor task. Similarly, Zhu et al. (2010) found increased frontal theta power as improvements in the learning of a sequential finger-tapping task began to plateau. Baumeister et al. (2008a) compared the cortical activity of expert golfers with novice golfers during a four-minute self-paced golf-putting task. Relatively higher frontal theta power was found in experts, indicative of more focused attention and the suppressed involvement of working memory in the putting stroke. Furthermore, Baumeister

---

1 An alternative interpretation is that increased T3 alpha power reflects suppression of lower-order motor processes. Kerick et al. (2001) analyzed alpha power activity during shooting at the central regions (C3 and C4) associated with lower-order activation of the motor cortex, and the T3 and T4 regions. Consistent with the findings of Hatfield et al. (1984), expert marksmen exhibited heightened alpha power relative to control measures at the T3 site only, implying that increased T3 alpha power reflects suppression of verbal-analytical processes, not lower-order motor processes.

et al. (2008a) found relatively high parietal alpha2 (10–12 Hz) power in his expert golf putters. Likewise, Smith et al. (1999) found parietal alpha2 power also increased with video game motor learning. High parietal alpha2 power is thought to reflect low somatosensory information processing activity (Gevins et al., 1997), presumably by working memory.

Spectral power analysis is used to assess cortical activation associated with specific regions in the cerebral cortex; however, during motor performance the co-activation of different regions or the integration of different processes may occur to varying degrees. EEG coherence is a frequency-dependent measure of the degree of linear relatedness between signals simultaneously recorded from two separate regions on the scalp. High coherence implies 'synched' communication between particular regions, whereas low coherence indicates regional independence (Silverstein, 1995; Weiss & Mueller, 2003). A number of studies have performed EEG coherence analysis to examine functional cortico-cortical communication during cognitive (e.g., Petsche, 1996; Petsche, Kaplan, Von Stein & Filz, 1997), verbal (e.g., Razoumnikova, 2000; Sheppard & Boyer, 1990; Volf & Razumnikova, 1999), and language tasks (e.g., Reiterer, Berger, Hemmelmann & Rappelsberger, 2005; Reiterer, Hemmelmann, Rappelsberger & Berger, 2005; see Weiss & Mueller, 2003, for a review). Previous work has also used this measure to examine the cortical networking related to neural adaptation in motor learning (e.g., Bell & Fox; 1996; Busk & Galbraith; 1975; Chen, Hung, Lin et al.; 2005; Deeny et al.; 2003; Deeny, Haufler, Saffer & Hatfield; 2009; Hatfield, Haufler & Contreras-Vital, 2009; Zhu et al., 2010). Specifically, co-activation of the left temporal region (T3) associated with verbal-analytical activity (Haufler et al., 2000; Kerick et al., 2001), and the frontal midline region (Fz) associated with motor planning (which has a direct connection to the left temporal region, Kaufer & Lewis, 1999) in the alpha frequency bandwidth has been recommended as an informative index of the involvement of verbal-analytical processing in motor control and learning. In support of this proposition, Deeny et al. (2003) found that expert marksmen displayed predictably lower EEG T3–Fz coherence than less skilled shooters.

Reduction in cortical activity associated with increased motor skill expertise has been described by Hatfield and Hillman (2001) as a marker of 'psychomotor efficiency', the principle of which holds that 'as skill level increases, the organism becomes increasingly capable of performing more psychomotor work with relatively less neuromuscular activity' (Janelle, Duley & Coombes, 2004: 291). Hatfield and Hillman (2001) proposed that a more efficient activation pattern is the result of pruning of neural regions that are nonessential to proficient performance of the task. Motor learning is characterized by the gradual withdrawal, or pruning, of verbal-analytical processes operated by working memory. Neurophysiological studies have begun to map out the neural network that reflects working memory's contribution to motor control and learning. In particular, the medial temporal lobe, anterior cingulate cortex, the left temporal region, the parietal region and the frontal region have been suggested. Collectively, a more efficient activation pattern is shown in these regions as working memory becomes less essential to motor control and learning. An understanding of the neural network that represents the verbal-analytical processing that supports motor control and learning facilitates psychophysiological investigation into implicit motor learning and the mechanisms that underpin skill failure and skill resilience.

Suppressing working memory activity from the onset of motor learning

Masters and colleagues (e.g., Masters, 1992; Masters & Maxwell, 2004; Masters & Poolton 2012; Maxwell, Masters & Eves, 2000, 2003; Maxwell, Masters, Kerr & Weedon, 2001) have argued that, by suppressing working memory's involvement in motor learning, a different kind of learning (i.e., implicit motor learning) takes place, which challenges the traditional stage theories of (explicit) motor learning (e.g., Anderson, 1983; Fitts & Posner, 1967; Schneider & Shiffrin, 1977). Implicit motor learning avoids the serial transition of explicit declarative knowledge to implicit procedural knowledge by encouraging implicit processes to support performance and implicit procedural knowledge to develop from the onset of learning.

In the cognitive psychology literature, tasks with simple motor components, such as finger tapping or continuous motor tracking, have been used to demonstrate that knowledge pertaining to a task can be extracted without an individual's awareness of what has been learnt or even a conscious attempt to learn. For example, the serial reaction time (SRT) task asks participants to respond as quickly as possible to stimuli presented in one of four positions on a computer monitor by pressing one of four keys with the right hand (see Nissen & Bullemer, 1987). Unbeknownst to participants, the order in which the stimuli are presented is a fixed sequence that repeats many times during practice. Most participants do not become aware of the repeating sequence, yet shortened response times suggest that parts of the sequence order have been learned. This is interpreted as evidence of implicit learning (e.g., Cohen, Ivry & Keele, 1990; Jiminez & Mendez, 1999; Nissen & Bullemer, 1987). Continuous motor tracking tasks ask participants to track a stimulus that follows a waveform pattern horizontally across a computer screen by manipulating a hand held device. Embedded within two random segments of the waveform pattern is a segment that recurs on every practice trial. Similar to the SRT task, participants are typically unaware of a repeating element within the task and are unable to recognize or describe the pattern of the waveform when prompted, yet superior tracking accuracy of the repeated segment compared to the random segments implies that parts of the waveform have been learned implicitly (e.g., Pew, 1974; Wulf & Schmidt, 1997).

In an advancement of this work, over the past two decades, Masters and his colleagues have set about curtailing verbal-analytical aspects of working memory activity when the to-be-learnt motor skill requires the coordination of multiple body segments, and when learners are aware of the specific learning objective, as is the case when golfers practice their short game or when surgeons prepare for a technically complex laparoscopic procedure. At some stage, the motor skill will be performed under challenging conditions, such as psychological pressure, that can result in skill failure. Implicit motor leaning was proposed by Masters (1992) as a means of effectively combating such challenges. Empirical evidence has consistently shown that, compared to explicitly learned motor skills, implicitly learned motor skills are characterized by stable motor performance under psychological stress (e.g., Hardy, Mullen & Jones, 1996; Lam, Maxwell & Masters, 2009; Law, Masters, Bray, Eves & Bardswell, 2003; Liao & Masters, 2001; Masters, 1992; Mullen, Hardy & Oldham, 2007), when multitasking (e.g., Maxwell et al., 2003; Masters, Poolton, Maxwell & Raab, 2008b; Maxwell et al., 2001; Poolton, Masters & Maxwell, 2006), and after prolonged and acute exhaustive exercise (e.g., Masters, Poolton & Maxwell, 2008a; Poolton, Masters & Maxwell, 2007).

The initial attempt to suppress verbal-analytical processing utilized a dual-task learning paradigm (Masters, 1992). Novice golfers were required to practice a golf putt while simultaneously performing a random letter generation task known to place high demands on the central executive and phonological components of working memory (Baddeley, 1986). Putting performance improved, but learners' task-relevant declarative knowledge appeared limited, which was interpreted by Masters as evidence of implicit motor learning. However, suppressing working memory in this manner resulted in slower learning rates than traditional explicit learning techniques (Hardy et al. 1996; MacMahon & Masters, 2002; Maxwell et al., 2000, 2003). In order to solve this problem, several more practical learning paradigms have been developed.

The errorless learning paradigm, for example, was developed upon the premise that verbal-analytical processes are active in error correction. Learning environments are, therefore, designed to significantly reduce the number of outcome errors. For example, Maxwell et al. (2001) structured golf putting practice so that it was virtually impossible to miss early in learning (25cm, 50cm, 75cm putts). Without errors to correct, verbal-analytical aspects of working memory play a subordinate role in the motor learning process. As a result, and as evidence of implicit motor learning, learners that follow an errorless learning paradigm have the residual working memory capacity to proficiently complete cognitively demanding secondary tasks without them impinging on motor performance (Capio, Poolton, Sit, Eguia & Masters, 2012; Capio, Poolton, Sit, Holmstrom & Masters, 2011; Lam, Maxwell & Masters, 2010; Maxwell et al., 2001; Poolton, Masters & Maxwell, 2005; Poolton et al., 2007).

A second paradigm, analogy learning, is designed to avoid working memory activity by packaging rules related to the mechanics of a skill in a single analogical instruction. For example, table-tennis forehands are coached by instructing novices to strike the ball as the bat moves up the hypotenuse of an imaginary right-angle triangle (Liao & Masters, 2001; Masters, 2000). The fundamental mechanics of the skill are thus conveyed to the learner effectively without the need for additional verbal instruction or hypothesis testing (i.e., without additional working memory activity), freeing processing capacity for other task-relevant tasks, if required (e.g. decision making; Masters et al., 2008b; Poolton et al., 2007).

Promoting psychomotor efficiency from the onset of motor learning

Since complex motor skills associated with sports activities are difficult to study by neuro-imaging technology, neuroscientists have used more fundamental motor tasks, such as the SRT task, over the past two decades to examine the different neural mechanisms that underpin implicit and explicit motor learning. For example, Grafton, Hazeltine and Ivry (1995) used positron emission tomography (PET) to examine learning-related changes in neural activation following SRT task practice. In a within-subject design, participants completed both an explicit learning session, in which they were made aware of the recurring sequence, and an implicit learning session, in which they were not. Following the implicit learning session, learning-related increases in blood flow were present in contralateral motor effector areas, including the motor cortex, supplementary motor area and putamen. During the explicit learning session, learning-related increases in blood flow were displayed in the right dorsolateral prefrontal cortex, right premotor cortex, right ventral putamen and biparieto-occipital cortex. Hazeltine, Grafton and Ivry (1997) reported similar findings in an SRT study that cued stimuli by color rather than spatial position. This work implies that implicit learning of finger-tapping sequences occurs in a different cerebral network than explicit learning (see also Rauch, Savage, Brown et al., 1995). However, Willingham, Salidis and Gabrieli (2002) argued that the dissociating pattern of neural activity found might be confounded by an order effect of measuring implicit and explicit learning-related activation sequentially. Willingham et al. (2002), therefore designed an SRT task paradigm that permitted implicit and explicit learning of response sequences to occur more concurrently, and so allowed a more direct comparison of learning-related changes in neural activation. Participants were explicitly informed that a repeating sequence would be presented in red circles and that a random sequence would be presented in black circles. Unbeknownst to participants, a second repeating sequence was actually hidden in the black circles. Functional magnetic resonance imaging (fMRI) analysis showed that implicit learning and explicit learning exhibited activation overlap among some brain regions, such as the left inferior frontal gyrus, right putamen, and left inferior parietal cortex. Nevertheless, many areas were relatively more active when participants were performing the repeating sequence that they were aware of, including the bilateral superior parietal cortex, bilateral cerebellar cortex, brain stem, bilateral middle frontal and inferior frontal gyri, cingulate gyrus, right caudate, and bilateral premotor cortex. Greater activation was interpreted to reflect conscious control over motor sequencing performance or a greater cognitive load.

Schendan, Searl, Melrose and Stern (2003) conducted an fMRI comparison of the medial temporal lobe (MTL) of participants who had learnt an SRT sequence in an implicit learning session or an explicit learning session. The MTL is associated with the working memory dependent retrieval of task-relevant facts and events from long-term memory (e.g., Clark & Squire, 1998; Schacter, 1997; Squire, 1992) and, therefore, Schendan et al. (2003) hypothesized that MTL would be more active when learning was explicit compared to when it was implicit. The fMRI analysis was consistent with this hypothesis, suggesting that processes governed by working memory were less active when learning was implicit. Further empirical evidence that implicit and explicit learning can be differentiated by MTL activity is provided by neuropsychological studies of patients with neurological illness or brain injury. For example, patients with MTL dysfunction (e.g., amnesic patients) demonstrate impaired explicit learning of SRT task stimulus sequences, yet, implicit learning remains intact (e.g, Nissen & Bullemer, 1987; Reber & Squire, 1994, 1998). In contrast, patients with basal ganglia dysfunction (e.g., Parkinson's disease and Huntington's disease) demonstrate impaired implicit learning of stimulus sequences, signaling the significance of the basal ganglia to implicit motor learning (e.g., Ferraro, Balota & Connor, 1993; Knopman & Nissen, 1991; Smith & McDowall, 2004; Willingham & Koroshetz, 1993). Such studies provide strong evidence to suggest that implicit and explicit learning of motor sequences is dependent on different neural structures.

The electroencephalographic investigation of cortico-cortical networks associated with implicit and explicit motor learning has largely centered on co-activation or 'synched' communication between the left temporal region (T3), associated with verbal-analytical processing (e.g., Haufler et al., 2000; Kerick et al., 2001), and the frontal midline region (Fz), associated with motor planning (Kaufer & Lewis, 1999). Rietschel, King, Pangelinan, Clark and Hatfield (2008) conducted a T3–Fz coherence analysis during a visuo-motor task, in which there was a 30-degree distortion between visual feedback and actual movement. Participants who were explicitly informed about the distortion and how to compensate their movements displayed higher T3–Fz coherence during the task after practice than a group who were not given this information, implying that the provision of instructions increases verbal-analytical involvement in motor performance.

Zhu et al. (2011a) reported comparable findings using a continuous tracking task paradigm (e.g., Pew, 1974; Wulf & Schmidt, 1997) that had been modified for surgical skills training. Participants tracked a target dot that followed a waveform pattern horizontally across a monitor screen. Tracking required bimanual manipulation of laparoscopic graspers to move a stylus across a tablet housed in a laparoscopic training box (Figure 8.2b). An explicit learning group was informed about the existence of a repeating middle segment (Figure 8.2a) and was frequently shown a pictorial representation of the pattern (e.g., Boyd & Winstein, 2006). An implicit learning group was not informed about the existence of a repeating segment and gave no indication of awareness of the repeating segment at the end of the experiment. In the retention test that followed practice, cortical activity was recorded and alpha band coherence between Fz and both T3 and T4 (the right temporal region associated with visuo-spatial processing) was calculated. The implicit and explicit motor learning groups showed comparable T4–Fz coherence,

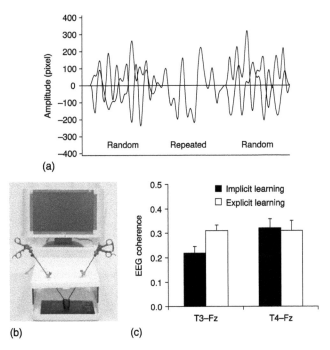

Figure 8.2 (a) A visual representation of waveform patterns from two exemplar trials that participants tracked during practice trials. The middle (repeated) segment was recurring in every trial and embedded between two (random) segments that differed in every trial. (b) Laparoscopic skills training set-up. Participants tracked a target dot across a monitor screen by bimanually moving a pen (represented on the screen by a cursor) held by laparoscopic graspers across a pen tablet housed in a laparoscopic training box. (c) Electroencephalographic alpha T3–Fz and T4–Fz coherence in the retention test for participants in the implicit motor learning group and participants in the explicit motor learning group (mean ± standard error) [adapted from Zhu et al. (2011a)].

suggesting that visuo-spatial processing was essential for tracking performance. However, the T3–Fz coherence exhibited by the explicit motor learning group was significantly higher than that of the implicit learning group (Figure 8.2c), suggesting that verbal-analytical processes contributed less to motor control of movement patterns that had been learnt implicitly.

In the first neurophysiological examination of the cortical activity underlying paradigms developed by Masters and his colleagues for the implicit learning of more 'real-world' motor skills, Zhu et al. (2011b) analyzed cortico-cortical activity during golf putts (Figure 8.3b) that were made in a retention test a day after a practice session. Implicit motor learning was encouraged by an established errorless learning protocol for golf putting that ensured errors were infrequent, particularly during the initial period of learning. Learners began very close to the target hole, just 25 cm, and moved sequentially to 50, 75, 100, 125 and 150 cm after 50 practice putts at each distance. Learners in the explicit motor learning condition practiced the same number of putts from the same distances but in a pseudorandom order (i.e., 125, 100, 25, 75, 50, and 150 cm), which allowed errors to more frequently occur. Coherence analysis between the T3–Fz and the T4–Fz cortical regions in alpha1 (8–10 Hz) and alpha2 (10–12 Hz) frequency bandwidths was computed across the four second period prior to ball strike. Compared to explicit motor learning (frequent errors) the analysis indicated that implicit motor learning (infrequent errors) resulted in a pattern of lowered co-activation between the regions. More

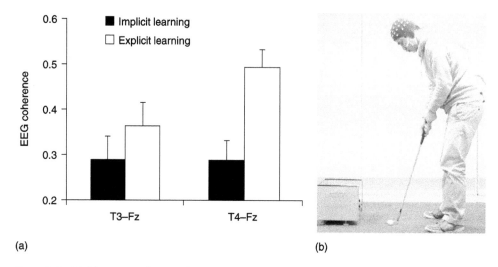

Figure 8.3 (a) Electroencephalographic alpha2 T3–Fz and T4–Fz coherence across the four second
period prior to ball strike in the retention test for participants in the errorless (implicit)
motor learning group and participants in the errorful (explicit) motor learning group
(mean ± standard error). (b) Golf putting set-up. Participants wore a stretchable elec-
trode-cap (ElectroCap Inc., USA) that was physically connected to a NeuroTop EEG
system (Symtop Instruments, China). A golf ball dispenser (Hono Golf, China) presented
standard white golf balls and was fitted with an infrared sensor that detected each ball strike
and marked it on the EEG record [adapted from Zhu et al. (2011b)].

specifically, compared to explicit motor learning, implicit motor learning resulted in lowered
alpha2 T4–Fz coherence, implying attenuated visual-spatial involvement in motor performance and lowered alpha2 T3–Fz coherence, implying attenuated verbal-analytical activity,
consistent with the conceptualization of implicit motor learning paradigms (Figure 8.3a).

Collectively, the neurophysiological study of implicit motor learning suggests that implicit
and explicit learning paradigms cause different adaptations at a neural level. Implicit motor
learning appears to result in greater 'psychomotor efficiency' (see Hatfield & Hillman, 2001),
in the sense that verbal-analytical processes essential for more traditional forms of learning
are used more sparingly and have residual capacity with which to handle other challenges
associated with motor performance (e.g., decision making, see Masters et al., 2008b; Poolton
et al., 2007). Thus, the pruning of neural regions that are nonessential to proficient motor
performance, which is normally the product of extended practice and expertise, begins from
the onset of learning. Furthermore, advocators of implicit motor learning propose that the
neural adaptations that go on to support motor control are less susceptible to changes in the
pattern of cortical activity potentially caused by common performance contingencies (e.g.,
psychological stress) and so implicit motor learning promotes robust motor performances.

The reintroduction of verbal-analytical processes to motor performance

Although traditional stage-models of learning argue that verbal-analytical aspects of working
memory activity are gradually withdrawn during learning, the declarative knowledge accrued
remains available for retrieval, manipulation and application by working memory (see Baumeister,
1984; Masters, 1992; Masters, Polman & Hammond, 1993; Norman & Shallice, 1986;
Willingham, 1998). The retrieval of declarative knowledge might be considered useful if the

performer wants to make strategic adjustments to his or her motor skill, such as to make whole-sale alterations to a technique (e.g., re-build a golf swing), or protect an injury (Beek, 2000; Masters & Maxwell, 2004). However, when the motor skill is relatively automated the reintro-duction of working memory dependent processes, such as the retrieval and application of declar-ative knowledge, to support on-line motor control, may be less desirable (Baumeister, 1984; Masters, 1992; Masters et al., 1993; Masters & Maxwell, 2004, 2008; Masters & Poolton 2012;).

Masters and Maxwell (2008) describe this phenomenon in the Theory of Reinvestment and detail wide-ranging empirical evidence that demonstrates the disruptive influence that the reintroduction of verbal-analytical processes to motor control has on performance (e.g., Hardy, Mullen & Martin, 2001; Jackson, Ashford & Nosworthy, 2006; Maxwell, Masters & Poolton, 2006; Mullen, Hardy & Tattersall, 2005; Mullen et al., 2007; Pijpers, Oudejans & Bakker, 2005; Poolton, Maxwell & Masters, 2004; Wan & Huon, 2005). For example, Gray (2004) showed that monitoring the position of the bat caused expert baseball batters to make greater temporal timing errors (see also Beilock, Bertenthal, McCoy & Carr, 2004; Ford, Hodges & Williams, 2005; Jackson et al., 2006). Raised anxiety induced by performance pressure is one contingency believed to evoke increased working memory involvement in motor control (e.g., Baumeister, 1984; Beilock & Carr, 2001; Masters, 1992; Masters et al., 1993). The importance of proficient performance causes individuals to allocate all available resources to the task at hand in what becomes an ironic attempt to ensure success (Baumeister, 1984). For example, the requirement to traverse a 5 m high section of a climbing wall, as opposed to a 0.4 m high section, caused changes in climbing behavior indicative of working memory involvement, including an increased number of exploratory movements, slower movements and longer grasping of the hand-holds, and culminated in increased traverse time (Pijpers et al., 2005).

The Theory of Reinvestment also argues that the propensity to reintroduce verbal-analytical processes into motor control or reinvestment is a dimension of personality that can be quantified psychometrically using specifically developed Reinvestment Scales (Masters et al., 1993; Masters, Eves & Maxwell, 2005), which are predictive of skill breakdown under pressure (Chell, Graydon, Crowley & Child, 2003; Jackson et al., 2006; Masters et al., 1993; Maxwell et al., 2006; Poolton et al., 2004). Besides psychological pressure, many other contingencies are proposed to trigger reinvestment, including the availability of too much time for movement preparation (Singer, 2002; Liao & Masters, 2002; Wilson, Smith & Holmes, 2007), injury or accident (Laufer, Roten-Lehrer, Ronen, Khayutin & Rozenberg, 2007; Wong, Masters, Maxwell & Abernethy, 2008), aerobic and anaerobic fatigue (Masters et al., 2008a; Poolton et al., 2007), movement disorders (Masters, Pall, MacMahon & Eves, 2007; Orrell, Masters & Eves, 2009), unexpected events prior to or during performance (Rotheram, Thomas, Bawden & Maynard, 2006), performance history (Gray, 2004), and the need to adapt to different conditions (Beilock & Carr, 2001). Recent neurophysiological studies that have examined the cortical activity changes brought about by the introduction of such contingencies are generally supportive of the Theory of Reinvestment (e.g., Baumeister, Reinecke, Schubert, Schade & Weiss, 2012; Baumeister, Reinecke, Schubert & Weiss, 2011; Baumeister, Reinecke & Weiss, 2008b; Chen et al., 2005; Zhu et al., 2011a, 2011b).

Cortico-cortical coherence changes in response to psychological stress manipulations have been shown in far-aiming motor tasks. Chen et al. (2005), for example, showed that disrupted dart throwing performance under pressure was accompanied by increased co-activation or increased communication, between verbal-analytical processing and motor planning cortical regions (higher T3–Fz coherence). Similarly, Hatfield et al. (2009) found that, compared to a low-stress condition, a competitive high-stress condition resulted in higher T3–Fz coherence levels within a group of skilled marksmen. Zhu et al. (2011b) subjected trained golf putters to an anxiety invoking evaluation apprehension manipulation. The manipulation had no significant effect on alpha2 T4–Fz coherence, implying that anxiety did not induce changes in visual-spatial involvement in motor planning. Nevertheless, the manipulation significantly heightened alpha2 T3–Fz coherence, supporting predictions by the Theory of Reinvestment that stress increases verbal-analytical involvement in motor planning. Of significant note was the finding that golf putters trained via an errorless (implicit) motor learning paradigm showed no such change in alpha2 T3–Fz coherence in response to the stress manipulation (Figure 8.4). The critical practical implication of this finding is that implicit motor learning installs resilience to cortical activity pattern changes associated with the reintroduction of verbal-analytical processes under pressure. One explanation for this resilience is that implicit motor learning results in limited declarative knowledge stored in long-term memory; thus, implicit learners are without a declarative knowledge pool from which to retrieve and apply knowledge to control movement should they choose to. Alternatively, suppressed verbal-analytical involvement in the learning process may be co-related with the suppression of what can be a strong inclination to reintroduce verbal-analytical process to motor control, especially under conditions that demand successful outcomes. This inherent ability to combat reinvestment is likely a decisive factor in the consistent performance robustness under pressure that has been shown for implicitly learned motor skills (e.g., Hardy et al., 1996; Koedijker, Ouedejans & Beek, 2007; Lam et al., 2009; Law, Masters, Bray, Eves & Bardswell, 2003; Liao & Masters, 2001; Masters, 1992).

Neurophysiological evidence suggesting that the propensity to (re)introduce verbal-analytical processes into motor control (or reinvest) is a dimension of personality was also reported by Zhu et al. (2011b). Novice golfers were invited to participate in the study if they scored one standard deviation above or below the population mean (over 200 responses) on the *conscious motor processing* factor of the Movement Specific Reinvestment Scale (MSRS, Masters et al., 2005; Masters & Maxwell, 2008). Cortical activity patterns were recorded during 10 putts. Zhu et al. (2011b) showed that high and low conscious motor processors could not be differentiated by alpha2 T4–Fz coherence, but as expected, could be differentiated by alpha2 T3–Fz coherence (Figure 8.5). Compared to participants with low scores, and as predicted by the Theory of Reinvestment, participants with high scores displayed the

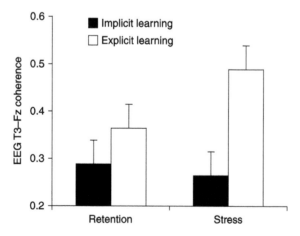

Figure 8.4 Electroencephalographic alpha2 T3–Fz coherence across the four second period prior to ball strike in the retention test and psychological stress transfer test for participants in the errorless (implicit) motor learning group and participants in the errorful (explicit) motor learning group (mean ± standard error) [adapted from Zhu et al. (2011b)].

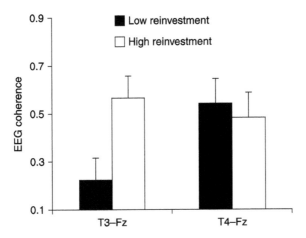

Figure 8.5 Electroencephalographic alpha2 T3–Fz and T4–Fz coherence during putting performance for participants with low and high conscious motor processing scores [adapted from Zhu et al. (2011b)].

higher alpha2 T3–Fz that implies greater verbal-analytical processing involvement in motor control and learning.

Baumeister et al. (2012) examined the effect that exercise-induced fatigue had on cortical activation patterns during a knee angle reproduction task. Cortical activity was measured using EEG methodology prior to, immediately after, and 60 minutes after prolonged (> 1 hour) and exhaustive exercise. The exercise protocol negatively impacted knee angle reproduction accuracy. Moreover, sensorimotor deficiencies were accompanied by cortical activity changes, including a decrease in frontal theta power and a decrease in parietal alpha2 power. A decrease in frontal theta power implies ineffective attention control (Doppelmayr, Finkenzeller & Sauseng, 2008; Gevins et al., 1997; Posner & Rothbart, 2007) and heightened working memory activity (Baumeister et al., 2008a), while a decrease in parietal alpha2 power implies increased information processing in the somatosensory areas to react to changes in afferent feedback caused by fatigue (Neuper, Wortz & Pfurtscheller, 2006; Pfurtscheller, 1992). Cortical activity patterns and sensorimotor function were restored to pre-exercise baseline levels after a 60-minute recovery period. Heightened working memory activity caused by exhaustive exercise might reflect the reintroduction of verbal-analytical processes to motor control that can be detrimental to motor performance. Recently, Masters and his colleagues showed that implicit motor learning curtailed the negative impact of fatigue on performance (Masters et al., 2008a; Poolton et al., 2007), possibly because fatigue did not trigger an increase in working memory's direct involvement in motor performance (Poolton et al., 2007). Participants practiced a rugby passing accuracy task following an infrequent error (implicit) or frequent error (explicit) learning paradigm and then passing accuracy was assessed before and after an anaerobic fatigue protocol (Poolton et al., 2007). In a second study, accuracy was assessed before and after an aerobic fatigue protocol (Masters et al., 2008a). On both occasions, fatigue resulted in detrimental passing accuracy when learning was explicit (frequent errors). In contrast, when learning was 'errorless' (infrequent errors), fatigue had no significant impact on passing accuracy.

In an earlier study, Baumeister and his colleagues (2008b) used the knee angle reproduction task to identify cortical activity adaptations following anterior cruciate ligament (ACL) reconstruction. Compared with healthy controls, ACL patients showed larger knee angle reproduction error when using their reconstructed limb, whereas there was no between-group difference when ACL patients used their non-reconstructed knee. During angle reproduction, ACL patients exhibited higher frontal theta power than healthy controls when the task was performed with the reconstructed and non-reconstructed limb, and lower parietal alpha2 power when using the reconstructed limb. Taken together, the pattern of cortical activity was considered to reflect heighted central executive activity in working memory, possibly reflecting the high demands associated with coordinating a rehabilitating body segment and processing afferent feedback. Interestingly, the greater neuronal resources devoted to the motor task by ACL patients were not restricted to the reconstructed limb, implying that the process of injury, surgical reconstruction and rehabilitation has a psycho-motor impact beyond overcoming altered coordination of the involved limb. Subseqently, Baumeister et al. (2011) also showed higher frontal theta power in ACL patients, compared to healthy controls, when using the reconstructed limb to perform a force reproduction task. These findings suggest that the treatment of an injury should take into account not only func-tional muscle deficits, but also changes in cortical activity patterns caused by the injury. This contention holds for the treatment of the elderly after a fall, for rehabilitation programs for stroke patients and for management of Parkinson's disease. Each of these groups has shown a tendency for raised predispositions for conscious control of movement (implied psychometri-cally using the Movement Specific Reinvestment Scale, Masters & Maxwell, 2008; Wong et al., 2008 etc.), while having to overcome cognitive deficits (e.g., Hochstenbach, Van Spaendonck, Cools, Horstink & Mulder, 1998; Hoyer & Verhaeghen, 2006). Implicit motor learning may rapidly return cortical activity patterns to pre-incident configurations by suppressing working memory activity and promoting psychomotor efficiency. Implicit motor learning research has been extended to the elderly and to patients with movement disorders (e.g., Parkinson's disease or stroke). For the most part, implicit motor learning paradigms have transferred well, overcoming age-related deficits in motor learning (Chauvel, Maquestiaux, Hartley, Joubert, Didierjean & Masters, 2011) and freeing the working memory resources of Parkinson's disease patients (Masters, MacMahon & Pall, 2004) and stroke patients (Orrell, Eves & Masters, 2006) to allow proficient multitasking (e.g., walking and listening). Therapists might consider incorporating implicit motor learning elements into rehabilitation programs designed for patients recovering from ACL reconstruction, in an effort to deter changes to cortical activity patterns (Baumeister et al., 2008b) that may reduce the efficiency of activities of daily living.

Summary and future directions

Evidence shows that motor skills can be acquired implicitly without the performer becoming consciously aware of much of the knowledge that underlies effective performance. Implicit motor learning approaches are designed to suppress working memory involvement in the skill acquisition process so that declarative knowledge about the skill is not created or at least is inaccessible to conscious awareness. Compared to explicitly learned motor skills, implicitly learned motor skills appear to be characterized by stable performance despite stress, fatigue and multitasking. Recently, this evidence base has been strengthened by neuroscientific find-ings in which cortical activity during motor performance has been measured using EEG technology. The findings suggest that implicit motor learning can result in different cortical activity compared to explicit motor learning. Additionally, there is a small amount of neuro-scientific evidence that hints at the possibility that cortical activity, in the form of coherence between motor and verbal-analytical regions of the brain, differs as a function of the proclivity that a person has for conscious motor processing (reinvestment) during performance. Recent advances in wireless EEG technology present the possibility to examine implicit aspects of motor performance in domains beyond sport, in which performers must also cope with conditions of psychological stress, multi-tasking or physiological fatigue. For example, the opportunity to examine cortical aspects of motor performance during live surgical procedures, in which such conditions are commonplace, is made feasible by the unobtrusive nature of wireless EEG technology. In the very near future, surgical educators in the operating theatre may be able to see the cortical impact of multi-tasking on the brain activity of trainees during a procedure, or coaches may be able to objectively gauge the level of conscious motor processing (reinvestment) of players during the heat of the game, perhaps even when making a penalty-kick in the World Cup finals, and physiotherapists may be able to assess the degree to which their rehabilitation techniques achieve motor (re)learning that is explicit or implicit.

References

Adams, J.A. (1971) 'A closed-loop theory of motor learning', *Journal of Motor Behavior*, 3: 111–50.

Anderson, J.R. (1983) *The architecture of cognition*. Cambridge, MA: Harvard University Press.

Andersen, J.R. and Lebiere, C.J. (1998) *The atomic components of thought*. Mahwah, NJ: Lawrence Erlbaum Associates.

Baddeley, A.D. (1986) *Working memory*. Oxford: Oxford University Press.

Baddeley, A.D. (1996) 'Exploring the central executive', *Quarterly Journal of Experimental Psychology Section A: Human Experimental Psychology*, 49: 5–28.

Baddeley, A.D. (1999) *Essentials of human memory*. Hove, England: Psychological Press.

Baddeley, A.D. (2000) 'The episodic buffer: A new component of working memory?' *Trends in Cognitive Sciences*, 4: 417–23.

Baddeley, A.D. (2007) *Working memory, thought and action*. Oxford: Oxford University Press.

Baddeley, A.D. and Hitch, G.J. (1974) 'Working memory', in G.H. Bower (Ed.) *The psychology of learning and motivation* (8, 47–89), New York: Academic Press.

Baumeister, R.F. (1984) 'Choking under pressure: Self-consciousness and paradoxical effects of incentives on skillful performance', *Journal of Personality and Social Psychology*, 46: 610–20.

Baumeister, J., Reinecke, K., Liesen, H. and Weiss, M. (2008a) 'Cortical activity of skilled performance in a complex sports related motor task', *European Journal of Applied Physiology*, 104: 625–31.

Baumeister, J., Reinecke, K., Schubert, M., Schade, J. and Weiss, M. (2012) 'Effects of induced fatigue on brain activity during sensorimotor control', *European Journal of Applied Physiology*, 112: 2475–82.

Baumeister, J., Reinecke, K., Schubert, M. and Weiss, M. (2011) 'Altered electrocortical brain activity after ACL reconstruction during force control', *Journal of Orthopaedic Research*, 29: 1383–9.

Baumeister, J., Reinecke, K. and Weiss, M. (2008b) 'Changed cortical activity after anterior cruciate ligament reconstruction in a joint position paradigm: An EEG study', *Scandinavian Journal of Medicine and Science in Sports*, 18: 473–84.

Beek, P.J. (2000) 'Toward a theory of implicit learning in the perceptual-motor domain', *International Journal of Sport Psychology*, 31: 547–54.

Beilock, S.L., Bertenthal, B.I., McCoy, A.M. and Carr, T.H. (2004) 'Haste does not always make waste: Expertise, direction of attention, and speed versus accuracy in performing sensorimotor skills', *Psychonomic Bulletin and Review*, 11: 373–9.

Beilock, S.L. and Carr, T.H. (2001) 'On the fragility of skilled performance: What governs choking under pressure?' *Journal of Experimental Psychology: General*, 130: 701–25.

Bell, M.A. and Fox, N.A. (1996) 'Crawling experience is related to changes in cortical organization during infancy: Evidence from EEG Coherence', *Developmental Psychobiology*, 29: 551–61.

Botvinick, M.M., Carter, C.S., Braver, T.S., Barch, D.M. and Cohen, J.D. (2001) 'Conflict monitoring and cognitive control', *Psychological Review*, 108: 624–52.

Boyd, L.A. and Winstein, C.J. (2006) 'Explicit information interferes with implicit motor learning of both continuous and discrete movement tasks after stroke', *Journal of Neurologic Physical Therapy*, 30: 46–57.

Busk, J. and Galbraith, G.C. (1975) 'EEG correlates of visual-motor practice in man', *Electroencephalography and Clinical Neurophysiology*, 38: 415–22.

Capio, C.M., Poolton, J.M., Sit, C.H.P., Eguia, K.F. and Masters, R.S.W. (2012) 'Reduction of errors during practice facilitates fundamental movement skill learning in children with intellectual disabilities', *Journal of Intellectual Disability Research*, doi: 10.III/j.1365-2788.2012.01535.x.

Capio, C.M., Poolton, J.M., Sit, C.H.P., Holmstrom, M. and Masters, R.S.W. (2011) 'Reducing errors benefits the field-based learning of a fundamental movement skill in children', *Scandinavian Journal of Medicine and Science in Sports*. doi: 101111/j.1600-0838.2011.01368.x.

Carter, C.S., Braver, T.S., Barch, D.M., Botvinick, M.M., Noll, D. and Cohen, J.D. (1998) 'Anterior cingulate cortex, error detection, and the online monitoring of performance', *Science*, 280: 747–9.

Chauvel, G., Maquestiaux, F., Hartley, A.A., Joubert, S., Didierjean, A. and Masters, R.S.W. (2012) 'Age effects shrink when motor learning is predominantly supported by nondeclarative, automatic memory processes: evidence from golf putting', *Quarterly Journal of Exerimental Psychology*, 65: 25–38.

Chell, B.J., Graydon, J.K., Crowley, P.L. and Child, M. (2003) 'Manipulated stress and dispositional reinvestment in a wall-volley task: An investigation into controlled processing', *Perceptual and Motor Skills*, 97: 435–48.

Chen, J., Hung, T., Lin, J., Lo, L., Kao, J., Hung, C., Chen, Y. and Lai, Z. (2005) 'Effects of anxiety on EEG coherence during dart throwing', paper presented at International Society of Sport Psychology World Congress, Sydney, August.

Clark, R.E. and Squire, L.R. (1998) 'Classical conditioning and brain systems: The role of awareness', *Science*, 280: 77–81.

Cohen, R.A. (1993) *The neuropsychology of attention*. New York: Plenum Press.

Cohen, A., Ivry, R.I. and Keele, S.W. (1990) 'Attention and structure in sequence learning', *Journal of Experimental Psychology: Learning, Memory, and Cognition*, 16: 17–30.

Crews, D.J. and Landers, D.M. (1993) 'Electroencephalographic measures of attentional patterns prior to the golf putt', *Medicine and Science in Sports and Exercise*, 25: 116–26.

Deeny, S.P., Haufler, A.J., Saffer, M. and Hatfield, B.D. (2009) 'Electroencephalographic coherence during visuomotor performance: A comparison of cortico-cortical communication in experts and novices', *Journal of Motor Behavior*, 41: 106–16.

Deeny, S.P., Hillman, C.H., Janelle, C.M. and Hatfield, B.D. (2003) 'Cortico-cortical communication and superior performance in skilled marksmen: An EEG coherence analysis', *Journal of Sport and Exercise Psychology*, 25: 188–204.

Doppelmayr, M., Finkenzeller, T. and Sauseng, P. (2008) 'Frontal midline theta in the pre-shot phase of rifle shooting: Differences between experts and novices', *Neuropsychologia*, 46: 1463–7.

Ferraro, F.R., Balota, D.A. and Connor, L.T. (1993) 'Implicit memory and the formation of new associations in nondemented parkinson's disease individuals and individuals with senile dementia of the Alzheimer type: A Serial reaction time (SRT) investigation', *Brain and Cognition*, 21: 163–80.

Fitts, P.M. and Posner, M.I. (1967) *Human Performance*. Belmont, CA: Brooks/Cole.

Ford, P., Hodges, N.J. and Williams, A.M. (2005) 'Online attentional-focus manipulations in a soccer-dribbling task: Implications for the proceduralization of motor skills', *Journal of Motor Behavior*, 37: 386–94.

Gevins, A., Smith, M.E., McEvoy, L. and Yu, D. (1997) 'High-resolution EEG mapping of cortical activation related to working memory: Effects of task difficulty, type of processing, and practice', *Cerebral Cortex*, 7: 374–85.

Grafton, S.T., Hazeltine, E. and Ivry, R. (1995) 'Functional mapping of sequence learning in normal humans', *Journal of Cognitive Neuroscience*, 7: 497–510.

Gray, R. (2004) 'Attending to the execution of a complex sensorimotor skill: Expertise differences, choking, and slumps', *Journal of Experimental Psychology: Applied*, 10: 42–54.

Hardy, L., Mullen, R. and Jones, G. (1996) 'Knowledge and conscious control of motor actions under stress', *British Journal of Psychology*, 86: 621–36.

Hardy, L., Mullen, R. and Martin, N. (2001) 'Effect of task-relevant cues and state anxiety on motor performance', *Perceptual and Motor Skills*, 92: 943–6.

Hatfield, B., Haufler, A. and Contreras-Vidal, J. (2009) 'Brain processes and neurofeedback for performance enhancement of precision motor behavior', *Lecture Notes in Computer Science*, 5638: 810–17.

Hatfield, B.D. and Hillman, C.H. (2001) 'The psychophysiology of sport: A mechanistic understanding of the psychology of superior performance', in R. Ringer, H. Hausenblas and C. Janelle (Eds.) *Handbook of sport psychology*, New York: Wiley & Sons.

Hatfield, B.D., Landers, D.M. and Ray, W.J. (1984) 'Cognitive processes during self-paced motor performance: An electroencephalographic profile of skilled marksmen', *Journal of Sport Psychology*, 6: 42–59.

Haufler, A.J., Spalding, T.W., Santa Maria, D.L. and Hatfield, B.D. (2000) 'Neuro-cognitive activity during a self-paced visuospatial task: Comparative EEG profiles in marksmen and novice shooters', *Biological Psychology*, 53: 131–60.

Hazeltine, E., Grafton, S.T. and Ivry, R. (1997) 'Attention and stimulus characteristics determine the locus of motor-sequence encoding. A PET study', *Brain*, 120: 123–40.

Hochstenbach, J., Van Spaendonck, K.P.M., Cools, A.R., Horstink, M.W.I.M. and Mulder, T. (1998) 'Cognitive deficits following stroke in the basal ganglia', *Clinical Rehabilitation*, 12: 514–20.

Hoyer, W.J. and Verhaeghen, P. (2006) 'Memory aging', in J.E. Birren and K.W. Schaie (Eds.) *Handbook of the psychology of aging*, San Diego, CA: Academic Press.

Jackson, R.C., Ashford, K.J. and Norsworthy, G. (2006) 'Attentional focus, dispositional reinvestment, and skilled motor performance under pressure', *Journal of Sport and Exercise Psychology*, 28: 49–68.

Janelle, C.M., Duley, A.R. and Coombes, S.A. (2004) 'Psychophysiological and related indices of attention during motor skill acquisition', in A.M. Williams and N.J. Hodges (Eds.) *Skill acquisition in sport: Research, theory and practice*, London: Routledge.

Jiménez, L. and Méndez, C. (1999) 'Which attention is needed for implicit sequence learning?' *Journal of Experimental Psychology: Learning Memory and Cognition*, 25: 236–59.

Karni, A., Meyer, G., Rey-Hipolito, C., Jezzard, P., Adams, M.M., Turner, R. and Ungerleider, L.G. (1998) 'The acquisition of skilled motor performance: Fast and slow experience-driven changes in primary motor cortex', *Proceedings of the National Academy of Sciences of the United States of America*, 95: 861–8.

Kaufer, D.I. and Lewis, D.A. (1999) 'Frontal lobe anatomy and cortical connectivity', in B. Miller and J. Cummings (Eds.) *The human frontal lobes*, New York: Guilford Press.

Keele, S.W. (1968) 'Movement control in skilled motor performance', *Psychological Bulletin*, 70: 283–304.

Kerick, S.E., Douglass, L.W. and Hatfield, B.D. (2004) 'Cerebral cortical adaptations associated with visuomotor practice', *Medicine and Science in Sports and Exercise*, 36: 118–29.

Kerick, S.E., McDowell, K., Hung, T.-M., Santa Maria, D.L., Spalding, T.W. and Hatfield, B.D. (2001) 'The role of the left temporal region under the cognitive motor demands of shooting in skilled marksmen', *Biological Psychology*, 58: 263–77.

Knopman, D. and Nissen, M.J. (1991) 'Procedural learning is impaired in Huntington's disease: Evidence from the serial reaction time task', *Neuropsychologia*, 29: 245–54.

Koedijker, J.M., Oudejans, R.R.D. and Beek, P.J. (2007) 'Explicit rules and direction of attention in learning and performing the table tennis forehand', *International Journal of Sport Psychology*, 38: 227–44.

Konttinen, N. and Lyytinen, H. (1993) 'Brain slow waves preceding time-locked visuo-motor performance', *Journal of Sports Sciences*, 11: 257–66.

Lam, W.K., Maxwell, J.P. and Masters, R. (2009) 'Analogy learning and the performance of motor skills under pressure', *Journal of Sport and Exercise Psychology*, 31: 337–57.

Lam, W.K., Maxwell, J.P. and Masters, R.S.W. (2010) 'Probing the allocation of attention in implicit (motor) learning', *Journal of Sports Sciences*, 28: 1543–54.

Landau, S.M., Schumacher, E.H., Garavan, H., Druzgal, T.J. and D'Esposito, M. (2004) 'A functional MRI study of the influence of practice on component processes of working memory', *NeuroImage*, 22: 211–21.

Landers, D.M., Han, M., Salazar, W., Petruzzello, S.J., Kubitz, K.A. and Gannon, T.L. (1994) 'Effects of learning on electroencephalographic and electrocardiographic patterns in novice archers', *International Journal of Sport Psychology*, 25: 313–30.

Laufer, Y., Rotem-Lehrer, N., Ronen, Z., Khayutin, G. and Rozenberg, I. (2007) 'Effect of attention focus on acquisition and retention of postural control following ankle sprain', *Archives of Physical Medicine and Rehabilitation*, 88: 105–8.

Law, J., Masters, R., Bray, S.R., Eves, F. and Bardswell, I. (2003) 'Motor performance as a function of audience affability and metaknowledge', *Journal of Sport and Exercise Psychology*, 25: 484–500.

Liao, C.M. and Masters, R.S.W. (2001) 'Analogy learning: A means to implicit motor learning', *Journal of Sports Sciences*, 19: 307–19.

Liao, C.M. and Masters, R.S.W. (2002) 'Self-focused attention and performance failure under psychological stress', *Journal of Sport and Exercise Psychology*, 24: 289–305.

Logie, R.H. (1995) *Visuo-spatial working memory.* Hove, England: Lawrence Erlbaum Associates, Inc.

Luks, T.L., Simpson, G.V., Feiwell, R.J. and Miller, W.L. (2002) 'Evidence for anterior cingulate cortex involvement in monitoring preparatory attentional set', *NeuroImage*, 17: 792–802.

Lungu, O.V., Liu, T., Waechter, T., Willingham, D.T. and Ashe, J. (2007) 'Strategic modulation of cognitive control', *Journal of Cognitive Neuroscience*, 19: 1302–15.

MacDonald, A.W., III, Cohen, J.D., Andrew Stenger, V. and Carter, C.S. (2000) 'Dissociating the role of the dorsolateral prefrontal and anterior cingulate cortex in cognitive control', *Science*, 288: 1835–8.

MacMahon, K.M.A. and Masters, R.S.W. (2002) 'The effects of secondary tasks on implicit motor skill performance', *International Journal of Sport Psychology*, 33: 307–24.

Masters, R.S.W. (1992) 'Knowledge, knerves and know-how: The role of explicit versus implicit knowledge in the breakdown of a complex motor skill under pressure', *British Journal of Psychology*, 83: 343–58.

Masters, R.S.W. (2000) 'Theoretical aspects of implicit learning in sport', *International Journal of Sport Psychology*, 31: 530–41.

Masters, R.S.W., Eves, F.F. and Maxwell, J.P. (2005) 'Development of a movement specific Reinvestment Scale', paper presented at International Society of Sport Psychology World Congress, Sydney, August.

Masters, R.S.W., MacMahon, K.M.A. and Pall, H.S. (2004) 'Implicit motor learning in Parkinson's disease', *Rehabilitation Psychology*, 49: 79–82.

Masters, R.S.W. and Maxwell, J.P. (2004) 'Implicit motor learning, reinvestment and movement disruption: What you don't konw won't hurt you?' in A.M. Williams and N.J. Hodges (Eds.) *Skill acquisition in sport: Research, theory and practice*, London: Routledge.

Masters, R.S.W. and Maxwell, J.P. (2008) 'The theory of reinvestment', *International Review of Sport and Exercise Psychology*, 1: 160–83.

Masters, R.S.W., Pall, H.S., MacMahon, K.M.A. and Eves, F.F. (2007) 'Duration of Parkinson disease is associated with an increased propensity for "reinvestment"', *Neurorehabilitation and Neural Repair*, Masters, R.S.W. and Poolton, J.M. (2012). Advances in implicit motor learning. In A.M Williams & N.J. Hodges (Eds.), Skill aquisition in Sport: Research, theory and practice (2nd ed., pp. 59–75). London, UK: 21: 123–6.

Masters, R.S.W., Polman, R.C.J. and Hammond, N.V. (1993) "Reinvestment': A dimension of personality implicated in skill breakdown under pressure', *Personality and Individual Differences*, 14: 655–66.

Masters, R.S.W., Poolton, J.M. and Maxwell, J.P. (2008a) 'Stable implicit motor processes despite aerobic locomotor fatigue', *Consciousness and Cognition*, 17: 335–8.

Masters, R.S.W., Poolton, J.M., Maxwell, J.P. and Raab, M. (2008b) 'Implicit motor learning and complex decision making in time-constrained environments', *Journal of Motor Behavior*, 40: 71–9.

Maxwell, J.P., Masters, R. and Eves, F. (2000) 'Explicit versus implicit motor learning: Dissociating selective and unselective modes of skill acquisition via feedback manipulation', *Journal of Sports Sciences*, 18: 559.

Maxwell, J.P., Masters, R.S.W. and Eves, F.F. (2003) 'The role of working memory in motor learning and performance', *Consciousness and Cognition*, 12: 376–402.

Maxwell, J.P., Masters, R.S.W., Kerr, E. and Weedon, E. (2001) 'The implicit benefit of learning without errors', *Quarterly Journal of Experimental Psychology Section A: Human Experimental Psychology*, 54: 1049–68.

Maxwell, J.P., Masters, R.S.W. and Poolton, J.M. (2006) 'Performance breakdown in sport: The roles of reinvestment and verbal knowledge', *Research Quarterly for Exercise and Sport*, 77: 271–6.

Mullen, R., Hardy, L. and Oldham, A. (2007) 'Implicit and explicit control of motor actions: Revisiting some early evidence', *British Journal of Psychology*, 98: 141–56.

Mullen, R., Hardy, L. and Tattersall, A. (2005) 'The effects of anxiety on motor performance: A test of the conscious processing hypothesis', *Journal of Sport and Exercise Psychology*, 27: 212–25.

Neuper, C., Wortz, M. and Pfurtscheller, G. (2006) 'ERD/ERS patterns reflecting sensorimotor activation and deactivation', *Progress on Brain Research*, 159: 211–22.

Nissen, M.J. and Bullemer, P. (1987) 'Attentional requirements of learning: Evidence from performance measures', *Cognitive Psychology*, 19: 1–32.

Norman, D.A. and Shallice, T. (1986) 'Attention and action: Willed and automatic control of behavior', in R.D. Davidson, G.E. Schwartz and D. Shapiro (Eds.) *Consciousness and self-regulation*, New York: Plenum.

Orrell, A.J., Eves, F.F. and Masters, R.S.W. (2006) 'Motor learning of a dynamic balancing task after stroke: Implicit implications for stroke rehabilitation', *Physical Therapy*, 86: 369–80.

Orrell, A.J., Masters, R.S.W. and Eves, F.F. (2009) 'Reinvestment and movement disruption following stroke', *Neurorehabilitation and Neural Repair*, 23: 177–83.

Petsche, H. (1996) 'Approaches to verbal, visual and musical creativity by EEG coherence analysis', *International Journal of Psychophysiology*, 24: 145–59.

Petsche, H., Kaplan, S., Von Stein, A. and Filz, O. (1997) 'The possible meaning of the upper and lower alpha frequency ranges for cognitive and creative tasks', *International Journal of Psychophysiology*, 26: 77–97.

Pew, R.W. (1974) 'Levels of analysis in motor control', *Brain Research*, 71: 393–400.

Pfurtscheller, G. (1992) 'Event-related synchronization (ERS): An electrophysiological correlate of cortical areas at rest', *Electroencephalography and Clinical Neurophysiology*, 83: 62–9.

Pijpers, J.R., Oudejans, R.R.D. and Bakker, F.C. (2005) 'Anxiety-induced changes in movement behaviour during the execution of a complex whole-body task', *Quarterly Journal of Experimental Psychology Section A: Human Experimental Psychology*, 58: 421–45.

Poolton, J.M., Masters, R.S.W. and Maxwell, J.P. (2005) 'The relationship between initial errorless learning conditions and subsequent performance', *Human Movement Science*, 24: 362–78.

Poolton, J.M., Masters, R.S.W. and Maxwell, J.P. (2006) 'The influence of analogy learning on decision-making in table tennis: Evidence from behavioural data', *Psychology of Sport and Exercise*, 7: 677–88.

Poolton, J.M., Masters, R.S.W. and Maxwell, J. P. (2007) 'Passing thoughts on the evolutionary stability of implicit motor behaviour: Performance retention under physiological fatigue.', *Consciousness and Cognition*, 16(2): 456–468.

Poolton, J., Maxwell, J. and Masters, R. (2004) 'Rules for reinvestment', *Perceptual and Motor Skills*, 99: 771–4.

Posner, M.I. and Rothbart, M.K. (2007) 'Research on attention networks as a model for the integration of psychological science', *Annual Review of Psychology*, 58: 1–23.

Radlo, S.J., Janelle, C.M., Barba, D.A. and Frehlich, S.G. (2001) 'Perceptual decision making for baseball pitch recognition: Using P300 latency and amplitude to index attentional processing', *Research Quarterly for Exercise and Sport*, 72: 22–31.

Rauch, S.L., Savage, C.R., Brown, H.D., Curran, T., Alpert, N.M., Kendrick, A., Fischman, A.J. and Kosslyn, S.M. (1995) 'A PET investigation of implicit and explicit sequence learning', *Human Brain Mapping*, 3: 271–86.

Razoumnikova, O.M. (2000) 'Functional organization of different brain areas during convergent and divergent thinking: An EEG investigation', *Cognitive Brain Research*, 10: 11–18.

Reber, P.J. and Squire, L.R. (1994) 'Parallel brain systems for learning with and without awareness', *Learning and memory*, 1: 217–29.

Reber, P.J. and Squire, L.R. (1998) 'Encapsulation of implicit and explicit memory in sequence learning', *Journal of Cognitive Neuroscience*, 10: 248–63.

Reiterer, S., Berger, M.L., Hemmelmann, C. and Rappelsberger, P. (2005) 'Decreased EEG coherence between prefrontal electrodes: a correlate of high language proficiency?' *Experimental Brain Research*, 163: 109–13.

Reiterer, S., Hemmelmann, C., Rappelsberger, P. and Berger, M.L. (2005) 'Characteristic functional networks in high- versus low-proficiency second language speakers detected also during native language processing: An explorative EEG coherence study in 6 frequency bands', *Cognitive Brain Research*, 25: 566–78.

Rietschel J. R., King B. R., Pangelinan M. M., Clark J. E. and Hatfield B. D. (2008) 'EEG coherence associated with using an explicit strategy during a motor task', paper presented at *North American Society for the Psychology of Sport and Physical Activity Conference*, Niagara Falls, Ontario, Canada, June.

Rotheram, M.J., Thomas, O.T., Bawden, M.A., & Maynard, I.W. (2006) 'Understanding the 'yips' in sport: A grounded theory interview study', *Journal of Sports Sciences*, 24: 323–4.

Salazar, W., Landers, D.M., Petruzzello, S.J., Han, M., Crews, D.J. and Kubitz, K.A. (1990) 'Hemispheric asymmetry, cardiac response, and performance in elite archers', *Research Quarterly for Exercise and Sport*, 61: 351–9.

Salmoni, A.W. (1989) 'Motor skill learning', in D. Holding (Ed.) *Human skills*, Chichester: John Wiley.

Schacter, D.L. (1997) 'The cognitive neuroscience of memory: Perspectives from neuroimaging research', *Philosophical Transactions of the Royal Society B: Biological Sciences*, 352: 1689–95.

Schendan, H.E., Searl, M.M., Melrose, R.J. and Stern, C.E. (2003) 'An fMRI study of the role of the medial temporal lobe in implicit and explicit sequence learning', *Neuron*, 37: 1013–25.

Schmidt, R.A. (1977) 'Schema theory: Implications for movement education', *Motor Skills: Theory into Practice*, 2: 36–8.

Schmidt, R.A. (1988) *Motor control and learning: A behavioral emphasis*, 2nd edn, Champaign, IL: Human Kinetics.

Schneider, W. and Schiffrin, R.M. (1977) 'Controlled and automated human information processing: Detection, search and attention.' *Psychological Review*, 44: 627–44.

Sheppard I, W.D. and Boyer, R.W. (1990) 'Pretrial EEG coherence as a predictor of semantic priming effects', *Brain and Language*, 39: 57–68.

Silverstein, R. (1995) 'Neuromodulation of neocortical dynamics', in P.L. Nunez (Ed.) *Neocortical dynamics and human EEG rhythms*, New York: Oxford University Press.

Singer, R.S. (2002) 'Preperformance state, routines and automaticity: What does it take to realise expertise in self-paced events?' *Journal of Sport and Exercise Psychology*, 24.

Smith, J.G. and McDowall, J. (2004) 'Impaired higher order implicit sequence learning on the verbal version of the serial reaction time task in patients with Parkinson's disease', *Neuropsychology*, 18: 679–91.

Smith, M.E., McEvoy, L.K. and Gevins, A. (1999) 'Neurophysiological indices of strategy development and skill acquisition', *Cognitive Brain Research*, 7: 389–404.

Squire, L.R. (1992) 'Memory and the hippocampus: A synthesis from findings with rats, monkeys, and humans', *Psychological Review*, 99: 195–231.

Volf, N.V. and Razumnikova, O.M. (1999) 'Sex differences in EEG coherence during a verbal memory task in normal adults', *International Journal of Psychophysiology*, 34: 113–22.

Wan, C.Y. and Huon, G.F. (2005) 'Performance degradation under pressure in music: an examination of attentional processes', *Psychology of Music*, 33: 155–72.

Weiss, S. and Mueller, H.M. (2003) 'The contribution of EEG coherence to the investigation of language', *Brain and Language*, 85: 325–43.

Willingham, D.B. (1998) 'A neuropsychological theory of motor skill learning', *Psychological Review*, 105: 558–84.

Willingham, D.B. and Koroshetz, W.J. (1993) 'Evidence for dissociable motor skills in Huntington's disease patients', *Psychobiology*, 21: 173–82.

Willingham, D.B., Salidis, J. and Gabrieli, J.D.E. (2002) 'Direct comparison of neural systems mediating conscious and unconscious skill learning', *Journal of Neurophysiology*, 88: 1451–60.

Wilson, M., Smith, N.C. and Holmes, P.S. (2007) 'The role of effort in influencing the effect of anxiety on performance: Testing the conflicting predictions of processing efficiency theory and the conscious processing hypothesis', *British Journal of Psychology*, 98: 411–28.

Wong, W.L., Masters, R.S.W., Maxwell, J.P. and Abernethy, A.B. (2008) 'Reinvestment and falls in community-dwelling older adults', *Neurorehabilitation and Neural Repair*, 22: 410–14.

Wulf, G. and Schmidt, R.A. (1997) 'Variability of practice and implicit motor learning', *Journal of Experimental Psychology: Learning Memory and Cognition*, 23: 987–1006.

Zhu, F.F., Maxwell, J.P., Hu, Y., Zhang, Z.G., Lam, W.K., Poolton, J.M. and Masters, R.S.W. (2010) 'EEG activity during the verbal-cognitive stage of motor skill acquisition', *Biological Psychology*, 84: 221–7.

Zhu, F., Poolton, J., Wilson, M., Hu, Y., Maxwell, J. and Masters, R. (2011a) 'Implicit motor learning promotes neural efficiency during laparoscopy', *Surgical Endoscopy*, 25: 2950–5.

Zhu, F.F., Poolton, J.M., Wilson, M.R., Maxwell, J.P. and Masters, R.S.W. (2011b) 'Neural co-activation as a yardstick of implicit motor learning and the propensity for conscious control of movement', *Biological Psychology*, 87: 66–73.

9

MIRROR NEURONS AND IMITATION

Stefano Rozzi,[1,2] *Giovanni Buccino*[3] *and Pier F. Ferrari*[1,2,4]

[1] DIPARTIMENTO DI NEUROSCIENZE, UNIVERSITÀ DI PARMA
[2] ISTITUTO ITALIANO DI TECNOLOGIA, UNIVERSITÀ DI PARMA
[3] DIPARTIMENTO DI SCIENZE MEDICHE E CHIRURGICHE, UNIVERSITÀ MAGNA GRECIA
[4] DIPARTIMENTO DI BIOLOGIA EVOLUTIVA E FUNZIONALE, UNIVERSITÀ DI PARMA

The mirror-neuron system in monkeys

The motor cortex and the posterior parietal cortex are formed by a mosaic of anatomically and functionally distinct areas (Figure 9.1), reciprocally connected in specialized circuits working in parallel (see Rizzolatti and Luppino, 2001), allowing the transformation of sensory information into action. These circuits represent the basic elements of the motor system.

The description of this type of anatomo-functional organization, together with the finding that several areas of the motor cortex contain neurons responsive to visual stimuli (Kurata and Wise, 1988; Rizzolatti et al., 1988) strongly boosted the idea that the motor system is not a mere executor of the commands generated by the associative areas, but is also involved in cognitive functions such as space coding, motor learning, action understanding and imitation. In this context, the concept of goal coding and the discovery of mirror neurons is particularly important and will be described in the present chapter.

Coding of motor goals at the single neuron level

Electrophysiological studies revealed that most of ventral premotor area F5 and inferior parietal area PFG neurons code specific motor acts, rather than individual movements (Figure 9.2a; Rizzolatti et al., 1988, Rozzi et al., 2008). Using the effective motor act as the classification criterion, F5 neurons were subdivided into various classes (e.g. 'grasping', 'holding', 'tearing', and 'manipulating'). Neurons of a given class do not respond or respond weakly when similar movements are executed for achieving different goals. For example, the same neurons that discharge during finger movements performed to grasp an object are not discharging during finger movements performed for scratching. Furthermore, many neurons code specific types of grips (e.g. precision grip, whole hand prehension, finger prehension). Finally, some neurons discharge during the whole motor act (e.g. opening and closure of the hand), while others fire only during a certain part of the act (e.g. the last part of the grasping). On the basis of these findings it was suggested that F5 contains a 'vocabulary' of motor acts (Rizzolatti et al.,

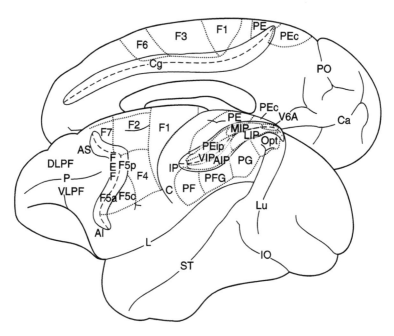

Figure 9.1 Lateral and mesial views of the monkey brain showing the parcellation of the agranular frontal and posterior parietal cortices. Intraparietal, arcuate and cingulated sulci are shown unfolded. Agranular frontal areas are labelled with 'F', similar to the von Economo classification for human frontal cortex (von Economo and Koskinas, 1925) and defined according to Matelli et al. (1985, 1991) and Belmalih et al. (2009). Frontal eye fields (FEF) are defined according to Bruce et al. (1985). All posterior parietal areas except those buried inside the intraparietal sulcus are defined according to Pandya and Seltzer (1982) and Gregoriou et al. (2006). The areas buried inside the intraparietal sulcus and V6A are defined according to functional criteria (for references see Rizzolatti et al. 1998). AI: inferior arcuate sulcus; AS: superior arcuate sulcus; C: central sulcus; Ca: calcarine fissure; Cg: cingulate sulcus; DLPF: dorsolateral prefrontal cortex; IO: inferior occipital sulcus; L: lateral fissure; Lu: lunate sulcus; P: principal sulcus; PO: parieto-occipital sulcus; ST: superior temporal sulcus; VLPF: venrolateral prefrontal cortex.

1988), with each 'word' of this vocabulary representing a population of F5 neurons. Some 'words' code the general goal of a motor act (e.g. grasping), others code how, within a general goal, a specific motor act must be executed (e.g. precision grip) and other specify the temporal aspects of the motor act to be executed (e.g. opening of the hand).

Further studies clearly supported the idea that F5 motor neurons code the goal of a motor act. In a recent experiment (Umiltà et al., 2008) monkeys were trained to grasp objects using 'normal' pliers, requiring hand closure in order to take possession of the object, and 'reverse' pliers requiring hand opening to achieve the same goal. Most of area F5 neurons fired in relation to goal achievement independent of hand opening or closure movements (Figure 9.2b). This evidence strongly supports the idea that F5 neurons code motor acts rather than individual movements.

Mirror neurons and understanding of motor acts

Goal coding is also the key feature of a peculiar class of visuomotor neurons named *mirror neurons* (MN) that activate both when the monkey performs a goal-directed motor act (e.g.

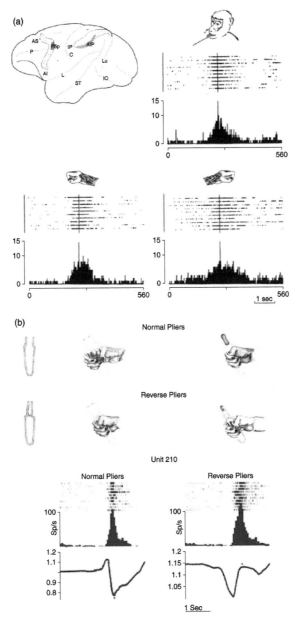

Figure 9.2 Goal coding in F5 motor neurons. (a) Top-left: lateral view of the monkey brain showing the grasping circuit areas AIP and F5. Top-right: discharge of an F5 neuron active during execution of a grasping act with the mouth, the right or the left hand. Abscissae: time; ordinates: spikes per bin; bin width: 20 ms. (Modified from Rizzolatti et al., 1988). (b) Example of an F5 neuron active during execution of grasping with normal and reverse pliers. (modified from Umiltà et al., 2008). Rasters and histograms are aligned with the end of the grasping closure phase of the plier (asterisks). The traces below each histogram indicate the hand position, recorded with a potentiometer, expressed as function of the distance between the pliers' handles. When the trace goes down, the hand closes; when it goes up, the hand opens. The values on the vertical axes indicate the voltage change measured with the potenziometer. The firing rate is expressed as spikes per second.

grasping) and when it observes the same, or a similar, motor act performed by another individual (Figure 9.3a; Di Pellegrino et al., 1992; Gallese et al., 1996; Rizzolatti et al., 1996; Ferrari et al., 2003; Gallese et al., 2002; Rozzi et al., 2008). MN were recorded in specific sectors of the ventral premotor (area F5) and in sectors of the inferior parietal cortex (area PFG). Generally, they do not respond to object presentation or during the observation of biological movements mimicking the motor act devoided of a goal. Although the visual response of most MN is invariant with respect to many visual aspects, some of them show specificity for the direction of the hand movement, the space sector in which the observed motor act is presented or the hand (left or right) used by the observed agent.

A recent study systematically investigated whether the discharge of MN is modulated by the distance at which the observed act is performed (Caggiano et al., 2009). In this study, the same motor act was performed by the experimenter inside the monkey's reaching space (peripersonal space) or outside it (extrapersonal space). The results showed that 50 percent of the studied MN discharged differently in the two conditions. Half of these neurons discharged stronger when the experimenter grasped a piece of food within the monkey's peripersonal space, while the other half responded better when the same motor act was performed in the extrapersonal space. However, when introducing a transparent barrier obstructing the monkey's peripersonal space, extrapersonal MN started to discharge within the previous peripersonal space (i.e. at a distance that could be grasped by the monkey in the absence of the barrier). These data suggest that MN code other's actions in relation to the potential interaction that might occur between the two individuals.

One of the most important properties of MN is the congruence in the neuronal discharge between the observed and executed motor act. The congruence can be strict or broad. In the 'strictly congruent' neurons (about one-third of MN), the executed and observed motor acts correspond both in terms of goals (e.g. grasping) and the means to achieve the goal (e.g. whole hand grip). In the other two-thirds ('broadly congruent'), the congruence is more in terms of the goals of the observed and executed motor act, rather than in the details of the movements necessary to achieve the target (e.g. precision grip on the motor side, unspecific grasping on the visual side).

It has been proposed that the property shown by mirror neurons of matching the visual description of a goal-directed act with its motor representation allows the observer to understand *what* another individual is doing, because the observation of a motor act determines an automatic retrieval of its motor representation from the motor repertoire of the observer. Note, however, that this retrieval does not normally produce an overt motor output. In fact, observers typically do not mimic the observed motor acts, indicating that an inhibitory mechanism is at work. In agreement with this observation, Kraskov et al. (2009) demonstrated that half of F5 mirror neurons characterized as cortico-spinal neurons, were excited during grasping execution but inhibited during grasping observation. Kraskov et al. (2009) speculated that these neurons may play a role in providing a signal as to whether the current state involves action observation (suppressed discharge) or active grasp (elevated discharge). Indeed, there are other important mechanisms that could contribute to disambiguate F5 movement discharge from their mirror discharge (e.g. command to perform voluntarily grasping coming from the prefrontal lobe or the proprioceptive signals arriving at the cortex during arm movements).

Various electrophysiological studies support the hypothesis that mirror neurons have an important role in understanding the motor acts of others. In particular, it has been shown that mirror neurons discharge both when the monkey observes a grasping act and when it sees only part of it, the hand–target interaction being hidden behind a screen (Umiltà et al., 2001).

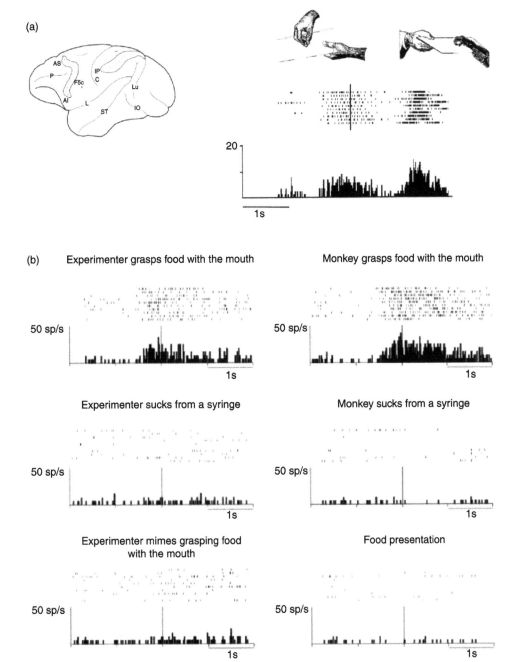

(a)

(b)   Experimenter grasps food with the mouth   Monkey grasps food with the mouth

50 sp/s   50 sp/s

Experimenter sucks from a syringe   Monkey sucks from a syringe

50 sp/s   50 sp/s

Experimenter mimes grasping food   Food presentation
with the mouth

50 sp/s   50 sp/s

Figure 9.3 Examples of F5 mirror neurons. (a) Left. Location of area F5c, where mirror neurons were recorded. Right. Mirror neuron responding during observation and execution of a hand grasping motor act. (Modified from Rizzolatti et al., 1996). (b) Mirror neuron responding during observation and execution of a specific mouth motor act. (Modified from Ferrari et al., 2003). Conventions as in Figure 9.2.

Interestingly, the discharge was absent when the monkey knew that no object was present behind the screen, suggesting that mirror neurons use prior information to retrieve the motor representation of the observed motor act. In another study (Kohler et al., 2002), it was demonstrated that a subset of mirror neurons ('audio-visual mirror neurons') discharge not only during execution and observation of a noisy act (e.g. breaking a peanut), but also during the simple audition of the typical sound produced by that act, in the absence of visual information. This finding suggests that the meaning of a motor act can be accessed through different sensory modalities.

Mouth mirror neurons

The hand is one of the most important effectors for interacting with the outer world, and is most often adopted in the behavioral paradigms used in electrophysiological experiments. A less studied, but not less important effector for primate and human social life is the mouth. Interestingly, the lateral part of area F5 hosts mouth MN (Ferrari et al., 2003). Most of them respond to the observation and execution of ingestive motor acts such as biting, sucking, and licking (Figure 9.3b). They do not respond to object presentation or to mimed mouth motor acts. A smaller but significant number of mouth mirror neurons respond specifically to the observation of mouth communicative gestures, such as lips-smacking or tongue protrusion. Mouth mirror neurons of this sub-category do not respond, or respond very weakly, to the observation of ingestive motor acts. This sub-category of mirror neurons is interesting because, considering a possible homology between area F5 and human area 44 (see Rizzolatti and Arbib, 1998, Fogassi and Ferrari, 2007), they could constitute an evolutionary ancient oro-facial system of communication, evolved from neurons coding other types of goal (ingestion).

Mirror neurons responding to the observation of motor acts performed with tools

Although in the original reports on mirror neuron properties (Gallese et al., 1996; Rizzolatti et al., 1996) no instances of visual responses to motor acts performed by the observed agent with a tool were described, more recently two studies showed two categories of mirror neurons with these properties. In the first study (Ferrari et al., 2005) it has been reported that while a higher number of the studied mirror neurons showed sensitivity only for the observation of hand-related motor acts, a smaller amount showed a clear preference for motor acts performed with tools, such as sticks or pliers, to which the monkey was visually exposed for a long time period although unable to use it. On the motor side these neurons responded best during hand and mouth motor acts, thus showing a high degree of generalization of the motor goal. In the second study (Rochat et al., 2010), after monkeys were trained to use tools such as normal or inverted pliers (which required two opposite finger movements, flexion of extensions, in order to grasp an object), mirror neurons responded to the observation of hand-related as well as tool-related motor acts, although the response to tool-motor acts was normally weaker.

Mirror neurons and intention coding

Recent studies showed that mirror neurons, besides playing a role in understanding motor acts, are also involved in coding the motor intention of an action performed by another individual (Fogassi et al., 2005; Bonini et al., 2010). Our everyday actions are formed by sequences of fluently linked motor acts that, as final outcome, lead to the achievement of an ultimate goal (e.g., reaching a piece of food, grasping it and bringing it to the mouth leads to the final goal of eating it). Note that each motor act belonging to a sequence has its own intrinsic goal (e.g. grasping aims to take possession). Often the same motor act is part of different action sequences having different final goals. A series of experiments were carried out to assess whether neurons discharging during the execution of grasping were influenced by the type of action in which they were embedded (Fogassi et al., 2005; Bonini et al., 2010). For this purpose, grasping neurons were recorded from areas PFG and F5 while the monkey executed a motor task and observed the same task, performed by an experimenter (Figure 9.4), in which the same motor act (grasping) was embedded into two different actions (grasping to eat and grasping to place). The results showed that a high percentage of parietal and premotor neurons discharged differently during execution of the grasping act, depending on the final goal of the action (either eating or placing) in which the act was embedded. On the basis of

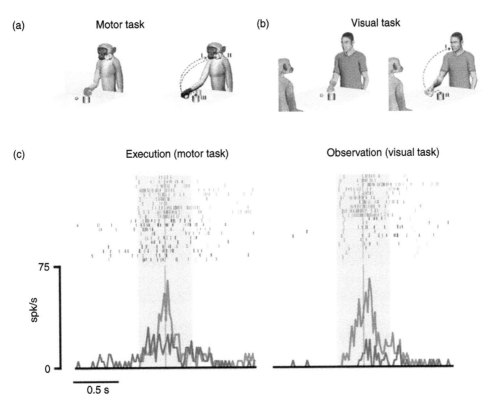

Figure 9.4 (a) Paradigm used for the motor task. The monkey, starting from a fixed position, reaches and grasps a piece of food or an object, then it brings the food to the mouth and eats it (I, grasp-to-eat), or places it into a container (II/III, grasp-to-place). (b) Paradigm used for the visual task. The experimenter, starting from a fixed position, reaches and grasps a piece of food or an object, then brings the food to the mouth and eats it (I, grasp-to-eat) or places it into a container (II, grasp-to-place). (c) Discharge of the neuron during execution (left) and observation (right) of the two actions. Rasters and histograms are aligned on the contact between the monkey or experimenter's hand and the object. First ten trials and light gray line in the histogram: neuron discharge during grasp-to-eat condition; second ten trials and dark gray line in the histogram: neuron discharge during grasp-to-place condition. (Modified from Bonini et al. 2010)

these results, it has been proposed that parietal and premotor neurons form pre-wired chains, in which a neuron coding a given motor act is facilitated by the neuron coding the previously executed one (Fogassi et al., 2005). Thus any time an agent has the intention (final goal) to perform an action, specific neuronal chains are activated. Note that this model would also account for the fluidity with which the different motor acts of an action are executed one after another, as shown by kinematic studies (Jeannerod, 1988).

During the observation of sequential actions it has been found that most recorded mirror neurons discharged differently during observation of grasping, when this act was embedded into different actions. Since in this case the grasping act was performed by the observed agent, it was suggested that the neuronal selectivity for the action goal during grasping observation represents a prediction of the action outcome. The possibility for the neuron to predict the action goal depended on contextual cues (e.g. presence or absence of a container for placing the objects) and the memory of the previous observed actions. Thus these neurons integrate several types of information that is used to understand the underlying motor intention of the observed agent. More recent investigations have shown similar response properties also in F5 mirror neurons (Bonini et al., 2010). In terms of mechanism involved and in agreement with the 'chain' interpretation of the results of the motor task, the observation of a motor act embedded in an action would activate a chain corresponding to a specific intention. This pattern of activation is considered as evidence that the monkey understands the motor intentions of others.

The mirror-neuron circuits in the monkey

An important issue about mirror neurons concerns their main source of sensory information. It is well known that the anterior region of the superior temporal sulcus (STSa) contains visual neurons responding to the observation of biological movements (Perrett, 1989; Barraclough et al., 2006). It is also known that IPL (inferior parietal lobule) areas hosting mirror neurons (AIP and PFG) are anatomically connected with specific sectors of STS (Seltzer and Pandya, 1984; Borra et al., 2008; Rozzi et al., 2006). In a combined fMRI and anatomy experiment in monkeys we recently showed that different areas of the superior temporal sulcus (STS), both in its upper and lower banks, become active during the observation of grasping motor acts and are anatomically connected with areas AIP and PFG (Nelissen et al., 2011). Thus IPL appears to be the first node of an IPL-ventral premotor (PMV) circuit processing biologically meaningful movements both visually and in motor terms. Interestingly, recent anatomical evidence indicated that a specific region of VLPF (rostral 46VC and intermediate area12r) is reciprocally connected with both parieto-premotor and temporal 'mirror' sectors (Borra et al., 2011; Gerbella et al., 2011). This prompted the idea that the prefrontal cortex could play a role in higher order functions relying on the mirror system, such as intention coding.

The mirror-neuron system in humans

A mirror-neuron system does also exist in humans. So far, however, there is little evidence for this at single neuron level. A recent work using extracellular recordings in patients who underwent neurosurgery has demonstrated the presence of neuron responding both to execution and observation of various actions within the supplementary motor area (SMA) and the middle temporal gyrus (Mukamel et al., 2010). These areas do not correspond to the monkey brain areas where mirror neurons have been found and do not overlap with those in human brain normally recruited during observation and/or execution of actions. Thus, the relevance of these data needs to be clarified by further experiments. Evidence of the presence of mirror neurons in humans in this sense has been indirectly provided by a series of experiments carried out with various techniques such as electroencephalography (EEG), magnetoencephalography (MEG), transcranial magnetic stimulation (TMS) and brain imaging techniques including fMRI and PET (see Rizzolatti et al., 2001).

By means of non-invasive techniques, it has also been possible to investigate the human brain and many studies have been conducted that confirm the existence of mirror properties also in specific regions of the human cortex. A pilot study of Fadiga and colleagues (1995), applying TMS pulses over hand motor cortex, showed that the excitability of this region is enhanced when subjects observed hand actions with respect to a control condition. This is considered as one of the first pieces of evidence that even in humans there exists a system similarly involved in the observation of actions and in performing actions. Comparable results were obtained by Strafella and Paus (2000). Furthermore, using the same technique, it has been found that during the observation of hand actions there is not only a general increase of the motor evoked potentials (MEP) size in the muscles involved in the actual execution of the observed action, but MEPs are modulated in a fashion strictly resembling the time-course of the observed action (Gangitano et al., 2001). For example, the amplitude of MEPs recorded from the first dorsal interosseus muscle was modulated according to the degree of aperture of the observed finger over time. Taken together, these TMS data support the notion that the mirror neuron system matches action execution and action observation both in terms of the muscles involved and the temporal sequence of the action.

Hari and co-workers (1998), using MEG, found a suppression of 15-25 Hz activity, known to originate from the precentral motor cortex, during the execution and, to a lesser extent, during the observation of object manipulation. Similar results were obtained by means of quantified electroencephalography (Cochin et al., 1998). In this study, muscle activity was blocked during both the observation and execution of various hand actions when compared to rest. As an historical note, it is worth recalling that already Gastaut and Bert (1954) noted the suppression of muscle activity during the presentation of video-clips depicting actions. All the studies quoted so far provide little, if any, insight on the localization of the mirror neuron system in humans. This issue has been addressed by a number of brain imaging studies. In an early positron emission tomography (PET), an experiment aimed at identifying the brain areas active during action observation. Rizzolatti and colleagues (1996), comparing hand action observation with the observation of an object, found activation of Broca's area, the middle temporal gyrus and the superior temporal sulcus region, providing the first evidence on the anatomical localization of the mirror neuron system for hand actions in humans. Broca's area is classically considered as an area devoted to speech production. Recently, however, it has been demonstrated that in this area a motor representation of hand actions is also present (Binkofski et al., 1999; Ehrsson et al., 2001; Binkofski and Buccino, 2004). Using the fMRI technique, different authors (Decety et al., 1997; Grèzes et al., 1998) searched for the brain foci active during the observation of meaningful and meaningless hand actions. Subjects performed either a passive observation or a delayed imitation task; both these conditions were compared to mere observation of static hand images. During the passive observation task the same cortical circuits were active during both the observation of meaningful and meaningless actions, the active foci being the superior occipital gyrus, the occipito-temporal junction bilaterally and the inferior parietal lobule and the precentral gyrus in the left hemisphere. During the delayed imitation task, the observation of both meaningful and meaningless actions led to the activation of common circuits involving bilaterally the inferior and the superior parietal lobules and the premotor cortex. Besides these common areas, observation of meaningful actions additionally evoked activity in the SMA and in the orbitofrontal cortex. Some years later, Buccino and co-workers (2001) showed that in humans the mirror neuron system is related to different body actions performed not only with the hand but also with the foot and the mouth. These authors asked participants to observe video sequences presenting different actions performed with the hand, the mouth and the foot, respectively. The actions shown could be either object directed (the action performed by a biological effector was acted upon a specific object) or non-object directed (the mouth/hand/foot action was mimicked, that is performed without an object). The following actions were presented: biting an apple, grasping a cup, grasping a ball, kicking a ball and pushing a brake. As a control, participants were asked to observe a static image of each action. The observation of both object directed and non object directed mimicked actions compared to the observation of a static image of the same action, led to the activation of different regions in the premotor cortex and Broca's area, depending on the effector involved in the observed action. The different sectors within the premotor cortex largely overlapped those where classical studies (Penfield and Rasmussen, 1952) had shown a somatotopically organized motor representation of the different effectors. Moreover, during the observation of object directed actions, distinct sectors in the inferior parietal lobule were active, including areas inside and around the intraparietal sulcus, with the localization depending on the effector involved in the observed action. In conclusion, this study supports the claim that like the actual execution of action, different somatotopically organized fronto-parietal circuits are recruited during action observation. These fronto-parietal regions are possibly the human homologue of monkey fronto-parietal regions where mirror neurons have been found (Gallese et al., 1996, 2002; Rizzolatti et al., 1996; Bonini and Ferrari, 2011). Furthermore, different from previous studies, the results of this experiment showed an involvement of the mirror sectors during a mere observation task, in the absence of higher order requirements for the subjects, suggesting that the mirror system is indeed operating independently of the observer's cognitive strategy.

Some studies have stressed the role of the observer's motor experience in the recruitment of the mirror neuron system during action observation. In particular, one study addressed the issue whether we can recognize actions performed by individuals belonging to other species (non conspecifics) using the same neural structures involved in the recognition of actions performed by conspecifics (Buccino et al., 2004). In this study, normal participants were asked to carefully observe different mouth actions performed by a man, a monkey and a dog, respectively. Two types of mouth actions were visually presented: ingestive actions (biting) or communicative actions (human silent speech, monkey lip-smacking and silent dog barking). The results showed that, during the observation of biting actions, there is a clear activation of the pars opercularis in the inferior frontal gyrus and of the inferior parietal lobule, regardless of the species doing the action. For oral communicative actions, a different pattern of activation was found, depending on the species performing the action. During the observation of silent speech there was an activation of Broca's region, with no clear asymmetry between the two hemispheres. Finally, during the observation of silent dog barking, no activation was found in Broca's region, but activation was present in the right superior temporal sulcus region. These findings suggest that recognition of actions may occur in two different ways: (a) for actions which are part of the observer's motor repertoire (biting, silent speech reading) there is a motor resonance within the mirror neuron system, and (b) for actions like barking, this resonance is missing. In the first case, there is a retrieval of a motor knowledge based on motor experience (I know what you are doing because I myself can do the same), while in the second case (barking) this personal motor knowledge is lacking and recognition cannot be based on motor resonance encoded by the mirror-neuron system. Similarly, the motor expertise of the observer affects the motor resonance system. In another study, the mirror circuit of expert dancers resonates more strongly when they observe another dancer performing exactly the same kind of dance they practice than in the condition when watching a different type of dance the subjects actually practiced (Calvo-Merino et al., 2005).

Being mirror neurons mostly activated by object directed actions in the monkey, little attention has been paid by authors to the neural substrates involved in the observation of non object directed actions in humans. However, a recent fMRI study addressed this issue (Lui et al., 2008). The study was aimed at assessing the cortical areas active when individuals observe different non-object directed actions (mimicked, symbolic emblems, meaningless) and imagine those same actions. During the observation, a fMRI signal increase was found in the premotor cortex and in a large region of the inferior parietal lobule. The premotor cortex activation largely overlapped that found during the observation of object directed actions, while in the inferior parietal lobule the active region extended in the supramarginal and angular gyri.

Altogether, these studies demonstrated that the human and monkey mirror-neuron systems share the capacity to match observed actions onto their motor representations. However, they also showed some important peculiarities of the human mirror system. In particular, as shown by fMRI studies (Buccino et al., 2001; Grèzes et al., 1998; Lui et al., 2008), the observation of meaningful hand actions without an object (mimed actions) in humans activates the mirror-neuron system. In agreement with these results, TMS experiments showed that there is a facilitation of the MEP size recorded from the muscles of the observer corresponding to those used by the actor performing an action, and this facilitation is also present when an individual observes non-object directed meaningless hand/arm gestures (Fadiga et al., 1995; Iacoboni et al., 1999; Maeda et al., 2002). Interestingly, the facilitation of MEPs recorded from the corresponding muscles follows the time-course of the observed action (Gangitano et al., 2001).

In conclusion, these properties indicate that the human mirror-neuron system, unlike the mirror-neuron system in the monkey, is able to describe both the goal of an action and the movements necessary to achieve it. This capacity is considered a necessary prerequisite for imitation.

Mirror-neuron system and imitation

The literature on imitation is as vast as it is heterogeneous and highly controversial, and the most debated issue concerns the identification of what should, and what should not be defined as 'imitation'. As a result, scientists have generated a plethora of new labels such as 'imitative learning', 'goal emulation', and 'true imitation' (Whiten et al., 1996). In fact, very often the term imitation is used to include several behavioral phenomena that could require different degrees of cognitive processing and intentionality (Byrne and Russon, 1998) and, very likely, different corresponding neurophysiological mechanisms.

From a behavioral and cognitive perspective, all the phenomena in which an agent displays a certain behavior similar in terms of goals or motor appearance to that previously or contingently displayed by another individual, share a common functional requirement: the capacity of recognizing the observed motor events. As described above, it has been proposed that this capacity is identified with the property of the observer's motor system to *'resonate'* with that of the observed agent by the activation of the mirror-neuron system (Jeannerod, 1994: Rizzolatti et al., 1999; Rizzolatti et al., 2002). This automatic embodied recognition of other's behaviors would render it possible for an observer to replicate the observed patterns of movements or the achievement of the same goals.

The involvement of the mirror-neuron system in imitation was demonstrated by a series of brain imaging studies. Using the functional magnetic resonance imaging (fMRI) technique,

Iacoboni and co-workers (1999) scanned normal human volunteers while they lifted a finger in response: (a) to the same action presented on a screen ('imitation'), (b) to a symbolic cue, or (c) to a spatial cue. The results show that the activation in the *pars opercularis* of the left inferior frontal gyrus (IFG), the right anterior parietal region, the right parietal operculum, and the right STS region was stronger during imitation than during the other motor conditions (see also Iacoboni et al., 2001). Other authors confirmed the importance of Broca's area, in particular when the action to be imitated had a specific goal (Grèzes et al., 1998).

Nishitani and Hari (2000, 2002) performed two studies, using MEG, in which they investigated imitation of grasping actions and of facial movements, respectively. The first study confirmed the importance of the left IFG (Broca's area) for imitation. In the second study (Nishitani and Hari, 2002), the authors asked volunteers to observe still pictures of verbal and non-verbal (grimaces) lip forms, to imitate them immediately after having seen them, or to make similar lip forms spontaneously. During lip forms observation, cortical activation progressed from the occipital cortex to the superior temporal region, the inferior parietal lobule, IFG (Broca's area), and finally to the primary motor cortex. The activation sequence during imitation of both verbal and non-verbal lip forms was the same as during observation.

In spite of some minor discrepancies, these data clearly show that the basic circuit underlying imitation coincides with that active during action observation, and indicate that, in the posterior part of IFG, a mapping of the observed action on its motor representation takes place. The importance of the 'pars opercularis' of IFG in imitation was further demonstrated (Heiser et al., 2003), using repetitive TMS, a technique that transiently disrupts the functions of the stimulated area. The task adopted in the study was, essentially, the same employed in the fMRI study by Iacoboni and colleagues (1999). The results showed that following stimulation of both left and right IFG, there was significant impairment in imitation of finger movements. The effect was absent when finger movements were made in response to spatial cues.

Important evidence that the mirror neuron system in humans is not only implicated in some imitative processes but also has a direct effect on the periphery of the motor system (nerves and muscles) is provided by neurophysiological experiments in which the observation of hand simple movements and actions induced specific muscles subliminal activations as measured with H-reflexes and transcranial magnetic stimulation (Borroni et al., 2005; 2008). It is important to note that these activations are modulated subliminally and reproduce, with high temporal fidelity, the motor commands needed to execute the observed movement.

Behavioral correlates of mirror-neuron activation

If MN have a motor output, one would expect that movements are produced during action observation, but this is not the case under normal circumstances. Actually, although most of the motor effects due to mirror-neuron activation in adults are suppressed, there are instances in which, following action observation, the motor output can reach the threshold and will be triggered. In these cases, we observe a phenomenon named 'automatic imitation' (Brass et al., 2001). Automatic imitation is characterized by a high speed of response, an activation of simple motor schema, and, more often, this involves the face.

Several studies document that people mimic emotional facial expressions of others within 1000 ms (Dimberg et al., 2000). This phenomenon, defined as *Rapid facial mimicry* (RFM), has been widely described in children (Jones, 2009) and adult humans (Dimberg and Thumberg, 1998), whose congruent reactions are elicited more frequently and rapidly in response to a dynamic facial expression compared to a static one (Sato and Yoshikawa, 2007). Recent works in nonhuman primate behavior seem to support the existence of such direct mirroring behavior.

Other examples suggest the presence of motor facilitation as a consequence of action observation, such as the synchronous behaviors displayed by young chimpanzees observing skilled individuals performing nut-cracking actions (Marshall-Pescini & Whiten, 2008). These types of behaviors, although not well investigated, are probably very common among primates and even other taxa. In highly social species, the necessity to coordinate activities among individuals of a group might have been a driving force to evolve a mechanism of sensori-motor coupling. For example, flocks of birds or groups of mammals tend to synchronize their movements during antipredatory responses or feeding activities. This synchronization could rely on automatic motor mirroring. Doing the same things at the same time has often undoubted advantages for survival as it helps group cohesiveness, increases defensive opportunities against predators, and tunes individuals in similar activities (Coussi-Korbel & Fragaszy, 1995).

One of the most convincing phenomena that seem to imply a direct involvement of the mirror system is neonatal imitation. First demonstrated in humans (Meltzoff & Moore, 1977; 1983) it has been subsequently shown in apes (Myowa, 1996; Myowa-Yamakoshi et al., 2004) and monkeys (Ferrari et al., 2006). Very likely, at the evolutionary origin of neonatal imitative response, there was the necessity to automatically establish and subsequently sustain the relation with the caregiver, typically the mother (Ferrari et al., 2006; Casile et al., 2011). Lip-smacking, for example, is an important facial communicative display which is used in affiliative contexts.

One of the problem raised by neonatal imitation is how does the infant translate the personal perspective of the model into her/his own perspective since, in both humans and monkeys, neonatal imitation involves typically effectors (tongue and mouth) that the neonate has never had the opportunity to visually access. We hypothesized that this phenomenon can rely on a mirror mechanism present at birth and capable of matching some facial features with an internal motor representation of these features (Ferrari et al., 2006; Casile et al., 2011; Gallese, 2010). Note that these motor representations are present at birth because foetuses in utero very often perform tongue and mouth movements (De Vries et al., 1984; Hata et al., 2005) similar to those shown in neonatal imitation experiments.

The possibility of a functional mirror system in the first stages of postnatal development has been investigated in human infants by means of electroencephalography (EEG). Similarly to adults, 6-month-old infants showed a desynchronization of the μ-rhythm during the observation of goal-directed movements. This result has been attributed to the fact that, in infants, motor cortical areas are active during action observation, an indirect evidence of an early mirror system activity. More recently, we explored EEG responses to facial gestures in one-week-old infant macaques and we observed significant suppression of the μ-rhythm (around 5-6Hz frequency band in infant macaques) compared to control stimuli (Ferrari et al., in press). This inhibition seems to reflect an activation of motor areas recorded in central-parietal sites. This preliminary data would constitute a more direct evidence of a 'premature' mirror system at birth that is sensitive to biological meaningful stimuli.

Functional mirror pathways

Recently, we proposed that MNs can exert an influence on the motor output through two distinct anatomo-functional pathways (Ferrari et al., 2009). One pathway is direct (*direct pathway*) and involves the parieto-premotor circuit, which, during action observation, can have an immediate influence on the motor output. A second pathway, named *indirect pathway*, would link parietal and premotor areas with the prefrontal cortex (for the anatomical bases of this pathway in the monkey see Borra et al., 2011; Gerbella et al., 2011; Nelissen et al., 2011). This latter part of the frontal cortex could exploit, during the observational phase of imitation, the sensory-motor representations provided by the mirror regions of the direct pathway in order to accomplish more complex cognitive and behavioral functions, such as those required for imitative behaviors in which the observation and execution phase of imitation are significantly segregated in time (order of several seconds or minutes). During ontogeny, neonatal imitation could be the consequence of a pre-wired but immature mirror system (Bonini and Ferrari, 2011; Casile et al., 2011), mainly based on a direct sensory-motor pathway having access to the descending traits. Subsequently, experience-mediated synaptic plasticity would allow suppression of this automatic motor activation assigning to the core mirror-neuron system essentially perceptuo-cognitive functions. The later emergence of more efficient and adaptive resonance behaviors would mainly depend on the indirect pathway allowing as complex mirroring behaviors as developed in the prefrontal cortex of the individual displaying them.

In summary, the direct mirror pathway is an uninhibited motor resonance phenomenon revealed by neonatal imitation and RFM, the functional meaning of which is probably that of promoting intersubjective exchanges between mother and infant. Its early disappearance is probably related to an increased voluntary motor control that would allow infants to exploit the representations coded by the mirror-neuron system in order to respond in a more flexible modality to the social environment solicitations. In humans, this capacity could enable individuals to intentionally imitate other's actions. The maintenance of forms of imitative behaviors in adulthood suggests that the mirror-neuron system could still affect directly, to a certain extent, the motor output, even though the biological relevance of this output could be, in some cases, partially lost.

The speed of the response crucially determines whether mirror neurons can have a direct impact on the motor output. The studies reported so far are related to behaviors in which imitation and mimicry are characterized by the automatic copying of a movement or a motor act, which is embedded in the behavioral repertoire of the observer. In addition, observation and copying are events closely linked in time. In contrast, more complex imitative processes are needed when copying is an off-line behavioral response. This requires not only the capacity to recognize the motor acts and the motor intention underlying the behavior, but also the capacity to re-enact motor programs similar to those observed after a considerable time lag. Other authors refer to this form of behavior as *response facilitation* or *social mirroring* (Byrne, 2005). Furthermore, such forms of imitation necessitate a sophisticate *working-memory system* to store the information related to the observed action subsequently needed for the organization of the correspondent motor plan.

Buccino and colleagues (2004) recently addressed the issue of which cortical areas become active in humans when individuals are required not to simply repeat an action or a posture present in their motor repertoire, but to produce, on the basis of action observation, a *novel motor pattern*. The basic task was imitation by naive participants of guitar chords played by an expert guitarist. By using an event-related fMRI paradigm, cortical activations were mapped during the following events: (a) observation of the chords made by an expert player, (b) pause (novel motor pattern formation and consolidation), (c) execution of the observed chords, and (d) rest. In addition to the imitation condition, there were three control conditions: observation of the guitar chords made by the player without any subsequent motor activity, observation of the chords followed by the execution of actions not related to guitar chord execution (grasp-release of the guitar neck, rhythmical covering, or gentle scratching of the fret-board), free execution of guitar chords.

The results show that, during the event observation in the imitation condition, there was activation of a cortical network formed by the inferior parietal lobule and the dorsal part of PMv plus the *pars opercularis* of IFG. This circuit was also active during the same event in the two control conditions in which participants merely observed the chords or observed them with the instruction to do subsequently an action not related to guitar chord execution. The strength of the activation was, however, much greater during imitation than during the control conditions. During observation in the imitation condition there was, in addition, activation of the anterior mesial areas, superior parietal lobule, and a modest activation of the middle frontal gyrus. The activations during the pause event in imitation condition involved the same basic circuit as in event observation, but with some important differences: increase of the superior parietal lobule activation, activation of PMd, and, most interestingly, a dramatic increase in extension and strength of the middle frontal cortex activation (area 46) and of the areas of the anterior mesial wall (Figure 9.5). Finally, during the execution event, the activation concerned, not surprisingly, the sensorimotor cortex contralateral to the acting hand.

These data show that the key centers for the formation of a novel motor pattern include the principal nodes of the mirror-neuron system. In addition, the specific involvement of the prefrontal cortex in imitation tasks (Buccino et al., 2004; Vogt et al., 2007) suggests that the

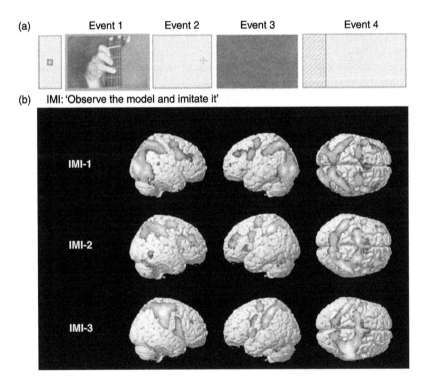

Figure 9.5 Cortical areas active during observation for imitation. (a) experimental paradigm. Each trial begins with a colored cue indicating that the subject had to perform the imitation condition (three other colored cues indicated three corresponding control conditions, not shown here). Each trial included four events. Event 1: Observation of a chord repetitively performed by a model; Event 2: Pause (preparation to imitate); Event 3: Imitation of the observed chord; Event 4: Blank screen during which the participants, after imitation, placed back their hand in rest position on the guitar neck. (b) fMRI activation of areas of the two hemispheres during the first three events of the imitation task. (Modified from Buccino et al., 2004.)

proposed subdivision of the mirror system in a *direct* and an *indirect* pathway could apply also to humans. These findings are compatible with the idea that the sensory-motor representations provided by the mirror-neuron system is exploited within a complex network that includes cortical areas involved not only in perceptuo-motor functions but also in planning, decision making, and working memory such as the prefrontal cortex. The role of this brain region appears to increase in phylogeny, since this region has undergone a significant expansion from monkeys to great apes and humans (Rilling & Insel, 1999; Semendeferi et al., 2002; Fuster, 2002). Moreover, prefrontal cortex is subject to deep changes and long-lasting development during ontogeny (Fuster, 2002). From the observed motor acts, this pathway would be fundamental for extracting the simple movements necessary to achieve its goal and to reassemble them into a new action. Therefore, this pathway is important for matching the observed movement not only in terms of action goal, but also of the specific motor patterns.

Conclusions

Several authors acknowledged the importance of distinguishing different levels among imitative processes (Byrne & Russon, 1998; Byrne, 2003; Visalberghi & Fragaszy, 2002; Meltzoff & Decety, 2003; Call & Tomasello, 2005; Brass & Heyes, 2005; Rizzolatti et al., 1999; Rizzolatti et al., 2001).

In our view the monkey mirror-neurons represent an important starting point to understand, based on its general properties and the neuroanatomical connections, the possible links with the behavioral phenomena that have been described in the monkey. It appears plausible that the mirror neuron system could be a core of a neural network that enables the monkey to code, starting from sensory information, others' action goals and motor acts already present in its motor repertoire. We propose that some behaviors (e.g. neonatal imitation) could be triggered by mirror neuron activity because of their direct influence on the motor output. Other behaviors, instead, are indirectly affected by the mirror-neuron activity as these latter provide important information about others' action goals and motor acts to other brain structures, to which they are connected. These brain structures integrate motivational and contextual factors in order to select the actions correspondent to the observed ones whenever the decision to re-enact the same motor plans is taken. The relative contribution of these two parallel systems in mirroring behaviors in the course of development depends on the level of control on intentional movements.

From a comparative perspective we can only make speculative considerations on what differentiates the human mirror system from that of the monkey in relation to imitative processes. As suggested recently by us (Ferrari & Fogassi, in press) and by other authors (Lyons et al., 2006), the capacity of the human system to encode intentions in a flexible fashion is highly dependent on the complexity of the motor system to segment movements independently and to organize complex behaviors. This possibility enlarged, in the course of human evolution, the capacity of the mirror system to encode behaviors not only in terms of goals and motor acts but also in terms of movements, even when they are devoid of a more general goal (Rizzolatti & Fabbri-Destro, 2008; Bonini & Ferrari, 2011). This clearly represents an enormous advantage to enable an individual to acquire new motor skills from others that are outside their own motor knowledge.

References

1. Barraclough, N.E., Xiao, D., Oram, M.W. and Perrett, D.I. (2006) 'The sensitivity of primate STS neurons to walking sequences and to the degree of articulation in static images', *Prog Brain Res*, 154: 135–148.
2. Belmalih, A., Borra, E., Contini, M., Gerbella, M., Rozzi, S. and Luppino, G. (2009) 'Multimodal architectonic subdivision of the rostral part (area F5) of the macaque ventral premotor cortex', *J Comp Neurol*, 512: 183–217.
3. Binkofski, F., Buccino, G., Posse, S., Seitz, R.J., Rizzolatti, G. and Freund, H.J. (1999) 'A fronto-parietal circuit for object manipulation in man: Evidence from an fMRI-study', *Eur J Neurosci*, 11: 3276–3286.
4. Binkofski, F. and Buccino, G. (2004) 'Motor functions of the Broca's region', *Brain Lang*, 89(2): 362–369.
5. Bonini, L., Rozzi, S., Ugolotti Serventi, F., Simone, L., Ferrari, P.F. and Fogassi, L. (2010) 'Ventral premotor and inferior parietal cortices make distinct contribution to action organization and intention understanding', *Cereb Cortex*, 20: 1372–1385.
6. Bonini, L. and Ferrari, P.F. (2011) 'Evolution of mirror systems: a simple mechanism for complex cognitive functions', *Ann N Y Acad Sci*, 1225: 166–175.
7. Borra, E., Belmalih, A., Calzavara, R., Gerbella, M., Murata, A., Rozzi, S. and Luppino, G. (2008) 'Cortical connections of the macaque anterior intraparietal (AIP) area', *Cereb Cortex*, 18(5): 1094–1111.
8. Borra, E., Gerbella, M., Rozzi, S. and Luppino, G. (2011) 'Anatomical evidence for the involvement of the macaque ventrolateral prefrontal area 12r in controlling goal-directed actions', *J Neurosci*, 31(34): 12351–12363.
9. Borroni, P., Montagna, M., Cerri, G. and Baldissera, F. (2005) 'Cyclic time course of motor excitability modulation during the observation of a cyclic hand movement', *Brain Res*, 1065(1–2): 115–124.
10. Borroni, P., Montagna, M., Cerri, G. and Baldissera, F. (2008) 'Bilateral motor resonance evoked by observation of a one-hand movement: role of the primary motor cortex', *Eur J Neurosci*, 28: 1427–1435.
11. Brass, M., Bekkering, H. and Prinz, W. (2001) 'Movement observation affects movement execution in a simple response task', *Acta Psychol*, 106: 3–22.
12. Brass, M. and Heyes, C. (2005) 'Imitation: is cognitive neuroscience solving the correspondence problem?', *Trends Cogn Sci*, 9: 489–495.
13. Bruce, C.J., Goldberg, M.E., Bushnell, M.C. and Stanton, G.B. (1985) 'Primate frontal eye fields. II. Physiological and anatomical correlates of electrically evoked eye movements', *J Neurophysiol*, 54: 714–734.
14. Buccino, G., Binkofski, F., Fink, G.R., Fadiga, L., Fogassi, L., Gallese, V., Seitz, R.J., Zilles, K., Rizzolatti, G. and Freund, H.J. (2001) 'Action observation activates premotor and parietal areas in a somatotopic manner: an fMRI study', *Eur J Neurosci*, 13: 400–404.
15. Buccino, G., Vogt, S., Ritzl, A., Fink, G.R., Zilles, K., Freund, H.J. and Rizzolatti, G. (2004) 'Neural circuits underlying imitation learning of hand actions: an event-related fMRI study', *Neuron*, 42: 323–334.
16. Byrne, R.W. and Russon, A.E. (1998) 'Learning by imitation: a hierarchical approach', *Behav Brain Sci*, 21: 667–684.
17. Byrne, R.W. (2003) 'Imitation as behavior parsing', *Philos Trans R Soc Lond B Biol Sci*, 358: 529–536.
18. Byrne, R.W. (2005) 'Social cognition: imitation, imitation, imitation', *Curr Biol*, 15: 498–500.
19. Caggiano, V., Fogassi, L., Rizzolatti, G., Thier, P. and Casile, A. (2009) 'Mirror neurons differentially encode the peripersonal and extrapersonal space of monkeys', *Science*, 324: 403–406.
20. Call, J., Carpenter, M. and Tomasello, M. (2005) 'Copying results and copying actions in the process of social learning: chimpanzees (*Pan troglodytes*) and human children (Homo sapiens)', *Anim Cogn*, 8: 151–163.
21. Calvo-Merino, B., Glaser, D.E., Grezes, J., Passingham, R.E. and Haggard, P. (2005) 'Action observation and acquired motor skills: an FMRI study with expert dancers', *Cereb Cortex*, 15: 1243–1249.
22. Casile, A., Caggiano, V. and Ferrari, P.F. (2011) 'The mirror neuro n system: a fresh view', *Neuroscientist*, 17(5): 524–538.

23. Cochin, S., Barthelemy, C., Lejeune, B., Roux, S. and Martineau, J. (1998) 'Perception of motion and qEEG activity in human adults', *Electroencephalogr Clin Neurophysiol*, 107: 287–295.

24. Coussi-Korbel, S. and Fragaszy, D. (1995) 'On the relation between social dynamics and social learning', *Anim Behav*, 50: 1441–1453.

25. Decety, J., Grezes, J., Costes, N., Perani, D., Jeannerod, M., Procyk, E., Grassi, F. and Fazio, F. (1997) 'Brain activity during observation of actions – Influence of action content and subject's strategy', *Brain*, 120: 1763–1777.

26. De Vries, J.I.P., Visser, G.H.A. and Prechtl, H.F.R. (1984) 'Fetal motility in the first half of pregnancy'. In Prechtl, H.F.R. (ed.) *Continuity of Neural Functions from Prenatal to Postnatal Life*. Spastics International Medical Publications.

27. Dimberg, U. and Thunberg, M. (1998) 'Rapid facial reactions to different emotionally relevant stimuli', *Scand J Psychol*, 39: 39–45.

28. Dimberg, U., Thunberg, M. and Elmehed, K. (2000) 'Unconscious facial reactions to emotional facial expressions', *Psycholog Sci*, 11: 86–89.

29. Di Pellegrino, G., Fadiga, L., Fogassi, L., Gallese, V. and Rizzolatti, G. (1992) 'Understanding motor events: a neurophysiological study', *Exp Brain Res*, 91: 176–180.

30. Ehrsson, H.H., Fagergren, E. and Forssberg, H. (2001) 'Differential fronto-parietal activation depending on force used in a precision grip task: an fMRI study', *J Neurophysiol*, 85(6): 2613–2623.

31. Fadiga, L., Fogassi, L., Pavesi, G. and Rizzolatti, G. (1995) 'Motor facilitation during action observation: a magnetic stimulation study', *J Neurophysiol*, 73: 2608–2611.

32. Ferrari, P.F., Gallese, V., Rizzolatti, G. and Fogassi, L. (2003) 'Mirror neurons responding to the observation of ingestive and communicative mouth actions in the monkey ventral premotor cortex', *Eur J Neurosci*, 17: 1703–1714.

33. Ferrari, P.F., Rozzi, S. and Fogassi, L. (2005) 'Mirror neurons responding to observation of actions made with tools in monkey ventral premotor cortex', *J Cogn Neurosci*, 17: 212–226.

34. Ferrari, P.F., Visalberghi, E., Paukner, A., Fogassi, L., Ruggiero, A. and Suomi, S.J. (2006) 'Neonatal imitation in rhesus macaques', *PLoS Biol*, 4: e302.

35. Ferrari, P.F. and Fogassi, L. 'Mirror neurons and primate social cognition. An evolutionary perspective'. In Platt, M.L. and Ghazanfar, A.A. (eds.) *Primate Neuroethology*, Oxford University Press, Oxford, UK, In press.

36. Ferrari, P.F., Bonini, L. and Fogassi, L. (2009) 'From monkey mirror neurons to primate behaviors: possible "direct" and "indirect" pathways', *Philos Trans R Soc Lond B Biol Sci*, 364(1528): 2311–2323.

37. Fogassi, L., Ferrari, P.F., Gesierich, B., Rozzi, S., Chersi, F. and Rizzolatti, G. (2005) 'Parietal lobe: from action organization to intention understanding', *Science*, 308: 662–667.

38. Fogassi, L. and Ferrari, P.F. (2007) 'Mirror neurons and the evolution of embodied language', *Curr Dir Psych Sci*, 16: 136–141.

39. Fuster, J.M. (2002) 'Frontal lobe and cognitive development', *J Neurocytol*, 31: 373–385.

40. Gallese, V., Fadiga, L., Fogassi, L. and Rizzolatti, G. (1996) 'Action recognition in the premotor cortex', *Brain*, 119: 593–609.

41. Gallese, V., Fadiga, L., Fogassi, L. and Rizzolatti, G. (2002) 'Action representation and the inferior parietal lobule'. In Prinz, W. and Hommel, B. (eds.) *Common Mechanisms in Perception and Action: Attention and Performance*. Oxford: Oxford University Press, pp. 334–355.

42. Gallese, V. (2010) 'Embodied Simulation and its role in Intersubjectivity'. In: Fuchs, T., Sattel, H.C. and Henningsen, P. (eds.) *The Embodied Self. Dimensions, Coherence and Disorders*. Stuttgart: Schattauer.

43. Gangitano, M., Mottaghy, F.M. and Pascual-Leone, A. (2001) 'Phase-specific modulation of cortical motor output during movement observation', *Neuroreport*, 12: 1489–1492.

44. Gastaut, H.J. and Bert, J. (1954) 'EEG changes during cinematographic presentation', *Electroencephalogr Clin Neurophysiol*, 6: 433–444.

45. Gerbella, M., Borra, E., Rozzi, S., Tonelli, S. and Luppino, G. (2011) 'Topographic organization of the parietal connections of the macaque ventral area 46 Program No. 914.09/QQ14', Neuroscience Meeting Planner. Washington, DC: Society for Neuroscience, 2011. Online.

46. Gregoriou, G.G., Borra, E., Matelli, M. and Luppino, G. (2006) 'Architectonic organization of the inferior parietal convexity of the macaque monkey', *J Comp Neurol*, 496(3): 422–451.

47. Grèzes, J., Costes, N. and Decety, J. (1998) 'Top down effect of strategy on the perception of human biological motion: A PET investigation', *Cognitive Neuropsychol*, 15: 553–82.

48. Heiser, M., Iacoboni, M., Maeda, F., Marcus, J. and Mazziotta, J.C. (2003) 'The essential role of Broca's area in imitation', *Eur J Neurosci Mar*, 17(5): 1123–1128.

49. Hari, R., Forss, N., Avikainen, S., Kirveskari, E., Salenius, S. and Rizzolatti, G. (1998) 'Activation of human primary motor cortex during action observation: a neuromagnetic study', *Proc Natl Acad Sci USA*, 95: 15061–15065.

50. Hata, T., Kanenishi, K., Akiyama, M., Tanaka, H. and Kimura, K. (2005) 'Real-time 3-D sonographic observation of fetal facial expression', *J Obstet Gynaecol Res*, 31: 337–340.

51. Iacoboni, M., Woods, R.P., Brass, M., Bekkering, H., Mazziotta, J.C. and Rizzolatti, G. (1999) 'Cortical mechanisms of human imitation', *Science*, 286: 2526–2528.

52. Iacoboni, M., Koski, L.M., Brass, M., Bekkering, H., Woods, R.P., Dubeau, M.C., Mazziotta, J.C. and Rizzolatti, G. (2001) 'Reafferent copies of imitated actions in the right superior temporal cortex', *Proc Natl Acad Sci USA*, 98(24): 13995–13999.

53. Iacoboni, M. (2009) 'Imitation, empathy and mirror neurons', *Ann Rev Psychol*, 60: 653–670.

54. Jeannerod, M. (1988) *The Neural and Behavioral Organization of Goaldirected Movements*. Oxford: Oxford University Press.

55. Jeannerod, M. (1994) 'The representing brain: neural correlates of motor intention and imagery', *Behav Brain Sci*, 17: 187–245.

56. Jones, S.S. (2009) 'The development of imitation in infancy', *Phil Trans R Soc B*, 364: 2325–2335.

57. Kohler, E., Keysers, C., Umiltà, M.A., Fogassi, L., Gallese, V. and Rizzolatti, G. (2002) 'Hearing sounds, understanding actions: action representation in mirror neurons', *Science*, 297: 846–848.

58. Kraskov, A., Dancause, N., Quallo, M.M., Shepherd, S. and Lemon, R.N. (2009) 'Corticospinal neurons in macaque ventral premotor cortex with mirror properties: A potential mechanism for action suppression?', *Neuron*, 64: 922–930.

59. Kurata, K. and Wise, S.P. (1988) 'Premotor cortex of rhesus monkeys: set-related activity during two conditional motor tasks', *Exp Brain Res*, 69(2): 327–343.

60. Lui, F., Buccino, G., Duzzi, D., Benuzzi, F., Crisi, G., Baraldi, P., Nichelli, P., Porro, C.A. and Rizzolatti, G. (2008) 'Neural substrates for observing and imagining non-object-directed actions', *Soc Neurosci*, 3(3–4): 261–275.

61. Lyons, D.E., Santos, L.R. and Keil, F.C. (2006) 'Reflections of other minds: how primate social cognition can inform the function of mirror neurons', *Curr Opin Neurobiol*, 16: 230–234.

62. Maeda, F., Kleiner-Fisman, G. and Pascual-Leone, A. (2002) 'Motor facilitation while observing hand actions: specificity of the effect and role of observer's orientation', *J Neurophysiol* 87: 1329–1335.

63. Marshall-Pescini, S. and Whiten, A. (2008) 'Social learning of nut-cracking behavior in East African sanctuary-living chimpanzees (*Pan troglodytes schweinfurthii*)', *J Comp Psychol*, 122: 186–194.

64. Matelli, M., Luppino, G. and Rizzolatti, G. (1985) 'Patterns of cytochrome oxidase activity in the frontal agranular cortex of the macaque monkey. Behavioral', *Brain Research*, 18: 125–136.

65. Matelli, M., Luppino, G. and Rizzolatti, G. (1991) 'Architecture of superior and mesial area 6 and the adjacent cingulate cortex in the macaque monkey', *J Comp Neurol*, 311: 445–462.

66. Meltzoff, A.N. and Decety, J. (2003) 'What imitation tells us about social cognition: a rapprochement between developmental psychology and cognitive neuroscience', *Philos Trans R Soc Lond B Biol Sci*, 358: 491–500.

67. Meltzoff, A.N. and Moore, M.K. (1977) 'Imitation of facial and manual gestures by human neonates', *Science*, 198: 75–78.

68. Meltzoff, A.N. and Moore, M.K. (1983) 'Newborn infants imitate adult facial gestures', *Child Dev*, 54: 702–709.

69. Mukamel, R., Ekstrom, A.D., Kaplan, J., Iacoboni, M. and Fried, I. (2010) 'Single-neuron responses in humans during execution and observation of actions', *Curr Biol*, 20(8): 750–756.

70. Myowa-Yamakoshi, M., Tomonaga, M., Tanaka, M. and Matsuzawa, T. (2004) 'Imitation in neonatal chimpanzees (*Pan troglodytes*)', *Dev Sci*, 7: 437–442.

71. Myowa, M. (1996) 'Imitation of facial gestures by an infant chimapnzee', *Primates*, 37: 207–213.

72. Nelissen, K., Borra, E., Gerbella, M., Rozzi, S., Luppino, G., Vanduffel, W., Rizzolatti, G. and Orban, G.A. (2011) 'Action observation circuits in the macaque monkey cortex', *J Neurosci*, 31(10): 3743–3756.

73. Nishitani, N. and Hari, R. (2000) 'Temporal dynamics of cortical representation for action', *Proc Natl Acad Sci USA*, 97: 913–918.

74. Nishitani, N. and Hari, R. (2002) 'Viewing lip forms: cortical dynamics', *Neuron*, 36(6): 1211–1220.

75. Pandya, D.N. and Seltzer, B. (1982) 'Intrinsic connections and architectonics of posterior parietal cortex in the rhesus monkey', *J Comp Neurol*, 204: 196–210.

76. Penfield, W. and Rasmussen, T. (1952) *The Cerebral Cortex of Man.* New York: Macmillan.
77. Perrett, D.I. (1989) 'Frameworks of analysis for the neural representation of animate objects and actions', *J Exp Biol*, 146: 87–113.
78. Rilling, J.K. and Insel, T.R. (1999) 'The primate neocortex in comparative perspective using magnetic resonance imaging', *J Hum Evol*, 37: 191–223.
79. Rizzolatti, G., Camarda, R., Fogassi, L., Gentilucci, M., Luppino, G. and Matelli, M. (1988) 'Functional organization of inferior area 6 in the macaque monkey. II. Area F5 and the control of distal movements', *Exp Brain Res*, 71: 491–507.
80. Rizzolatti, G., Fadiga, L., Gallese, V. and Fogassi, L. (1996) 'Premotor cortex and the recognition of motor actions', *Brain Res Cogn Brain Res*, 3: 131–141.
81. Rizzolatti, G. and Arbib, M.A. (1998) 'Language within our grasp', *Trends Neurosci*, 21: 188–194.
82. Rizzolatti, G., Luppino, G. and Matelli, M. (1998) 'The organization of the cortical motor system: new concepts', *Electroencephalogr Clin Neurophysiol*, 106(4): 283–296.
83. Rizzolatti, G., Fadiga, L., Fogassi, L. and Gallese, V. (1999) 'Resonance behaviors and mirror neurons', *Arch Ital Biol*, 137: 85–100.
84. Rizzolatti, G. and Luppino, G. (2001) 'The cortical motor system', *Neuron*, 31(6): 889–901.
85. Rizzolatti, G., Fogassi, L. and Gallese, V. (2001) 'Neurophysiological mechanisms underlying the understanding and imitation of action', *Nat Rev Neurosci*, 2: 661–670.
86. Rizzolatti, G., Fogassi, L. and Gallese, V. (2002) 'Motor and cognitive functions of the ventral premotor cortex', *Curr Opin Neurobiol*, 12: 149–154.
87. Rizzolatti, G. and Fabbri-Destro, M. (2008) 'The mirror system and its role in social cognition', *Curr Opin Neurobiol*, 18(2): 179–184.
88. Rochat, M.J., Caruana, F., Jezzini, A., Escola, L., Intskirveli, I., Grammont, F., Gallese, V., Rizzolatti, G. and Umiltà, M.A. (2010) 'Responses of mirror neurons in area F5 to hand and tool grasping observation', *Exp Brain Res*, 204: 605–616.
89. Rozzi, S., Calzavara, R., Belmalih, A., Borra, E., Gregoriou, G.G., Matelli, M. and Luppino, G. (2006) 'Cortical connections of the inferior parietal cortical convexity of the macaque monkey', *Cereb Cortex*, 16: 1389–1417.
90. Rozzi, S., Ferrari, P.F., Bonini, L., Rizzolatti, G. and Fogassi, L. (2008) 'Functional organization of inferior parietal lobule convexity in the macaque monkey: electrophysiological characterization of motor, sensory and mirror responses and their correlation with cytoarchitectonic areas', *Eur J Neurosci*, 28: 1569–1588.
91. Sato, W. and Yoshikawa, S. (2007) 'Spontaneous facial mimicry in response to dynamic facial expressions', *Cognition*, 104: 1–18.
92. Seltzer, B. and Pandya, D.N. (1984) 'Further observations on parieto-temporal connections in the rhesus monkey', *Exp Brain Res*, 55(2): 301–312.
93. Semendeferi, K., Lu, A., Schenker, N. and Damasio, H. (2002) 'Humans and great apes share a large frontal cortex', *Nat Neurosci*, 5: 272–276.
94. Strafella, A.P. and Paus, T. (2000) 'Modulation of cortical excitability during action observation: a transcranial magnetic stimulation study', *Neuroreport*, 11: 2289–2292.
95. Umiltà, M.A., Escola, L., Intskirveli, I., Grammont, F., Rochat, M., Caruana, F., Jezzini, A., Gallese, V. and Rizzolatti, G. (2008) 'When pliers become fingers in the monkey motor system', *Proc Natl Acad Sci USA*, 105: 2209–2213.
96. Umiltà, M.A., Kohler, E., Gallese, V., Fogassi, L., Fadiga, L., Keysers, C. and Rizzolatti, G. (2001) 'I know what you are doing. A neurophysiological study', *Neuron*, 31: 155–165.
97. Visalberghi, E. and Fragaszy, D.M. (2002) 'Do monkey apes? Ten years after. In Dautenhan, K. and Nehaniv, C.L. (eds.) *Imitation in Animals and Artifacts.* Cambridge, MA: MIT Press.
98. Vogt, S., Buccino, G., Wohlschlager, A.M., Canessa, N., Shah, N.J., Zilles, K., Eickhoff, S.B., Freund, H.J., Rizzolatti, G. and Fink, G.R. (2007) 'Prefrontal involvement in imitation learning of hand actions: effects of practice and expertise', *Neuroimage*, 37: 1371–1383.
99. von Economo, K. and Koskinas, G. (1925) *Die Cytoarchitektonik der Hirnrinde des erwachsenen Menschen*, Wien: Springer.
100. Whiten, A., Custance, D.M., Gomez, J.C., Teixidor, P. and Bard, K.A. (1996) 'Imitative learning of artificial fruit processing in children (*Homo sapiens*) and chimpanzees (*Pan troglodytes*)', *J Comp Psychol*, 110: 3–14.

PART III

Motor control and learning in locomotion and posture

10
NEURAL CONTROL OF WALKING

Michael J. Grey,[1] *Laurent Bouyer*[2] *and Jens Bo Nielsen*[3]

[1]COLLEGE OF LIFE & ENVIRONMENTAL SCIENCES, UNIVERSITY OF BIRMINGHAM
[2]CENTRE INTERDISCIPLINAIRE DE RECHERCHE EN RÉADAPTATION ET INTÉGRATION SOCIALE
(CIRRIS), QUEBEC
[3]DEPARTMENT OF NEUROSCIENCE AND PHARMACOLOGY, UNIVERSITY OF COPENHAGEN

Introduction

When observing an able-bodied individual walking across a room, the movement appears fluid and effortless. This graceful movement is achieved through a complex neural control system in which dozens of muscles must be activated and coordinated in the midst of an inherently unstable multiarticular system. The central nervous system integrates control commands that may be generated at cortical and spinal levels; it makes adjustments to these motor commands based on sensory feedback from a variety of sources including vestibular, visual, and somatosensory components; and it controls multi joint limbs using a variety of muscles, each of which have very different intrinsic dynamic mechanical properties.

Animal models have been fundamental in the development of our understanding of this control system. They have, for example, been crucial in demonstrating that control is distributed across the central nervous system (CNS), involving structures at all levels of complexity, from the spinal cord to the posterior parietal cortex. While research in animal models has been conducted for hundreds of years, work on the neural circuitry underlying human walking is only very recent. For example, technology for the advanced intracellular recording required to study walking has only been available in the last century. Even more recently, focal transcranial magnetic stimulation (TMS) has provided a non-invasive tool to excite or inhibit the brain during movement, thus allowing the exploration of cortical contributions during human walking. While we believe we have a good theoretical understanding of the control of human walking, much remains to be investigated. Furthermore, one must keep in mind that the bipedal nature of human locomotion will, by necessity, have required evolutionary changes in the role of some structures or in the processing of sensorimotor information compared with quadrupedal gait. The goal of this chapter is not to compare these two modes of progression, but rather to present an integrated view of our current knowledge of the neural control of *human* walking. Due to the presence of knowledge gaps, evidence from animal work is used to complement the human data, and to point in the direction of future research in the field. As several in-depth reviews have been published on this topic (Dietz and Duysens, 2000; Pearson, 1993; Pearson, 1995; Rossignol et al., 2006; Rossignol et al., 2011), the reader is referred to them for a more detailed discussion.

The locomotor pattern

When looking at the lower limb and trunk muscle activation pattern present during walking (Figure 10.1), it is impressive to see: (1) the large number of muscles activated; (2) the precise moment of activation and termination of these muscles; and (3) the differences in activation profiles (EMG envelopes) exhibited by each muscle. In fact, each muscle group seems to have a different signature, allowing the expert eye to tell them apart simply by looking at the raw signal. This complex muscle activation profile requires a precise level of control that is mostly produced without the need for conscious control.

Historical perspective and concept of a basic locomotor pattern generator

The first major advance in our understanding of the neural control of walking was made more than 100 years ago, when it was shown that the brainstem and spinal cord can produce basic rhythmic patterns of muscle activity in decerebrate cats (Bickel and Ewald, 1897; Hering, 1897; Brown, 1911). In 1911 Graham Brown demonstrated that rhythmic contractions of cat hind leg muscles could be generated following complete transection of the spinal cord (spinal-isation), thus leading to the hypothesis that the spinal cord itself contained an intrinsic mechanism that could produce alternating flexion and extension movements. This was commonly referred to as the 'half-centre' model. Much later, in the 1960s, Jankowska and Lundberg described how the application of noradrenergic precursors to acute spinal cats activated a network of spinal neurones that most likely represented the mechanism originally described by Graham Brown (Jankowska et al., 1965). Following a series of novel investigations into spinal control mechanisms, Grillner (1975) subsequently reformulated the half-centre model, and proposed that the spinal cord contains a network of cells that are capable of generating the basic locomotor pattern he called the 'central pattern generator' (CPG). Numerous studies have shown that the existence of a CPG is beyond doubt in lower animals such as the lamprey, and in quadruped mammals (reviewed in Rossignol et al., 1996). While evidence for the existence of a similar CPG in the spinal cord of higher animals such as primates and humans remains elusive (see below), these studies have nevertheless set the foundation for our current knowledge of the neural control of walking.

Using low spinal cats (i.e., animals with transection of the spinal cord at low thoracic level) walking on a treadmill, Rossignol and colleagues have demonstrated that the spinal control of locomotion can be quite sophisticated (Rossignol et al., 2002; Rossignol et al., 2004; Rossignol et al., 1996). Indeed, after a complete spinal cord transection at low thoracic level (i.e., above the lumbar enlargement), spinal cats can generate an alternating movement of the hind limbs with proper paw placement, bear the weight of their hindquarters, recover from a stumble (Forssberg et al., 1977), adjust walking speed (Barbeau and Rossignol, 1987), and compensate for added load (Duysens and Pearson, 1980; Pearson, 1993; Pearson, 1995). These movements are very similar to those made by intact animals (Belanger et al., 1996). However, spinal animals also have major deficits. For example, they cannot initiate walking voluntarily, have no postural control, and are unable to anticipate obstacles placed on their walking path. Their ability to walk with assistance on a treadmill is, however, very compatible with the expression of a functional spinal CPG.

To date, it has not been possible to elicit walking in humans with complete spinal cord transection and it is not possible to conclude definitively whether or not CPG circuitry remains in the human spinal cord. While it is clear that some level of automatic control similar to that exhibited in quadrupeds must also be present in human walking, it is quite

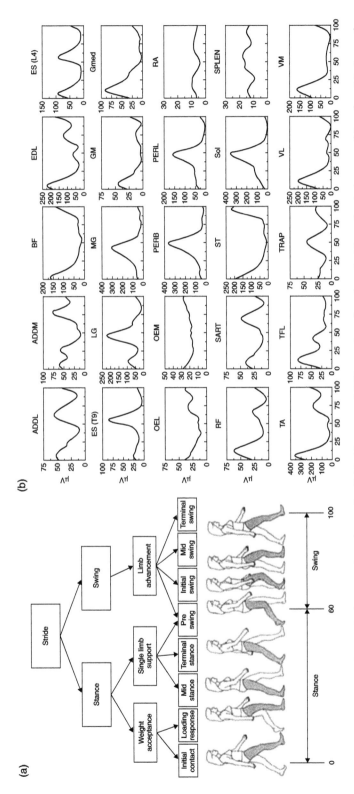

Figure 10.1 Human walking. The human gait cycle may be divided into several different phases based on the biomechanical events taking place (a). A complex pattern of EMG activity underlies these events (b) (from Winter and Yack, 1987). The relative amplitude and duration of activity in each of the muscles are presented as shaded areas. All activities were scaled in relation to the maximal EMG activity and the total duration of one gait cycle. Swing and stance phases are indicated by vertical lines.

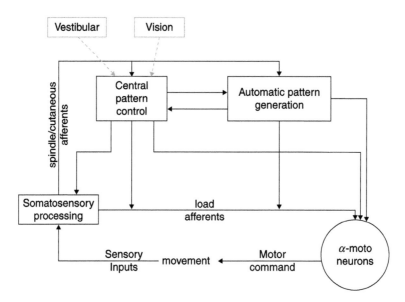

Figure 10.2 Schematic view of the neural control of walking.

likely that the evolutionary demands for bipedal walking have changed either the nature of this circuitry or the mechanism by which it operates.

Based on evidence from animal models and human studies, we can start putting together a schematic view of how the neural control of locomotion operates (Figure 10.2). One box generates the 'basic locomotor pattern'. One role of this part of the circuitry is to time the different muscles groups and to produce the different muscle activation profiles.

Role of somatosensory inputs

Sensory feedback is thought to be important for the maintenance of a stable locomotor pattern in the face of a wide range of externally imposed conditions. In particular, it has been suggested that somatosensory afferents are responsible for modulating the basic locomotor programs and/or the output from these programs; for controlling phase transitions; and for reinforcing the locomotor muscle activity (Pearson, 1993; Pearson, 1995). The discussion in this section will therefore focus on the role of somatosensory input to the neural control system.

State-dependent modulation of sensory feedback

In addition to its involvement in the timing of muscle activation, animal models have shown that the CPG also modulates sensory afferent activity and interneuron excitability, and is therefore actively involved in the sensorimotor integration that occurs during walking (Grillner et al., 1981; Rossignol et al., 1996; Rossignol et al., 2006). By modulating afferent activity, the CPG acts as a gate for sensory information. The same sensory stimulation occurring during walking or other tasks such as standing will therefore have a different influence on CNS neurons.

In humans, the Hoffmann reflex, or H-reflex, has proven to be a very useful technique in the investigation of spinal excitability during different tasks (Hoffmann, 1910; Hoffmann, 1918). For example, in a series of influential studies conducted in the 1980s, the H-reflex was used to compare task dependency between walking and standing (Morin et al., 1982; Capaday and Stein, 1986), walking and stepping (Crenna and Frigo, 1987), and to compare walking with running (Capaday and Stein, 1987). These studies showed that the ankle plantar flexor H-reflex magnitude is largest during standing; that it decreases during walking; and that it is lowest during running.

As a result of Sherrington's important investigations on the stretch reflex in the early 1900s, most human studies have focussed, either directly or indirectly, on the contribution from the monosynaptic group Ia pathway. For example, the H-reflex studies in the 1980s and 1990s led to the conclusion that the group Ia pathway provided the feedback signal that integrated with descending drive to reinforce locomotor activity. This notion was supported in subsequent stretch reflex experiments (Dietz et al., 1987; Sinkjaer et al., 1996; Yang et al., 1991). For

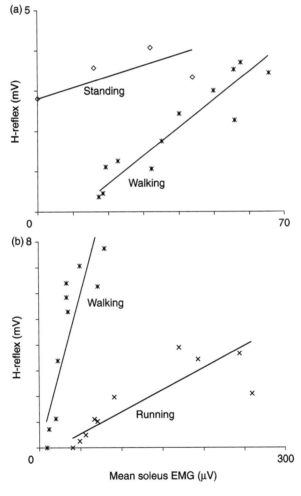

Figure 10.3 State dependent reflex modulation. Comparison in two participants of the changes in the H-reflex (at a constant M-wave) for (a) standing and walking and (b) walking and running. The difference between the H-reflexes during standing and walking involved a large change in y-intercept (a), whereas the difference between walking and running was mainly a change in slope (b). Straight lines were fitted by the standard least mean squares method. From Stein and Capaday (1988).

example, rapid dorsiflexion perturbations produced a reflex response in the soleus EMG characterised by a group Ia mediated short latency response (Sinkjaer et al., 1996) and a group II mediated medium latency response (Grey et al., 2001), again suggesting that spindle afferents contributed to the spinally mediated muscle activity. This was an attractive notion that remains popular in a large segment of the current locomotion literature, and forms the basis for numerous ideas of locomotor control including the stretch-shortening cycle.

An extreme example of state-dependent modulation occurs when a reflex changes from excitation in one task, to inhibition in another, a process called reflex reversal. In the cat, state-dependent reflex reversals have been demonstrated during the transition from standing to walking. During standing, stimulation of group Ib afferents produces an inhibition in the homonymous motoneurons, whereas stimulation of the same pathway during locomotion produced an excitation in the same motoneurons (Conway et al., 1987; Duysens and Pearson, 1980; Pearson and Collins, 1993).

Similar to the cat, state-dependent group Ib reflex reversal may also occur in humans. To study this pathway non-invasively, recent studies in humans have used a different approach consisting in removing rather than enhancing the afferent traffic to the CNS (af Klint et al., 2010; Grey et al., 2004; Nielsen and Sinkjaer, 2002; Sinkjaer et al., 2000). For the ankle extensor muscles, this can be achieved by imposing a rapid plantar flexion rotation on the ankle when the extensor muscles are active; thereby unloading the triceps surae muscle–tendon complex and transiently removing its proprioceptive feedback (Grey et al., 2004; Sinkjaer et al., 2000). This type of response has been called an unload response (Angel, 1987; Angel et al., 1965; Sinkjaer et al., 2000). Sinkjaer et al. (2000) showed this response represents the removal of afferent feedback from the locomotor muscle activity rather than autogenic group Ia activity or reciprocal inhibition from a stretch reflex of the antagonist tibialis anterior muscle. In a subsequent study, Grey et al. (2004) demonstrated that the unload response is caused by the removal of proprioceptors rather than cutaneous and muscle afferents from the foot and/or ankle, and concluded that the afferent-mediated reinforcement of the locomotor muscle activity must be effected via proprioceptors. A careful analysis of onset latencies revealed that the onset of this depression was delayed with respect to the short latency component of stretch reflex, but preceded its medium latency component, suggesting that the feedback-mediated locomotor activity might not be mediated by muscle spindle output (Grey et al., 2004). Later studies have demonstrated that the magnitude of the unload response varies with load (af Klint et al., 2010; Grey et al., 2007). In summary, the evidence in both cat and human suggests that positive load feedback, via a group Ib afferent pathway, is likely to provide the afferent-mediated feedback that reinforces the locomotor muscle activity, in contrast to the H-reflex and stretch reflex studies which implicated spindle afferents. Spindle feedback, through the groups Ia and II pathways, is more likely to provide a corrective response either by direct production of resistive force or as a trigger for supraspinal responses.

Studies designed to investigate state-dependent reflex modulation have led to the important conclusion that the CNS interprets sensory feedback differently depending on the task being performed. This leads to the conclusion that it is only through experiments performed during walking that we truly gain an understanding of the mechanisms by which sensory feedback contributes to its control.

Phase-dependent modulation of sensory feedback

In the previous section, we described how the CNS interprets sensory feedback across different tasks (e.g. standing and walking). In this section, we will look at the modulation of

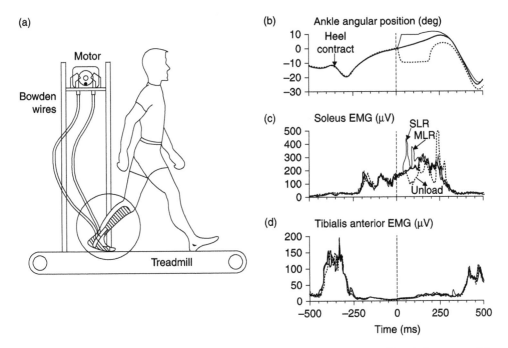

Figure 10.4 Soleus EMG responses evoked by sudden perturbations during walking in human participants. Sudden perturbations of the position of the ankle joint were induced in the stance
phase while participants were walking on a treadmill (a). This was done by a computer-
controlled servomotor through Bowden wires attached to a light-weight cast fitted around
the ankle joint of the participant. The perturbations (8 deg; 300 deg/s) were induced 300
ms after heel strike and moved the joint in either plantar flexion direction (dashed lines in
(b) and (c)) or dorsiflexion direction (full thin lines in (b) and (c)). Whereas the dorsiflexion perturbation induced two clear reflex responses at latencies of around 40 and 60 ms
(M1 and M2 respectively), the plantar flexion perturbation induced a drop in the EMG at
a latency of approximately 60 ms. Adapted from Grey et al. (2004).

sensory feedback 'within' the walking task. The step cycle consists of several phases as illustrated in Figure 10.1.

In the cat, reflexes elicited by cutaneous stimulation or by direct stimulation of the afferent
nerve are modified depending on the phase of the step cycle (Forssberg et al., 1975; Duysens
and Pearson, 1976; Forssberg et al., 1977; Duysens and Stein, 1978). Similarly, mechanically
evoked stretch reflexes evoked during different phases of the step cycle of the walking cat
produce different response amplitudes (Akazawa et al., 1982). This modulation varies
depending on the pathway studied, from simple amplitude modulation that parallels the locomotor activity of the muscle (with a small phase lead), to complete response reversal (see
below). It must be noted that phase lead and response reversals show that the reflex modulation cannot be attributed simply to changes in motoneuron excitability or phasic afferent
feedback, but must be influenced by central control. Furthermore, studies involving intracellular recordings during fictive locomotion in cats also show reflex modulation and reversals,
reinforcing the idea of a central control (Gossard et al., 1994).

Several studies in humans have also demonstrated that the amplitude of the H-reflex
elicited in a particular muscle varies in a consistent manner over the step cycle (Capaday
and Stein, 1986; Dietz et al., 1990; Dyhre-Poulsen and Simonsen, 2002; Simonsen et al.,

2002). For example, the soleus H-reflex is greatest during the stance phase and absent during the swing phase of the gait cycle (Capaday and Stein, 1986). Similarly, stretch reflexes evoked during walking are phase-modulated in the soleus (Sinkjaer et al., 1996; Yang et al., 1991), and tendon tap reflexes are modulated in biceps femoris (Faist et al., 1999a; Van de Crommert et al., 1996) and rectus femoris (Dietz et al., 1990; Faist et al., 1999b) muscles.

A particularly extreme example of phase-dependent reflex modulation is the response to skin stimulation during walking observed in both cat and human. If the dorsal surface of a cat's paw is stimulated during early swing, an excitatory reflex response is produced in tibialis anterior, an ankle flexor muscle in the cat (Rossignol & Jones, 1976). If the same stimulation is applied at the end of swing, an inhibition is produced instead (phase-dependent reflex reversal), thereby preventing the foot from being dorsiflexed just before heel strike, a situation that could compromise balance or even cause a fall. Similar observations have been demonstrated with non-noxious electrical stimulation of the cutaneous nerves supplying the foot in humans (Duysens et al., 1992; Zehr et al., 1997; Van Wezel et al., 1997; Zehr et al., 1998a; Zehr et al., 1998b) and with stretch reflexes (Dietz et al., 1986; Eng et al., 1994). As discussed further below, a significant part of these reflexes are likely mediated by transcortical reflex pathways in humans.

In summary, sensory input provides a rich variety of information that is used by the CNS to modulate its output according to the task and the phase of movement. Sensory feedback is, therefore, used to adapt to the environment and provide *functionally* meaningful output.

Role of supraspinal structures

While an automatic neural circuit produces the locomotor rhythm and determines the pattern of muscle activation, it is generally thought that supraspinal structures are responsible for the initiation and termination of the motor programs that produce walking (Armstrong, 1988; Rossignol et al., 1996). At about the same time that the nature of the CPG was being determined, the Russian researchers, Shik, Orlovsky and Severin (1966), demonstrated that continuous stimulation of a set of neurones in the brainstem and the lateral hypothalamus, just superior to the brainstem, called the mesencephalic locomotor region (MLR), could evoke organized locomotion in decerebrate cats by activating the spinal locomotor centers. Electrical stimulation of the MLR initiates stepping and the speed of stepping can be controlled by the intensity of the electrical stimulus. Like the locomotor CPG circuitry, a human MLR has not yet been conclusively demonstrated although it is widely believed that such control circuitry exists. Human studies have failed to demonstrate activation in the brainstem during actual locomotion tasks, but in a single fMRI study activation was observed in the subthalamic nucleus, the pontomesencephalic tegmentum and the paramedian reticular formation during imagined locomotion (Jahn et al., 2008; la Fougere et al., 2010). It was suggested that this could reflect activation of a circuitry involved in planning and initiation of locomotion rather than the actual execution of locomotion, but it also cannot be excluded that the activation reflects suppression of activity in the brainstem areas to prevent the participants from executing the movements while imagining the locomotion.

Numerous studies have demonstrated activity in corticospinal cells during uncomplicated overground walking in the cat (Armstrong and Drew, 1984; Beloozerova and Sirota, 1985). Consistent with this, human studies have demonstrated increased metabolic activity in the primary motor cortex during gait (Fukuyama et al., 1997; Iseki et al., 2008; la Fougere et al., 2010), simulation of gait (Jahn et al., 2008; la Fougere et al., 2010) and during rhythmic bicycling resembling gait (Christensen et al., 2000a). Motor evoked potentials (MEPs) in leg and

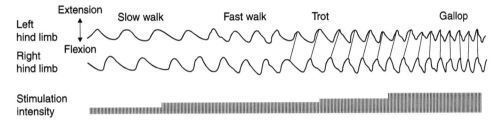

Figure 10.5 Gait pattern evoked in the cat by MLR stimulation. The graph shows extension-flexion movements of the right hind limb in a cat during stimulation of the MLR at increasing intensity. It is seen that the frequency of movement increases with the stimulation intensity reflecting a change from slow to fast walking, trot and finally gallop. (Modified from Keir Pearsons (in Kandel et al. 2000)).

arm muscles elicited by transcranial magnetic stimulation (TMS) over the primary motor cortex are also rhythmically modulated during gait, but this may be explained simply by the modulation of excitability in spinal motoneurones (Schubert et al., 1997; Capaday et al., 1999). However, Petersen et al. (2001) and Barthelemy & Nielsen (2010) demonstrated that subthreshold TMS produces a suppression of EMG in leg and arm muscles during gait. Since this suppression is in all likelihood mediated by activation of intracortical inhibitory interneurones, this finding reflects a removal of the corticospinal contribution to the EMG activity and thus provides good evidence of a direct contribution of activity in the primary motor cortex to the EMG activity. Independent evidence of this has been provided from analysis of the coupling between EEG and EMG activity during gait (Petersen et al., 2012). It has been known for some time that tibialis anterior motor units show coupled activity in the 15–35 Hz frequency band in the swing phase during gait (Halliday et al., 2003). Petersen et al. (2012) provided evidence that some of this coupled activity is generated by the motor cortex since a coupling in a similar frequency band between EEG and EMG recordings was found.

There is thus little doubt that the primary motor cortex and the corticospinal tract contribute directly to the EMG activity during uncomplicated gait. This is also testified by the pronounced foot drop and generally impaired gait ability that is experienced following lesion of the corticospinal tract (Barthelemy and Nielsen, 2010).

In all likelihood the motor cortex and the corticospinal tract contributes even more importantly when the gait activity has to be modulated in order to avoid obstacles, when steering in specific directions or when walking in a complex environment. A long line of experiments have demonstrated that this is certainly the case in the cat. Whereas overground locomotion in an uncomplicated environment may be seen following motor cortical ablation (albeit with some paw drag), the cats are greatly impaired when trying to walk under more complex circumstances (Armstrong, 1988; Beloozerova and Sirota, 1985; Drew, 1991). Consistent with this, much greater corticospinal activity is seen under these circumstances than during uncomplicated overground or treadmill walking (Drew, 1991; Beloozerova and Sirota, 1986). Also, when the cats have to react to a visually observed obstacle by lifting the paw, very significant corticospinal activity is observed in relation to the paw lift (Beloozerova and Sirota, 1986; Marple-Horvat and Armstrong, 1999). The observations from the cat suggest that one important contribution of the motor cortex and the corticospinal tract during locomotion is in end-point control; i.e. ensuring proper foot placement. Observations from human experiments during treadmill locomotion are consistent with this in finding predominant corticospinal excitability towards the end of the swing phase just prior to heel strike (Petersen et al., 2012). In humans much less is known about corticospinal activity during gait modifications and gait under more challenging circumstances. Schubert et al. (1997) observed larger MEPs during visual steering of gait than during uncomplicated treadmill walking, which suggests that there may be a larger contribution of corticospinal activity to the gait activity in that situation, consistent with the idea that the corticospinal tract may play a specific role in foot placement during gait. With new advances in methodology and technology it is likely that much more knowledge is going to be available regarding this important issue in the next few years.

Transcortical reflexes

In humans, it is generally found that long-latency reflexes are larger than the shorter latency spinal reflexes during locomotion (Christensen et al., 2000a). A range of different experimental approaches have shown that these long-latency reflexes are in all likelihood at least partly mediated by transcortical reflexes with the corticospinal tract as the efferent limb of the reflex (Christensen et al., 2000b; van Doornik et al., 2004; Zuur et al., 2009). In the stance phase long-latency stretch responses dominate in all muscles, suggesting that they play a major role in ensuring stability of the supporting limb. In contrast long-latency cutaneous reflexes dominate in the swing phase, suggesting that they are likely involved in adaptive reactions to unexpected contact with obstacles (Christensen et al., 2000b; Zehr and Stein, 1999). As is the case for spinal reflexes in the cat, these long-latency cutaneous reflexes may show reflex reversal depending on the functional requirements at the specific time of the gait cycle (Nielsen and Sinkjaer, 1997; Yang et al., 1990). The advantage of integrating the sensory feedback primarily at the cortical level in humans rather than at lower levels is unclear, but may be related to the necessity of coordinating muscle activity throughout the body in order to maintain posture during human bipedal gait and/or the necessity of integrating visual and somatosensory control before issuing the motor commands to the muscles.

Motor planning in locomotion

This may tie in also to the planning involved in the various gait modifications which are necessary during gait in response to visual information of upcoming obstacles (Cinelli and Patla, 2008; Drew et al., 2008). Little is known about the structures involved in this planning, but it is conceivable that circuitries involving premotor cortex (PMC), supplementary motor area (SMA), basal ganglia, cerebellum and posterior parietal cortex (PPC) are involved as has been shown for reaching movements in both monkeys and humans (Rizzolatti and Luppino, 2001). In line with this, experiments in the cat have demonstrated significant activity in PPC neurones during locomotion especially when involving obstacle avoidance (Beloozerova and Sirota, 2003) and PPC lesion produces significant errors in foot placement in front of an advancing obstacle and causes problems in forelimb–hindlimb coordination (Drew et al., 2008). PPC in addition plays a significant role in the working memory of the coordinates of an obstacle that has recently been traversed (Lajoie et al., 2010). Imaging studies in humans have shown activity in each of PMC, SMA and PPC during locomotion, indicating that cortical processing and planning takes place in relation to locomotion in much the same way as for any other voluntary motor task, but the precise role of this activity is unclear and requires further research.

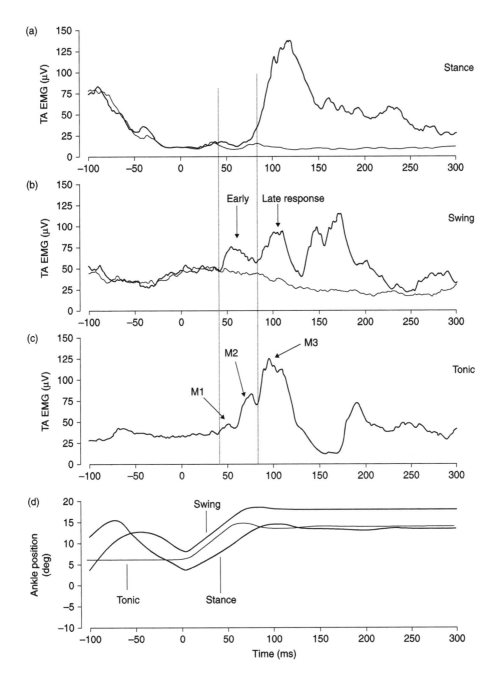

Figure 10.6 Comparison of TA stretch responses during the walking cycle and during tonic dorsi-flexion in a sitting participant. In (a) the response in the TA EMG evoked by stretch in the early stance phase (100 ms after heel contact) is shown, whereas (b) and (c) show the responses to stretch in the early swing phase (700 ms after heel contact) and during static dorsiflexion. The positional records are shown in (d). The static dorsiflexion was recorded with the participant sitting down and performing a dorsiflexion against a resistance. The participant was asked to produce a similar amount of TA EMG activity to that in early swing. Time zero corresponds to stretch onset. Modified from Christensen et al. (2001).

Cerebellum and gait

The significance of the cerebellum for locomotion is clearly evidenced from the postural problems and ataxic gait in patients with cerebellar disorders (Mariotti et al., 2005). Such observations suggest that the cerebellum is chiefly important for the generation of appropriate movement patterns and the feedforward control of locomotion based on error-feedback (Morton and Bastian, 2007). In cats, the CPG sends an efference copy of its motor command to the cerebellum through axons ascending in the ventral spinocerebellar tract. The cerebellum can therefore compare this information with the limb sensory feedback arriving through axons ascending in the dorsal spinocerebellar tract and fulfil its role as a comparator of expected and actual movement trajectory, even in this complex control situation where both voluntary and automatic drive reach limb motoneurons. In line with this, cerebellar neurons discharge rhythmically during walking in the cat (Armstrong, 1988) and significant activation is seen in the cerebellum during locomotion in humans (Fukuyama et al., 1997; Jahn et al., 2008; la Fougere et al., 2010). Also, locomotor adaptation is greatly influenced by transcranial magnetic stimulation applied over the cerebellum (Jayaram et al., 2011) consistent with the idea that the cerebellum is mainly involved in the feedforward adaptation and planning of gait.

Basal ganglia and gait

In vertebrates including the lamprey and the cat, the basal ganglia has a major role in the initiation of locomotion and in the selection of goal-directed locomotion (Grillner et al., 2008). The severe gait deficits observed in Parkinson patients including freezing of gait are consistent with a similar role in humans, but it should also be remembered that the loss of dopaminergic innervation in Parkinson's disease affects many levels of the nervous system (Shine et al., 2011). The pathophysiological mechanisms involved in freezing of gait are therefore far from clarified. Human imaging studies have in general found activation in the basal ganglia (Iseki et al., 2008; Fukuyama et al., 1997), but little is known about the exact role of this activity.

In summary, supraspinal structures are involved in the volitional control of gait (initiation/termination), interlimb coordination (walking vs running), obstacle avoidance (visuomotor control), and feedforward adaptation of gait. They can also contribute to within step modifications by means of transcortical reflexes. While a large body of literature exists in animal models, a lot remains to be documented in humans, and understood with respect to our unique bipedal gait pattern.

References

af Klint R, Mazzaro N, Nielsen JB, et al. (2010) Load rather than length sensitive feedback contributes to soleus muscle activity during human treadmill walking. *J Neurophysiol* 103: 2747–2756.

Akazawa K, Aldridge JW, Steeves JD, et al. (1982) Modulation of stretch reflexes during locomotion in the mesencephalic cat. *J Physiol* 329: 553–567.

Angel RW. (1987) Unloading reflex of hand muscle. *Electroencephalogr Clin Neurophysiol* 67: 447–451.

Angel RW, Eppler W and Iannone A. (1965) Silent period produced by unloading of muscle during voluntary contraction. *Physiol Lond* 180: 864–870.

Armstrong DM. (1988) The supraspinal control of mammalian locomotion. *J Physiol Lond* 405: 1–37.

Armstrong DM and Drew T. (1984) Locomotor-related neuronal discharges in cat motor cortex compared with peripheral receptive fields and evoked movements. *J Physiol* 346: 497–517.

Barbeau H and Rossignol S. (1987) Recovery of locomotion after chronic spinalization in the adult cat. *Brain Res* 412: 84–95.

Barthelemy D and Nielsen JB. (2010) Corticospinal contribution to arm muscle activity during human walking. *J Physiol* 588: 967–979.

Belanger M, Drew T, Provencher J, et al. (1996) A comparison of treadmill locomotion in adult cats before and after spinal transection. *J Neurophysiol* 76: 471–491.

Beloozerova IN and Sirota MG. (1985) [Activity of neurons of the motosensory cortex during natural locomotion in the cat]. *Neirofiziologiia* 17: 406–408.

Beloozerova IN and Sirota MG. (1986) [Activity of neurons of the motor-sensory cortex of the cat during natural locomotion while stepping over obstacles]. *Neirofiziologiia* 18: 546–549.

Beloozerova IN and Sirota MG. (2003) Integration of motor and visual information in the parietal area 5 during locomotion. *J Neurophysiol* 90: 961–971.

Bickel A and Ewald JR. (1897) Ueber den Einfluss der sensibelen Nerven und der Labyrinthe auf die Bewegungen der Thiere. *Pflügers Arch Eur J Physiol* 67: 299–344.

Brown TG. (1911) The Intrinsic Factors in the Act of Progression in the Mammal. *Proc R S Lond B* 84: 308–319.

Capaday C, Lavoie BA, Barbeau H, et al. (1999) Studies on the corticospinal control of human walking. I. Responses to focal transcranial magnetic stimulation of the motor cortex. *J Neurophysiol* 81: 129–139.

Capaday C and Stein RB. (1986) Amplitude modulation of the soleus H-reflex in the human during walking and standing. *J Neurosci* 6: 1308–1313.

Capaday C and Stein RB. (1987) Difference in the amplitude of the human soleus H reflex during walking and running. *J Physiol* 392: 513–522.

Christensen LO, Andersen JB, Sinkjaer T, et al. (2001) Transcranial magnetic stimulation and stretch reflexes in the tibialis anterior muscle during human walking. *J Physiol* 531: 545–557.

Christensen LO, Johannsen P, Sinkjaer T, et al. (2000a) Cerebral activation during bicycle movements in man. *Exp Brain Res* 135: 66–72.

Christensen LO, Petersen N, Andersen JB, et al. (2000b) Evidence for transcortical reflex pathways in the lower limb of man. *Prog Neurobiol* 62: 251–272.

Cinelli ME and Patla AE. (2008) Task-specific modulations of locomotor action parameters based on on-line visual information during collision avoidance with moving objects. *Hum Mov Sci* 27: 513–531.

Conway BA, Hultborn H and Kiehn O. (1987) Proprioceptive input resets central locomotor rhythm in the spinal cat. *Exp Brain Res* 68: 643–656.

Crenna P and Frigo C. (1987) Excitability of the soleus H-reflex arc during walking and stepping in man. *Exp Brain Res* 66: 49–60.

Dietz V, Discher M, Faist M, et al. (1990) Amplitude modulation of the human quadriceps tendon jerk reflex during gait. *Exp Brain Res* 82: 211–213.

Dietz V and Duysens J. (2000) Significance of load receptor input during locomotion: a review. *Gait Posture* 11: 102–110.

Dietz V, Quintern J and Berger W. (1986) Stumbling reactions in man: release of a ballistic movement pattern. *Brain Res* 362: 355–357.

Dietz V, Quintern J and Sillem M. (1987) Stumbling reactions in man: significance of proprioceptive and pre-programmed mechanisms. *J Physiol* 386: 149–163.

Drew T. (1991) Visuomotor coordination in locomotion. *Curr Opin Neurobiol* 1: 652–657.

Drew T, Andujar JE, Lajoie K, et al. (2008) Cortical mechanisms involved in visuomotor coordination during precision walking. *Brain Res Rev* 57: 199–211.

Duysens J and Pearson KG. (1976) The role of cutaneous afferents from the distal hindlimb in the regulation of the step cycle of thalamic cats. *Exp Brain Res* 24: 245–255.

Duysens J and Pearson KG. (1980) Inhibition of flexor burst generation by loading ankle extensor muscles in walking cats. *Brain Res* 187: 321–332.

Duysens J and Stein RB. (1978) Reflexes induced by nerve stimulation in walking cats with implanted cuff electrodes. *Exp Brain Res* 32: 213–224.

Duysens J, Tax AA, Trippel M, et al. (1992) Phase-dependent reversal of reflexly induced movements during human gait. *Exp Brain Res* 90: 404–414.

Dyhre-Poulsen P and Simonsen EB. (2002) H reflexes recorded during locomotion. *Adv Exp Med Biol* 508: 377–383.

Eng JJ, Winter DA and Patla AE. (1994) Strategies for recovery from a trip in early and late swing during human walking. *Exp Brain Res* 102: 339–349.

Faist M, Blahak C, Duysens J, et al. (1999a) Modulation of the biceps femoris tendon jerk reflex during human locomotion. *Exp Brain Res* 125: 265–270.

Faist M, Ertel M, Berger W, et al. (1999b) Impaired modulation of quadriceps tendon jerk reflex during spastic gait: differences between spinal and cerebral lesions. *Brain* 122 (Pt 3): 567–579.

Forssberg H, Grillner S and Rossignol S. (1975) Phase dependent reflex reversal during walking in chronic spinal cats. *Brain Res* 85: 103–107.

Forssberg H, Grillner S and Rossignol S. (1977) Phasic gain control of reflexes from the dorsum of the paw during spinal locomotion. *Brain Res* 132: 121–139.

Fukuyama H, Ouchi Y, Matsuzaki S, et al. (1997) Brain functional activity during gait in normal subjects: a SPECT study. *Neurosci Lett* 228: 183–186.

Gossard JP, Brownstone RM, Barajon I, et al. (1994) Transmission in a locomotor-related group Ib pathway from hindlimb extensor muscles in the cat. *Exp Brain Res* 98: 213–228.

Grey MJ, Ladouceur M, Andersen JB, et al. (2001) Group II muscle afferents probably contribute to the medium latency soleus stretch reflex during walking in humans. *J Physiol* 534: 925–933.

Grey MJ, Mazzaro N, Nielsen JB, et al. (2004) Ankle extensor proprioceptors contribute to the enhancement of the soleus EMG during the stance phase of human walking. *Can J Physiol Pharmacol* 82: 610–616.

Grey MJ, Nielsen JB, Mazzaro N, et al. (2007) Positive force feedback in human walking. *J Physiol* 581: 99–105.

Grillner S. (1975) Locomotion in vertebrates: central mechanisms and reflex interaction. *Physiol Rev* 55: 247–304.

Grillner S, McClellan A, Sigvardt K, et al. (1981) Activation of NMDA-receptors elicits 'fictive loco-motion' in lamprey spinal cord in vitro. *Acta Physiol Scand* 113: 549–551.

Grillner S, Wallen P, Saitoh K, et al. (2008) Neural bases of goal-directed locomotion in vertebrates – an overview. *Brain Res Rev* 57: 2–12.

Halliday DM, Conway BA, Christensen LO, et al. (2003) Functional coupling of motor units is modulated during walking in human subjects. *J Neurophysiol* 89: 960–968.

Hering HE. (1897) Ueber Bewegungsstörungen nach centripetaler Lähmung. *Naunyn Schmiedebergs Arch Pharmacol* 38: 266–283.

Hoffmann P. (1910) Beiträge zur Kenntnis der menschlichen Reflexe mit besonderer Berücksichtigung der electrischen Erscheinungen. *Arch Anat Physiol* 1: 223–246.

Hoffmann P. (1918) Concerning the connections of tendon reflexes for deliberate movement and tonus. *Zeitschrift für Biologie* 68: 351–370.

Iseki K, Hanakawa T, Shinozaki J, et al. (2008) Neural mechanisms involved in mental imagery and observation of gait. *Neuroimage* 41: 1021–1031.

Jahn K, Deutschlander A, Stephan T, et al. (2008) Imaging human supraspinal locomotor centers in brainstem and cerebellum. *Neuroimage* 39: 786–792.

Jankowska E, Jukes MGM, Lund S, et al. (1965) Reciprocal innervation through interneuronal inhibition [29]. *Nature* 206: 198–199.

Jayaram G, Galea JM, Bastian AJ, et al. (2011) Human locomotor adaptive learning is proportional to depression of cerebellar excitability. *Cereb Cortex* 21: 1901–1909.

Kandel ER, Schwartz JH, Jessel, TM. (2000) Principle of neural science. 4th edn. McGraw-Hill.

la Fougere C, Zwergal A, Rominger A, et al. (2010) Real versus imagined locomotion: a [18F]-FDG PET-fMRI comparison. *Neuroimage* 50: 1589–1598.

Lajoie K, Andujar JE, Pearson K, et al. (2010) Neurons in area 5 of the posterior parietal cortex in the cat contribute to interlimb coordination during visually guided locomotion: a role in working memory. *J Neurophysiol* 103: 2234–2254.

Mariotti C, Fancellu R and Di Donato S. (2005) An overview of the patient with ataxia. *J Neurolo* 252: 511–518.

Marple-Horvat DE and Armstrong DM. (1999) Central regulation of motor cortex neuronal responses to forelimb nerve inputs during precision walking in the cat. *J Physiol* 519 Pt 1: 279–299.

Morin C, Katz R, Mazieres L, et al. (1982) Comparison of soleus H reflex facilitation at the onset of soleus contractions produced voluntarily and during the stance phase of human gait. *Neurosci Lett* 33: 47–53.

Morton SM and Bastian AJ. (2007) Mechanisms of cerebellar gait ataxia. *Cerebellum* 6: 79–86.

Nielsen JB and Sinkjaer T. (2002) Afferent feedback in the control of human gait. *J Electromyogr Kinesiol* 12: 213–217.

Nielsen JF and Sinkjaer T. (1997) Long-lasting depression of soleus motoneurons excitability following repetitive magnetic stimuli of the spinal cord in multiple sclerosis patients. *Mult Scler* 3: 18–30.

Pearson KG. (1993) Common principles of motor control in vertebrates and invertebrates. *Annu Rev Neurosci* 16: 265–297.

Pearson KG. (1995) Proprioceptive regulation of locomotion. *Curr Opin Neurobiol* 5: 786–791.

Pearson KG and Collins DF. (1993) Reversal of the influence of group Ib afferents from plantaris on activity in medial gastrocnemius muscle during locomotor activity. *J Neurophysiol* 70: 1009–1017.

Petersen NT, Butler JE, Marchand-Pauvert V, et al. (2001) Suppression of EMG activity by transcranial magnetic stimulation in human subjects during walking. *J Physiol* 537: 651–656.

Petersen TH, Willerslev-Olsen M, Conway BA, et al. (2012) The motor cortex drives the muscles during walking in human subjects. *J Physiol*. May 15; 590(10): 2443–2452.

Rizzolatti G and Luppino G. (2001) The cortical motor system. *Neuron* 31: 889–901.

Rossignol S, Bouyer L, Barthelemy D, et al. (2002) Recovery of locomotion in the cat following spinal cord lesions. *Brain Res Brain Res Rev* 40: 257–266.

Rossignol S, Bouyer L, Langlet C, et al. (2004) Determinants of locomotor recovery after spinal injury in the cat. *Prog Brain Res* 143: 163–172.

Rossignol S, Chau C, Brustein E, et al. (1996) Locomotor capacities after complete and partial lesions of the spinal cord. *Acta Neurobiol Exp (Wars)* 56: 449–463.

Rossignol S, Dubuc R and Gossard JP. (2006) Dynamic sensorimotor interactions in locomotion. *Physiol Rev* 86: 89–154.

Rossignol S, Frigon A, Barriere G, et al. (2011) Chapter 16 – spinal plasticity in the recovery of loco-motion. *Prog Brain Res* 188: 229–241.

Rossignol S, Jones GM. (1976) Audio-spinal influence in man studied by the H-reflex and its possible role on rhythmic movements synchronized to sound. *Electroencephalogr Clin Neurophysiol* 41:83–92.

Schubert M, Curt A, Jensen L, et al. (1997) Corticospinal input in human gait: modulation of magneti-cally evoked motor responses. *Exp Brain Res* 115: 234–246.

Shik ML, Orlovskii GN and Severin FV. (1966) [Organization of locomotor synergism]. *Biofizika* 11: 879–886.

Shine JM, Naismith SL and Lewis SJ. (2011) The pathophysiological mechanisms underlying freezing of gait in Parkinson's Disease. *J Clin Neurosci* 18: 1154–1157.

Simonsen EB, Dyhre-Poulsen P, Alkjaer T, et al. (2002) Interindividual differences in H reflex modu-lation during normal walking. *Exp Brain Res* 142: 108–115.

Sinkjaer T, Andersen JB, Ladouceur M, et al. (2000) Major role for sensory feedback in soleus EMG activity in the stance phase of walking in man. *J Physiol* 523 Pt 3: 817–827.

Sinkjaer T, Andersen JB and Larsen B. (1996) Soleus stretch reflex modulation during gait in humans. *J Neurophysiol* 76: 1112–1120.

Stein RB and Capaday C. (1988) The modulation of human reflexes during functional motor tasks. *Trends Neurosci* 11: 328–332.

Van de Crommert HW, Faist M, Berger W, et al. (1996) Biceps femoris tendon jerk reflexes are enhanced at the end of the swing phase in humans. *Brain Res* 734: 341–344.

van Doornik J, Masakado Y, Sinkjaer T, et al. (2004) The suppression of the long-latency stretch reflex in the human tibialis anterior muscle by transcranial magnetic stimulation. *Exp Brain Res* 157: 403–406.

Van Wezel BM, Ottenhoff FA and Duysens J. (1997) Dynamic control of location-specific information in tactile cutaneous reflexes from the foot during human walking. *J Neurosci* 17: 3804–3814.

Winter DA and Yack HJ. (1987) EMG profiles during normal human walking: stride-to-stride and inter-subject variability. *Electroencephalogr Clin Neurophysiol* 67: 402–411.

Yang JF, Stein RB and James KB. (1991) Contribution of peripheral afferents to the activation of the soleus muscle during walking in humans. *Exp Brain Res* 87: 679–687.

Yang JF, Stein RB, Jhamandas J, et al. (1990) Motor unit numbers and contractile properties after spinal cord injury. *Ann Neurol* 28: 496–502.

Zehr EP, Fujita K and Stein RB. (1998a) Reflexes from the superficial peroneal nerve during walking in stroke subjects. *J Neurophysiol* 79: 848–858.

Zehr EP, Stein RB, Komiyama T. (1998b) Function of sural nerve reflexes during human walking. *J Physiol* 507: 305–314.

Zehr EP, Komiyama T and Stein RB. (1997) Cutaneous reflexes during human gait: electromyographic and kinematic responses to electrical stimulation. *J Neurophysiol* 77: 3311–3325.

Zehr EP and Stein RB. (1999) What functions do reflexes serve during human locomotion? *Prog Neurobiol* 58: 185–205.

Zehr EP, Stein RB and Komiyama T. (1998b) Function of sural nerve reflexes during human walking. *J Physiol* 507 (Pt 1): 305–314.

Zuur AT, Christensen MS, Sinkjaer T, et al. (2009) Tibialis anterior stretch reflex in early stance is suppressed by repetitive transcranial magnetic stimulation. *J Physiol* 587: 1669–1676.

11

ADAPTIVE PLASTICITY OF GAIT

Laurent Bouyer,[1] *Michael J. Grey*[2] *and Jens Bo Nielsen*[3]

[1]DEPARTMENT OF REHABILITATION, UNIVERSITÉ LAVAL, QUEBEC CITY
[2]COLLEGE OF LIFE & ENVIRONMENTAL SCIENCES, UNIVERSITY OF BIRMINGHAM
[3]DEPARTMENT OF NEUROSCIENCE AND PHARMACOLOGY, UNIVERSITY OF COPENHAGEN

Why is it necessary to adapt gait?

The previous chapter described the mechanisms by which the sensory feedback is integrated with CNS control systems to produce smooth and fluid locomotion. There are many situations that will challenge this homeostasis: growing, aging, injury to the central or peripheral nervous system, musculoskeletal injury, walking on different terrains, etc. As locomotion is one of the key rhythmical movements for survival (together with breathing, chewing and reproduction), the nervous system requires mechanisms by which gait can remain functionally adapted. Globally, these mechanisms can be divided into two main categories: movement recalibration and skilled learning. Both require goal-directed modifications in the neural control of walking, a process called adaptive plasticity.

Adjusting rhythmic motor output over the lifespan: An example of movement recalibration

Growing from children to adults, gait fluidity remains regardless of the transformations in body mass and length of body segments. Therefore, for movements to remain fluid despite the changes in musculoskeletal metrics and dynamics, the stereotyped muscle activation pattern producing walking (motor output) must be continuously adjusted, a process called *recalibration*. Movement recalibration can be considered a form of continuous motor learning, with the goal of keeping movement optimally 'tuned'. Its theoretical limit is the capacity for change of the neural control circuitry generating the rhythmic behaviour, i.e. adaptive plasticity. Understanding the extent of adaptive plasticity is of great interest in the field of motor control and rehabilitation, as it can provide a framework for developing more efficient locomotor rehabilitation protocols.

Combined feedforward and feedback control: Implications for adaptive capacity

Before further considering the mechanisms of locomotor movement plasticity, it is important to understand how this movement is controlled by the central nervous system. The generation of the locomotor muscle activation pattern involves voluntary commands, central pattern generators and sensory feedback (see previous chapter). Schematically, locomotion can therefore be considered to involve two types of neural control: feedforward and feedback. Feedforward control is what the central nervous system components (voluntary drive and pattern generators) provide to the muscle activation pattern. This includes the 'predictive' or planned aspect of the movement. Motor learning occurs through adaptive modifications in feedforward control. Feedback control is what sensory feedback components provide to motor output. In the case of locomotion, both positive and negative feedback loops contribute to muscle activation pattern shaping.

It should be noted that positive feedback is not only a powerful mode of control, but also of movement adaptation to changes in the environment. It must, therefore, not be neglected when considering movement adaptation. Its *reactive* nature will make it come 'on' and 'off' as needed. However, being reactive to sensory feedback limits its versatility: for example, it cannot be used to preload a muscle in the expectation of an upcoming dynamic event such as landing from a fall/jump. Furthermore, motor output adjustments generated by positive feedback will only transfer to other situations associated with similar sensory feedback, and as such are not very useful to train new motor patterns for rehabilitation purposes. On the contrary, feedforward adaptation, requiring modifications in central predictive elements, will have a better chance of transferring to other tasks due to a reduced need for continuous sensory feedback in this mode of control.

Central pattern generators: Implications for adaptive capacity

While changing feedforward control to update the locomotor muscle activation pattern may seem to be a relatively simple task for the central nervous system (CNS), the fact that central pattern generators (CPGs) are involved in movement control could potentially complicate the situation and reduce adaptive capacity. Indeed, CPGs are designed to produce autonomously a stereotyped muscle activation pattern. Such interconnected interneuronal networks create specific restricted synergies at different moments of the step cycle (McCrea and Rybak, 2007; Grillner et al., 1998). While this coupling is not absolute (e.g. burst deletions can occur; (Lafreniere-Roula and McCrea, 2005; Rybak et al., 2006), it is also not random (Trank et al., 1996; Carlson-Kuhta et al., 1998; Smith et al., 1998). For the sake of the present review, it is important to keep in mind that CPGs probably follow simple rules of activation (Yakovenko et al., 2002; Prochazka et al., 2002), and therefore changing the drive to one motoneuron pool for adaptive reasons may have consequences on other, maybe more remote motoneuron pools due to network properties. As a consequence, adaptation in central drive must take this factor into consideration.

Assessing the relative contribution of positive feedback and feedforward control mechanisms

As motor learning requires modifications in feedforward (i.e. predictive) control, separating feedforward control from positive feedback control is essential when attempting to understand underlying neural mechanisms. To do so, sensory feedback must be unexpectedly removed, thereby isolating the feedforward contribution to motor output. A good example of how this can be readily done is the 'foot in the hole' protocol (Fig. 11.1). In freely walking cats, when motor output from ankle extensors measured during normal gait is compared to a situation where the support surface is rapidly removed just before the paw is expected to contact the

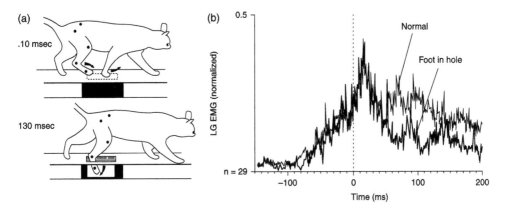

Figure 11.1   Foot in the hole protocol. (a) Line drawings illustrating the procedure. A force sensor built into the trap door detects forelimb contacts and trigger door opening at the end of fore-limb stance. (b) Average rectified EMG activity of lateral gastrocnemius when the cat steps into the hole (thick line; n=29) superimposed on EMG obtained during normal stepping (thin line; n=29). The initial part of the burst is the same in both situations. Modified from Gorassini et al. (1994).

ground, the initial part of ankle extensor activity normally involved in weight acceptance is identical in both cases. However, 50 ms after expected ground contact, ankle extensor activity rapidly decreases when the support surface is absent. The initial part of the electromyogram (EMG) burst is therefore considered to be centrally generated (feedforward control), while the rest of the burst is largely due to positive feedback from load receptors (feedback control). A similar phenomenon has been observed in human walking (af Klint et al., 2009; af Klint et al., 2010). A support surface drop of 8 cm during mid and late stance phase produced a decrease in soleus EMG at a latency between 45–49 ms and was attributed to the removal of ankle extensor force feedback (af Klint et al., 2009). A subsequent study used high-speed ultrasonography to investigate fascicle and tendon movement during the drop (af Klint et al., 2010). This study confirmed that the transient reduction in soleus EMG associated with the platform drop could be attributed to the removal of sensory feedback via the group Ib pathway.

"Evidence from animal experiments"

One approach to the study of adaptive capacity of the control of rhythmic movements that has received considerable attention over the past century is the model of partial denervation (reviewed in Bouyer and Rossignol, 2001). In this model, one (or more) peripheral nerves are transected and prevented from growing back, in an otherwise intact animal. The rhythmic movement is documented before and repeatedly after the denervation. Surgically sectioning a mixed nerve interrupts sensory afferents (reducing sensory feedback) and motor axons (producing partial paralysis), effectively changing the way the affected body part can be controlled. A motor deficit (movement error) is observed when the animal tries to move after the denervation, as the usual motor command is no longer appropriate to execute the movement. For recovery of function to be observed, modifications in feedforward control must be implemented by the nervous system, a process referred to as adaptive capacity/plasticity.

Several denervation models have been used over the years (see Bouyer and Rossignol, 2001; Rossignol et al., 2008; Rossignol et al., 2004: for reviews) including cutaneous and motor nerve transections, and nerve transpositions. Examples and limitations of this approach are presented below.

Extent and limits of adaptive capacity

Denervation experiments have shown that quite extensive recovery can be seen after an experimentally-induced modification in sensorimotor processing (Pearson et al., 1999; Carrier et al., 1997; Whelan and Pearson, 1997; Bouyer and Rossignol, 2003a). The time course of locomotor recovery occurs over a few days, i.e. faster than the time needed for muscle growth (Bouyer et al., 2001), thereby demonstrating that recovery results from neural modifications.

Adaptive plasticity has limits, however. Muscle/nerve crossing experiments have shown that antagonistic muscles cannot have their locomotor activation pattern fully reversed. For example, even three years after MG/TA nerve crossing, the output of MG during walking remains that of an extensor, discharging during stance, while its mechanical action is now into swing (Forssberg and Svartengren, 1983). In an extensive review of the older literature, Sperry (1945) demonstrated that this was also true for other movements, and that the reported recovery at the time may in fact have been due to compensation by non-denervated agonists rather than by a change in activation patterns in the experimentally transposed muscles. Examples of this limit have also been reported very early in the human literature (Close and Todd, 1959; Sutherland et al., 1960).

Underlying mechanisms of adaptive plasticity

In addition to being easy to implement and to control, the denervation model is very well suited to study some of the neural mechanisms underlying adaptive capacity/plasticity. Using successive peripheral and central lesions, animal models have also shown that adaptation involves structures distributed across the nervous system both at supraspinal and spinal sites (Whelan and Pearson, 1997; Carrier et al., 1997). Furthermore, reduced preparations such as spinal cats demonstrate that substantial adaptive capacity remains even in the 'disconnected' spinal cord (Bouyer and Rossignol, 2003b; Bouyer et al., 2001). An extreme example is shown in Figure 11.2, where a chronic spinal cat trained to walk on a treadmill undergoes a lateral gastrocnemius and soleus nerve section and fully recovers walking within a week (Bouyer et al., 2001). This recovery time course is similar to that reported in non-spinal cats (Whelan and Pearson, 1997), and is implemented in the complete absence of supraspinal structures. Interestingly, the central modifications associated with locomotor recovery seem to be different between intact and spinal cats (Bouyer et al., 2001), a phenomenon that has since been observed in other situations (Frigon and Rossignol, 2009).

Limits of the denervation model

While the denervation model is very useful to study some of the mechanisms underlying the adaptive capacity of central command to remodel during rhythmic behaviours, it also has limits. For example, after a muscle nerve transection, the fact that both the sensory and the motor components of the nerve have been interrupted imposes an additional constraint. In this case, limits in locomotor recovery could be due to at least three non-mutually exclusive factors: (1) limits in actual central adaptive capacity; (2) limits in motor output due to muscle paralysis subsequent to the transection of motor axons; (3) limits in error detection, as part of the richness of sensory feedback is removed.

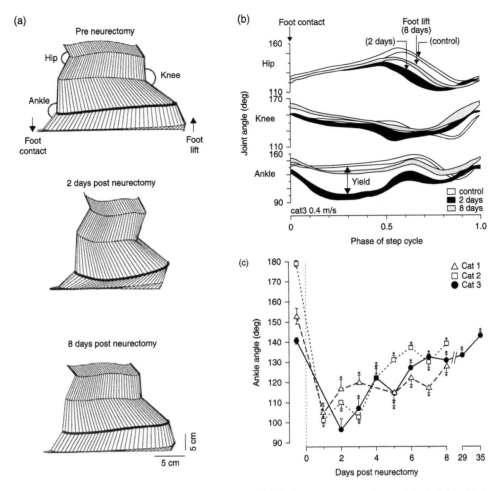

Figure 11.2 LGS denervation in the spinal cat. (a) Stick figure reconstructions of the left hindlimb before (top), 2 days (middle) and 8 days (bottom) after the neurectomy. (b) Details of the angular excursions during the step cycle obtained before (n=23; pale gray traces), 2 days (n=11; black traces) and 8 days (n=9; dark gray traces) after the LGS neurectomy for cat 3. (c) Time course of ankle angle (±SD) at a phase of 0.3 (representing maximum yield) superimposed for all cats. ★p < 0.05. Modified from Bouyer et al. (2001).

Modifying the environment to study adaptive capacity and the concept of force field adaptation

Over the past 20 years, a new approach involving experimental manipulation of the force environment in which movements are made ('*force fields*') has been used to study the adaptive capacity of movement control without a need to alter the integrity of the nervous system of the individual under study. This approach was largely inspired by pioneer work originating in non-rhythmic movement control (Lackner and DiZio, 1994; Shadmehr and Mussa-Ivaldi, 1994).

Original studies in the upper limb

When naïve subjects make an initial reaching movement in a force field, their arm is deviated from the planned movement trajectory, thereby creating a *'movement error'* (difference between perturbed and planned trajectories). Classical theories of motor control suggest that this error is interpreted by the nervous system as a signal to recalibrate the movement (Kawato, 1999; Loeb et al., 2000; Wolpert et al., 2001; Lackner and DiZio, 2000; Ghez et al., 2000). Several movements are required before it becomes completely recalibrated. Interestingly, modifications in the motor pattern persist temporarily after the force field is removed, demonstrating that feedforward control is modified as part of the motor adaptation (Lackner and DiZio, 1994; Shadmehr and Mussa-Ivaldi, 1994). Movement can be adapted to a large variety of force field directions (Ghez et al., 2000; Shadmehr and Mussa-Ivaldi, 1994; Shadmehr and Moussavi, 2000), and repeated exposure to the same force field leads to a more rapid compensation by the subject, i.e. the motor system can 'learn' new motor patterns using force fields (Thoroughman and Shadmehr, 2000; Shadmehr and Holcomb, 1997; Joiner and Smith, 2008). When people adapt to a force field in one region of the workspace, there is carry over of this adaptation to movements performed outside of the training region (Shadmehr and Mussa-Ivaldi, 1994) when the two regions of the workspace require similar changes in muscle activity (Thoroughman and Shadmehr, 1999; Shadmehr and Moussavi, 2000).

Studies during rhythmic movement

Recent studies show that when healthy, stroke, and spinal cord injured participants walk on a treadmill while a force field is applied to the leg, motor adaptations are observed as previously shown for reaching. While the present chapter focuses on force field adaptation, it must be noted that the latter represent only one of the available means of manipulating the environment in which the locomotor movement is executed. Other protocols exist, such as walking on split-belt treadmills and on rotating platforms, and produce similar results. In all cases, a movement error is initially present and gradually compensated for over the next several strides (e.g. Figure 11.3). Once the force is removed, aftereffects are present and followed by a gradual return to baseline values (Jensen et al., 1998; Prokop et al., 1995; Dietz et al., 1994; Blanchette and Bouyer, 2009; Fortin et al., 2009; Noel et al., 2009; Kao and Ferris, 2009; Choi and Bastian, 2007; Gordon and Ferris, 2007; Lam et al., 2006; Lam et al., 2008; Noble and Prentice, 2006; Emken and Reinkensmeyer, 2005; Reisman et al., 2009; Reisman et al., 2007; Sawicki et al., 2006; Weber et al., 1998; Gordon et al., 1995; Bouyer et al., 2003). These findings are indicative that the adaptive capacity of the neural control of rhythmic movement can be studied in humans using non-invasive protocols that simply manipulate the walking environment in a controlled manner.

By effectively altering limb dynamics, force field adaptation protocols are in many ways similar to the experience faced by animals and humans during development (e.g. changes in body weight, in limb segment length, etc.) and also when aging (e.g. reduction in muscle force, etc.), and it is therefore believed to tap into similar underlying adaptive mechanisms.

Using modern robotics to create tailored force fields

Within the past few years, several robotic devices and exoskeletons (Emken and Reinkensmeyer, 2005; Andersen and Sinkjaer, 1995; Gordon and Ferris, 2007; Noel et al., 2008; Colombo et al., 2000; Blaya and Herr, 2004) have been developed to study human locomotor

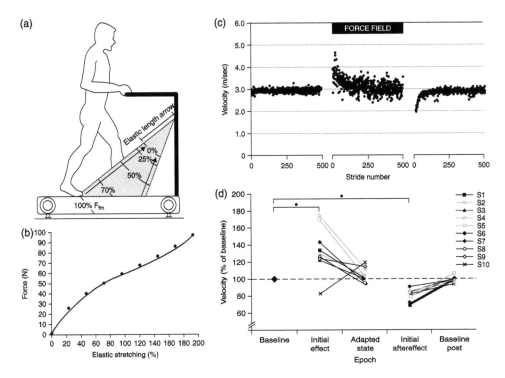

Figure 11.3 Elastic force field adaptation protocol. (a) Experimental set-up to produce the force field: elastic tubing was attached between the right foot and the frame of the treadmill. This arrangement generated a force that pulled the foot forward and up as shown schematically by the arrows. Force intensity was related to foot antero-posterior position, being largest around toe off. In addition, elastic stretching during stance did not require active work, but resulted from the backward movement of the loaded lower limb with the treadmill belt. (b) Force-stretching relationship for a 40-cm elastic tubing i.e., the average elastic length used in the study. (c) Peak foot velocity during swing for each stride of the three walking periods in a representative subject (*S1*). (d) Normalized (percent of baseline) peak right foot velocity during swing for the five epochs analyzed. Asterisk, $P < 0.05$. Modified from Blanchette & Bouyer (2009).

adaptation/learning (reviewed in Ferris, 2009). With advances in robotics, there is now little limit to the types and number of altered dynamics (i.e. force fields) that can be created to test this rhythmic movement. While most of the force field adaptation protocols have so far only been tested in humans, walking robots also exist for animal models such as rodents. DeLeon et al. (2002) have shown motor adaptation in rats, opening a new field for future neurophysiological studies.

Using catch trials to identify modifications in feedforward control

While force field exposure can be used to induce motor adaptations in rhythmic movements, the relative contribution of feedforward and feedback mechanisms to this process cannot be quantified by simply looking at muscle activation patterns during movement execution in the force field. Indeed, the mere presence of the force field will modify muscle loading, and hence also positive feedback contribution to motor output. While complete separation between feedforward and feedback control cannot be achieved using non-invasive methods, the modification in feedforward control can be estimated relatively easily by unexpectedly removing the force field for individual movement cycles ('*catch trials*'). In this situation, motor output used during the catch trial will involve the modified central command, but the altered muscle loading will not be available for positive feedback to influence motor output. Therefore, using a similar reasoning as was presented earlier for the 'foot in the hole' protocol, by comparing the catch trial motor output to control movement, the differences will reveal modifications in feedforward control (Lam et al., 2006; Noel et al., 2009).

Phase-dependent feedforward adaptive capacity

Using a velocity-dependent resistance applied against hip and knee movement during the swing phase, Lam et al. (2006) compared the EMG activity of several muscles of the lower limb during catch trials (unexpected force removal during a stride) to baseline walking. They concluded that adapted muscle activation patterns in rectus femoris (knee extensor/hip flexor) and tibialis anterior (ankle dorsiflexor) during swing were controlled by feedback mechanisms (catch EMG activity was not different from baseline), while modifications in biceps femoris and medial hamstring (knee flexors/hip extensors) activity during pre-swing involved modifications in feedforward control (catch EMG was different from baseline). The authors suggested that the mechanisms underlying force field adaptation during locomotion were therefore different from one muscle group to another.

However, there is an alternative hypothesis that could explain their results: considering that the resistance applied to knee flexion and knee extension did not occur at the same moment in the gait cycle, the difference in feedforward adaptation between the muscle groups studied may have depended on the timing in the gait cycle where adaptation had to occur. Indeed, neurophysiological studies have shown that central and peripheral drive to lower limb motoneurons is modulated as a function of the timing in the gait cycle (Nielsen, 2002; Capaday and Stein, 1987; Duysens et al., 1992; Sinkjaer et al., 1996; Yang and Stein, 1990; Zehr et al., 1997). Therefore, it is very likely that feedforward and feedback contributions to the adapted motor output may not be constant throughout a stride. To test this hypothesis, Blanchette and Bouyer (2009) used a position-dependent force field that modified the force environment over a wide portion of the gait cycle (pre-swing and swing phases). Their specific objectives were to describe the modifications in the walking pattern during and after force field exposure, and to relate the muscle activation aftereffects to the modifications that occurred in the presence of the force field. In the adapted state, hamstring EMG activity started earlier and remained elevated throughout swing. After force field exposure, aftereffects in hamstring EMGs consisted of increased activity around toe off, but contrary to the adapted state, this increase was not maintained during the rest of swing (Figure 11.4). These results suggest that adapted hamstring EMG activity may rely more on feedforward mechanisms around toe off and more on feedback mechanisms during the rest of swing, thereby supporting the hypothesis of phase-dependent adaptive capacity.

In another recent study, Noel and colleagues (2009) used a robotized ankle exoskeleton (Noel et al., 2008) to apply short duration force fields during the stance phase of human walking, and applied catch trials at pseudo-randomly selected gait cycles within a 5 min adaptation period (Figure 11.5). When exposed to a mid-stance force field, participants initially showed a large movement error (increased ankle dorsiflexion velocity). Catch trials applied early into the force field exposure period presented a kinematic pattern identical to baseline, suggesting that feedforward control had not been modified yet. The participants gradually

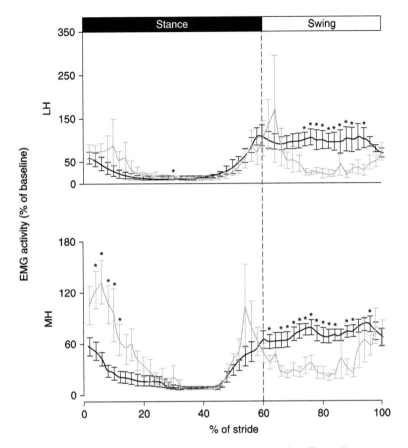

Figure 11.4 Phase-specific carry-over of force field adaptation aftereffects. Group average (n=7) Lateral Hamstring (LH) and Medial Hamstring (MH) rectified EMG activity. Comparisons between adapted motor output (thick black line) and initial aftereffect (thick gray line) show that carry over after force field removal varied depending on the phase of the gait cycle. Error bars correspond to SE. Data were time-normalized to 60 percent for stance and 40 percent for swing. Data were also amplitude-normalized to the peak locomotor EMG measured during baseline. Asterisks, $P < 0.05$. From Blanchette & Bouyer (2009).

adapted by returning ankle velocity to baseline over ~50 strides. Catch trials applied thereafter showed decreased ankle velocity when compared to baseline, indicating the presence of feedforward adaptation. These results are compatible with the idea of a gradual adaptation of feedforward control based on recent experience. Noel and colleagues (2009) also exposed the same participants to a similar force field, but this time presented during push-off (end of stance). Plantar flexion velocity was initially reduced at the moment of force field application, but no adaptation occurred over the 5 min exposure period. Catch trials kinematics remained similar to baseline at all times, suggesting that the same participants who could easily adapt to a mid-stance force field had no adaptive capacity during the push-off phase of gait, at least over these short duration exposures.

Taken together, these studies suggest that while the neural control of rhythmic movement can adapt to force field exposure, the mechanisms underlying this adaptation may vary according to the timing in the gait cycle. In addition, these results show that robotized

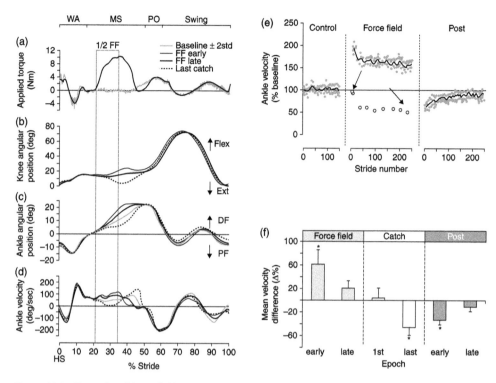

Figure 11.5 Example of force field adaptation with pseudorandomly inserted catch trials. (a) Torque applied on a subject's ankle by an electro-hydraulic ankle foot orthosis during mid-stance (Noel et al., 2008). Baseline (grey band) and force field late (thick black line). Outside of the force field application zone, the robotized orthosis applied a null field to minimize its influence on the subject's walking pattern. (b) Knee angular displacements superimposed for baseline (grey band), force field early (thin black line), force field late (thick black line), and last catch (dashed line). (c) Ankle angular displacements superimposed for baseline (grey band), force field early (thin black line), force field late (thick black line), and last catch (dashed line). (d) Ankle angular velocity for the same traces as in 'C'. Grey box: zone used for velocity measurement. Grey bands represent mean value ± 2 STD. For all conditions, data were synchronized on heel strike. Abbrev. *WA*: weight acceptance; *MS*: mid-stance; *PO*: push-off; *DF*: dorsiflexion; *PF*: plantar flexion; *HS*: heel strike. (e) Time course of ankle velocity across walking conditions. Each grey symbol represents a stride. Black symbols represent an 11 points moving average. Open symbols represent catch strides; arrows focus on early and late catch trials velocity. (f) Group results (n = 11) expressed as percent difference from control for the two epochs in each walking condition. Error bars represent 95 percent confidence intervals. ★:Epochs statistically different from baseline (P < 0.05; repeated measure ANOVA with Bonferonni correction). Modified from Noel et al. (2009).

exoskeletons such as the Lokomat (Lam et al., 2006) and the EHO (Noel et al., 2009) are useful tools to study phase-dependent adaptive control of movement.

Potential limits of adaptation of human gait

In the protocols presented thus far, the demand on movement reorganization (i.e. extent of pattern reshaping) has been relatively modest. In the section on 'evidence from animal experiments' above, it was discussed that functional antagonists such as TA and MG cannot be trained to become activated in the opposite phase of their normal activation after nerve transposition. However, due to the detrimental effects of nerve transposition on sensory feedback, these experiments are not conclusive. Using a very simple force field that overassists foot dorsiflexion during swing, Barthelemey et al. (2012) have recently provided initial evidence that even in the intact human subject, the ability to produce a swing/stance reversal of muscle activation may be very limited. Indeed, after 10 min. of walking with this force field, their subjects showed a drastic reduction in TA activation, but no increase in SOL activity. The reduction in TA activation shows that the locomotor system is capable of adaptive plasticity. However, only reducing TA without inducing a concomitant increase in SOL was insufficient to counteract the force field applied at the ankle, and a large ankle movement error persisted. This effect was not seen when the same subjects were exposed to a force field that instead *resisted* ankle dorsiflexion during swing. In the latter, TA activation was increased until movement error was minimal. It therefore seems that the lack of movement error reduction observed with the force field overassisting ankle dorsiflexion was due to the inability of the CNS to activate SOL (the antagonist muscle) during swing, similarly to what was reported after nerve transposition in the cat, but here without the experimental limitations associated with nerve surgery. Further studies will be needed to explore this phenomenon further.

Effects of force field training on aftereffects, next day performance, and retention of feedforward control parameters

A review of the plasticity of gait could not be complete without touching upon two additional questions: (1) how long should one be exposed to force fields to obtain motor adaptation? (2) How much retention can be expected on subsequent exposures to the same force field?

These questions were recently addressed by Fortin and colleagues (2009). Participants (n=17) were randomly assigned to one of several predetermined force field exposure durations (range 1–30 min, i.e. 49–1629 strides); the effects of exposure duration were measured on aftereffect duration and amplitude, and on next day performance in the same force field. This study demonstrated that while aftereffect duration was correlated to exposure duration (Figure 11.6), all subjects performed equally better on a 24 hour retest. These results suggest that during walking, even short daily exposures to a force field (≥ 49 strides) lead to significant retention of the adapted feedforward parameters. They also suggest that neural consolidation mechanisms triggered by force field exposure continue to act after the end of exposure.

An example of the extent of memory of the adapted feedforward parameters and of our ability to rapidly recall the adapted muscle activation pattern is demonstrated in Figure 11.7, where a non-naïve subject is tested using a catch trial protocol similar to that of Figure 11.5. This time, catch trial performance was measured as early as 2–3 strides into the force field walking, a period way too short for a novel adaptation to occur.

Development of gait

In contrast to other animals, humans are not capable of independent walking until an average age of 1 year (Berger, 2001; Forssberg, 1985; Ivanenko et al., 2007). This relates to the postponed development of the human brain as compared to other animals and the larger demands on balance and motor control during human bipedal gait (requiring that an erect lean body

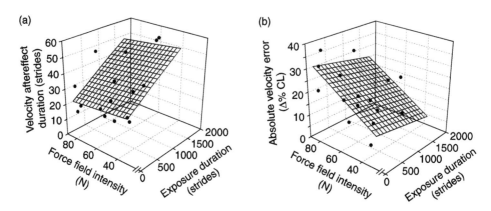

Figure 11.6 Effects of elastic force field adaptation duration and force field intensity on aftereffects duration. Each point represents one subject. It can be seen that aftereffect duration is proportional to force field exposure duration (a), and that aftereffects magnitude is proportional to force field intensity (b). From Fortin et al. (2009).

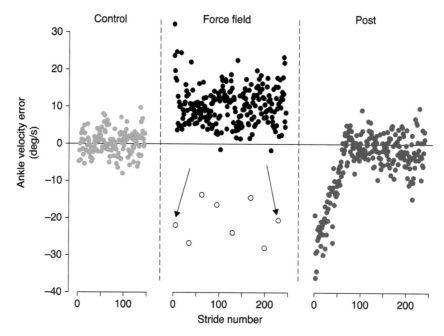

Figure 11.7 Force field adaptation with catch trials (open circles) in a non-naïve subject. When compared to Figure 11.5(e), it can be seen that in this case, ankle velocity error on the first catch trial (left arrow) was already as large as after 5 min. of force field adaptation (right arrow). The non-naïve subject anticipated well how to perform during force field walking. This is an indication of storage capacity of the previously adapted feedforward control parameters, i.e. a sign of motor learning (Bouyer, 2011).

has to be balanced on top of a small supporting surface that alternates with every step). However, newborns show stepping movements, which has been proposed as a sign of the existence of a spinal CPG in human subjects (Forssberg, 1985; Yang et al., 1998). The fact that such stepping movements may also be seen in anencephalic infants is strong support for this (Forssberg, 1985), whereas the argument that supraspinal fibres have not reached the lumbar spinal cord at the time of birth does not seem to hold true (Eyre et al., 2000). The disappearance of stepping movements in infants older than 1 month and the subsequent appearance of independent bipedal gait around the age of 1 year has been interpreted as a replacement of the spinal rhythm generating network by supraspinal control centres (Forssberg, 1985). However, later studies have shown that stepping movements may be seen also in infants older than 1 month under the right circumstances (reduction of gravity and/or when providing sensory input from a moving treadmill (Yang et al., 1998)); and corticospinal fibres seem to innervate all of the human spinal cord already from birth, although the maturation of these connections appear to continue for a number of years (Eyre et al., 2000; Fietzek et al., 2000; Muller and Homberg, 1992; Muller et al., 1997). From early on, spinal and supraspinal networks thus appear to interact and the development of independent bipedal gait around the age of 1 year should be seen as the result of a gradual adaptation of the involved spinal and supraspinal networks to the Newtonian reality that the infant has been placed in. Bipedal gait requires coordination in muscles throughout the body based on predictions of the consequences of moving the centre of gravity away from a relatively safe position between the two legs. This is a task that requires considerable computation based on information from the skin, proprioceptors, the vestibular system and the visual system sampled from all the successful and less successful attempts by the infant at reaching a desired object in the environment. Rather than focusing on an intrinsic maturation process and a switch between different hierarchies of control at a specific point in time, we thus should acknowledge that our motor control already from birth relies heavily on integrated control at multiple levels and that this control from the very outset is continuously optimized based on interaction with the environment.

One of the defining features of human bipedal gait is the heel strike which initiates each stance phase (Forssberg, 1985). A proper heel strike requires tibialis anterior muscle activation to lift the forefoot when the leg is at its maximal forward position and is lowered so that the foot makes contact with the ground. Following heel strike, TA remains active to ensure a controlled lowering of the foot. In most children, heel strike is seen already at the age of 2, but it continues to be adapted and optimized for at least another 10 years (Forssberg, 1985; Norlin et al., 1981). This is also the case with many other features of human gait (Forssberg, 1985; Hirschfeld and Forssberg, 1992; Norlin et al., 1981; Petersen et al., 2010). Although the basic movement repertoire is in place from relatively early on, there is a continued adaptation and optimization taking place up until the age of around 10–12 years where a fully adult gait pattern is seen. Among the gait features that have been investigated are the step-to-step variability in the trajectory of the foot during swing phase, the precision of foot placement (heel strike), the lateral deviation of the foot and the extent of foot lift prior to heel strike (Forssberg, 1985; Hirschfeld and Forssberg, 1992; Norlin et al., 1981; Petersen et al., 2010). We know little about the adaptations in specific parts of the control network which are responsible for these behavioural adaptations. In one recent study, Petersen et al. (2010) found that the common central drive to tibialis anterior motor units in the swing phase of gait gradually increases until the age of 10–12 years and that these changes are correlated to a reduction in the step-to-step variability of the lift of the foot in the swing phase. Since the common central drive requires corticospinal transmission it seems likely that these developmental changes reflect adaptations in the corticospinal contribution to the control of the ankle joint during gait. This would be consistent with evidence showing on one hand that corticospinal transmission continues to develop and mature until at least the age of 11–12 years (Muller and Homberg, 1992; Muller et al., 1997) and that one severe consequence of lesion of corticospinal input to the spinal motor network often is a lasting reduction in dorsiflexion muscle force and a functional foot drop during gait (Barthelemy et al., 2010).

Gait and ageing

It is one of the most obvious signs of old age that our gait becomes slower and more unstable with a marked reduction in step length and an increased risk of falling (Rubino, 1993; Sternberg et al., 2011). Characteristically, the elderly show reduced ability to perform dual-tasks during walking, signifying increased conscious involvement in gait control (Beauchet et al., 2009). Many interrelated factors contribute to these changes. Whether primarily caused by inactivity or by a physiological ageing process, we lose muscle mass already from a relatively early age (according to some, already from the age of 35–40 years) and at such a high rate after the age of 60–65 that functional problems during daily life activities easily occur (Berger and Doherty, 2010; Roubenoff, 2000). Muscular weakness definitely contributes to the unsteadiness of gait in the ageing population and is also most likely involved in the high incidence of falls (Visser and Schaap, 2011). Changes in the compliance of connective tissue, joints and tendons also contribute to reduced function of the musculo-skeletal system and lead to restrictions in gait speed and versatility (McGibbon, 2003). Reduced proprioception is well documented in the elderly (Boisgontier et al., 2012), but it is not clarified if and how this contributes to reduced gait ability although it seems reasonable to assume that diminished sensory information will leave the central neuronal circuitries with suboptimal updating of the motor commands leading to reduced automaticity of gait (i.e. forcing subjects to move more slowly under increased conscious control). Reduced vision may also be a contributing factor (Hallemans et al., 2009; Hallemans et al., 2010; Huitema et al., 2005), but otherwise the most important effect of reduced vision in relation to gait is probably a reduced ability to adequately predict upcoming obstacles in the walking path and thereby modulate the gait adequately in advance (Hallemans et al., 2009; Hallemans et al., 2010). Degeneration of hair cells in the vestibular organs contributes to dizziness and falls in elderly subjects, and impaired integration of vestibular signals with other sensory signals will also make it more difficult for elderly subjects to maintain balance during gait and force them to take shorter steps and walk more slowly under increased conscious control (Lord and Sturnieks, 2005).

References

af Klint R, Cronin NJ, Ishikawa M, et al. (2010) Afferent contribution to locomotor muscle activity during unconstrained overground human walking: an analysis of triceps surae muscle fascicles. *J Neurophysiol* 103: 1262–1274.

af Klint R, Nielsen JB, Sinkjaer T, et al. (2009) Sudden drop in ground support produces force-related unload response in human overground walking. *J Neurophysiol* 101: 1705–1712.

Andersen JB and Sinkjaer T. (1995) An actuator system for investigating electrophysiological and biomechanical features around the human ankle joint during gait. *IEEE Trans Rehabil Eng* 3 299–306.

Barthelemy D, Alain S, Grey MJ, et al. (2012) Rapid changes in corticospinal excitability during force field adaptation of human walking. *Exp Brain Res* 217: 99–115.

Barthelemy D, Willerslev-Olsen M, Lundell H, et al. (2010) Impaired transmission in the corticospinal tract and gait disability in spinal cord injured persons. *J Neurophysiol* 104: 1167–1176.

Beauchet O, Annweiler C, Dubost V, et al. (2009) Stops walking when talking: a predictor of falls in older adults? *Eur J Neurol* 16: 786–795.

Berger MJ and Doherty TJ. (2010) Sarcopenia: prevalence, mechanisms, and functional consequences. *Interdiscip Top Gerontol* 37: 94–114.

Berger W. (2001) Normal and impaired development of gait. *Adv Neurol* 87: 65–70.

Blanchette A and Bouyer LJ. (2009) Timing-specific transfer of adapted muscle activity after walking in an elastic force field. *J Neurophysiol* 102: 568–577.

Blaya JA and Herr H. (2004) Adaptive control of a variable-impedance ankle-foot orthosis to assist drop-foot gait. *IEEE Trans Neural Syst Rehabil Eng* 12: 24–31.

Boisgontier MP, Olivier I, Chenu O, et al. (2012) Presbypropria: the effects of physiological ageing on proprioceptive control. *Age (Dordr)* 34: 1179–1194.

Bouyer L and Rossignol S. (2001) Spinal cord plasticity associated with locomotor compensation to peripheral nerve lesions in the cat. In: Patterson MM and Grau JW (eds) *Spinal cord plasticity: Alterations in reflex function*. Boston, Massachusetts: Kluwer Academic Publishers, 207–224.

Bouyer LJ. (2011) Chapter 8–challenging the adaptive capacity of rhythmic movement control: from denervation to force field adaptation. *Prog Brain Res* 188: 119–134.

Bouyer LJ, DiZio P and Lackner JR. (2003) Adaptive modification of human locomotion by Coriolis force. 33rd Annual Meeting of Society for Neuroscience.

Bouyer LJ and Rossignol S. (2003a) Contribution of cutaneous inputs from the hindpaw to the control of locomotion. I. Intact cats. *J Neurophysiol* 90: 3625–3639.

Bouyer LJ and Rossignol S. (2003b) Contribution of cutaneous inputs from the hindpaw to the control of locomotion. II. Spinal cats. *J Neurophysiol* 90: 3640–3653.

Bouyer LJ, Whelan PJ, Pearson KG, et al. (2001) Adaptive locomotor plasticity in chronic spinal cats after ankle extensors neurectomy. *J Neurosci* 21: 3531–3541.

Capaday C and Stein RB. (1987) Difference in the amplitude of the human soleus H reflex during walking and running. *J Physiol* 392: 513–522.

Carlson-Kuhta P, Trank TV and Smith JL. (1998) Forms of forward quadrupedal locomotion. II. A comparison of posture, hindlimb kinematics, and motor patterns for upslope and level walking. *J Neurophysiol* 79: 1687–1701.

Carrier L, Brustein E and Rossignol S. (1997) Locomotion of the hindlimbs after neurectomy of ankle flexors in intact and spinal cats: model for the study of locomotor plasticity. *J Neurophysiol* 77: 1979–1993.

Choi JT and Bastian A. (2007) Adaptation reveals independent control networks for human walking. *Nat Neurosci* 10: 1055–1062.

Close JR and Todd FN. (1959) The phasic activity of the muscles of the lower extremity and the effect of tendon transfer. *J Bone Joint Surg* 41 189–235.

Colombo G, Joerg M, Schreier R, et al. (2000) Treadmill training of paraplegic patients using a robotic orthosis. *J Rehabil Res Dev* 37: 693–700.

De Leon RD, Kubasak MD, Phelps PE, et al. (2002) Using robotics to teach the spinal cord to walk. *Brain Res Brain Res Rev* 40: 267–273.

Dietz V, Zijlstra W and Duysens J. (1994) Human neuronal interlimb coordination during split-belt locomotion. *Exp Brain Res* 101: 513–520.

Duysens J, Tax AA, Trippel M, et al. (1992) Phase-dependent reversal of reflexly induced movements during human gait. *Exp Brain Res* 90: 404–414.

Emken J and Reinkensmeyer D. (2005) Robot-enhanced motor learning: accelerating internal model formation during locomotion by transient dynamic amplification. *IEEE Trans Neural Syst Rehabil Eng* 13: 33–39.

Eyre JA, Miller S, Clowry GJ, et al. (2000) Functional corticospinal projections are established prenatally in the human foetus permitting involvement in the development of spinal motor centres. *Brain* 123 (Pt 1): 51–64.

Ferris DP. (2009) The exoskeletons are here. *J Neuroeng Rehabil* 6: 17.

Fietzek UM, Heinen F, Berweck S, et al. (2000) Development of the corticospinal system and hand motor function: central conduction times and motor performance tests. *Dev Med Child Neurol* 42: 220–227.

Forssberg H. (1985) Ontogeny of human locomotor control. I. Infant stepping, supported locomotion and transition to independent locomotion. *Exp Brain Res* 57: 480–493.

Forssberg H and Svartengren G. (1983) Hardwired locomotor network in cat revealed by a retained motor pattern to gastrocnemius after muscle transposition. *Neurosci Lett* 41: 283–288.

Fortin K, Blanchette A, McFadyen BJ, et al. (2009) Effects of walking in a force field for varying durations on aftereffects and on next day performance. *Exp Brain Res* 199: 145–155.

Frigon A and Rossignol S. (2009) Partial denervation of ankle extensors prior to spinalization in cats impacts the expression of locomotion and the phasic modulation of reflexes. *Neuroscience* 158: 1675–1690.

Ghez C, Krakauer JW, Sainburg RL, et al. (2000) Spatial representations and internal models of limb dynamics in motor learning. In: Gazzaniga MS (ed.) *The new cognitive neuroscience*, 501–514.

Gorassini MA, Prochazka A, Hiebert GW, et al. (1994) Corrective responses to loss of ground support during walking. I. Intact cats. *J Neurophysiol* 71: 603–610.

Gordon CR, Fletcher WA, Melvill JG, et al. (1995) Adaptive plasticity in the control of locomotor trajectory. *Exp Brain Res* 102: 540–545.

Gordon KE and Ferris DP. (2007) Learning to walk with a robotic ankle exoskeleton. *J Biomechanics* 40: 2636–2644.

Grillner S, Ekeberg El, Manira A, et al. (1998) Intrinsic function of a neuronal network – a vertebrate central pattern generator. *Brain Res Brain Res Rev* 26: 184–197.

Hallemans A, Beccu S, Van Loock K, et al. (2009) Visual deprivation leads to gait adaptations that are age- and context-specific: I. Step-time parameters. *Gait Posture* 30: 55–59.

Hallemans A, Ortibus E, Meire F, et al. (2010) Low vision affects dynamic stability of gait. *Gait Posture* 32: 547–551.

Hirschfeld H and Forssberg H. (1992) Development of anticipatory postural adjustments during locomotion in children. *J Neurophysiol* 68: 542–550.

Huitema RB, Brouwer WH, Mulder T, et al. (2005) Effect of ageing on the ability to adapt to a visual distortion during walking. *Gait Posture* 21: 440–446.

Ivanenko YP, Dominici N and Lacquaniti F. (2007) Development of independent walking in toddlers. *Exerc Sport Sci Rev* 35: 67–73.

Jensen L, Prokop T and Dietz V. (1998) Adaptational effects during human split-belt walking: influence of afferent input. *Exp Brain Res* 118: 126–130.

Joiner WM and Smith MA. (2008) Long-term retention explained by a model of short-term learning in the adaptive control of reaching. *J Neurophysiol* 100: 2948–2955.

Kao PC and Ferris DP. (2009) Motor adaptation during dorsiflexion-assisted walking with a powered orthosis. *Gait Posture* 29: 230–236.

Kawato M. (1999) Internal models for motor control and trajectory planning. *Curr Opin Neurobiol* 9: 718–727.

Lackner JR and DiZio P. (1994) Rapid adaptation to Coriolis force perturbations of arm trajectory. *J Neurophysiol* 72: 299–313.

Lackner JR and DiZio PA. (2000) Aspects of body self-calibration. *Trends Cogn Sci* 4: 279–288.

Lafreniere-Roula M and McCrea DA. (2005) Deletions of rhythmic motoneuron activity during fictive locomotion and scratch provide clues to the organization of the mammalian central pattern generator. *J Neurophysiol* 94: 1120–1132.

Lam T, Anderschitz M and Dietz V. (2006) Contribution of feedback and feedforward strategies to locomotor adaptations. *J Neurophysiol* 95: 766–773.

Lam T, Wirz M, Lunenburger L, et al. (2008) Swing phase resistance enhances flexor muscle activity during treadmill locomotion in incomplete spinal cord injury. *Neurorehabilitation Neural Repair* 22: 438–446.

Loeb EP, Giszter SF, Saltiel P, et al. (2000) Output units of motor behavior: an experimental and modeling study. *J Cogn Neurosci* 12: 78–97.

Lord SR and Sturnieks DL. (2005) The physiology of falling: assessment and prevention strategies for older people. *J Sci Med Sport* 8: 35–42.

McCrea DA and Rybak IA. (2007) Modeling the mammalian locomotor CPG: insights from mistakes and perturbations. *Prog Brain Res* 165: 235–253.

McGibbon CA. (2003) Toward a better understanding of gait changes with age and disablement: neuromuscular adaptation. *Exerc Sport Sci Rev* 31: 102–108.

Muller K and Homberg V. (1992) Development of speed of repetitive movements in children is determined by structural changes in corticospinal efferents. *Neurosci Lett* 144: 57–60.

Muller K, Kass-Iliyya F and Reitz M. (1997) Ontogeny of ipsilateral corticospinal projections: a developmental study with transcranial magnetic stimulation. *Ann Neurol* 42: 705–711.

Nielsen JB. (2002) Motoneuronal drive during human walking. *Brain Res Brain Res Rev* 40: 192–201.

Noble J and Prentice S. (2006) Adaptation to unilateral change in lower limb mechanical properties during human walking. *Exp Brain Res* 169: 482–495.

Noel M, Cantin B, Lambert S, et al. (2008) An electrohydraulic actuated ankle foot orthosis to generate force fields and to test proprioceptive reflexes during human walking. *IEEE Trans Neural Syst Rehabil Eng* 16: 390–399.

Noel M, Fortin K and Bouyer LJ. (2009) Using an electrohydraulic ankle foot orthosis to study modifications in feedforward control during locomotor adaptation to force fields applied in stance. *J Neuroeng Rehabil* 6: 16.

Norlin R, Odenrick P and Sandlund B. (1981) Development of gait in the normal child. *J Pediatr Orthop* 1: 261–266.

Pearson KG, Fouad K and Misiaszek JE. (1999) Adaptive changes in motor activity associated with functional recovery following muscle denervation in walking cats. *J Neurophysiol* 82: 370–381.

Petersen TH, Kliim-Due M, Farmer SF, et al. (2010) Childhood development of common drive to a human leg muscle during ankle dorsiflexion and gait. *J Physiol* 588: 4387–4400.

Prochazka A, Mushahwar V and Yakovenko S. (2002) Activation and coordination of spinal motoneuron pools after spinal cord injury. *Prog Brain Res* 137: 109–124.

Prokop T, Berger W, Zijlstra W, et al. (1995) Adaptational and learning processes during human split-belt locomotion: interaction between central mechanisms and afferent input. *Exp Brain Res* 106: 449–456.

Reisman DS, Wityk R, Silver K, et al. (2007) Locomotor adaptation on a split-belt treadmill can improve walking symmetry post-stroke. *Brain* 130: 1861–1872.

Reisman DS, Wityk R, Silver K, et al. (2009) Split-belt treadmill adaptation transfers to overground walking in persons poststroke. *Neurorehabilitation and Neural Repair.*

Rossignol S, Barriere G, Frigon A, et al. (2008) Plasticity of locomotor sensorimotor interactions after peripheral and/or spinal lesions. *Brain Res Rev* 57: 228–240.

Rossignol S, Brustein E, Bouyer L, et al. (2004) Adaptive changes of locomotion after central and peripheral lesions. *Can J Physiol Pharmacol* 82: 617–627.

Roubenoff R. (2000) Sarcopenia and its implications for the elderly. *Eur J Clin Nutr* 54 Suppl 3: S40–47.

Rubino FA. (1993) Gait disorders in the elderly. Distinguishing between normal and dysfunctional gaits. *Postgrad Med* 93: 185–190.

Rybak IA, Shevtsova NA, Lafreniere-Roula M, et al. (2006) Modelling spinal circuitry involved in loco-motor pattern generation: insights from deletions during fictive locomotion. *J Physiol* 577: 617–639.

Sawicki GS, Domingo A and Ferris DP. (2006) The effects of powered ankle-foot orthoses on joint kinematics and muscle activation during walking in individuals with incomplete spinal cord injury. *J Neuroeng Rehabil* 3: 3.

Shadmehr R and Holcomb HH. (1997) Neural correlates of motor memory consolidation. *Science* 277: 821–825.

Shadmehr R and Moussavi ZM. (2000) Spatial generalization from learning dynamics of reaching movements. *J Neurosci* 20: 7807–7815.

Shadmehr R and Mussa-Ivaldi FA. (1994) Adaptive representation of dynamics during learning of a motor task. *J Neurosci* 14: 3208–3224.

Sinkjaer T, Andersen JB and Larsen B. (1996) Soleus stretch reflex modulation during gait in humans. *J Neurophysiol* 76: 1112–1120.

Smith JL, Carlson-Kuhta P and Trank TV. (1998) Forms of forward quadrupedal locomotion. III. A comparison of posture, hindlimb kinematics, and motor patterns for downslope and level walking. *J Neurophysiol* 79: 1702–1716.

Sperry RW. (1945) The problem of central nervous reorganization after nerve regeneration and muscle transposition. *Q Rev Bio* 20 311–369.

Sternberg SA, Wershof Schwartz A, Karunananthan S, et al. (2011) The identification of frailty: a systematic literature review. *J Am Geriatr Soc* 59: 2129–2138.

Sutherland DH, Bost FC and Schottstaedt ER. (1960) Electromyographic study of transplanted muscles about the knee in poliomyelitic patients. *J.Bone Joint Surg* 42 919–939.

Thoroughman KA and Shadmehr R. (1999) Electromyographic correlates of learning an internal model of reaching movements. *J Neurosci* 19: 8573–8588.

Thoroughman KA and Shadmehr R. (2000) Learning of action through adaptive combination of motor primitives. *Nature* 407: 742–747.

Trank TV, Chen C and Smith JL. (1996) Forms of forward quadrupedal locomotion. I. A comparison of posture, hindlimb kinematics, and motor patterns for normal and crouched walking. *J Neurophysiol* 76: 2316–2326.

Visser M and Schaap LA. (2011) Consequences of sarcopenia. *Clin Geriatr Med* 27: 387–399.

Weber KD, Fletcher WA, Gordon CR, et al. (1998) Motor learning in the 'podokinetic' system and its role in spatial orientation during locomotion. *Exp Brain Res* 120: 377–385.

Whelan PJ and Pearson KG. (1997) Plasticity in reflex pathways controlling stepping in the cat. *J Neurophysiol* 78: 1643–1650.

Wolpert DM, Ghahramani Z and Flanagan JR. (2001) Perspectives and problems in motor learning. *Trends Cogn Sci* 5: 487–494.

Yakovenko S, Mushahwar V, VanderHorst V, et al. (2002) Spatiotemporal activation of lumbosacral motoneurons in the locomotor step cycle. *J Neurophysiol* 87: 1542–1553.

Yang JF and Stein RB. (1990) Phase-dependent reflex reversal in human leg muscles during walking. *J Neurophysiol* 63: 1109–1117.

Yang JF, Stephens MJ and Vishram R. (1998) Infant stepping: a method to study the sensory control of human walking. *J Physiol* 507 (Pt 3): 927–937.

Zehr EP, Komiyama T and Stein RB. (1997) Cutaneous reflexes during human gait: electromyographic and kinematic responses to electrical stimulation. *J Neurophysiol* 77: 3311–3325.

12

MOTOR CONTROL
AND MOTOR LEARNING
IN STRETCH-SHORTENING
CYCLE MOVEMENTS

Wolfgang Taube,[1] *Christian Leukel,*[1,2] *and Albert Gollhofer*[2]

[1] DEPARTMENT OF MEDICINE, MOVEMENT AND SPORT SCIENCE, UNIVERSITY OF FRIBOURG
[2] DEPARTMENT OF SPORT SCIENCE, UNIVERSITY OF FREIBURG

Introduction

By definition, the stretch-shortening cycle (SSC) describes a natural muscle function in which the pre-activated muscle–tendon complex is lengthened in the eccentric phase followed by muscle–tendon shortening in the concentric phase (Ishikawa & Komi, 2008; Komi & Gollhofer, 1997). Animal and human locomotion as well as many other movements, such as hopping and throwing, are organized in a SSC. Mechanically, the major advantage of the SSC compared to isolated concentric and eccentric muscle activation is considered to be the storage and subsequent release of kinetic energy leading to enhanced power and/or greater economy (Dietz et al., 1979; Gollhofer et al., 1992; Voigt et al., 1998b). The efficiency of the SSC is assumed to depend largely on the ability to transfer energy from the pre-activated and eccentrically stretched muscle–tendon complex to the concentric push-off phase (Bosco et al., 1981; Bosco & Komi, 1979). The elastic properties of the tendo-muscular system therefore greatly influence the efficiency of the SSC.

The stiffness of the tendo-muscular complex during SSC is determined by the stiffness of the muscle(s) attached to the tendon. Task specific activation of the muscle(s) is precisely adjusted by the central nervous system (CNS). In regard to the complexity of the SSC several functional phases have been distinguished. First, the pre-programmed muscular activation prior to touch down (the pre-activation) is adapted, presumably to provide an appropriate stiffness before touch down, i.e. to ensure adequate mechanical loading of the tendo-muscular complex. Second, reflex activity after touch down can be modulated dependent on the task and the context. These reflex contributions induced by stretching of the antigravity muscles during touch down (eccentric phase) have been proposed to influence the muscular stiffness, too (Dietz et al., 1979; Gollhofer et al., 1992; Voigt et al., 1998b). Third, the pre-programmed muscular activity after touch down, which is not primarily influenced by stretch reflex activity (Zuur et al., 2010), might be adapted according to the

231

distinct requirements of the motor task (for instance drop height) and to the status of training (Taube et al., 2011b).

This shows that the neural control of the SSC is complex as both feedforward (pre-programmed) and feedback mechanisms have to be highly specific to ensure the balance between a 'powerful system' and possible risks of overload. Kjaer (2004) reported the maximal loading for several tendons and demonstrated that the Achilles and patellar tendons work in ranges relatively close to their point of failure. Especially during jumping, very high forces are exerted on the tendons (Fukashiro et al., 1995; Kjaer et al., 2005). Thus, muscular activation during SSC-movements has to be adjusted both task- and phase-specifically. For this purpose, multiple hierarchical levels of the CNS have to accurately interact to ensure the appropriate muscular activation.

Longitudinal studies, with training over several weeks, have demonstrated that plyometric training (i.e. training of stretch-shortening cycle movements) can change functional properties of serial and parallel elastic tissue (e.g. Kubo et al., 2007). As stress–strain characteristics of tendons adapt by long-term plyometric training, it appears logically that the CNS also adapts in order to control the altered biomechanical properties. Short-term SSC-training (four weeks) was also shown to influence the neural control of SSC-movements, which was reflected in phase-specific adaptations of the muscular activity (Taube et al., 2011b). Depending on the training regime, different aspects of the drop jump, such as the time of ground contact or the rebound height, can be specifically addressed.

The first section of this chapter describes the current knowledge of the neural organization of SSC-movements and highlights in particular the interaction of spinal and cortical levels. Information provided in this section was previously highlighted in our review article 'How Neurons Make Us Jump – The Neural Control of Stretch-Shortening-Cycle Movements' (Taube et al., 2011a). The second main section of the chapter concentrates on neural plasticity as a function of SSC-training.

Phases of SSC-movements

Due to the complexity of the movement, the SSC is classically divided into several phases. Some authors distinguish between the activation before ground contact (the pre-activation or pre-innervation) and the eccentric (tendo-muscular lengthening) and the concentric (tendo-muscular shortening) phase (Komi, 1984). Slightly different approaches separated the activity of the SSC into pre-activity (or pre-activation; before touch down), 'reflex induced activation' (i.e. activation dominated by stretch reflex contributions (Marsden et al., 1978; Gollhofer & Schmidtbleicher, 1989) and 'late muscular responses' occurring after 120 ms following ground contact (Gollhofer, 1987). These phase separations were not based on the behavior of the tendo-muscular system but on the muscular activation profile. When considering the muscular activation in more detail, the reflex induced activation phase can be further differentiated into the first peak of activation, the so-called short-latency response (SLR), and subsequent peaks which were called medium-latency response (MLR), long-latency response (LLR) and long-latency response 2 (LLR2) (Taube et al., 2008). However, as it is often difficult to discriminate between those peaks, the same authors proposed another kind of phase division, which is based on reflex latencies: the pre-activation (the 100 ms before ground contact), the early reflex phase (20 to 70 ms after touch down which incorporates the SLR), the late reflex phase (70 to 120 ms incorporating long-latency reflexes) and the push-off phase (from 120 to 170 ms) (Taube et al., 2011b). Whereas research concentrating on the behavior of the muscle–tendon complex favors the first approach listed here, the neuromuscular control of SSC-movements is best described when applying a model based on reflex latencies. In this chapter, the separation proposed by Taube et al. (2011b) will be used as all other approaches can easily be integrated into this phase-division.

The pre-activation phase

It was shown in landing movements of monkeys that the pre-activation EMG pattern was uninfluenced by turning off the light during the fall and by opening a collapsible platform delaying the time of ground contact (Dyhre-Poulsen & Laursen, 1984). Thus, it seems most likely that this muscular activity is pre-programmed as Jones and Watt (1971) have already suggested. These authors also proposed that the pre-activation is programmed and dispatched from higher centers as a single entity before landing.

From a functional point of view, it is generally accepted that the activity prior to touch down is pre-programmed in order to provide an appropriate stiffness for the period after touch down. In this respect, Arampatzis and co-workers demonstrated that the change in leg stiffness was related to the level of pre-activation but not to its duration (Arampatzis et al., 2001a). Furthermore, Schmidtbleicher and Gollhofer (1982) have observed a relationship between both the duration and the amount of the muscle pre-activation and the drop height, whereas the drop height is known to influence the ankle stiffness resulting in reduced stiffness when jumping from higher drop heights (Taube et al., 2011b). Similarly, the amount of pre-activation has been shown to increase with increasing running speed (Komi et al., 1987) and increased running speed is considered to be associated with a greater leg stiffness (Arampatzis et al., 1999; Gollhofer et al., 1984).

Comparing pre-activation of the leg extensors, Gollhofer and Kyrölainen (1991) demonstrated that both integrated EMG as well as pre-activation duration changed drastically when subjects were functionally loaded or unloaded during drop jumps from identical heights. They concluded that according to the load condition, the main neuromuscular regulation takes part in the pre-activation phase, adjusting the tendo-muscular system for the mechanical condition. In a recent study, stiffness adjustments during running were investigated in response to changes in the step height (Muller et al., 2010). Most interestingly, the pre-activation of the gastrocnemius medialis correlated as highly significant with the activity of the muscle after touch down as well as with the kinematics and dynamics. This is strong evidence in favor of a pre-programmed pre-activation, which in turn influences the leg stiffness after touch down to achieve situation-specific leg properties.

The early reflex phase

When considering the early reflex phase we have to verify in a first step that SSC-movements indeed incorporate reflex responses. In the first part of this section, the question is therefore whether spinal reflexes occur during the SSC. In the next section, the contribution of supraspinal centers to the muscular activity of the early reflex phase is discussed. Finally, it is highlighted how the activity within the early reflex phase may be of functional significance for the performance of a successful SSC-movement.

Do spinal (stretch) reflexes occur during SSC?

The human central nervous system (CNS) responds instantaneously to stretches of a relaxed muscle. The muscle spindle detects changes in the muscle length and alters firing frequencies in Ia afferent fibers proportionate to the velocity of the change in length (Gollhofer & Rapp, 1993; Gottlieb & Agarwal, 1979) and to a smaller degree in relation to the amplitude

(De Doncker et al., 2003). The increased activity of Ia afferents after muscle stretch depolarizes α-motoneurons at the spinal level, which elicit a stretch reflex called short-latency response (SLR). If the muscle is preactivated before stretching, not only an SLR can be observed but also medium (MLR) and long-latency responses (LLR) (Lee & Tatton, 1975; Petersen et al., 1998; Taube et al., 2006; Toft et al., 1989). Based on these observations it seems conceivable that stretch reflexes are elicited in the eccentric phase of SSC as one may argue that the mechanism of stretching the extensor muscles during touch down is similar to other tasks like rotating the ankle joint. Otherwise it may be proposed that the ankle rotation and the consequent muscle stretch during a voluntary initiated drop jump is organized differently than after externally driven ankle rotations due to the capacity of the CNS to accurately predict the instance of ground contact and thus the onset of the muscle stretch (Dyhre-Poulsen & Laursen 1984; McDonagh & Duncan, 2002; Santello & McDonagh, 1998). Potentially, the CNS could inhibit muscular activation caused by afferent excitation precisely at the time of the SLR. Different experimental setups were used to clarify whether muscular activity at the time of the SLR during SSC-movements was influenced by spinal stretch reflexes. Observations strengthening the hypothesis of an integrated stretch reflex were the following: (a) the latency of the first muscular activation peak after ground contact corresponded to the latency of the SLR elicited by stretching of the relaxed muscle (Dietz et al., 1979; Leukel et al., 2008b); (b) the muscular activation peaks increased with increasing stretch-velocity (Komi & Gollhofer 1997); (c) the maximum EMG amplitude during the contact phase of running was two to three times higher than the activity during maximum voluntary contractions (Dietz et al., 1979); (d) the activation peak at the time of the SLR decreased during running after partial blockage of Ia afferents by ischemia (Dietz et al., 1979); (e) vibration of the Achilles tendon, which is known to decrease primarily the efficacy of Ia afferent activity, led to a significant decrease of the SLR during running (Cronin et al., 2011).

Recently, two further observations strongly supported the assumption of stretch reflex generated muscular activity during SSC-movements. Leukel et al. (2009) introduced a new methodology to investigate stretch reflex responses by means of a pneumatic cuff surrounding the lower leg. Immediately after inflation of the cuff, a selective reduction of the SLR could be seen, which was elicited by a dorsiflexion of the foot in an ankle ergometer. Changes in the stretch velocity but not the stretch amplitude affected the size of the SLR pointing towards the Ia afferent pathway being primarily responsible for this response. As the effect was seen immediately after inflation of the cuff, the time was too short to cause ischemia. Therefore, it was postulated that inflation restricted the stretching of the muscles under the cuff so that most of the changes in length probably occurred in the series elastic structures of the muscle–tendon complex distal to the cuff. As a consequence, the muscle spindles embedded within the muscle may be less excited, resulting in a reduced SLR. When the cuff was applied during hopping, the muscular activity at the phase of the SLR was reduced, too (Leukel et al., 2009). Thus, it seems likely that the Ia afferent pathway is important to generate the SLR in hopping whereas other structures like the Golgi tendon organs or cutaneous receptors are probably much less involved.

A recent and convincing argument that spinal stretch reflexes contribute to SSC-movements was derived from the observation of a time-locked occurrence of the SLR with respect to the instant of ground contact (Zuur et al., 2010). Zuur and co-workers altered the time of ground contact during hopping by changing the height of the landing surface while subjects were airborne. The authors hypothesized that if a stretch reflex indeed contributes to the early EMG burst, then advancing or delaying the touch down without the subject's knowledge should similarly advance or delay the SLR-burst. This was indeed the case when touch down was advanced or delayed by shifting the height of a programmable platform up or down between two hops and this resulted in a corresponding shift of the SLR (Figure 12.1). These

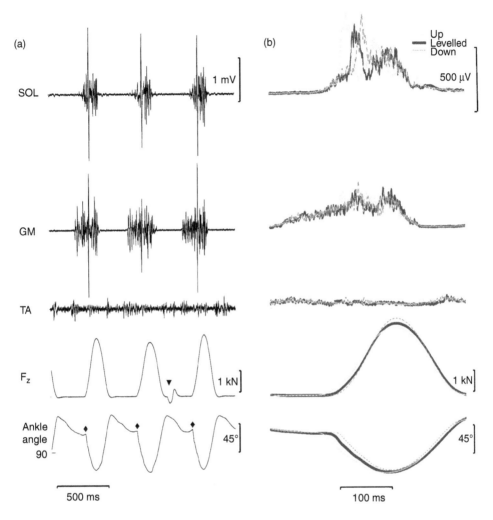

Figure 12.1 Evidence of spinal stretch reflex activity during hopping on a moveable platform. (a) raw EMG and kinematics from three hops. ▼ indicates an artefact in the force signal due to acceleration and deceleration of the platform when moving up or down. ◆ indicates the point where the ankle angle shows a maximal dorsiflexion acceleration. (b), ensemble average of 25 sweeps for the conditions 'Up', 'Level' and 'Down'. 'Up', 'Level' and 'Down' refer to the positions of the platform. During the period the subject was in the air the platform either stayed in the leveled position (Level) or moved 2.5 cm up or down in a randomized fashion. In the control trials the platform made a lateral movement, but returned to the level position before the subject touched down so that in all three conditions the same sound was made and no audible cues were given about the position of the platform. All averaged trials are aligned to the crossing of the light barrier, which was positioned 3 cm above the moveable force platform. Note that the platform in the 'Up' position causes a shift in the signals for ankle angle, ground reaction force and the short-latency EMG burst in SOL ahead in time, whereas the 'Down' position delays these signals in time. The shift in the short-latency response dependent on the position of the landing surface indicates that this burst is caused by peripheral feedback due to the impact at touch down. In contrast, the initial rise in EMG activity (before the sharp peak of the short-latency response) is comparable in all three conditions. Thus, it can be assumed that this rise is not caused by a feedback process but is rather pre-programmed (feedforward control); (from Zuur et al. (2010)).

results are in line with observations from landing where the EMG burst that appears shortly after landing disappeared when the subjects fell through a false floor, thus confirming that the burst results from a feedback loop, i.e. from a stretch reflex (Duncan & McDonagh, 2000).

In summary, there is good evidence for spinal stretch reflex activity at the time of the SLR during SSC-movements. The most likely source of this reflex activity is the excitation of primary muscle spindle endings. However, there is an ongoing discussion about how muscle spindles are activated: some authors propose muscle fascicle stretches to be the relevant stimuli (Ishikawa & Komi, 2008) while others argue that mechanical vibrations in response to the ground contact trigger the response (Cronin et al., 2011). Furthermore, it is likely that during functional movements several other pathways may influence the SLR including Golgi tendon organ Ib afferents, cutaneous receptors and mechanoreceptors in other muscles, as well as pre-programmed input from supraspinal centers (see next section).

Motor cortical contribution during the early reflex phase

In contrast to sudden and unexpected perturbations, drop jumping and hopping allow supraspinal centers to accurately predict the time of ground contact and thus the instant of muscular stretch (Dyhre-Poulsen & Laursen, 1984; Jones & Watt, 1971; McDonagh & Duncan 2002; Santello & McDonagh, 1998). Therefore, it has been speculated that during SSC, pre-programmed activation of supraspinal structures may contribute to the muscular activity at any time during the movement. Previous studies assumed that such a centrally pre-programmed muscular activity was important for the pre-innervation, the reflex modulation and the stiffness regulation during stretch-shortening cycle and landing movements in humans and animals (Dyhre-Poulsen & Laursen, 1984; Horita et al., 1996; Horita et al., 2002; Laursen et al., 1978; Leukel et al., 2008b). In other words, it was proposed that supraspinal centers not only initiate jumping and landing movements, but also modulate or influence at least part of the muscular activation pattern after touch down. However, the question of whether this type of modulation is organized by the motor cortex or by other centers is not well understood. So far, only two studies have investigated corticospinal activity during drop jumping and hopping by means of transcranial magnetic stimulation (TMS) (Taube et al., 2008; Zuur et al., 2010). In the first study, modulation of the TMS-evoked motor potential (MEP) was assessed during different phases after ground contact in drop jumps (Taube et al., 2008). The MEPs were small and non-augmented shortly after ground contact (at the times of the SLR, MLR, and LLR = early and late reflex phase), but were significantly facilitated after approximately 120 ms (LLR2; the activity 120 ms after touch down was labeled 'LLR2' in this study = late reflex phase and push-off phase). As this modulation was reciprocal to the modulation of the H-reflex (providing information about changes at the spinal level), which was high at SLR and then progressively declined towards the push-off phase, the authors argued that it is conceivable that corticomotoneurons (neurons within the motor cortex) enhanced their excitability at the time of the LLR2. Consequently, the cortical influence was supposed to be minor in the early contact phase, which was proposed to be dominated by spinal reflex activity. However, based on the methodology applied in this study, no information about the cortical involvement is available, i.e. whether the motor cortex itself contributed to the muscular activation. To assess motor cortical contribution, another TMS-method is needed, which was introduced by Davey et al. (1994). These authors were the first to demonstrate that a single transcranial magnetic stimulus below the threshold to elicit an MEP can produce a suppression in the electromyogram (EMG) of a voluntary contracted muscle without prior facilitation. Several control experiments suggested that this TMS-evoked EMG suppression is due to the activation of intracortical inhibitory interneurons which reduce the output from the motor cortex (Davey et al., 1994; Petersen et al., 2001). Thus, motor cortical output during movement is degraded by subthreshold magnetic stimulation, resulting in a decreased muscular activation. Most interesting, with respect to SSC-movements, was whether the motor cortex contributed to the muscular activity at the time of the SLR. As described previously in this chapter (see 'Do spinal (stretch) reflexes occur during SSC?'), the SLR during hopping is influenced by a stretch reflex response (Zuur et al., 2010). Dyhre-Poulsen et al. (1991) suggested that this stretch reflex response was set on top of the voluntary EMG activity. To test this assumption and to reveal motor cortical contribution at the time of the SLR, low-intensity magnetic stimuli to the motor cortex were applied during hopping (Zuur et al., 2010). The SLR was significantly reduced, indicating that the SLR in hopping is not only composed of activity resulting from sensory feedback but also influenced by a descending drive from the motor cortex (Figure 12.1).

Functional role of the stretch reflex in stretch-shortening cycle movements

As mentioned above, the efficiency of the stretch-shortening cycle is dependent on the energy transfer from the pre-activated and eccentrically stretched muscle–tendon complex to the concentric push-off phase. An appropriate stiffness regulation is considered to be one constitutive factor for a successful transfer (Komi, 2003). It is widely assumed that stretch reflexes may have a role in adjusting the leg stiffness. More specifically, reflex contributions induced by stretching of the antigravity muscles during touch down (eccentric phase) are thought to enhance muscular stiffness and therefore increase the performance during the concentric phase when compared to isolated concentric action (Dietz et al., 1979; Gollhofer et al., 1992; Voigt et al., 1998b). This reflex-induced enhancement of performance may be even more relevant in submaximal SSC-contractions as it was observed that the SLR after a rapid stretch of the isometrically contracted muscle is largest when the intrinsic muscle stiffness is low. Thus, the reflex may prevent muscle yielding in conditions where the muscle is not strongly (pre-) activated. In line with this assumption, vibration-induced reductions of the SLR were shown to increase muscle yielding while running with low to moderate speeds (7 to 12 km/h) but not at a faster speed (15 km/h) (Cronin et al., 2011).

The importance of stretch reflex responses to modulate the leg stiffness was first highlighted in animal studies demonstrating that muscle stiffness is dependent on an intact reflex system and is reduced in the a-reflexive state (Hoffer & Andreassen, 1981). Furthermore, animal studies indicated that reflex responses can change the form of the mechanical force response of the muscle from one dominated by viscosity to one dominated by elasticity (Lin & Rymer, 1993). Thus, reflex responses seem to ensure and to preserve the system's elasticity (Lin & Rymer, 1993). However, although it seems, based on these observations, beyond controversy that the mechanical stiffness of the leg can be influenced by spinal stretch reflexes, it remains a topic of debate how the neuromuscular system has to be adjusted to create optimal mechanical properties for energy storage and utilization during SSC (a thorough discussion of the mechanical properties determining the performance in SSC movements can be found in Brughelli & Cronin, 2008). Cross-sectional studies indicate that the stretch reflex circuit is adjusted according to the falling height by the central nervous system: when subjects were asked to perform drop jumps from different drop heights, the muscular activity at the SLR was lower in drop jumps from excessive (80 cm) than from low (30 cm) heights (Komi & Gollhofer 1997) (Figure 12.2). In a subsequent experiment, the size of the H-reflex was inversely related to the drop height (Leukel et al., 2008a; Leukel et al., 2008b). Recently, the

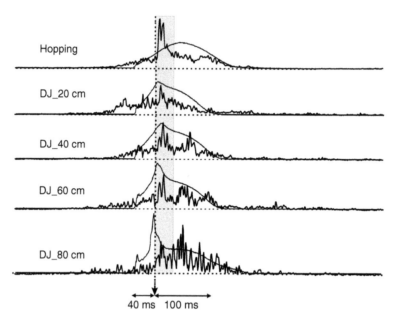

Figure 12.2 Changes in EMG pattern due to modulation of drop height. Rectified and averaged EMG pattern of the soleus muscle and vertical ground reaction force in various stretch-shortening cycle drop jumps with both legs. The figure illustrates the modulation in the pattern and in the force record with increasing stretch load (drop height). From top: Both-leg hopping in place; DJ_20 – DJ_80 cm, drop jumps from 20 to 80 cm drop height. It can be seen that the short-latency response (first peak in the EMG after the dotted vertical line) is rather increasing from 20 to 60 cm drop height but is drastically reduced when jumping from 80 cm (modified from Komi & Gollhofer (1997)).

decrease in H-reflex amplitude with increasing drop height was shown to be correlated with a reduction in the ankle joint stiffness (Taube et al., 2011b). It was speculated that this correlation supported and strengthened the long held assumption (Dietz et al., 1979; Dyhre-Poulsen et al., 1991; Gollhofer et al., 1992; Hobara et al., 2007; Komi, 2003; Voigt et al., 1998b) that spinal reflexes at the time of the SLR modulate the ankle joint stiffness during SSC-movements. It was hypothesized that from a functional point of view, reduction of the Ia afferent input (Leukel et al., 2008a; Leukel et al., 2008b) and the muscular activity (Komi, 2003; Komi & Gollhofer, 1997) at the time of the SLR could serve as a preventative method to compensate for the higher loads associated with greater drop heights (Figure 12.2). Diminished reflexes were thought to reduce the stiffness and therefore the peak stress of the tendo-muscular system (Dyhre-Poulsen et al., 1991; Hoffer & Andreassen, 1981). Furthermore, with the above-cited results of Lin and Rymer (1993) in mind, which demonstrated that tendo-muscular elasticity is dependent on intact reflex responses, it can be hypothesized that the mechanical force response of the muscles progressively changes from an elastic to a more viscous state when augmenting the drop height due to a reduction in reflex contribution. Such gradual switches in reflex contribution influencing the muscle properties should be most pronounced when subjects are asked to land instead of rebounding from the ground. Indeed, it was demonstrated that subjects drastically reduced their H-reflexes in the landing condition (Dyhre-Poulsen et al., 1991; Leukel et al., 2008a) when a viscous system was required.

Interestingly, the neuromuscular system adjusts muscular activity not only based on the drop height, but also depending on the characteristics of the landing surface. It was demonstrated that the stretch reflex response is diminished when rebounding from soft elastic surfaces (Gollhofer, 1987; Moritz & Farley, 2005). Moritz and Farley (2005) further observed that apart from the stretch component, the overall muscular activity was higher on soft surfaces than on solid surfaces despite similar joint moments and mechanical leg work. Additionally, the leg kinematics changed from the normal pattern known from solid surfaces where, during the contact phase, the legs are first flexed and subsequently extended to a reversed pattern. Therefore, the authors assumed that a higher overall muscular activation might be needed to compensate for the loss of the normal extensor muscle stretch–shortening cycle i.e. to compensate for the loss of the stretch reflex contribution.

In the preceding paragraphs it has been argued that the Ia afferent transmission and the muscular activity at the time of the SLR are task-specifically modulated. However, although there is a clear decrease in the muscular activity at the time of the SLR from drop jumps from low drop heights to drop jumps from high drop heights and finally to landings, it is difficult or even impossible to conclude that the elastic properties of the muscles are best when the reflexes are largest and thus, the stiffness is highest. For instance, although the leg stiffness was shown to be higher in experienced jumpers (French elite long and triple jumpers) compared to untrained subjects of the same age, the elite jumpers demonstrated a strong but negative correlation between the maximal height reached during hopping and the corresponding leg stiffness (Rabita et al., 2008). Similarly, Laffaye and co-workers (Laffaye et al., 2005) reported that elite handball, basketball and volleyball players as well as high-jumpers and novice jumpers decreased their leg stiffness when they augmented their rebound height in a one-leg jump task. Furthermore, it was demonstrated during drop jumps that the greatest power production but not the maximal rebound height was related to a highly stiff system (Arampatzis et al., 2001b). Recently, Cronin and co-workers (Cronin et al., 2011) showed that vibration of the Achilles tendon led to a significant decrease of the size of the SLR during running and this decrease was accompanied by a reduction in ankle joint stiffness (evidenced as ankle yielding) at low to intermediate running speeds (7 to 12 km/h). However, at 15 km/h, vibration still reduced the SLR but had no effect on the stiffness (Cronin et al., 2011). These observations demonstrate that it is in all likelihood wrong (i) to assume that maximal performance is correlated with maximal stiffness (Arampatzis et al., 2001b; Laffaye et al., 2005; Rabita et al., 2008) and (ii) to suggest that the stiffness is directly related to the magnitude of the SLR-peak (Cronin et al., 2011). Both adjustment of stiffness and changes of the SLR may therefore rather follow an optimum function with a u-shape than be linear.

In line with those assumptions, longitudinal training studies also emphasized that training-related increases in rebound height may not be primarily associated with changes in the lower leg stiffness, but may be more strongly dependent on the compliance of the tendo-muscular system (Hunter & Marshall, 2002; Taube et al., 2011b) (for further details please see section 'Influence of training on the neural control in SSC-movements'). Based on these observations and according to Rabita et al. (2008), it may therefore be proposed that in order to control SSC movements the central nervous system is challenged to find the right balance between a stiff 'high power system' and a more compliant system, which is probably better suited to store elastic energy. As a consequence, stretch reflex contributions have probably to be gated specifically to adjust the tendo-muscular stiffness and thus, to meet the criteria of context (e.g. duration of ground contact, rebound height etc.) and environment (e.g. low versus high drop height; surface characteristics etc.). In conclusion, the appropriate neuromuscular activation at the time of the SLR in SSC-movements is therefore most likely influenced by multiple factors and can only be detected by studies combining neurophysiological with biomechanical measurements so that interrelations of neuronal control and tendo-muscular properties can be identified. Furthermore, the muscular activity at the SLR seems important to adjust the tendo-muscular stiffness but it is certainly not the only parameter determining physical outcomes.

The late reflex phase and the push-off phase

Studies investigating the Ia afferent transmission by means of peripheral nerve stimulation during hopping and drop jumping revealed a phase-dependent modulation of the H-reflex: the H-reflex excitability was high during the contact phase, decreased just before push-off and remained low during the flight phase (Dyhre-Poulsen et al., 1991; Moritani et al., 1990; Taube et al., 2008; Voigt et al., 1998b). The functional significance of enhanced Ia-afferent transmission in the early contact phase may be that impulses from the Ia-afferents may enhance motoneuron activation on top of the ongoing EMG activity. Furthermore, it has been argued that afferent feedback can be used to produce a peak impulse by synchronizing muscular activation in the already active SOL muscle at the time of ground contact (Dyhre-Poulsen et al., 1991). The progressive decline of the H-reflex amplitude during the contact phase implies that muscular activity in the later contact phase may be less dependent on Ia-afferent input than in the early phase, suggesting other sources of neural activity are becoming more involved. In this context, Taube et al. (2008) demonstrated that despite a reduction in the H-reflex amplitude, the MEP was augmented towards push-off. Therefore, the authors speculated that the motor cortex might be more strongly involved during the later period of ground contact. Together with the results of Zuur and co-workers (2010) showing the contribution of the motor cortex to the early reflex phase of hopping, it can be assumed that motor cortical contribution is present throughout the entire SSC-movement, i.e. from the initiation of the jump throughout the pre-activation, the time of ground contact (SLR, MLR, LLR) until push-off.

Interaction of feedforward and feedback control during the SSC

Based on the current understanding of the neural control of SSC movements, we conceptualized and introduced a model showing how feedforward and feedback control interact in order to allow appropriate motor function during plyometric actions (Taube et al., 2011a). In general, 'feedforward control' (or 'predictive control') refers to the portion of the movement that is planned in advance and is not altered by online peripheral feedback (Bastian, 2006). In contrast, 'feedback' or 'reactive control' involves in-flight integration of peripheral feedback into the current movement in order to provide on-line reinforcement and/or correction (Bastian, 2006). Most of our natural movements involve interaction of feedforward and feedback control (e.g. Lam et al., 2006) (Figure 12.3). The feedforward component in walking could, for instance, be shown by demonstrating after-effects following a period of training in a new environment. Similarly, repeated jumping on an elastic surface leads to after-effects when subjects are tested on solid ground afterwards (Marquez et al., 2010). These examples indicate that for a given task the motor output has to be aligned and accordingly calibrated in order to meet the requirements of the modified environmental conditions. Therefore, the model that the CNS possesses an internal representation of the dynamics of the limbs and the body to compute the necessary motor output for a desired task in a given setting is highly attractive. When the biomechanics of the limb or task are changed, movement errors will

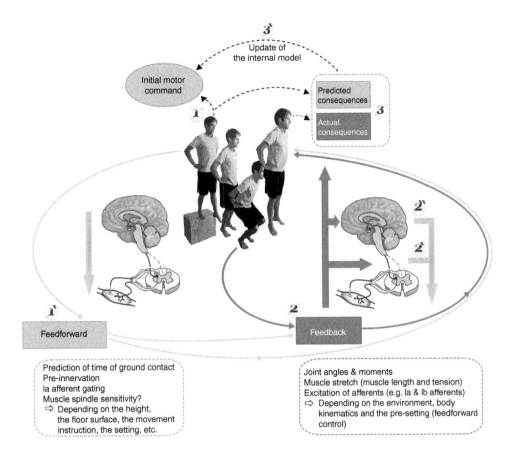

Figure 12.3 Interaction of feedforward and feedback control during stretch-shortening cycle movement.

(1) Initial motor command to initiate the movement and to adjust the system in accordance with the expected environmental setting (1'): The feedforward or predictive motor control refers to the portion of the movement that is planned in advance and is not altered by online peripheral feedback. In case of the drop jump, the instant of ground contact can be estimated and factors like floor surface, aim of the movement (for instance 'to rebound as fast as possible' or 'to rebound as high as possible') and the stability of the environment (e.g. opponent) can be given consideration. Dependent on the situation (Leukel et al., 2012), the central nervous system will adjust its activity, like for instance the amount and duration of pre-innervation or the Ia afferent gating.

(2) At touch down, peripheral feedback will be generated and can be integrated into the current movement in order to provide on-line reinforcement (e.g. activity of the short-latency stretch reflex (2') on top of a supraspinally caused baseline activity (Zuur et al., 2010)) and/or correction (for instance if the CNS miscalculated the instant of touch down or the properties of the landing surface (Marquez et al., 2010)). The feedback loop can involve spinal structures (2') or can be travelling via supraspinal centres (2'').

(3) The predicted and the actual consequences of the movement are compared. If there are differences between predicted and actual consequences, the internal model has to be updated. This may be the case when the biomechanics of the limb or task have been changed. Consequently, the internal model has to adjust the motor output to the new setting. It has been shown that for the update of the internal model during a series of jumps, information about one single miscalculated jump is sufficient to appropriately recalibrate the model (Marquez et al., 2010). Most probably, subjects use the error between the predicted and the actual sensory feedback to recalibrate their internal jump model. With permission from American College of Sports Medicine (Taube et al., 2011a).

occur (Shadmehr & Mussa-Ivaldi, 1994), which are needed for updating the internal model in order to adjust the motor output to the new setting (Kawato et al., 1987). The structural need for feedforward control is basically a consequence of the time delays of sensoriomotor loops that limit the rapidity with which the motor system can respond to sensory events (Wolpert & Flanagan, 2010). Thus, particularly fast movements depend on pre-programmed muscular activity. This is also the case in SSC movements and can easily be demonstrated in blindfolded subjects who perform drop jumps from a certain drop height. When the drop height remains unchanged, the movement pattern after a few attempts resembles the one with open eyes (Rapp & Gollhofer, 1993). However, when the instant of ground contact is earlier than expected (for example advising the subject to jump from 60 cm height, but offering only 20 cm), subjects are not able to perform a drop jump any more due to a wrongly timed muscular activation (own unpublished observations). A feedforward strategy was also demon-strated in landing movements of monkeys as the onset of the pre-activation EMG pattern was uninfluenced by turning off the light during the fall and by opening a collapsible platform delaying the time of ground contact (Dyhre-Poulsen & Laursen 1984). In a similar approach during human hopping, it was illustrated that lifting and lowering of the landing surface shifted the stretch reflex component forward and backward, respectively (Zuur et al., 2010). However, the initial rise in the soleus background EMG activity always occurred at the same time independent of the position of the landing surface indicating a pre-programmed feed-forward control (Figure 12.1). It was shown in several studies that the CNS accurately predicts the time of ground contact when jumping down from an elevated platform with open eyes (Dyhre-Poulsen & Laursen, 1984; McDonagh & Duncan, 2002; Santello & McDonagh, 1998). During hopping, the prediction of the time of ground contact seems to be similarly precise but it may be speculated that the visual information may not be as important as during landings or drop jumps due to the repetitive character of this movement. In summary, a considerable part of the muscular activity seen in SSC-movements seems to be pre-programmed. Thereby, the feedforward control in all likelihood not only affects supraspinal motor commands, but predetermines the integration of afferent feedback too. This may be assumed based on adapted stretch and H-reflex activities in drop jumps in response to modu-lations of the drop height (Leukel et al., 2008a; Leukel et al., 2008b). In particular, when subjects were asked to jump from high drop heights (i.e. over 60 cm) reflex responses were reduced. If the CNS had not decreased the spinal excitability and/or the susceptibility of the fusimotor system, the faster stretch velocities going along with increases in drop height should have resulted in an increased reflex activity (Gollhofer & Rapp, 1993). As this was not the case, Leukel et al. (2008b) speculated that the CNS pre-programmed the level of reflex inhibition based on the drop height to avovid abundant spinal reflex activation, which other-wise might have resulted in extensive muscle-tendon-stiffness, potentially causing overload injuries. A typical example of where the CNS has been misleadingly pre-programmed is where the neuromuscular activity in SSC-movements could be seen in subjects who got accustomed to jumping on an elastic surface and were asked afterwards to perform the same kind of jump on solid ground (Marquez et al., 2010). The after-effects included an increase in leg stiffness, decrease in jump height, and perceptual misestimation of the jump height. The authors proposed that the after-effects were due to an erroneous internal model acquired on the elastic surface. A further study highlighted that the parameterization of the internal model is dependent not only on previous experience, but also on the setting of the task (Leukel et al., 2012) (Figure 12.4). Leukel et al. (2012) instructed their subjects in the first condition to perform drop jumps from 50 cm ('no switch condition'). In the second condi-tion, subjects also performed drop jumps from 50 cm, but when a tone was presented prior to

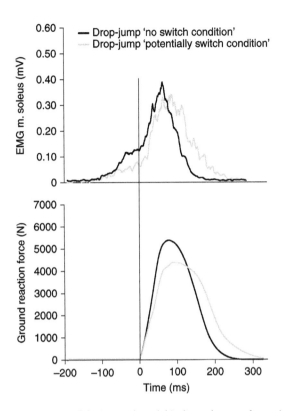

Figure 12.4 The parameterization of the internal model is dependent on the setting of the task. Shown are grand mean values of the m. soleus EMG and the ground reaction force for two conditions. In the first condition (termed 'no switch condition'), subjects performed drop jumps from 50 cm and knew that there would be no sign (auditory cue) indicating that they had to change their movement (i.e. to land). In the second condition, called 'potentially switch condition', the subjects performed the instructed drop jump and no switch of the movement had to be performed either. However, subjects were aware that an auditory cue could be presented in this condition. Zero on the x-axis refers to ground contact. Data are truncated at 340 ms following ground contact. It can be seen that the muscular activation and the resulting ground reaction forces were clearly modulated depending on the performed condition despite biomechanical identical conditions (modified from Leukel et al. (2012)).

ground contact they had to switch from jumping to landing ('potentially switch condition'). The authors were interested in the two conditions where the subject did exactly the same but with different settings, i.e. when subjects performed drop jumps from 50 cm and potentially had to switch the movement from a drop jump to a landing ('potentially switch condition') and in the other case when subjects performed the drop jump, but could be absolutely sure that no switch signal would occur ('no switch condition'). As the two tasks were biomechanically identical, every change in the movement execution in the 'potentially switch condition' must have been caused by a change in the pre-programmed (feedforward) motor command. The results demonstrated that the feedforward control indeed changed so that the muscular activity was somewhere between the muscular activity observed during drop jumps in the 'no switch condition' and the muscular activity in pure landings ('no switch landing condition'). The changes in the feedforward control were evidenced by a reduced muscular activity of the extensor muscles at the time of the SLR and augmentation of activity towards push-off

(Figure 12.4). Such a strategy may have allowed a more flexible task execution because the decision whether to apply a drop jump or to land could be 'postponed' as the muscular activity shortly before take-off decided whether subjects rebounded or landed. When rebounding was required, subjects displayed a strong activation towards take-off whereas landing was accompanied by a suppression of this activity (Leukel et al., 2012). This study demonstrates that depending on the setting ('no switch condition' versus 'potentially switch condition') the same task may be differently pre-programmed in order to accomplish the requirements of the situation. At the same time this study proposes that the integration of afferent feedback may also change depending on the setting. The H-reflexes recorded at the time of the SLR were reduced as soon as subjects had to potentially alter their movement. Although it could not be clarified whether this reduction was due to post or presynaptic mechanisms, it nevertheless demonstrates that modulation of the feedforward control affects the integration of spinal reflexes.

In summary, centrally pre-programmed muscular activity can be considered as being extremely important for the organization of SSC-movement, especially for the control of the pre-innervation (pre-activation), the reflex modulation and hence the stiffness regulation (Dyhre-Poulsen & Laursen 1984; Funase et al., 2001; Horita et al., 1996; Horita et al., 2002; Leukel et al., 2008b; Zuur et al., 2010). However, as already discussed in the paragraph entitled 'Spinal mechanisms contributing to the SSC', feedback mechanisms are also integrated into SSC-movements. Although they have the disadvantage of being time lagged, they do allow the system to provide specific responses to certain sensory events. The sensory consequences may either be expected – like the impact of touch down and the resulting muscle stretch – or be unexpected due to changes in the environment or errors in the feedforward program, for example. It may therefore be speculated that the benefit of using a feedback system during SSC-movements relies on the precise timing of the muscular activity through given sensorimotor loops. For instance, the activation of the monosynaptic stretch reflex circuit leads to contraction of the triceps surae complex some 35 to 45 ms after touch down (Funase et al., 2001; Taube et al., 2008). The delay may vary from subject to subject but within a subject is as precise as a few milliseconds (Gollhofer & Rapp, 1993). This time-locked reflex response may ensure an appropriate source of activation for reinforced calibration of the tendo-muscular stiffness due to the synchronous activation of the motoneuron pool. Moreover, in cases where the internal model wrongly predicts the time of ground contact, a feedback mechanism has the advantage to generate the muscular activity with reference to the instant of ground contact. When the miscomputation takes place during a series of jumps or during cyclic SSC-tasks like hopping or running, the detection of the movement error will be used to update the internal model in order to adjust the motor output for the next jump. It was shown that, for the updating, only information about one single miscalculated jump is necessary (Marquez et al., 2010). Most probably, subjects use the error between the predicted and the actual sensory feedback to recalibrate their internal jump model (Wolpert & Flanagan, 2001).

Influence of plyometric training on the neural control in SSC-movements

In many training studies incorporating SSC movements, the training of these movements was investigated together with other training set-ups such as purely concentric or eccentric exercises. Thus, the effect of the SSC tasks cannot clearly be inferred from these studies. From training studies, which concentrated on 'pure' SSC movements, it is obvious that most of them were carried out with the intention to monitor alterations in performance related biomechanical parameters of the SSC (e.g. Cornu et al., 1997; de Villarreal et al., 2008). Markovic and Mikolic (2010) summarized in a recent review article that relatively little emphasis was laid on the detection of the underlying changes taking place within the CNS. Interestingly, although some studies observed adaptations of the muscular activity by means of surface EMG in response to SSC interventions, these adaptations were most often not considered to be the driving force behind improved task performance. For instance, Kubo and co-workers (2007) compared a plyometric training with a weight training program in order to assess training specific adaptations of muscular activity and changes in the mechanical properties of the muscle–tendon complex. After 12 weeks, in which subjects trained in hopping and drop jumping in a sledge apparatus with one leg and doing strength exercises with their other leg, muscular activity of the plantarflexor muscles was enhanced in both legs in the concentric phase of the SSC. At the same time, lower leg stiffness was increased in the plyometrically trained leg whereas the strength trained side revealed an enhanced tendon stiffness of the knee extensors. As only the plyometric training group significantly increased vertical rebound height and the resistance group did not, despite similar adaptations in the EMG activity, the authors concluded that the increases in rebound height were most likely caused by changes in the mechanical properties of the muscle–tendon complex.

Similarly, Kyrolainen et al. (2004) speculated that the improved jump performance after 15 weeks of plyometric training was not caused by an altered neuromuscular control because they could not detect concomitant changes in the muscular activity. On first view, it therefore seems that adaptations in neuronal control either do not occur during plyometric training or are not relevant with respect to the biomechanical alterations in jump performance. However, another explanation may be that the neuronal mechanisms are difficult to assess due to the difficulties of quantifying EMG signals. Unfortunately, neither of the two studies normalized the individual EMG activity. Thus, any differences in the muscular properties as well as in the recording conditions may have biased their results. This is a serious concern as the reproducibility of EMG signals and especially their quantitative comparability obtained during drop jumps was shown to be low (m. gastrocnemius) to moderate (m. soleus) when the data collection was separated by seven or more days (Gollhofer et al., 1990). Especially when the electrode position was not exclusively controlled, quantitative comparisons were no longer possible.

Furthermore, in the studies of Kubo and co-workers (2007) and Kyrolainen et al. (2004) interventions were accomplished over a considerable period of time (12 and 15 weeks, respectively). Thus, anatomical changes within the muscle–tendon complex, specifically in the increased stiffness properties of the Achilles tendon, are very likely and have even been shown in the study of Kubo et al. (2007). Thus, the main argument against the interrelation of adaptations in the neuronal control strategy and gains in rebound height in this study was the lack of improvement in jump performance after strength training despite a similar increase in muscular activity during the SSC task. A limiting factor in the study of Kubo et al. (2007) is that the effects of weight and plyometric training on the neuronal control of the involved muscles cannot be separated: after strength as well as after coordinative training, cross-education effects have been reported which indicate that the training of one limb improves performance in the other limb as well (Farthing, 2009). Considering these restrictions and the fact that measurement and training sessions were performed in a sledge apparatus, in which natural SCC-movements are difficult to reproduce as biomechanical constraints are different and contact times are longer (Kramer et al., 2010), it is obvious that the aforementioned studies cannot convincingly demonstrate a non-interrelation of neuronal adaptation and improved performance. However, Kubo and co-workers (2007) as well as other authors (e.g. Malisoux et al., 2006) have demonstrated that the tendo-muscular properties adapt in response to plyometric training. Theoretically, changes in the tendo-muscular properties could be sufficient to explain increases in motor performance. In this case, neural adaptations would not be necessary to alter motor behavior. However, it seems more likely that there are interactions between tendo-muscular adaptations and changes within the central nervous system. The question is how these factors can be differentiated. With respect to neural adaptations, it is preferential to consider studies with shorter training duration as the neuronal changes are thought to occur at a faster rate than tendo-muscular adaptations (Moritani & deVries, 1979).

Wu et al. (2010) analyzed different types of jumps for eight weeks and assessed training-related effects during isometric contractions after four and eight weeks. The EMG activity in the soleus was normalized to the electrically evoked maximal M-wave of the same muscle and was significantly increased after four and eight weeks of training. This adaptation in the isometric condition was accompanied by improvements in maximal jump height during countermovement jumps, which were not significant after four weeks (total gain of 8 % corresponding to 3.9 cm) but after eight weeks (total gain of 12 % corresponding to 5.9 cm; P = 0.023). Correlations could be detected between changes in tendon stiffness and gains in jump height as well as between the tendon stiffness and the electromechanical delay of the gastrocnemius medialis muscle. The authors proposed that gains in performance may be attributed to the accumulated effects of changes in EMG activity, tendon stiffness and elastic energy utilization, but also stated that only the increases in tendon stiffness were correlated with the performance. One serious limitation of the study of Wu and co-workers (2009) is that, except for the vertical jump height, none of the other parameters such as muscular activity and tendon stiffness were assessed during the performance of an SSC-movement. Instead, adaptations seen during isometric contractions were 'transferred' to the gains in performance seen in the vertical jump test. However, it is well known that training adaptations are highly task specific and cannot be generalized (e.g. Schubert et al., 2008). Based on these limitations, the study of Wu and co-workers could also not clarify whether neuronal adaptations may contribute to improved SSC-performance after plyometric training.

In contrast, one study which could clearly demonstrate neural adaptations in response to SSC-training was presented by Voigt and co-workers (1998a). The authors investigated the influence of four weeks of hopping training on the neuromuscular activity at the time of the early reflex phase, i.e. the SLR. The H-reflex elicited at the time of the SLR increased after training. However, no changes in the movement pattern could be detected. Furthermore, no performance related parameters were assessed other than the time to task failure of continuous hopping, which increased from under 5 minutes before training to over 20 minutes after the intervention. The latter observation, however, may be interpreted in the way that the subjects improved in their motor control strategy in order to enable an optimized elastic interaction in the hopping exercises (see Figure 12.5).

In line with Voigt et al. (1998a), one of our most recent drop jump studies (Taube et al., 2011b) also showed clear neural adaptations after four weeks of SSC-training. These neural adaptations were associated with changes in performance related parameters. Two training groups were compared: in group I subjects exercised drop jumps exclusively from low drop heights (30 cm) whereas in group II subjects trained from different drop heights (30, 50 and 75 cm). In both groups the subjects were instructed to 'rebound as high and as fast as possible from the ground' in each single drop jump. It was hypothesized that the drop height applied during training should influence the neuromuscular adaptation. This hypothesis was based on

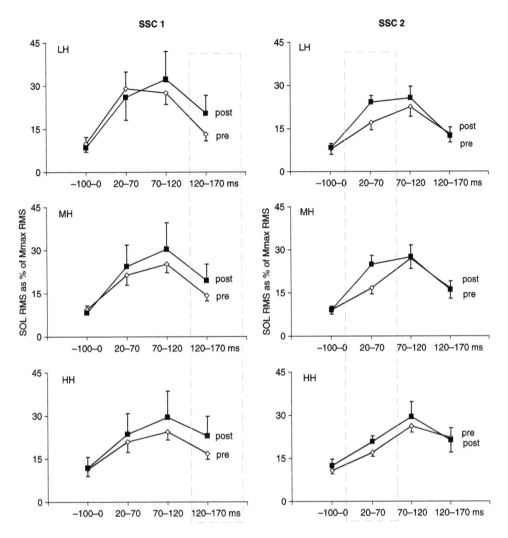

Figure 12.5 Drop height specific training adaptations of the neuromuscular activity. Muscular activity of the m. soleus during drop jumps from 30 cm (LH), 50 cm (MH), and 75 cm (EH) in pre- (◊) and post-measurement (■) of SSC1- and SSC2-trained subjects. After SSC1, in which subjects trained drop jumps from 30, 50 and 75 cm drop heights, muscular activity remained unchanged in the pre-activation phase (−100 to 0 ms before touch down) and the early duration of ground contact (20 to 70 ms = 'early reflex phase'). In later phases of ground contact (70 to 120 and 120 to 170 ms) muscular activity increased following SSC1-training. In contrast, changes after SSC2, in which subjects trained solely from 30 cm drop height, were most pronounced shortly after ground contact in the 'early reflex phase', but displayed no significant changes in any other phase. This demonstrates that the drop height chosen during training influenced the outcome of the neuromuscular adaptation (from Taube et al. (2011b)).

the height dependent modulation of muscular activity at the time of the SLR (please see 'Functional role of the stretch reflex in stretch-shortening cycle movements', above). The results confirmed the hypothesis and showed an increased muscular activity at the time of the SLR in the training regime using solely low level drop heights (30 cm) whereas the training incorporating high drop heights (75 cm) resulted in an enhanced muscular activity towards take-off. Interestingly, the differential adaptations in the neuromuscular control after the two training interventions were accompanied by distinct adaptations in performance related parameters: training from low drop heights led to shorter times of ground contact, while training incorporating excessive drop heights ameliorated the rebound height. These results indicate that changes in the neuromuscular control are indeed linked to the jump performance. However, this study could not clarify which structures of the CNS adapted in response to the two different plyometric training interventions. Based on the temporal occurrence (the phase of the drop jump) of the changes in muscular activity after training, it may vaguely be speculated that training from low drop heights may primarily affect the efficiency of the spinal reflex circuits whereas training incorporating excessive drop heights may act more strongly on supraspinal structures. Subsequent studies using electrophysiological methods have to clarify the sites of adaptation, which are responsible for the altered neuromuscular control after training from different drop heights.

Concluding this chapter, we can agree with Markovic and Mikulic (2010) who stated in their review article that 'our current knowledge about SSC-training induced changes in neural function is limited'. Considering the great impact of plyometric training on many kinds of sports disciplines, this lack of knowledge is surprising and has to be addressed with longitudinal training studies in the future.

Conclusions and prospects

This chapter illustrates the complex nature of SSC-movements, in which the CNS has to coordinate and adjust the contribution of anticipated (feedforward controlled) and reflexive (feedback controlled) neuromuscular activity in order to provide an appropriate tendo-muscular stiffness. To accomplish this task, cortical, subcortical and spinal levels have to closely interact. SSC-training can change the neuromuscular pattern indicating plasticity of the CNS. Future research is needed to elaborate the specific neuromuscular adaptations to SSC-long term training regimens. It is important to distinguish between different types of training set-ups (intensity, repetitions, duration) as well as between different subgroups with respect to gender, age and training status.

References

Arampatzis, A., Bruggemann, G. P., and Klapsing, G. M. (2001a), 'Leg stiffness and mechanical energetic processes during jumping on a sprung surface', *Med.Sci.Sports Exerc.*, 33: 923–931.

Arampatzis, A., Bruggemann, G. P., and Metzler, V. (1999), 'The effect of speed on leg stiffness and joint kinetics in human running', *J. Biomech.*, 32: 1349–1353.

Arampatzis, A., Schade, F., Walsh, M., and Bruggemann, G. P. (2001b), 'Influence of leg stiffness and its effect on myodynamic jumping performance', *J. Electromyogr. Kinesiol.*, 11: 355–364.

Bastian, A. J. (2006), 'Learning to predict the future: the cerebellum adapts feedforward movement control', *Curr. Opin. Neurobiol.*, 16: 645–649.

Bosco, C. and Komi, P. V. (1979), 'Potentiation of the mechanical behavior of the human skeletal muscle through prestretching', *Acta Physiol. Scand.*, 106: 467–472.

Bosco, C., Komi, P. V., and Ito, A. (1981), 'Prestretch potentiation of human skeletal muscle during ballistic movement', *Acta Physiol. Scand.*, 111: 135–140.

Brughelli, M. and Cronin, J. (2008), 'A review of research on the mechanical stiffness in running and jumping: methodology and implications', *Scand. J. Med. Sci. Sports*, 18: 417–426.

Cornu, C., Almeida Silveira, M. I., and Goubel, F. (1997), 'Influence of plyometric training on the mechanical impedance of the human ankle joint', *Eur. J. Appl. Physiol. Occup. Physiol.*, 76: 282–288.

Cronin, N. J., Carty, C. P., and Barrett, R. S. (2011), 'Triceps surae short latency stretch reflexes contribute to ankle stiffness regulation during human running', *PLoS One*, 6, e23917.

Davey, N. J., Romaiguere, P., Maskill, D. W., and Ellaway, P. H. (1994), 'Suppression of voluntary motor activity revealed using transcranial magnetic stimulation of the motor cortex in man', *J. Physiol.*, 477: 223–235.

De Doncker, L., Picquet, F., Petit, J., and Falempin, M. (2003), 'Characterization of spindle afferents in rat soleus muscle using ramp-and-hold and sinusoidal stretches', *J. Neurophysiol.*, 89: 442–449.

de Villarreal, E. S., Gonzalez-Badillo, J. J., and Izquierdo, M. (2008), 'Low and moderate plyometric training frequency produces greater jumping and sprinting gains compared with high frequency', *J. Strength Cond. Res.*, 22: 715–725.

Dietz, V., Schmidtbleicher, D., and Noth, J. (1979), 'Neuronal mechanisms of human locomotion', *J. Neurophysiol.*, 42: 1212–1222.

Duncan, A. and McDonagh, M. J. (2000), 'Stretch reflex distinguished from pre-programmed muscle activations following landing impacts in man', *J. Physiol.*, 526 Pt 2: 457–468.

Dyhre-Poulsen, P. and Laursen, A. M. (1984), 'Programmed electromyographic activity and negative incremental muscle stiffness in monkeys jumping downward', *J. Physiol.*, 350: 121–136.

Dyhre-Poulsen, P., Simonsen, E. B., and Voigt, M. (1991), 'Dynamic control of muscle stiffness and H reflex modulation during hopping and jumping in man', *J. Physiol.*, 437: 287–304.

Farthing, J. P. (2009), 'Cross-education of strength depends on limb dominance: implications for theory and application', *Exerc. Sport Sci. Rev.*, 37: 179–187.

Fukashiro, S., Komi, P. V., Jarvinen, M., and Miyashita, M. (1995), 'In vivo Achilles tendon loading during jumping in humans', *Eur. J. Appl. Physiol. Occup. Physiol.*, 71: 453–458.

Funase, K., Higashi, T., Sakakibara, A., Imanaka, K., Nishihira, Y., and Miles, T. S. (2001), 'Patterns of muscle activation in human hopping', *Eur. J. Appl. Physiol.*, 84: 503–509.

Gollhofer, A. (1987), 'Innervation characteristics of m. gastrocnemius during landing on different surfaces,' in *Biomechanics*, B. Jonsson, ed., IL: Human Kinetics Publishers, Champaign, IL, pp. 701–706.

Gollhofer, A., Horstmann, G. A., Schmidtbleicher, D., and Schonthal, D. (1990), 'Reproducibility of electromyographic patterns in stretch-shortening type contractions', *Eur.J . Appl. Physiol. Occup. Physiol.*, 60: 7–14.

Gollhofer, A. and Kyrölainen, H. (1991), 'Neuromuscular control of the human leg extensor muscles in jump exercises under various stretch-load conditions', *Int. J. Sports Med.*, 12: 34–40.

Gollhofer, A. and Rapp, W. (1993), 'Recovery of stretch reflex responses following mechanical stimulation', *Eur. J. Appl. Physiol. Occup. Physiol.*, 66: 415–420.

Gollhofer, A. and Schmidtbleicher, D. (1989), 'Stretch reflex responses of the human m. triceps surae following mechanical stimulation'. In R. J. Gregor, Zernicke, R. F. and Whiting, W. C. (eds), *Proceedings of the XII International Congress of Biomechanics in LosAngeles*, 219–220.

Gollhofer, A., Schmidtbleicher, D., and Dietz, V. (1984), 'Regulation of muscle stiffness in human locomotion', *Int. J. Sports Med.*, 5: 19–22.

Gollhofer, A., Strojnik, V., Rapp, W., and Schweizer, L. (1992), 'Behaviour of triceps surae muscle–tendon complex in different jump conditions', *Eur. J. Appl. Physiol. Occup. Physiol.*, 64: 283–291.

Gottlieb, G. L. and Agarwal, G. C. (1979), 'Response to sudden torques about ankle in man: myotatic reflex', *J. Neurophysiol.*, 42: 91–106.

Hobara, H., Kanosue, K., and Suzuki, S. (2007), 'Changes in muscle activity with increase in leg stiffness during hopping', *Neurosci. Lett.*, 418: 55–59.

Hoffer, J. A. and Andreassen, S. (1981), 'Regulation of soleus muscle stiffness in premammillary cats: intrinsic and reflex components', *J. Neurophysiol.*, 45: 267–285.

Horita, T., Komi, P. V., Nicol, C., and Kyrolainen, H. (1996), 'Stretch shortening cycle fatigue: interactions among joint stiffness, reflex, and muscle mechanical performance in the drop jump [corrected]', *Eur. J. Appl. Physiol. Occup. Physiol.*, 73: 393–403.

Horita, T., Komi, P. V., Nicol, C., and Kyrolainen, H. (2002), 'Interaction between pre-landing activities and stiffness regulation of the knee joint musculoskeletal system in the drop jump: implications to performance', *Eur. J. Appl. Physiol.*, 88: 76–84.

Hunter, J. P. and Marshall, R. N. (2002), 'Effects of power and flexibility training on vertical jump technique', *Med. Sci. Sports Exerc.*, 34: 478–486.

Ishikawa, M. and Komi, P. V. (2008), 'Muscle fascicle and tendon behavior during human locomotion revisited', *Exerc. Sport Sci. Rev.*, 36: 193–199.

Jones, G. M. and Watt, D. G. (1971), 'Observations on the control of stepping and hopping movements in man', *J. Physiol.*, 219: 709–727.

Kawato, M., Furukawa, K., and Suzuki, R. (1987), 'A hierarchical neural-network model for control and learning of voluntary movement', *Biol. Cybern.*, 57: 169–185.

Kjaer, M. (2004), 'Role of extracellular matrix in adaptation of tendon and skeletal muscle to mechanical loading', *Physiol. Rev.*, 84: 649–698.

Kjaer, M., Langberg, H., Miller, B. F., Boushel, R., Crameri, R., Koskinen, S., Heinemeier, K., Olesen, J. L., Dossing, S., Hansen, M., Pedersen, S. G., Rennie, M. J., and Magnusson, P. (2005), 'Metabolic activity and collagen turnover in human tendon in response to physical activity', *J. Musculoskelet. Neuronal. Interact.*, 5: 41–52.

Komi, P. V. (1984), 'Physiological and biomechanical correlates of muscle function: effects of muscle structure and stretch-shortening cycle on force and speed', *Exerc. Sport Sci. Rev.*, 12: 81–121.

Komi, P. V. (2003), 'Stretch-shortening cycle,' in *Strength and Power in Sport*, P. V. Komi, ed., Blackwell Science, Oxford, pp. 184–202.

Komi, P. V. and Gollhofer, A. (1997), 'Stretch reflex can have an important role in force enhancement during SSC-exercise.', *J. Appl. Biomech.*, 13: 451–460.

Komi, P. V., Gollhofer, A., Schmidtbleicher, D., and Frick, U. (1987), 'Interaction between man and shoe in running: considerations for a more comprehensive measurement approach', *Int. J. Sports Med.*, 8: 196–202.

Kramer, A., Ritzmann, R., Gollhofer, A., Gehring, D., and Gruber, M. (2010), 'A new sledge jump system that allows almost natural reactive jumps', *J. Biomech.*, 43: 2672–2677.

Kubo, K., Morimoto, M., Komuro, T., Yata, H., Tsunoda, N., Kanehisa, H., and Fukunaga, T. (2007), 'Effects of plyometric and weight training on muscle-tendon complex and jump performance', *Med. Sci. Sports Exerc.*, 39: 1801–1810.

Kyrolainen, H., Avela, J., McBride, J. M., Koskinen, S., Andersen, J. L., Sipila, S., Takala, T. E., and Komi, P. V. (2004), 'Effects of power training on mechanical efficiency in jumping', *Eur. J. Appl. Physiol.*, 91: 155–159.

Laffaye, G., Bardy, B. G., and Durey, A. (2005), 'Leg stiffness and expertise in men jumping', *Med. Sci. Sports Exerc.*, 37: 536–543.

Lam, T., Anderschitz, M., and Dietz, V. (2006), 'Contribution of feedback and feedforward strategies to locomotor adaptations', *J. Neurophysiol.*, 95: 766–773.

Laursen, A. M., Dyhre-Poulsen, P., Djorup, A., and Jahnsen, H. (1978), 'Programmed pattern of muscular activity in monkeys landing from a leap', *Acta Physiol. Scand.*, 102: 492–494.

Lee, R. G. and Tatton, W. G. (1975), 'Motor responses to sudden limb displacements in primates with specific CNS lesions and in human patients with motor system disorders', *Can. J. Neurol. Sci.*, 2: 285–293.

Leukel, C., Gollhofer, A., Keller, M., and Taube, W. (2008a), 'Phase- and task-specific modulation of soleus H-reflexes during drop-jumps and landings', *Exp. Brain Res.*, 190: 71–79.

Leukel, C., Lundbye-Jensen, J., Gruber, M., Zuur, A. T., Gollhofer, A., and Taube, W. (2009), 'Short-term pressure induced suppression of the short-latency response: a new methodology for investigating stretch reflexes', *J. Appl. Physiol.*, 107: 1051–1058.

Leukel, C., Taube, W., Gruber, M., Hodapp, M., and Gollhofer, A. (2008b), 'Influence of falling height on the excitability of the soleus H-reflex during drop-jumps', *Acta Physiol. (Oxf.)*, 192: 569–576.

Leukel, C., Taube, W., Lorch, M., and Gollhofer, A. (2012), 'Changes in predictive motor control in drop-jumps based on uncertainties in task execution', *Hum. Mov. Sci.*, 31: 152–160.

Lin, D. C. and Rymer, W. Z. (1993), 'Mechanical properties of cat soleus muscle elicited by sequential ramp stretches: implications for control of muscle', *J. Neurophysiol.*, 70: 997–1008.

Malisoux, L., Francaux, M., Nielens, H., and Theisen, D. (2006), 'Stretch-shortening cycle exercises: an effective training paradigm to enhance power output of human single muscle fibers', *J. Appl. Physiol.*, 100: 771–779.

Markovic, G. and Mikulic, P. (2010), 'Neuro-musculoskeletal and performance adaptations to lower-extremity plyometric training', *Sports Med.*, 40: 859–895.

Marquez, G., Aguado, X., Alegre, L. M., Lago, A., Acero, R. M., and Fernandez-del-Olmo, M. (2010), 'The trampoline aftereffect: the motor and sensory modulations associated with jumping on an elastic surface', *Exp. Brain Res.*, 204: 575–584.

Marsden, C. D., Merton, P. A., Morton, H. B., Adam, J. and Hallett, M. (1978), 'Automatic and voluntary reponses to muscle stretch in man'. In J. E. Desmedt (ed.), *Cerebal Motor Control in Man: Long loop mechanisms*. Basel: Karger, 167–177.

McDonagh, M. J. and Duncan, A. (2002), 'Interaction of pre-programmed control and natural stretch reflexes in human landing movements', *J. Physiol.*, 544: 985–994.

Moritani, T. and deVries, H. A. (1979), 'Neural factors versus hypertrophy in the time course of muscle strength gain', *Am. J. Phys. Med.*, 58: 115–130.

Moritani, T., Oddsson, L., and Thorstensson, A. (1990), 'Differences in modulation of the gastrocnemius and soleus H-reflexes during hopping in man', *Acta Physiol. Scand.*, 138: 575–576.

Moritz, C. T. and Farley, C. T. (2005), 'Human hopping on very soft elastic surfaces: implications for muscle pre-stretch and elastic energy storage in locomotion', *J. Exp. Biol.*, 208: 939–949.

Muller, R., Grimmer, S., and Blickhan, R. (2010), 'Running on uneven ground: leg adjustments by muscle pre-activation control', *Hum. Mov. Sci.*, 29: 299–310.

Petersen, N., Christensen, L. O., Morita, H., Sinkjaer, T., and Nielsen, J. (1998), 'Evidence that a transcortical pathway contributes to stretch reflexes in the tibialis anterior muscle in man', *J. Physiol.*, 512: 267–276.

Petersen, N. T., Butler, J. E., Marchand-Pauvert, V., Fisher, R., Ledebt, A., Pyndt, H. S., Hansen, N. L., and Nielsen, J. B. (2001), 'Suppression of EMG activity by transcranial magnetic stimulation in human subjects during walking', *J. Physiol.*, 537: 651–656.

Rabita, G., Couturier, A., and Lambertz, D. (2008), 'Influence of training background on the relationships between plantarflexor intrinsic stiffness and overall musculoskeletal stiffness during hopping', *Eur. J. Appl. Physiol.*, 103: 163–171.

Rapp, W. & Gollhofer, A. (1993), 'Motor control in reactive drop jump condition with and without visual control', Biomechanics, Proceedings of the XIVth Biomechanics-Congress, Paris, pp. 1104–1105.

Santello, M. and McDonagh, M. J. (1998), 'The control of timing and amplitude of EMG activity in landing movements in humans', *Exp. Physiol.*, 83: 857–874.

Schmidtbleicher, D. and Gollhofer, A. (1982), 'Neuromuskuläre Untersuchungen zur Bestimmung individueller Belastungsgrössen für ein Tieftsprungtraining', *Leistungsport*, 12: 298–307.

Schubert, M., Beck, S., Taube, W., Amtage, F., Faist, M., and Gruber, M. (2008), 'Balance training and ballistic strength training are associated with task-specific corticospinal adaptations', *Eur. J. Neurosci.*, 27: 2007–2018.

Shadmehr, R. and Mussa-Ivaldi, F. A. (1994), 'Adaptive representation of dynamics during learning of a motor task', *J. Neurosci.*, 14: 3208–3224.

Taube, W., Leukel, C., and Gollhofer, A. (2011a), '"How Neurons Make Us Jump" The Neural Control of Stretch-Shortening Cycle Movements', *Exerc. Sport Sci. Rev.* 40(2): 106–115.

Taube, W., Leukel, C., Lauber, B., and Gollhofer, A. (2011b), 'The drop height determines neuromuscular adaptations and changes in jump performance in stretch-shortening cycle training', *Scand J Med Sci Sports*, doi: 10.1111/j.1600–0838.2011.01293.x.

Taube, W., Leukel, C., Schubert, M., Gruber, M., Rantalainen, T., and Gollhofer, A. (2008), 'Differential modulation of spinal and corticospinal excitability during drop jumps', *J. Neurophysiol.*, 99: 1243–1252.

Taube, W., Schubert, M., Gruber, M., Beck, S., Faist, M., and Gollhofer, A. (2006), 'Direct corticospinal pathways contribute to neuromuscular control of perturbed stance', *J. Appl. Physiol.*, 101: 420–429.

Toft, E., Sinkjaer, T., and Andreassen, S. (1989), 'Mechanical and electromyographic responses to stretch of the human anterior tibial muscle at different levels of contraction', *Exp. Brain Res.*, 74: 213–219.

Voigt, M., Chelli, F., and Frigo, C. (1998a), 'Changes in the excitability of soleus muscle short latency stretch reflexes during human hopping after 4 weeks of hopping training', *Eur. J. Appl. Physiol. Occup. Physiol.*, 78: 522–532.

Voigt, M., Dyhre-Poulsen, P., and Simonsen, E. B. (1998b), 'Modulation of short latency stretch reflexes during human hopping', *Acta Physiol. Scand.*, 163: 181–194.

Wolpert, D. M. and Flanagan, J. R. (2001), 'Motor prediction', *Curr. Biol.*, 11: R729–R732.

Wolpert, D. M. and Flanagan, J. R. (2010), 'Motor learning', *Curr. Biol.*, 20: R467–R472.

Wu, Y. K., Lien, Y. H., Lin, K. H., Shih, T. T., Wang, T. G., and Wang, H. K. (2010), 'Relationships between three potentiation effects of plyometric training and performance', *Scand. J. Med. Sci. Sports*, 20, e80–e86.

Zuur, A. T., Lundbye-Jensen, J., Leukel, C., Taube, W., Grey, M. J., Gollhofer, A., Nielsen, J. B., and Gruber, M. (2010), 'Contribution of afferent feedback and descending drive to human hopping', *J. Physiol.*, 588: 799–807.

13

POSTURAL CONTROL AND BALANCE TRAINING

Wolfgang Taube[1] and Albert Gollhofer[2]

[1] DEPARTMENT OF MEDICINE, MOVEMENT AND SPORT SCIENCE, UNIVERSITY OF FRIBOURG, SWITZERLAND
[2] DEPARTMENT OF SPORT SCIENCE, UNIVERSITY OF FREIBURG, GERMANY

Introduction

Human stance and locomotion are basically related to the ability of balance and the control of posture. From a behavioral point of view the goal of postural control may therefore be defined as to orient body parts to one another and the external environment without loss of equilibrium. In order to accomplish this task and to achieve an appropriate muscular activation, the postural system has to take into account gravitational forces. Theoretically, 'static equilibrium' is achieved when all moments acting on the body are balanced and a specific body position is maintained. However, even static seeming postures like upright stance in humans involve a great deal of neural and muscular activity and the term 'static equilibrium' has – despite its widespread occurrence – to be used with caution (see 'The Organization of Un-Disturbed Stance – Reactive or Feedforward Control?', below, pp. 260–2).

The counterpart of static equilibrium constitutes 'dynamic equilibrium' where balance has to be maintained despite movement of the body. Although we are usually not aware of the complexity of this task, our postural system is greatly challenged to balance a centre of gravity with a height of approximately 1 m over the small area of support provided by our feet. The devastating effects easily observable when balance is impaired due to vestibular deterioration after accidents or due to selective loss of sensory feedback under pathological conditions impressively underline the importance of our balance controlling system. In order to meet the biomechanical requirements of stance and locomotion, it is the major task of the muscles encompassing the ankle, knee and hip joints to provide adequate forces in order to generate appropriate joint torques. However, postural control is not only relevant to avoid loss of balance but also to provide a fundamental basis for the execution of other movements. An alpine skier, for example, must rapidly compensate internal and external forces under highly dynamic conditions in order to avoid loss of balance. In other situations, when rapid movement changes are important and when disturbances from external sources (such as opponents in various team sports disciplines) have to be compensated, an immediate control of posture is highly relevant to enable the execution of goal directed actions.

For many daily and sportive activities, postural control is one of the limiting factors of performance and subsequently, certain training interventions aim to improve this skill. However, it has been shown that training of balance is not only efficient to improve postural control but also to enhance the rate of force development, the jumping behavior, and the regeneration after injury. Furthermore, balance training can significantly reduce the incidence rates of joint injuries. In particular, sprain injuries at the ankle joint as well as cruciate ligament ruptures at the knee joint can effectively be reduced when subjects perform balance training in parallel to their normal athletic training. Thus, balance training seems to be beneficial to improve motor control in general so that intra- and intermuscular coordination avoids the occurrence of injuries in demanding situations in which the organism is prone to overload.

The present chapter highlights in a first step the interaction of different structures of the central nervous system (CNS) to allow an appropriate postural control. The second part of this chapter outlines the influence of balance training. More specifically, neural adaptations within the CNS are discussed which may be responsible for improving postural control, increasing explosive force and reducing the incidence of injuries after balance training.

Postural control depends on sensorimotor interaction

Adequate postural control strategies depend on the detection and integration of relevant sensory information to get an impression about the body in space and the surrounding environment. Moreover, sensory signals constitute feedback with a time delay, due to the neural loop characteristics. Thus, integrated postural control can only be achieved if sensory information is properly sensed, predicted and incorporated in the ongoing motor control strategies. Based on this sensory information, the CNS has to plan and execute appropriate muscular activation patterns. Impairment of both the sensory and the motor side affects balance skills. In the following, the systems which are responsible for the perception of balance-related sensory information are identified. Subsequently, the neurological centers, in which this sensory information is processed, are introduced. Some information given in this chapter was previously published in review articles (Taube et al., 2008a; Taube, 2012) or book chapters (Taube & Gollhofer, 2011).

Sensory information

Vision

It is well established that the control of upright posture depends on sensory information from proprioceptive (Fitzpatrick & McCloskey, 1994), vestibular (Nashner et al., 1989), cutaneous (Kavounoudias et al., 2001; Meyer et al., 2004b), and visual sources (Buchanan & Horak, 1999; Duarte & Zatsiorsky, 2002). With regard to vision, it was shown that postural sway increases when the visual field is restricted or visual motion feedback is deprived (Hafstrom et al., 2002). Similarly, motion of the visual scene evokes postural responses (Guerraz et al., 2001; Mergner et al., 2005). Lee & Aronson (1974) were some of the first to highlight the importance of visual cues by testing infants in a 'moving room'. While the children stood in upright stance on solid non-moveable ground, the walls of the room were shifted either back or forward and the infants' postural reactions were monitored. In 82 percent of all trials, the subjects clearly showed a compensatory reaction in the direction of the moving room resulting in increased sway (26 percent), stagger (23 percent) or even fall (33 percent).

Almost 40 years after Lee & Aronson (1974) introduced the 'moving room' paradigm, the brain activity during such visual field motions was monitored in a virtual reality setting implemented into an fMRI-scanner (Slobounov et al., 2006). It was shown that visual field motions were accompanied by significant activation of specific brain structures, including prefrontal, parietal cortices and bilateral cerebellum. More specifically, when subjects had experienced actual postural responses (ego-motion) or reported the sense of self-motion (vection) in the virtual reality environment, perceptual areas, such as the visual cortical area V5/MT and the superior temporal sulcus (STS) were activated. The authors therefore proposed that these areas together with the anterior cingulate cortex (ACC) may represent at least in part the neural substrate of self-motion.

Vision influences not only supraspinal brain areas, which are important for postural control, but also spinal reflex pathways. For instance, it was demonstrated that H-reflexes in the soleus muscle were reduced when subjects closed their eyes (Earles et al., 2000; Hoffman & Koceja, 1995). In both studies, the influence of vision on the spinal reflex system was evaluated while subjects stood in bipedal stance on (i) a stable surface and (ii) an unstable surface (foam surface and mini-trampoline, respectively). As a common finding, occluded vision increased postural sway and reduced the gain of the soleus H-reflex (Earles et al., 2000; Hoffman & Koceja 1995). However, there is also evidence that postural control can be improved when visual cues are reinforced and this influences spinal reflexes. In this context, Taube et al. (2008b) provided subjects with a laser pointer, which should be directed at a target on the wall. On the behavioural level it was observed that subjects instantly reduced their centre of pressure (COP) displacement when using the laser pointer independently whether they stood on a stable or unstable support surface. On the neurophysiological level, an increase in the amplitude of the H-reflex was observed when subjects were provided with the enhanced (augmented) visual feedback (laser pointer).

One of the most important connections of the visual system in order to ensure appropriate gaze and postural control lies in the close interaction with the vestibular system. Before addressing this close interaction, the vestibular system is briefly outlined.

Vestibular system

In brief, the vestibular system gathers sensory signals about three-dimensional head rotations and translations. Vestibular information arises either from the semi-circular canals sensing rotations in three-dimensional space or from the otolithic system, which references linear accelerations. This information is subsequently transferred to the brain and serves to enable postural and oculomotor control as well as spatial and bodily perception (for details see Chapter 2 of this volume). As already mentioned in the introduction, the perception of gravitational forces is important to organize upright posture with regard to the ground. The vestibular receptors accomplish this task and at the same time account for a variety of reflexes in order to activate muscles responsible for gaze stabilization as well as muscles to stabilize body posture in space (Lopez & Blanke, 2011). The vestibular influence on balance control includes modulation of the muscle tonus and the antigravity reflexes, which are needed to maintain and adapt the configuration and stance of the body with respect to gravity without conscious endeavor (Lackner & DiZio, 2005). These vestibulo-spinal reflexes interact with reflexes related to the neck muscles, the limbs, and the overall disposition of forces acting on the body.

The vestibular structures are also important to distinguish self-motion and object-motion (Straube & Brandt, 1987). These complex actions are only possible because the vestibular nuclei are very well connected to other brain structures such as the cerebellum, the thalamus, and different cortical areas such as the parietal, frontal, cingulate, striate and extrastriate visual cortex, and the hippocampus (Lopez & Blanke 2011).

The parietal-insular cortex is involved in the perception of verticality and self-motion and interacts with the visual cortex in terms of a reciprocal visual-vestibular inhibition. This mechanism of an inhibitory interaction allows a shift of the dominant sensorial weighting from the visual to the vestibular modality and vice versa (Brandt & Dieterich, 1999). Vestibular stimuli are always associated to the sensation that the body is moving. However, the problem that our CNS faces is that visual stimuli of motion can always have two perceptual interpretations: either object-motion or self-motion. For instance, visual self-motion is often perceived while sitting in a stationary train while another train moves on an adjacent track. Our CNS therefore has to shift its concentration depending on what kind of stimulation prevails: body accelerations are associated with a dominant sensory weighting of the vestibular input while constant velocity motion favors the visual input (Brandt & Dieterich 1999). When misinterpretation or mis-weighting occurs, sensory illusions come into play, which affect postural control.

Apart from the strong linkage of visual and vestibular information processing, there is integration of other information, which also influences postural control. It is striking that comorbidity of balance disorders and anxiety disorders are widespread. Evidence suggests that information processing of vestibular and of emotional information share some common features (Balaban, 2002; Balaban et al., 2011). In this respect, a region in the brainstem, the parabrachial nucleus, seems to be an important node in a network that integrates vestibular, somatic, and visceral information. Moreover, noradrenergic and serotonergic projections to the vestibular nuclei have parallel connections with anxiety pathways. This close interaction is therefore thought to be responsible for the close association of balance disorders and anxiety.

Tactile information

In patients, for example, with loss of plantar cutaneous sensation due to diabetic neuropathy, impairments of postural control can be observed. Many authors therefore discussed postural instability in those patients primarily to be associated with deterioration of cutaneous tactile input. However, as sensory neuropathy is not restricted to the decreased sensation of the plantar surface of the feet but may also impair sensory information from muscle spindles and Golgi tendon organs, the significance of these findings is difficult to estimate (van Deursen & Simoneau, 1999). Based on studies showing the great importance of muscle spindle contributions to the control of upright posture (e.g. Fitzpatrick et al., 1994) it might even be speculated that the influence of cutaneous receptors of the feet is minor for the control of unperturbed stance. More recent studies support this view as anaesthetizing the foot sole had only a moderate influence on stance stability even when subjects were asked to balance on one leg or had to close their eyes (Meyer et al., 2004b). However, the same authors demonstrated that tactile information of the feet is important to control posture after lateral perturbations of the support surface (Meyer et al., 2004a). When plantar sensitivity was reduced, subjects displayed a relative shift in compensatory torque production from the ankles and trunk to the hips. This finding is in line with earlier observations, where foot and ankle ischemia produced also a shift from the 'ankle' to the 'hip strategy' (Horak et al., 1990).

The sensation of the foot sole is not the only tactile feedback which influences balance control. Light touch of the hand is very helpful to stabilize body posture even when the hand touches the reference surface at mechanically non-supportive force levels (Lackner & DiZio 2005). This phenomenon was shown in healthy subjects as well as in all kinds of patients such as people with labyrinthine deficits, diabetic neuropathy, cerebellar disease, or alcoholism. However, if the support surface is not stationary but moves, subjects will increase their body sway (Lackner & DiZio 2005). Thus, the support surface obviously serves as a reference system for body sway relative to its surrounding environment.

As vestibular and visual information is integrated to achieve adequate postural control, tactile and visual signals are also combined in the brain to ensure appropriate interactions with the surrounding space (Macaluso & Maravita, 2010). Tactile and visual signals converge in different regions of the brain such as the parietal and premotor cortices. This integration allows orienting of spatial attention and the generation of a cortical representation of the space around the body.

Proprioception

Previous studies have shown that postural control of stance relies considerably on proprioception. For instance, disrupting proprioceptive information processing increases postural sway (Bergin et al., 1995; Inglis et al., 1994). Complementary to this observation it was shown that when other sensory information is no longer available, postural control can be ensured by proprioceptive information. For instance, studies have proposed that standing with closed eyes is mainly organized by the somatosensory system with the ankle joint proprioception as the most important source regarding the detection of postural sway (Fitzpatrick & McCloskey 1994; Peterka & Benolken, 1995; Winter et al., 1998). Fitzpatrick et al. (1994) demonstrated that the control of upright stance is even possible after the exclusion of visual, vestibular, and cutaneous sensory information. In this condition, there remain three main candidates which may be selected to achieve proper undisturbed stance: group Ia and II afferents from muscle spindles, as well as group Ib afferents from Golgi tendon organs. The importance of these afferents for an organized balanced stance is supported by findings from patients with polyneuropathy, in whom the degradation of proprioception was related to the extent of their body oscillations (Bergin et al., 1995). More recent studies have particularly emphasized the relevance of the group II afferents for the control of upright stance (Nardone et al., 2000; Nardone et al., 2001; Nardone & Schieppati, 2004).

In elderly people, decline of proprioception is well documented (Goble et al., 2009) and it is further known that this decline is associated with deficits in postural control (Lord et al., 1991). Recently, neural correlates were found for the differences in proprioception between young and older individuals: when proprioceptors (i.e., muscle spindles) were activated by means of tendon vibration, fMRI showed reduced activity of the right putamen (basal ganglia) in elderly people (Goble et al., 2012). In a subsequent study, the same authors highlighted that the neural activity in response to stimulation of foot proprioceptors (i.e., muscle spindles) was correlated with their balance ability (Goble et al., 2011). In particular, greater balance performance was associated with greater activity in parietal, frontal, and insular cortical areas, as well as structures within the basal ganglia. It might therefore be proposed that processing in those structures is important to ensure balance control.

To come back to the proprioceptors, not only the muscle spindles of the leg play a prominent role to ensure adequate postural control but also those of the neck. In people with neck pain, coordination of head movements, inter-segmental coordination of the vertebrae of the cervical spine, and postural balance were shown to be impaired (Childs et al., 2008). The sensorimotor dysfunction resulted, for instance, in a reduced ability to reposition the head in a certain predefined position (Revel et al., 1991; Treleaven et al., 2005) and an impaired oculomotor control (Tjell & Rosenhall, 1998). Furthermore, when measured during unperturbed upright stance, enhanced postural sway was observed (Field et al., 2008; Stapley et al., 2006). These observations indicate that proprioceptive information of the neck muscles is crucial to ensure an appropriate control of posture and gaze. In line with this assumption, neck coordination exercises as intervention in patients with chronic neck pain have been shown to reduce postural sway, alleviate pain and improve performance in sensorimotor tasks (Jull et al., 2007; Revel et al., 1994; Roijezon et al., 2008; Taimela et al., 2000). Thus, proprioception of the cervical spine seems to greatly influence posture in general.

Information processing, motor planning and execution

In the preceding sections, the focus was primarily set on stimulus reception and the importance of visual, vestibular, tactile, and proprioceptive information with respect to postural control. In the following part, the integration and processing of those sensory signals is discussed. The relevant structures of the CNS with regard to balance control are introduced and it is illustrated how these structures affect motor commands. Thereafter, a more comprehensive view is provided by discussing the organization of undisturbed and disturbed upright stance.

Spinal Cord

The quickest and probably simplest afferent information processing takes place in the spinal cord. Following a fast calf muscle stretch (e.g. slipping on a slick surface), the muscle spindles perceive information about the type of change of muscular length and release action potentials to the spinal cord via Ia afferents. The activation of homonymous α-motoneurons then leads to a reflex response following around 40 to 50 ms after the muscle stretch (Gollhofer & Rapp, 1993). Although the stretch reflex is monosynaptic, i.e., only one synapse interposes afferent and efferent pathways, this does not exclude interaction of other parts of the neural system. For instance, during postural demanding situations, reduced reflex amplitudes can be observed, which most likely are caused by an increase in presynaptic inhibition (Hoffman & Koceja 1995; Llewellyn et al., 1990). Presynaptic inhibition is described as an inhibited release of transmitter at the synaptic cleft. In cases where demanding postural balance tasks lead to spinal reflex inhibition, the excitation of the Ia afferents is not fully transmitted to the postsynaptic neuron (the α-motoneuron). This means that the presynaptic transmitter release is reduced without affecting the postsynaptic side, which is still susceptible to other inputs. During balance control, presynaptic inhibition therefore allows the reduction of spinal reflexes without affecting the input of supraspinal sites to the α-motoneuron pool. This may allow movements to be controlled less by reflexes, but rather by higher centres.

Brainstem

The spinal cord connects to the brain stem, which is composed of the medulla oblongata, pons, and midbrain. From earlier experiments involving animals, it has been deduced that this area of the brain plays an important role in regulating balance. At the beginning of the twentieth century, research showed that mammals were able to compensate postural disturbances, utilizing reflex mechanisms involving the spinal cord and brainstem even if the connections to higher neural centres were dissected (Magnus, 1924; Sherrington, 1906). Later on, these observations were confirmed and expanded. Thereby, the reticular formation, which passes through the medulla oblongata, pons, and midbrain, was identified as playing an important role in controlling balance. In this area of the brain, information from the vestibular apparatus, the proprioceptive and the visual systems converges and is integrated into cortical motor commands. Luccarini et al. (1990), for instance, showed that postural balance adjustments could be suppressed by inhibiting the activity of the brainstem. Preventing the development of neural projections from the brainstem to the spinal cord consequently impairs the development of motor skills and postural control mechanisms (Vinay et al., 2005).

Cerebellum

The importance of the cerebellum for postural control could first be proven through studies involving patients with cerebral defects. As early as 1891, Luciani described the following consequences from cerebral lesions: atony (loss of muscular strength), asthenia (muscle exhaustion), astasia (inability to stand), dysmetry (movement disturbances, from either over-shooting or undershooting goal directed movements) (Luciani, 1891). Later on, the descrip-tion of the symptoms was expanded and the role of the cerebellum in adjusting agonist/antagonist movements, as well as coordinating successive movements was highlighted (Diener & Dichgans, 1992).

Due to the close connection of nodulus and flocculus with the vestibular system, this part of the cerebellum was called 'vestibulocerebellum'. The area accommodates important func-tions concerning the control of eye and axial muscles. Impairment of this structure therefore causes balance and gait to deteriorate and causes pathological nystagmus (Dietrichs, 2008). Regarding the two other parts of the cerebellum, the spino- and the cerebrocerebellum, the spinocerebellum in particular is also highly important to secure appropriate postural control. Impairments of the spinocerebellum are often associated with ataxias, which can impair postural control.

In general, the cerebellum seems to be crucial for the selection and memorization of appropriate compensation reactions in the specific disturbance situation. Patients with cere-bral lesions are often unable to alter their movement pattern when the situation is changing. Unlike healthy people, they cannot adapt the amplitudes of their reflex responses ('long-latency response' with a latency of around 120 ms) situation specifically (Nashner, 1976). Morton and Bastian (2007) further inferred by observation that a healthy person's cerebellum 'learns' from previous movement mistakes, thereby developing a different form of postural reaction by adjusting feedforward control.

Basal ganglia

Postural impairment is also observable following basal ganglia lesions. A dysfunctional Substantia nigra, as for example in Parkinson patients, is commonly associated with insecure stance, numerous falls and gait disabilities (Bloem et al., 2004; Bloem & Roos, 1995). Visser and Bloem (2005) claim that the basal ganglia facilitate postural control through two mecha-nisms: one being *postural flexibility* and the other *sensorimotor integration*. Postural flexibility is the ability to constantly adapt to environmental changes. Similar to the effects after cerebral lesions, damage of the basal ganglia leads to the inability to adapt muscular responses to specific disturbance stimuli (Bloem & Bathia, 2004). Moreover, during dual tasks incor-porating cognitive and motor demands, basal ganglia patients are unable to prioritize the task execution. Instead of placing retention of balance as the highest priority, they attempt to accomplish all tasks simultaneously, increasing the risk of falls (Bloem et al., 2001).

Apart from the loss of postural flexibility – or the ability to flexibly weigh resources – damage to basal ganglia accompanies an impairment to integrate sensorimotor information. When Parkinson patients are asked to close their eyes and then reposition their arm in the same way as their other arm, which has been passively or actively moved, they are often unable to do so (Klockgether et al., 1995; Zia et al., 2000). The dysfunctional capability of integrating sensorimotor information could possibly explain why Parkinson patients fail to realize their own body oscillations, and are often surprised when they see themselves oscillating in front of a mirror (Visser & Bloem, 2005). The role of the basal ganglia as a sensory analyzer could also be shown in healthy animals and humans. In monkeys, neurons in certain structures of the basal ganglia, the putamen and the globus pallidus, were shown to code for passive joint displacement and this strengthens the idea that those structures are important to process proprioceptive information relevant for postural control (Goble et al., 2011). In humans, the level of activation of the basal ganglia after stimulation of leg proprioceptors was shown to predict their performance in a postural task (Goble et al., 2011). Thus, sensory processing in the basal ganglia seems to be extremely important to ensure appropriate postural control.

Motor cortex

Initially, the motor cortex was not considered to play a major role in the maintenance of balance. However, recent studies present experimental data, which highlight its relevance with respect to postural reactions. Many of these studies are summarized in two recently published review articles (Jacobs & Horak, 2007a; Taube et al., 2008a). In rabbits and cats the role of the motor cortex in postural regulation was demonstrated by a correlation of cortical and muscular activity during balance reactions while directly recording electrical physiological currents from their cortical neurons (Beloozerova et al., 2003; Beloozerova et al., 2005). With humans, evidence of cortical involvement during balance reactions was provided using non-invasive electro-physiological techniques such as transcranial magnetic stimulation (TMS), positron emission topography (PET), or electroencephalography (EEG) (Jacobs et al., 2008; Ouchi et al., 1999; Taube et al., 2006). For instance during walking, the central nervous system is responsible for balancing numerous dynamic body segments with a high center of gravity over a small base of support. During such tasks TMS made it possible to demonstrate that despite highly automated movements, control was still relying on cortical centers (Nielsen, 2003).

However, cortical involvement is not confined to walking. Cortical control was also apparent following rapid perturbations during upright bipedal stance resulting in so-called transcortical long-loop reflexes of the lower extremity muscles with latencies around 85–100 ms (Taube et al., 2006). When the surface underneath a person is taken away, such as when abrupt accelerations or decelerations are experienced in buses or trams, these events are perceived by the muscle spindles of the calf muscles. Under lab conditions, these events are investigated by quick and unpredicted fast forward or backward displacements while the subjects are in upright stance. When the length change of the muscle is sufficiently fast, a short latency stretch reflex (occurring after 40 to 50 ms) is evident. The perception of stretch not only triggers the short latency response but is also transmitted to higher centers where it meets the sensory cortex following approximately 50 ms. After a short central processing (~ 10 ms), a motor command from the motor cortex to the muscles is sent and arrives there an additional 30 to 35 ms later (Petersen et al., 1998; Taube et al., 2006). The cortical compensation response following a balance disturbance appears later (around 85–100 ms) as the response generated at the spinal level (40–50 ms; short latency response). However, the cortical responses are better adaptable to the specific situation.

It appears plausible that when people find themselves in situations where they may potentially lose their balance, cortical structures are conceptually producing counteractive measures. However, in recent studies it was further shown that the motor cortex is also active in stable situations such as during undisturbed two legged stance (Soto et al., 2006; Tokuno et al., 2009). In this situation, no correlation between the modulation of motor cortex excitability and the observed body oscillations could be demonstrated (Tokuno et al., 2009). This could mean that the motor cortex is either shifted to a readiness position in order to quickly intervene in case of lost balance, or that other – non cortical areas – of the brain (i.e. cerebellum) are dependent on (tonic) motor cortex output to become self active.

The organization of upright posture

The term postural control describes the ability to balance the body over the base of support as well as providing a situation specific body position to enable the execution of other tasks. Generally, two modes of postural control are discussed: the 'feedback mode', which refers to the postural system aiming at re-establishing balance by compensatory reactions and the 'anticipation mode' or 'feedforward mode' that functions when potential 'disturbances' are cognitively anticipated in order to sustain postural control. The second mode, the 'anticipation mode', is characterized by its ability to foresee potentially destabilizing moments and sustains itself through self initiated adequate counter-movements. However, as postural control may not always be clearly differentiated into either 'feedback' or 'anticipation' modes during certain movements but may rather be a hybrid from both, the following section distinguishes between motor control in 'non-disturbed' and 'externally influenced' postural tasks. The contributions of anticipated and feedback actions are highlighted within this broader frame.

The organization of un-disturbed stance – reactive or feedforward control?

As compared to quadruped animals, humans have a more challenging task during bi-pedal stance. This is due to our uniquely high positioned centre of gravity when standing upright, the small stance supporting surface of our feet, and the low joint stiffness at the ankle (Casadio et al., 2005; Loram & Lakie, 2002). Humans are normally not aware of the demands for such a sensorimotor achievement. When 'uncertainty' during stance is experienced, caused for example by an 'unstable support surface' or a 'degraded sensorimotor system' due to pathological processes, the complexity of postural control is revealed. Usually, the standing posture is conceived as the initial state of walking or running, and thus locomotion is considered to be more complex than upright stance. However, it is just the other way round as standing requires a continuous stabilization process, which is not relevant in the same way during cyclic movements (Morasso et al., 1999).

Mechanically, the major task of the central nervous system (CNS) is to avoid falls and thus to keep the body in balance over its supporting surface. The perception of 'unbalance' through either optical (Buchanan & Horak 1999), proprioceptive (Fitzpatrick & McCloskey 1994), tactile (Holden et al., 1994; Lackner et al., 1999), or vestibular sensors (Nashner et al., 1989) is essential to secure muscular activation in order to produce adequate corrective strategies. The information received from various receptors is not separately processed but rather integrated and processed differently according to the situational needs. An example would be to stand straight maintaining balance in darkness or with the eyes closed. In this situation the body is unable to utilize visual feedback and must rely on other sensory information to complete the task. The task of standing undisturbed with closed eyes is then mainly organized by the somatic sensory system. This was shown in several experiments (Fitzpatrick &

McCloskey 1994; Peterka & Benolken 1995; Winter et al., 1998). In one of them, Fitzpatrick et al. (1994) demonstrated that the control of upright stance is even possible after the exclusion of visual, vestibular, and cutaneous sensory information so that only proprioceptive cues could be utilized. Conversely, the example of Ian Waterman, who has lost his proprioception from the neck downwards, demonstrates that postural control can be achieved without proprioception when other sensory information such as vision is available. These two examples demonstrate that loss of sensory information can normally be compensated by other sensory sources. However, as a common feature it can be stated that whenever certain sensory cues are withdrawn, postural sway increases. This was shown in a variety of experiments as for instance when subjects were asked to close their eyes (Dornan et al., 1978), when proprioceptive inputs were altered through cooling (Asai et al., 1992), or ischemic block (Mauritz & Dietz, 1980), or when vestibular inputs were (made) unreliable (Brandt et al., 1981). Traditional theories argue that this increase in postural sway is due to insufficient sensory information available to correct for the 'unbalance'. Thus, they assume that a loss of sensory information leads to greater sway oscillations as detection of uncertainties may be delayed or inaccurate. However, there is an alternative explanation why postural sway increases when sensory cues are withdrawn. Several studies postulated that postural sway may serve an 'exploratory' function (e.g. Van Emmerik & Van Wegen, 2002). In this respect, postural sway is not considered to reflect error or noise in the postural control system but may constitute an active strategy to gain essential information about the environment. Recently, Carpenter and colleagues (Carpenter et al., 2010; Murnaghan et al., 2011) conceptualized a design where they could reduce the movements of the body, i.e. the movements of the center of mass (COM), without the subject's awareness. At the same time they measured the ground reaction forces (center of pressure = COP) as these forces are traditionally considered to be caused by motor actions with the aim to stabilize the COM. If this traditional view was correct one would have assumed reduced COP-values when the COM is artificially stabilized. However, subjects showed exactly the contrary: when the movement of the COM was stabilized, deflections of the ground reaction forces (COP) were increased (Carpenter et al., 2010). Based on this observation the authors proposed that the COP displacement is increased purposefully by the CNS in order to ensure that the CNS receives a certain quality and/or quantity of sensory information from the surrounding environment. This hypothesis was further supported by subsequent experiments of the same group, in which potential confounding effects such as sensory illusions or motor drift were ruled out by providing the subjects with actual feedback about the movement of the COM (Murnaghan et al., 2011). Despite the knowledge that their COM was stable, subjects augmented their COP-displacements. Thus, postural sway may indeed serve – at least to a certain extent – to explore the surrounding environment and may therefore even be actively initiated.

The idea that stance regulation is not simply a reactive task but involves task planning and anticipation is not new. Based on computational frameworks, Morasso & Schieppati (1999) calculated that even very small delays of about 50 ms in the feedback loop were sufficient to destabilize the whole system. Therefore, they suggested a central computational process that carries out at least two main functions: (a) integration of the multisensory information in order to estimate the actual postural state and (b) compensation of transmission delays by anticipatory (feedforward) actions. A recent observation further strengthens the assumption that stance control is not reactive but is pre-planned and anticipated. Based on ultrasound measurements, Loram and co-workers (Loram et al., 2005a; Loram et al., 2005b) observed that the calf muscles do not follow a spring-like behavior during stance, meaning that the muscles do not lengthen during forward sway and shorten during backwards sway. Instead, calf muscles showed a shortening during forward sway and a lengthening during backwards sway, pointing towards a pre-programmed anticipatory muscular activity and not a reflex controlled activity. Fitzpatrick & Gandevia (2005) named this process a 'catch-and-throw-strategy' of the body and compared this behavior to the action of keeping a balloon in the air by repeated hits. The postural system may therefore use a simple pattern of anticipated cyclic muscle activations in order to stabilize the body instead of trying to achieve a static equilibrium by continuous feedback muscle activation.

Nowadays, there exists more and more evidence as to how the CNS may be able to ensure such an anticipatory movement control. In general, the cerebellum seems to play an important role (Bastian, 2006; Ramnani, 2006). It is assumed that the cerebellum receives a copy of the motor command – a so-called 'efference copy' – from cortical centers (Ramnani 2006). In line with this it was shown that the cerebellar Purkinje cells fire in advance of changes in the kinematics so that it seems likely that the cerebellum indeed predicts the forthcoming motor state based on this efference copy (Roitman et al., 2005). Beyond that, recent observations suggest that Ia-afferents anticipate the upcoming state of the extrafusal muscle fibers, too (Dimitriou & Edin, 2010). Thus, it is proposed that muscle spindles may also act as a sensory feedforward model. The advantage of predictive signals is that corrections can be made much faster. However, predictions make sense only if they can be compared with each other. In case of proprioceptive information it was assumed that an area in the brainstem, the inferior olive, may overtake the comparative function as it is closely linked with the cerebellum (target of the efference copy) and the spinal cord (source of proprioceptive information) (Ramnani, 2006).

Another point underlying the complex neural organization of unperturbed stance may be the observation that the cortical excitability is enhanced when subjects had to switch from supported to normal (free) standing (Tokuno et al., 2009) (Figure 13.1). At the same time, presynaptic inhibition acting upon the spinal reflex system was shown to increase when switching from supported to normal standing (Katz et al., 1988). Thus, it may be assumed that spinal reflex contributions are reduced while cortical involvement is enhanced during normal upright stance. This also suggests that motor control of upright stance is a complex pre-programmed motor activity involving higher motor centers rather than a set of purely reflex-controlled compensation reactions.

Postural compensation when stance is disturbed – reactive or feedforward control?

To compensate balance disturbances the CNS has to activate the muscles appropriately, according to disturbance stimuli and biomechanical conditions. It seems logical that unforeseen perturbations can only be compensated by reactive motor actions. However, anticipation may play a greater role than initially expected. Historically, compensatory postural responses were thought to be controlled by the brainstem and spinal reflex circuits as Magnus (1924) and Sherrington (1910) observed that animals were able to maintain and correct posture despite transection at the midbrain, eliminating input from the cerebral cortex. In the meantime, many experiments have demonstrated that cortical centers are also important to control posture (Jacobs & Horak, 2007b; Taube et al., 2008a). The importance of higher centers may especially be related to their ability to adjust the postural actions context specifically and to predefine a specific 'postural setting' (or 'central set'). For example, Horak et al. (1989) demonstrated that prior experience as well as knowledge about the magnitude of the postural disturbance affected the compensatory response. When subjects expected a larger perturbation than what actually took place, they over-responded. The opposite was true when a smaller perturbation was expected. Therefore it was argued that both peripheral and central

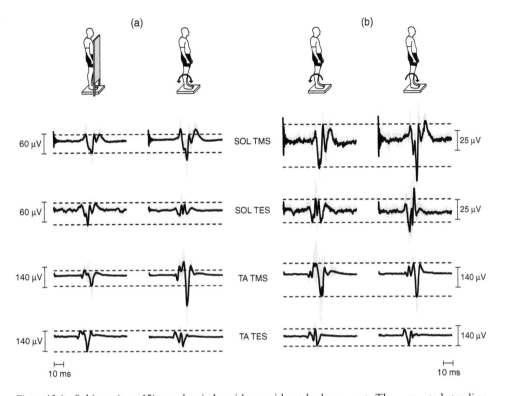

Figure 13.1 Subjects (n = 15) stood quietly with or without body support. The supported standing condition enabled subjects to stand with a reduced amount of postural sway. Transcranial magnetic stimulation (TMS) or transcranial electrical stimulation (TES) was applied to elicit responses in the soleus (SOL) and the tibialis anterior muscle (TA). During normal when compared to supported standing (a), the TMS-evoked MEPs from the SOL and TA were facilitated. However, TES-evoked MEPs in the SOL and TA were not different between the normal and supported standing conditions. When regarding the influence of the sway direction (b), the SOL TMS- and TES evoked MEPs were all greater during forward when compared to backward sway. In contrast, the TMS- and TES-evoked MEPs from the TA were smaller when swaying forward as compared to backward. These results indicate that cortical excitability is enhanced in the free standing condition compared to the situation with body support. However, based on the data analyzed with respect to sway direction it seems unlikely that the modulation of cortical excitability is directly involved in the on-going control of postural sway. Modified from Tokuno et al. (2009).

mechanisms modulate actual and anticipated perturbation characteristics. Some years later, evidence was provided by means of electroencephalography (EEG) that the cortical activity is changed prior to anticipated postural perturbations, supporting the abovementioned idea that cortical activity may be crucial to optimize postural responses by changing the postural pre-setting (central set) (Jacobs et al., 2008). Interestingly, the authors found that changes in the cortical activity shortly before the perturbation correlated with changes in postural stability, such that the subjects with the largest changes in readiness potentials between the anticipated and the non-anticiapted perturbations demonstrated the largest improvements in postural stability.

When disturbance of posture cannot be anticipated, sensory information about the modality and the magnitude of the perturbation is essential to successfully program adequate compensatory responses. Depending on the actual situation and the specificity of the disturbance stimulus, information from participating sensory systems will be weighed differently. Due to the close interaction between the different sensory modalities, it is impossible to designate relative contributions from each system (visual, vestibular, somatosensory). However, disturbance specific preferences have been identified. For an undisturbed upright stance, the importance of the somatosensory system was already emphasized in the preceding paragraph. This system also exhibits a dominate role for compensating fast direction changes from the supporting surface (Dietz et al., 1988). However, not only the somatosensory system (Horak et al., 1990) but also the vestibular system has the potential to trigger and modulate motor balance reactions utilizing its sensory information (Allum et al., 1998; Boyle et al., 1992). The vestibular system is thereby ascribed to control the slower body oscillations beginning around 1 Hz (Mauritz & Dietz 1980).

It seems meaningful to differentiate the disturbing stimuli into fast and slow perturbations but also to distinguish between rotational and translational surface displacements. The vestibular-spinal mechanisms in all probability play a major role when the support surface rotates (Allum & Pfaltz, 1985) whereas fast translational disturbances are processed more strongly by the somatosensory system (Dietz et al., 1988). A comparison in patients with different pathological impairments emphasizes the relevance of the somatosensory system compensating translational disturbances. In this situation, people with vestibular deficiencies revealed unaffected postural reactions (Horak et al., 1990), while patients with sensory neural pathology demonstrated delayed onset latencies (Inglis et al., 1994).

The organization of postural compensation reactions is not only dependent on the speed and type of perturbation (rotation or translation) but also relies on the supporting surface (Runge et al., 1999). On a solid and stable surface, the ankle strategy is predominantly utilized. In rough approximation this would mean that the body acts as an inverted single segmented pendulum. However, when the supporting surface is very small or fails to support, than the 'hip strategy' is adopted. Here the body resembles an inverted pendulum with two segments, which are connected together at the hip. Horak et al. (1990) proposed that an intact somatosensory system is necessary to conduct the 'ankle strategy' as they noted a change to the 'hip strategy' when the foot and ankle underwent ischemia. On the other hand, patients with vestibular loss were reported to use exclusively the ankle strategy. Thus it may be assumed that the hip strategy relies on vestibular input (Horak et al., 1990).

The dependency of the type of reflex pattern on the extent of the mechanical perturbation has led to assuming the existence of 'load receptors' (Dietz et al., 1992). It has been shown that the body load influences the amplitude of the reflex responses independently from the stretch velocity. Based on these observations it was concluded that a load dependent receptor – probably the Golgi tendon organ – could mediate different resultant joint torques at the ankle joint. Load receptors have been described for the extensor muscles of the spinal cat during fictive locomotion (Conway et al., 1987; Duysens & Pearson, 1980) and it was shown that these receptor signals arise from Golgi tendon organs before being mediated by Ib afferents to the spinal locomotor generator.

Balance training

Behavioral adaptations in response to balance training

Posture

The ability to maintain balance is subject to strong fluctuations during life span. Small children need approximately one year in order to stand upright or to independently learn how to walk. Children require then at least seven years until they have developed balance strategies similar to those of adults. Throughout the course of maturation, postural control continuously improves before it starts to get impaired with increased age. Age related balancing ability is, however, not solely influenced by age but is strongly affected by the activity level of an individual. Balance training studies demonstrated that balance skills can be improved in healthy individuals in every period of life. Hadders-Algra and co-workers (1996) evaluated the impact of a two-month BT program on postural responses while sitting on a moveable platform for healthy infants aged 5–10 months. Balance training facilitated the execution of the most complete direction-specific postural response patterns and accelerated the development of response modulation. In young and middle-aged participants, a variety of studies have emphasized the beneficial effect of balance training on postural control (e.g. Schubert et al., 2008; Taube et al., 2007a). In elderly subjects between the ages of 60 and 80 years, improved balance control could also be induced by means of balance training (Granacher et al., 2006). Apart from studies with healthy individuals, balance training proved to be efficient to ameliorate postural control in a variety of patients suffering from pathologies like, for instance, stroke (Lubetzky-Vilnai & Kartin, 2010), Parkinson (Allen et al., 2011), cerebellar ataxia (Ilg et al., 2009) or cerebral palsy (Ledebt et al., 2005).

Strength and power production

The benefit of balance training is not confined to the execution of postural tasks. Previous studies demonstrated that balance exercises also improved strength and jumping abilities (Bruhn et al., 2004; Granacher et al., 2006; Gruber et al., 2007b; Gruber & Gollhofer, 2004; Heitkamp et al., 2001; Kean et al., 2006; Myer et al., 2006; Taube et al., 2007b). Balance training influenced most efficiently the maximum rate of force development (Bruhn et al., 2006; Granacher et al., 2006; Gruber et al., 2007a; Gruber & Gollhofer 2004). In untrained or impaired subjects balance training also enhanced maximal voluntary strength (Bruhn et al., 2006; Heitkamp et al., 2001; Hirsch et al., 2003). Highly relevant from a functional point of view is the observation that balance training was capable of improving vertical jump performance (Kean et al., 2006; Taube et al., 2007b). As balance training stresses the tendomuscular system only moderately, these training effects may be particularly beneficial for pubertal children and athletes who face a risk of overload or overuse injuries.

Health benefits

The preventive character of balance training was evidenced in a variety of prospective studies investigating team sports like handball (Myklebust et al., 2003; Myklebust et al., 2007; Petersen et al., 2005; Wedderkopp et al., 1999; Wedderkopp et al., 2003), soccer (Caraffa et al., 1996; McGuine & Keene, 2006), basketball (Emery et al., 2007; McGuine & Keene 2006), and volleyball (Bahr & Bahr, 1997; Verhagen et al., 2004). The occurrences of ankle and knee injuries in these disciplines could be reduced by approximately 50 percent when athletes performed an additional prevention program incorporating balance exercises (Hübscher et al., 2010).

Apart from its preventive character, balance training supports the regeneration of neuromuscular structures following injury (Eils & Rosenbaum, 2001; Freeman et al., 1965; Gauffin et al., 1988; Henriksson et al., 2001; Rozzi et al., 1999) and efficiently prevents their re-occurrence (Holme et al., 1999; McGuine & Keene 2006; Verhagen et al., 2004).

Furthermore, the positive effect of balance training on the pain perception of chronic neck pain patients was shown in a recent study (Beinert & Taube, 2012). Subjects in this study suffered from neck pain and displayed impairments in sensorimotor control of head movements as well as deterioration in balance performance. After five weeks of balance training, not only postural skills were improved but also the sensorimotor function of the cervical spine. Furthermore, neck pain perception could be significantly reduced.

Neurophysiological adaptations in response to balance training

This paragraph links behavioral changes after balance training with the underlying neurophysiological adaptations in order to establish a basic understanding, how the plasticity of the CNS may enable improving postural skills, strength and power production, as well as reducing the incident rate of injuries. This section is strongly based on a review article, recently published in the *German Journal for Sports Medicine* (Taube 2012).

How can balance training improve postural skills?

Instant reflex adaptations in response to changes of the postural setting Llewellyn and co-workers (1990) demonstrated in humans that changes in the support surface influence the spinal reflex processing. In their experiment, they asked subjects to either walk over ground or over a small beam. As soon as subjects had to perform the more challenging task of walking over the beam, they demonstrated reduced H-reflex responses. Subsequent experiments confirmed reduced H-reflexes going along with increased postural demands (Earles et al., 2000; Hoffman & Koceja 1995). The H-reflex was measured in those studies, as this reflex shares some similarities with the stretch reflex. Thus, excitation of the Ia-afferents is mediated to the spinal cord and is transmitted there onto the α-motoneuron, which in turn activates the muscle. Changes of the H-reflex size without concurrent changes of the background muscular activity are therefore thought to point towards a modified spinal processing of afferent information. The most likely mechanism responsible for the reflex inhibition observed in more challenging postural tasks is an enhanced presynaptic inhibition (Katz et al., 1988). Presynaptic inhibition allows that the excitation of the Ia afferents is not fully transmitted to the postsynaptic neuron (the α-motoneuron). This means that the presynaptic transmitter release is reduced without affecting the postsynaptic side, which is still susceptible to other inputs. During balance control, presynaptic inhibition therefore allows the reduction of spinal reflexes without affecting the input of supraspinal sites to the α-motoneuron pool. As a consequence, it may be speculated that movements are controlled less by reflexes but rather by higher centres (Solopova et al., 2003; Taube et al., 2007a).

The functional significance of reducing spinal reflex responses in more challenging postural settings was interpreted as a way of reducing unwanted joint oscillations (Koceja & Mynark, 2000; Llewellyn et al., 1990). Keller et al. (2012) described this scenario in the following way: 'The unwanted and uncontrollable joint oscillations are assumed to originate from muscle stretch reflexes that occur when a fast deflection of the slackline or a fast tilt of a classical balance training device takes place and the counteracting stretch reflex results in an overshooting joint repositioning. This overcorrection may trigger subsequent stretch reflexes that are probably responsible for the build up of the described joint oscillations.'

Interestingly, reflex responses are not only inhibited as postural complexity increases, but they are also facilitated once postural demands become less challenging, for instance by additional mechanical support (Katz et al., 1988; Tokuno et al., 2009) or augmented visual feedback (Taube et al., 2008b). It is therefore argued that presynaptic inhibition is intensified or weakened in order to allow a flexible and adequate reflex adaptation to meet the requirements of the task.

Spinal adaptations after longer periods of balance training   Based on the abovementioned studies showing reduced reflex excitability when increasing postural demands, it may be speculated that balance training may also target these spinal reflex responses in order to improve postural control. This seems indeed to be the case as several studies demonstrated that improved balance skills went along with reduced H-reflexes after several weeks of balance training (Gruber et al., 2007b; Taube et al., 2007a; Taube et al., 2007c; Trimble & Koceja, 1994). Similarly, four weeks of slackline training proved to be efficient to ameliorate performance on the rope and at the same time reduced the H-reflex (Keller et al., 2012). Thus, it seems likely that the CNS learns during balance training to adequately adjust spinal reflex responses so that reflex mediated joint oscillations can be avoided.

However, it should be mentioned that these reflex adaptations are not a phenomenon which can be generalized to all movements, but are strictly related to the task in which they are elicited. In this respect, six weeks of balance training reduced H-reflexes exclusively during a postural task on the treadmill but not during unperturbed stance (Taube et al., 2007c). In another study, evidence was provided that not only the task itself may be crucial for the training related H-reflex modulation but also the phase of the movement (Taube et al., 2007a). After few weeks of balance training, subjects were tested while compensating for a rapid posterior displacement of the support surface. During the early compensation phase (around 50 ms after the perturbation) the H-reflex remained unchanged, whereas the H-reflex was significantly inhibited after training when measured in the late compensation phase (around 120 ms after the perturbation). Furthermore, spinal reflexes may even be facilitated when they are relevant to counteract the perturbation (Granacher et al., 2006). Thus, balance training does not change the spinal reflex behavior per se but rather seems to improve the ability to find the right 'reflex setting' for certain postural conditions. In other words, balance training improves the task-specific reflex modulation.

Adaptations within the brain in response to balance training   The compensatory reaction in response to a sudden perturbation is composed of activity, which results from activation of spinal reflex circuits followed by activity processed by a transcortical loop. The spinal contributions are generally termed short- (SLR) and medium-latency responses (MLR) and occur after approximately 50 ms and 70 ms, respectively. The subsequent muscular activation peak around 90 to 120 ms is already influenced by motor cortical centers and is called long-latency response (LLR) (Taube et al., 2006). Apart from the motor cortex (for review: Jacobs & Horak 2007b; Taube et al., 2008a), multiple supraspinal structures such as the basal ganglia, the cerebellum, and the brainstem have important functions in the organization of posture (Lalonde & Strazielle, 2007; for review Visser & Bloem 2005). However, due to the better accessibility of cortical compared to subcortical structures, adaptations following balance training are best investigated for the motor cortex. In this respect, motor cortical activity was shown to decrease after four weeks of balance training (Beck et al., 2007; Schubert et al., 2008; Taube et al., 2007a). Most interestingly, subjects who showed the greatest adaptation of their motor cortical activity displayed also the most profound improvements in postural control (Figure 13.2). As there was no such correlation between spinal adaptations and changes

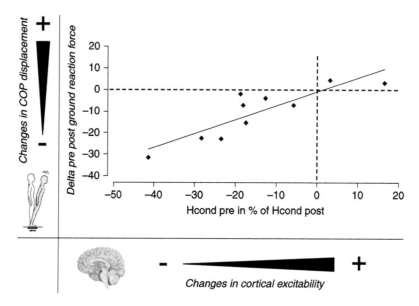

Figure 13.2   Interrelation of cortical plasticity and stance stability. Balance training related differences of pre and post measurement of the cortical excitability (expressed as H_{cond} (conditioned H-reflex) pre as % of H_{cond} post) are illustrated on the abscissa while changes in the vertical peak ground reaction force (center of pressure displacement) are displayed on the ordinate. The regression line demonstrates that subjects who have greater reductions of their cortical excitability after balance training demonstrate also greater reductions in postural sway (reduced ground reaction forces). Modified from Taube et al. (2007a).

in stance stability, the improvement in balance performance was argued to rely mostly on supraspinal plasticity (Taube et al., 2007a).

The reduction in cortical activity after several weeks of balance training displays some parallels with skill acquisition studies. These studies have shown (for example by means of fMRI) that motor cortical activity was large during the initial training phase (i.e. during skill acquisition) but decreased with progressive training (i.e. automatization). Conversely, activity in subcortical regions like the basal ganglia and the cerebellum increased with increasing task automatization (e.g. Puttemans et al., 2005). This scenario may also take place with balance training (Figure 13.3). Supporting evidence comes from a study of Nashner (1976), who compared short-term learning of balance strategies in cerebellar patients with the acquisition of healthy control subjects. Subjects were first tested while applying posterior directed translational platform displacements. After some trials, the perturbation was changed and the translation was replaced by a backward rotation of the foot. Thus, in both conditions, the triceps surae was stretched in a very similar way. However, during translational displacement, a compensatory contraction of the calf muscle (in this case a LLR) is appropriate to regain balance whereas during rotational displacement it is not. Thus, subjects had to adapt to the new postural perturbation by reducing the activity of the triceps surae. Healthy subjects adapted their LLR within three to five trials and drastically reduced the postural sway whereas the cerebellar patients were either significantly slower or unable to do so. Consequently, it can be argued that the cerebellum plays an important role in modulating the muscular activity task specifically. Furthermore, the importance of cerebellar structures for the acquisition of balance skills makes it reasonable to assume that balance training induces a similar 'shift in

(a)

(b)

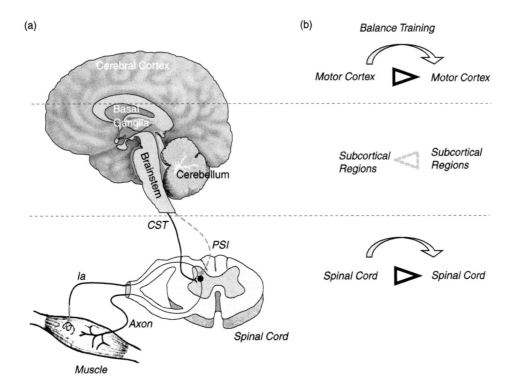

Figure 13.3 Balance training induced adaptations assessed during postural tasks. (a) Structures of the nervous system, which are considered to play an important role in the maintenance and recovery of balance. Sensory information from visual, vestibular, cutaneous and proprioceptive sources is integrated into postural control. With respect to the somatosensory system, changes in muscle length after perturbation are signalled by Ia and II (not illustrated) afferents originating from the muscle spindle. This information is transmitted to the spinal cord and to supraspinal centres (the latter connection is not displayed). The early part of the compensatory response (the short and the medium-latency response) is processed at the spinal level. After approximately 90–100 ms there is sufficient time for supraspinal sources to control muscular output (the LLR), e.g. via the corticospinal tract (CST) emanating from the motor cortex. (b) Balance training is thought to reduce spinal reflex excitability by increasing the supraspinal induced presynaptic inhibition (PSI). This reduction (▷) is schematically illustrated in the bottom section. Another well documented adaptation after balance training is the reduction in cortical involvement (first section). It is therefore assumed that training of balance skills and improved postural control after balance training strongly relies on subcortical structures (indicated as a grey dotted triangle in the middle line). Modified from Taube et al. (2008a).

movement control' from cortical to more subcortical and cerebellar structures (Figure 13.3) as described in the skill acquisition studies mentioned above (e.g. Puttemans et al., 2005).

It is reasonable to assume that balance training accomplished over a long period of time (several years) leads to fundamental structural plasticity within the CNS. In a recent study it was highlighted that the hippocampus formation of professional dancers and slackliners is differently structured compared to that of recreational sportspeople (Hüfner et al., 2011). However, it must be mentioned that these data were obtained in a cross-sectional study design so that it cannot be entirely ensured that the differences are really linked to the training or whether they are based on some genetic predispositions of the professional athletes or based on coincidental interindividual differences. However, the relatively great number of participants (21 athletes versus 20 controls) makes drawbacks rather unlikely.

Adaptation of proprioception   As highlighted earlier in this chapter, postural control seems to be closely related to the ability to gain and process proprioceptive information. A question which cannot easily be answered, is whether and – if yes – how proprioceptive abilities can be trained in order to improve balance. There is limited evidence that intrafusal muscle fibres (muscle spindle fibres) show metabolic and structural changes in response to exercise (Hutton & Atwater, 1992). Further, it is well known that the representation of sensory areas within the cortex increases following training of somatosensory demanding tasks like Braille reading (Pascual-Leone & Torres, 1993).

Concerning balance training, there are several studies proposing improved proprioception after balance exercises in young (Lee & Lin, 2008; Song et al., 2011) and elderly people (Waddington & Adams, 2004; Westlake et al., 2007). Song and co-workers (2011) demonstrated that balance training improved not only postural skills but also the repositioning accuracy of the trunk. Lee & Lin (2008) assessed active and passive repositioning sense of the ankle joint before and after training in patients with unilateral functional ankle instability. After postural training, active as well as passive repositioning values improved – however, only in the leg with functional ankle instability. In elderly subjects, ankle movement discrimination was shown to improve after five weeks of wobble board training (Waddington & Adams 2004). In a more recent study, Westlake and co-workers (2007) assessed changes in three measures of proprioceptive function (threshold to perception of passive movement, passive joint position sense, and velocity discrimination) after eight weeks of balance training. Significant improvements could only be found for the velocity discrimination test.

These observations make it reasonable to assume that proprioceptive function can be improved by balance training. However, further research is necessary to unambiguously relate sensorimotor improvements (for instance active repositioning accuracy) to adaptations of proprioceptive structures and/or changes in processing of proprioceptive information.

How can balance training enhance the rate of force development?

Schubert et al. (2008) assessed alterations in the excitability of direct corticospinal pathways following balance training. After training, cortical excitability was reduced during the execution of a postural task. However, when balance trained subjects were measured in a voluntary strength task where the same muscles were activated, enhanced cortical excitability was evident (Figure 13.4). The improved explosive force production may therefore be caused by an enhanced cortical drive to the agonistic muscles. It can be speculated that (synaptic) efficiency of direct corticospinal projections to the muscles encompassing the ankle joint had been increased by balance training and could be utilized in the voluntary contraction. A stronger drive to the ankle muscles after balance training could be quantified by means of surface EMG (Gruber et al., 2007a). The authors found significant increases in the frequency pattern of extensor muscle activation, especially in the early phase of the voluntary activation. They assumed that the extensive afferent input during balance training may play an important role in the strength adaptations because Grande & Cafarelli (2003) reported that recruitment thresholds and firing rates of MUs depend on the amount of afferent input.

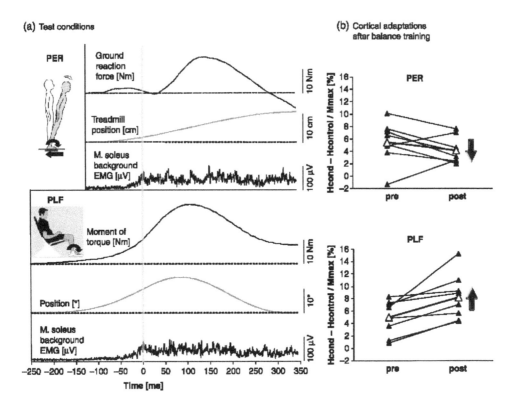

Figure 13.4 Motor tasks in which subjects were measured before and after balance training are displayed on the left hand side (a). Subjects were tested in a postural perturbation task and in a plantarflexion task in an ankle ergometer, where movements of a foot pedal had to be counteracted. Thus, in both cases (perturbation and ankle ergometer task) subjects activated their plantarflexors in a comparable way. During task execution, corticospinal adaptations were assessed by means of conditioning the H-reflex with transcranial magnetic stimulation (b). It is noteworthy that the responses to magnetic stimulation headed in opposing directions, i.e. when subjects were measured in the postural task, cortical excitability was reduced after balance training whereas measurements in the ankle ergometer task produced an augmentation of the cortical responses (modified from Schubert et al., 2008).

How can balance training prevent injuries?

There is an extensive body of evidence that balance training prevents the occurrence of injuries (some articles are mentioned in Hübscher et al., 2010). However, the mechanisms responsible for the reduction of knee and ankle injuries in particular are not known. Some authors assumed that balance training optimizes spinal reflex mechanisms (Gruber et al., 2006). However, the latency of spinal reflexes is too long to prevent fast traumata (Thacker et al., 1999). This observation also questions the assumption of other authors who propose that an improved sensory function may explain the preventive character (Panics et al., 2008). Thus, the question remains how balance training can be so effective. Probably, several factors contribute to the prophylactic character of balance training. However, the mechanism, which is additionally outlined in this chapter, is prevention by 'feedforward mechanisms'. It is speculated that balance trained people optimize the coordination of their movements in general so that they may be able to avoid critical joint positions. It is obvious that in order to improve the anticipatory reactions, improved sensory integration would be beneficial.

Can the efficiency of balance training be enhanced in the future?

Improvements in postural control are best evaluated in response to classical balance training with therapeutic devices (Taube et al., 2008a). However, there is more and more evidence that postural control can also be improved by other physical activities such as slacklining (Keller et al., 2012), skiing (Lauber et al., 2011) or inline skating (Taube et al., 2010). In another line of research, several studies examined the influence of augmented feedback on stance stability. In one of these studies, which was already mentioned in the preceding text, subjects were provided with additional feedback about their body sway by means of a hand held laser pointer (Taube et al., 2008b). It was demonstrated that subjects could significantly reduce their body sway when provided with this kind of additional feedback. However, so far it is not known whether the feedback from a laser pointer may also facilitate the adaptations taking place during balance training.

Similarly, it is not known how other forms of augmented feedback influence the efficiency of balance training. Recently, for instance, it was shown that the application of strips of adhesive bandage to the skin over and around the neck suppressed deteriorating effects of fatigue on postural sway (Pinsault & Vuillerme, 2010). The authors assumed that the adhesive tape may have served to reweigh the sensory information of neck muscles so that the somato-sensory information of the neck region could be better utilized.

In another experiment, Priplata and colleagues (2002) investigated the influence of subthreshold tactile stimuli, which were applied to the feet of the participants. Subthreshold means that the subjects could not consciously detect the stimuli. However, the stimuli nevertheless influenced motor behavior. In the first experiment, several hundred small nylon indentors were used that passed through the supporting ground and touched the sole of each foot (Priplata et al., 2002). It was demonstrated that postural sway was significantly reduced in elderly people and to a minor extent in young subjects, too. In the second experiment, stimuli to the feet of elderly people were applied by means of three vibrating elements in each insole (Priplata et al., 2003) and a reduction of postural sway could be seen in this experiment, too. Similarly, subthreshold stimuli to the soles of the feet proved efficient to reduce gait variability in elderly fallers (Galica et al., 2009). The information of the sole pressure distribution can also be decoded and used when the information is artificially transferred to other structures, which are highly sensitive to sensory input. In this way, Vuillerme and colleagues (2007) found improvements in sway parameters when feedback about the sole pressure distribution was provided to the tongue via electro-tactile stimulation.

All these observations point out that additional feedback and/or additional stimulation help to stabilize body posture. Longitudinal studies have to clarify, now, whether these applications may also facilitate motor learning during long-term balance training.

Conclusions

This chapter illustrates the complex sensori-motor interaction which is necessary to accomplish postural control. Not only has the CNS to integrate and weight multiple sources of afferent input, but it also has to use this information to adequately program anticipatory muscular activation (when the environment is predictable) or reactive muscle activity when unforeseen perturbations have to be counteracted. Thus, postural skills rely on the perception of sensory information, its integration and processing and the subsequent programming and execution of appropriate motor commands. If any of these steps is disturbed postural control is impaired. To counteract the risk of falling due to impaired postural skills, balance training proved efficient. In this context it was shown that balance training does not only improve balance per se but increases the rate of force development, the jumping abilities while at the same time reducing the incidence of lower limb injuries as well as the perception of neck pain. The neural adaptations underlying these behavioral changes are thought to involve both the sensory and motor side and are best understood with respect to spinal reflexes and activity of the primary motor cortex.

References

Allen, N.E., Sherrington, C., Paul, S.S., and Canning, C.G. (2011), 'Balance and falls in Parkinson's disease: a meta-analysis of the effect of exercise and motor training', *Mov. Disord.*, 26: 1605–1615.

Allum, J.H., Bloem, B.R., Carpenter, M.G., Hulliger, M., and Hadders-Algra, M. (1998), 'Proprioceptive control of posture: a review of new concepts', *Gait Posture*, 8: 214–242.

Allum, J.H. and Pfaltz, C.R. (1985), 'Visual and vestibular contributions to pitch sway stabilization in the ankle muscles of normals and patients with bilateral peripheral vestibular deficits', *Exp. Brain Res.*, 58: 82–94.

Asai, H., Fujiwara, K., Toyama, H., Yamashina, T., Tachino, K., & Nara, I. 1992, 'The influence of foot soles cooling on standing postural control analyzed by tracking the center of foot pressure.,' in *Posture and gait: control mechanisms II*, M. Woollacott & F. Horak, eds., University of Oregon Books, Eugene, pp. 151–154.

Bahr, R. and Bahr, I.A. (1997), 'Incidence of acute volleyball injuries: a prospective cohort study of injury mechanisms and risk factors', *Scand. J. Med. Sci. Sports*, 7: 166–171.

Balaban, C.D. (2002), 'Neural substrates linking balance control and anxiety', *Physiol. Behav.*, 77: 469–475.

Balaban, C.D., Jacob, R.G., and Furman, J.M. (2011), 'Neurologic bases for comorbidity of balance disorders, anxiety disorders and migraine: neurotherapeutic implications', *Expert Rev. Neurother.*, 11: 379–394.

Bastian, A.J. (2006), 'Learning to predict the future: the cerebellum adapts feedforward movement control', *Curr. Opin. Neurobiol.*, 16: 645–649.

Beck, S., Taube, W., Gruber, M., Amtage, F., Gollhofer, A., and Schubert, M. (2007), 'Task-specific changes in motor evoked potentials of lower limb muscles after different training interventions', *Brain Res.*, 1179: 51–60.

Beinert, K. & Taube, W. The effect of balance training on cervical sensorimotor function and neck pain. 2012. (Unpublished work.)

Beloozerova, I.N., Sirota, M.G., Orlovsky, G.N., and Deliagina, T.G. (2005), 'Activity of pyramidal tract neurons in the cat during postural corrections', *J. Neurophysiol.*, 93: 1831–1844.

Beloozerova, I.N., Sirota, M.G., Swadlow, H.A., Orlovsky, G.N., Popova, L.B., and Deliagina, T.G. (2003), 'Activity of different classes of neurons of the motor cortex during postural corrections', *J. Neurosci.*, 23: 7844–7853.

Bergin, P.S., Bronstein, A.M., Murray, N.M., Sancovic, S., and Zeppenfeld, D.K. (1995), 'Body sway and vibration perception thresholds in normal aging and in patients with polyneuropathy', *J. Neurol. Neurosurg. Psychiatry*, 58: 335–340.

Bloem, B.R. & Bhatia, K.P. (2004). 'Gait and balance in basal ganglia disorders.' *In Clinical disorders of balance, Posture and Gait*, eds. Bronstein, A., Brandt, T., Woollacott M.H. & Nutt, J.G., pp.173–206. Arnold: London.

Bloem, B.R., Hausdorff, J.M., Visser, J.E., and Giladi, N. (2004), 'Falls and freezing of gait in Parkinson's disease: a review of two interconnected, episodic phenomena', *Mov. Disord.*, 19: 871–884.

Bloem, B.R. & Roos, R.A. 1995, 'Neurotoxicity of designer drugs and related compounds,' in *Handbook of Clinical Neurology, Vol. 21: Intoxications of the nervous system, Part II*, F.A. De Wolff, ed., Elsevier, Amsterdam, Netherlands, pp. 363–414.

Bloem, B.R., Valkenburg, V.V., Slabbekoorn, M., and van Dijk, J.G. (2001), 'The multiple tasks test. Strategies in Parkinson's disease', *Exp. Brain Res.*, 137: 478–486.

Boyle, R., Goldberg, J.M., and Highstein, S.M. (1992), 'Inputs from regularly and irregularly discharging vestibular nerve afferents to secondary neurons in squirrel monkey vestibular nuclei. III. Correlation with vestibulospinal and vestibuloocular output pathways', *J. Neurophysiol.*, 68: 471–484.

Brandt, T. and Dieterich, M. (1999), 'The vestibular cortex. Its locations, functions, and disorders', *Ann. N.Y. Acad. Sci.*, 871: 293–312.

Brandt, T., Krafczyk, S., and Malsbenden, I. (1981), 'Postural imbalance with head extension: improvement by training as a model for ataxia therapy', *Ann. N. Y. Acad. Sci*, 374: 636–649.

Bruhn, S., Kullmann, N., and Gollhofer, A. (2004), 'The effects of a sensorimotor training and a strength training on postural stabilisation, maximum isometric contraction and jump performance', *Int. J. Sports Med.*, 25: 56–60.

Bruhn, S., Kullmann, N., and Gollhofer, A. (2006), 'Combinatory effects of high-intensity-strength training and sensorimotor training on muscle strength', *Int. J. Sports Med.*, 27: 401–406.

Buchanan, J.J. and Horak, F.B. (1999), 'Emergence of postural patterns as a function of vision and translation frequency', *J. Neurophysiol.*, 81: 2325–2339.

Caraffa, A., Cerulli, G., Projetti, M., Aisa, G., and Rizzo, A. (1996), 'Prevention of anterior cruciate ligament injuries in soccer. A prospective controlled study of proprioceptive training', *Knee Surg. Sports Traumatol. Arthrosc.*, 4: 19–21.

Carpenter, M.G., Murnaghan, C.D., and Inglis, J.T. (2010), 'Shifting the balance: evidence of an exploratory role for postural sway', *Neuroscience*, 171: 196–204.

Casadio, M., Morasso, P.G., and Sanguineti, V. (2005), 'Direct measurement of ankle stiffness during quiet standing: implications for control modelling and clinical application', *Gait Posture*, 21: 410–424.

Childs, J.D., Cleland, J.A., Elliott, J.M., Teyhen, D.S., Wainner, R.S., Whitman, J.M., Sopky, B.J., Godges, J.J., and Flynn, T.W. (2008), 'Neck pain: Clinical practice guidelines linked to the International Classification of Functioning, Disability, and Health from the Orthopedic Section of the American Physical Therapy Association', *J. Orthop. Sports Phys. Ther.*, 38: A1–A34.

Conway, B.A., Hultborn, H., and Kiehn, O. (1987), 'Proprioceptive input resets central locomotor rhythm in the spinal cat', *Exp. Brain Res.*, 68: 643–656.

Diener, H.C. and Dichgans, J. (1992), 'Pathophysiology of cerebellar ataxia', *Mov. Disord.*, 7: 95–109.

Dietrichs, E. (2008), 'Clinical manifestation of focal cerebellar disease as related to the organization of neural pathways', *Acta Neurol Scand*, 117: 6–11.

Dietz, V., Gollhofer, A., Kleiber, M., and Trippel, M. (1992), 'Regulation of bipedal stance: dependency on "load" receptors', *Exp. Brain Res.*, 89: 229–231.

Dietz, V., Horstmann, G., and Berger, W. (1988), 'Involvement of different receptors in the regulation of human posture', *Neurosci. Lett.*, 94: 82–87.

Dimitriou, M. and Edin, B.B. (2010), 'Human muscle spindles act as forward sensory models', *Current Biology*, 20: 1763–1767.

Dornan, J., Fernie, G.R., and Holliday, P.J. (1978), 'Visual input: its importance in the control of postural sway', *Arch. Phys. Med. Rehabil.*, 59: 586–591.

Duarte, M. and Zatsiorsky, V.M. (2002), 'Effects of body lean and visual information on the equilibrium maintenance during stance', *Exp. Brain Res.*, 146: 60–69.

Duysens, J. and Pearson, K.G. (1980), 'Inhibition of flexor burst generation by loading ankle extensor muscles in walking cats', *Brain Res.*, 187: 321–332.

Earles, D.R., Koceja, D.M., and Shively, C.W. (2000), 'Environmental changes in soleus H-reflex excitability in young and elderly subjects', *Int. J. Neurosci.*, 105: 1–13.

Eils, E. and Rosenbaum, D. (2001), 'A multi-station proprioceptive exercise program in patients with ankle instability', *Med. Sci. Sports Exerc.*, 33: 1991–1998.

Emery, C.A., Rose, M.S., McAllister, J.R., and Meeuwisse, W.H. (2007), 'A prevention strategy to reduce the incidence of injury in high school basketball: a cluster randomized controlled trial', *Clin. J. Sport Med.*, 17: 17–24.

Field, S., Treleaven, J., and Jull, G. (2008), 'Standing balance: a comparison between idiopathic and whiplash-induced neck pain', *Man. Ther.*, 13: 183–191.

Fitzpatrick, R. and McCloskey, D.I. (1994), 'Proprioceptive, visual and vestibular thresholds for the perception of sway during standing in humans', *J. Physiol.*, 478 (1): 173–186.

Fitzpatrick, R., Rogers, D.K., and McCloskey, D.I. (1994), 'Stable human standing with lower-limb muscle afferents providing the only sensory input', *J. Physiol.*, 480 (2): 395–403.

Fitzpatrick, R.C. and Gandevia, S.C. (2005), 'Paradoxical muscle contractions and the neural control of movement and balance', *J. Physiol.*, 564: 2.

Freeman, M.A., Dean, M.R., and Hanham, I.W. (1965), 'The etiology and prevention of functional instability of the foot', *J. Bone Joint Surg. Br.*, 47: 678–685.

Galica, A.M., Kang, H.G., Priplata, A.A., D'Andrea, S.E., Starobinets, O.V., Sorond, F.A., Cupples, L.A., and Lipsitz, L.A. (2009), 'Subsensory vibrations to the feet reduce gait variability in elderly fallers', *Gait Posture*, 30: 383–387.

Gauffin, H., Tropp, H., and Odenrick, P. (1988), 'Effect of ankle disk training on postural control in patients with functional instability of the ankle joint', *Int. J Sports Med.*, 9: 141–144.

Goble, D.J., Coxon, J.P., Van, I.A., Geurts, M., Doumas, M., Wenderoth, N., and Swinnen, S.P. (2011), 'Brain activity during ankle proprioceptive stimulation predicts balance performance in young and older adults', *J. Neurosci.*, 31: 16344–16352.

Goble, D.J., Coxon, J.P., Van, I.A., Geurts, M., Van, H.W., Sunaert, S., Wenderoth, N., and Swinnen, S.P. (2012), 'The neural basis of central proprioceptive processing in older versus younger adults: An important sensory role for right putamen', *Hum. Brain Mapp* Apr; 33(4): 895–908.

Goble, D.J., Coxon, J.P., Wenderoth, N., Van, I.A., and Swinnen, S.P. (2009), 'Proprioceptive sensibility in the elderly: degeneration, functional consequences and plastic-adaptive processes', *Neurosci. Biobehav. Rev.*, 33: 271–278.

Gollhofer, A. and Rapp, W. (1993), 'Recovery of stretch reflex responses following mechanical stimulation', *Eur. J. Appl. Physiol. Occup. Physiol.*, 66: 415–420.

Granacher, U., Gollhofer, A., and Strass, D. (2006), 'Training induced adaptations in characteristics of postural reflexes in elderly men', *Gait Posture*, 24: 459–466.

Grande, G. and Cafarelli, E. (2003), 'Ia afferent input alters the recruitment thresholds and firing rates of single human motor units', *Exp. Brain Res.*, 150: 449–457.

Gruber, M., Bruhn, S., and Gollhofer, A. (2006), 'Specific adaptations of neuromuscular control and knee joint stiffness following sensorimotor training', *Int. J. Sports Med.*, 27: 636–641.

Gruber, M. and Gollhofer, A. (2004), 'Impact of sensorimotor training on the rate of force development and neural activation', *Eur. J. Appl. Physiol.*, 92: 98–105.

Gruber, M., Gruber, S.B., Taube, W., Schubert, M., Beck, S.C., and Gollhofer, A. (2007a), 'Differential effects of ballistic versus sensorimotor training on rate of force development and neural activation in humans', *J. Strength Cond. Res.*, 21: 274–282.

Gruber, M., Taube, W., Gollhofer, A., Beck, S., Amtage, F., and Schubert, M. (2007b), 'Training-specific adaptations of H- and stretch reflexes in human soleus muscle', *J. Mot. Behav.*, 39: 68–78.

Guerraz, M., Thilo, K.V., Bronstein, A.M., and Gresty, M.A. (2001), 'Influence of action and expectation on visual control of posture', *Brain Res. Cogn. Brain Res.*, 11: 259–266.

Hadders-Algra, M., Brogren, E., and Forssberg, H. (1996), 'Training affects the development of postural adjustments in sitting infants', *J. Physiol.*, 493 (Pt 1): 289–298.

Hafstrom, A., Fransson, P.A., Karlberg, M., Ledin, T., and Magnusson, M. (2002), 'Visual influence on postural control, with and without visual motion feedback', *Acta Otolaryngol.*, 122: 392–397.

Heitkamp, H.C., Horstmann, T., Mayer, F., Weller, J., and Dickhuth, H.H. (2001), 'Gain in strength and muscular balance after balance training', *Int. J. Sports Med.*, 22: 285–290.

Henriksson, M., Ledin, T., and Good, L. (2001), 'Postural control after anterior cruciate ligament reconstruction and functional rehabilitation', *Am. J. Sports Med.*, 29: 359–366.

Hirsch, M.A., Toole, T., Maitland, C.G., and Rider, R.A. (2003), 'The effects of balance training and high-intensity resistance training on persons with idiopathic Parkinson's disease', *Arch. Phys. Med. Rehabil.*, 84: 1109–1117.

Hoffman, M.A. and Koceja, D.M. (1995), 'The effects of vision and task complexity on Hoffmann reflex gain', *Brain Res.*, 700: 303–307.

Holden, M., Ventura, J., and Lackner, J.R. (1994), 'Stabilization of posture by precision contact of the index finger', *J. Vestib. Res.*, 4: 285–301.

Holme, E., Magnusson, S.P., Becher, K., Bieler, T., Aagaard, P., and Kjaer, M. (1999), 'The effect of supervised rehabilitation on strength, postural sway, position sense and re-injury risk after acute ankle ligament sprain', *Scand. J. Med. Sci. Sports*, 9: 104–109.

Horak, F.B., Diener, H.C., and Nashner, L.M. (1989), 'Influence of central set on human postural responses', *J. Neurophysiol.*, 62: 841–853.

Horak, F.B., Nashner, L.M., and Diener, H.C. (1990), 'Postural strategies associated with somatosensory and vestibular loss', *Exp. Brain Res.*, 82: 167–177.

Hübscher, M., Zech, A., Pfeifer, K., Hansel, F., Vogt, L., and Banzer, W. (2010), 'Neuromuscular training for sports injury prevention: a systematic review', *Med. Sci. Sports Exerc.*, 42: 413–421.

Hüfner, K., Binetti, C., Hamilton, D.A., Stephan, T., Flanagin, V.L., Linn, J., Labudda, K., Markowitsch, H., Glasauer, S., Jahn, K., Strupp, M., and Brandt, T. (2011), 'Structural and functional plasticity of the hippocampal formation in professional dancers and slackliners', *Hippocampus*, 21: 855–865.

Hutton, R.S. and Atwater, S.W. (1992), 'Acute and chronic adaptations of muscle proprioceptors in response to increased use', *Sports Med.*, 14: 406–421.

Ilg, W., Synofzik, M., Brotz, D., Burkard, S., Giese, M.A., and Schols, L. (2009), 'Intensive coordinative training improves motor performance in degenerative cerebellar disease', *Neurology*, 73: 1823–1830.

Inglis, J.T., Horak, F.B., Shupert, C.L., and Jones-Rycewicz, C. (1994), 'The importance of somatosensory information in triggering and scaling automatic postural responses in humans', *Exp. Brain Res.*, 101: 159–164.

Jacobs, J.V., Fujiwara, K., Tomita, H., Furune, N., Kunita, K., and Horak, F.B. (2008), 'Changes in the activity of the cerebral cortex relate to postural response modification when warned of a perturbation', *Clin. Neurophysiol.*, 119: 1431–1442.

Jacobs, J.V. and Horak, F.B. (2007b), 'Cortical control of postural responses', *J. Neural Transm.*, 114: 1339–1348.

Jull, G., Falla, D., Treleaven, J., Hodges, P., and Vicenzino, B. (2007), 'Retraining cervical joint position sense: the effect of two exercise regimes', *J. Orthop. Res.*, 25: 404–412.

Katz, R., Meunier, S., and Pierrot-Deseilligny, E. (1988), 'Changes in presynaptic inhibition of Ia fibres in man while standing', *Brain*, 111: 417–437.

Kavounoudias, A., Roll, R., and Roll, J.P. (2001), 'Foot sole and ankle muscle inputs contribute jointly to human erect posture regulation', *J. Physiol.*, 532: 869–878.

Kean, C.O., Behm, D.G., and Young, W.B. (2006), 'Fixed foot balance training increases rectus femoris activation during landing and jump height in recreationally active women', *J. Sports Sci. Medic.*, 5: 138–148.

Keller, M., Pfusterschmied, J., Buchecker, M., Muller, E., and Taube, W. (2012), 'Improved postural control after slackline training is accompanied by reduced H-reflexes', *Scand. J. Med. Sci. Sports*, 22: 471–477.

Klockgether, T., Borutta, M., Rapp, H., Spieker, S., and Dichgans, J. (1995), 'A defect of kinesthesia in Parkinson's disease', *Mov. Disord.*, 10: 460–465.

Koceja, D.M. and Mynark, R.G. (2000), 'Comparison of heteronymous monosynaptic Ia facilitation in young and elderly subjects in supine and standing positions', *Int. J. Neurosci.*, 103: 1–17.

Lackner, J.R. and DiZio, P. (2005), 'Vestibular, proprioceptive, and haptic contributions to spatial orientation', *Ann. Rev. Psychol.*, 56: 115–147.

Lackner, J.R., DiZio, P., Jeka, J., Horak, F., Krebs, D., and Rabin, E. (1999), 'Precision contact of the fingertip reduces postural sway of individuals with bilateral vestibular loss', *Exp. Brain Res.*, 126: 459–466.

Lalonde, R. and Strazielle, C. (2007), 'Brain regions and genes affecting postural control', *Prog. Neurobiol.*, 81: 45–60.

Lauber, B., Keller, M., Gollhofer, A., Muller, E., and Taube, W. (2011), 'Spinal reflex plasticity in response to alpine skiing in the elderly', *Scand. J. Med. Sci. Sports*, 21 Suppl 1: 62–68.

Ledebt, A., Becher, J., Kapper, J., Rozendaalr, R.M., Bakker, R., Leenders, I.C., and Savelsbergh, G.J. (2005), 'Balance training with visual feedback in children with hemiplegic cerebral palsy: effect on stance and gait', *Motor Contr.*, 9: 459–468.

Lee, A.J. and Lin, W.H. (2008), 'Twelve-week biomechanical ankle platform system training on postural stability and ankle proprioception in subjects with unilateral functional ankle instability', *Clin. Biomech. (Bristol, UK)*, 23: 1065–1072.

Lee, D.N. and Aronson, E. (1974), 'Visual proprioceptive control of standing in human infants', *Percept. Psychophys.* 15: 529–532.

Llewellyn, M., Yang, J.F., and Prochazka, A. (1990), 'Human H-reflexes are smaller in difficult beam walking than in normal treadmill walking', *Exp. Brain Res.*, 83: 22–28.

Lopez, C. and Blanke, O. (2011), 'The thalamocortical vestibular system in animals and humans', *Brain Res. Rev.*, 67: 119–146.

Loram, I.D. and Lakie, M. (2002), 'Direct measurement of human ankle stiffness during quiet standing: the intrinsic mechanical stiffness is insufficient for stability', *J. Physiol*, 545: 1041–1053.

Loram, I.D., Maganaris, C.N., and Lakie, M. (2005a), 'Active, non-spring-like muscle movements in human postural sway: how might paradoxical changes in muscle length be produced?', *J. Physiol.*, 564: 281–293.

Loram, I.D., Maganaris, C.N., and Lakie, M. (2005b), 'Human postural sway results from frequent, ballistic bias impulses by soleus and gastrocnemius', *J. Physiol.*, 564: 295–311.

Lord, S.R., Clark, R.D., and Webster, I.W. (1991), 'Physiological factors associated with falls in an elderly population', *J. Am. Geriatr. Soc.*, 39: 1194–1200.

Lubetzky-Vilnai, A. and Kartin, D. (2010), 'The effect of balance training on balance performance in individuals poststroke: a systematic review', *J. Neurol. Phys. Ther.*, 34: 127–137.

Luccarini, P., Gahery, Y., and Pompeiano, O. (1990), 'Injection of a cholinergic agonist in the dorso-lateral pontine tegmentum of cats affects the posturokinetic responses to cortical stimulation', *Neurosci. Lett.*, 114: 75–81.

Luciani, L. 1891, *Il cervelletto. Nuovi studi di fisiologia normale e patologica. (Das Kleinhirn. Neue Studien zur normalen und pathologischen Physiologie)* Florence: Le Monnier.

Macaluso, E. and Maravita, A. (2010), 'The representation of space near the body through touch and vision', *Neuropsychologia*, 48: 782–795.

Magnus, R. 1924, *Körperstellung*, Springer Verlag, Berlin, Germany.

Mauritz, K.H. and Dietz, V. (1980), 'Characteristics of postural instability induced by ischemic blocking of leg afferents', *Exp. Brain Res.*, 38: 117–119.

McGuine, T.A. and Keene, J.S. (2006), 'The effect of a balance training program on the risk of ankle sprains in high school athletes', *Am. J. Sports Med.*, 34: 1103–1111.

Mergner, T., Schweigart, G., Maurer, C., and Blumle, A. (2005), 'Human postural responses to motion of real and virtual visual environments under different support base conditions', *Exp. Brain Res.*, 167: 535–556.

Meyer, P.F., Oddsson, L.I., and De Luca, C.J. (2004a), 'Reduced plantar sensitivity alters postural responses to lateral perturbations of balance', *Exp. Brain Res.*, 157: 526–536.

Meyer, P.F., Oddsson, L.I., and De Luca, C.J. (2004b), 'The role of plantar cutaneous sensation in unperturbed stance', *Exp. Brain Res.*, 156: 505–512.

Morasso, P.G., Baratto, L., Capra, R., and Spada, G. (1999), 'Internal models in the control of posture', *Neural Netw.*, 12: 1173–1180.

Morasso, P.G. and Schieppati, M. (1999), 'Can muscle stiffness alone stabilize upright standing?', *J. Neurophysiol.*, 82: 1622–1626.

Morton, S.M. and Bastian, A.J. (2007), 'Mechanisms of cerebellar gait ataxia', *Cerebellum.*, 6: 79–86.

Murnaghan, C.D., Horslen, B.C., Inglis, J.T., and Carpenter, M.G. (2011), 'Exploratory behavior during stance persists with visual feedback', *Neuroscience*, 195: 54–59.

Myer, G.D., Ford, K.R., Brent, J.L., and Hewett, T.E. (2006), 'The effects of plyometric vs. dynamic stabilization and balance training on power, balance, and landing force in female athletes', *J. Strength Cond. Res*, 20: 345–353.

Myklebust, G., Engebretsen, L., Braekken, I.H., Skjolberg, A., Olsen, O.E., and Bahr, R. (2003), 'Prevention of anterior cruciate ligament injuries in female team handball players: a prospective intervention study over three seasons', *Clin. J. Sport Med.*, 13: 71–78.

Myklebust, G., Engebretsen, L., Braekken, I.H., Skjolberg, A., Olsen, O.E., and Bahr, R. (2007), 'Prevention of noncontact anterior cruciate ligament injuries in elite and adolescent female team handball athletes', *Instr. Course Lect.*, 56: 407–418.

Nardone, A., Galante, M., Lucas, B., and Schieppati, M. (2001), 'Stance control is not affected by paresis and reflex hyperexcitability: the case of spastic patients', *J. Neurol. Neurosurg. Psychiatry*, 70: 635–643.

Nardone, A. and Schieppati, M. (2004), 'Group II spindle fibres and afferent control of stance. Clues from diabetic neuropathy', *Clin. Neurophysiol.*, 115: 779–789.

Nardone, A., Tarantola, J., Miscio, G., Pisano, F., Schenone, A., and Schieppati, M. (2000), 'Loss of large-diameter spindle afferent fibres is not detrimental to the control of body sway during upright stance: evidence from neuropathy', *Exp. Brain Res.*, 135: 155–162.

Nashner, L.M. (1976), 'Adapting reflexes controlling the human posture', *Exp. Brain Res.*, 26: 59–72.

Nashner, L.M., Shupert, C.L., Horak, F.B., and Black, F.O. (1989), 'Organization of posture controls: an analysis of sensory and mechanical constraints', *Prog. Brain Res.*, 80: 411–418.

Nielsen, J.B. (2003), 'How we walk: central control of muscle activity during human walking', *Neuroscientist.*, 9: 195–204.

Ouchi, Y., Okada, H., Yoshikawa, E., Nobezawa, S., and Futatsubashi, M. (1999), 'Brain activation during maintenance of standing postures in humans', *Brain*, 122: 329–338.

Panics, G., Tallay, A., Pavlik, A., and Berkes, I. (2008), 'Effect of proprioception training on knee joint position sense in female team handball players', *B. J. Sports Med.*, 42: 472–476.

Pascual-Leone, A. and Torres, F. (1993), 'Plasticity of the sensorimotor cortex representation of the reading finger in Braille readers', *Brain*, 116 (Pt 1): 39–52.

Peterka, R.J. and Benolken, M.S. (1995), 'Role of somatosensory and vestibular cues in attenuating visually induced human postural sway', *Exp. Brain Res.*, 105: 101–110.

Petersen, N., Christensen, L.O., Morita, H., Sinkjaer, T., and Nielsen, J. (1998), 'Evidence that a transcortical pathway contributes to stretch reflexes in the tibialis anterior muscle in man', *J. Physiol.*, 512: 267–276.

Petersen, W., Braun, C., Bock, W., Schmidt, K., Weimann, A., Drescher, W., Eiling, E., Stange, R., Fuchs, T., Hedderich, J., and Zantop, T. (2005), 'A controlled prospective case control study of a prevention training program in female team handball players: the German experience', *Arch. Orthop. Trauma Surg.*, 125: 614–621.

Pinsault, N. and Vuillerme, N. (2010), 'Vestibular and neck somatosensory weighting changes with trunk extensor muscle fatigue during quiet standing', *Exp. Brain Res.*, 202: 253–259.

Priplata, A., Niemi, J., Salen, M., Harry, J., Lipsitz, L.A., and Collins, J.J. (2002), 'Noise-enhanced human balance control', *Phys. Rev. Lett.*, Dec 2; 89(23): 238101.

Priplata, A.A., Niemi, J.B., Harry, J.D., Lipsitz, L.A., and Collins, J.J. (2003), 'Vibrating insoles and balance control in elderly people', *Lancet*, 362: 1123–1124.

Puttemans, V., Wenderoth, N., and Swinnen, S.P. (2005), 'Changes in brain activation during the acquisition of a multifrequency bimanual coordination task: from the cognitive stage to advanced levels of automaticity', *J. Neurosci.*, 25: 4270–4278.

Ramnani, N. (2006), 'The primate cortico-cerebellar system: anatomy and function', *Nature Reviews Neurosci.*, 7: 511–522.

Revel, M., Andre-Deshays, C., and Minguet, M. (1991), 'Cervicocephalic kinesthetic sensibility in patients with cervical pain', *Arch. Phys. Med. Rehabil.*, 72: 288–291.

Revel, M., Minguet, M., Gregoy, P., Vaillant, J., and Manuel, J.L. (1994), 'Changes in cervicocephalic kinesthesia after a proprioceptive rehabilitation program in patients with neck pain: a randomized controlled study', *Arch. Phys. Med. Rehabil.*, 75: 895–899.

Roijezon, U., Bjorklund, M., Bergenheim, M., and Djupsjobacka, M. (2008), 'A novel method for neck coordination exercise–a pilot study on persons with chronic non-specific neck pain', *J. Neuroeng. Rehabil.*, 5: 36.

Roitman, A.V., Pasalar, S., Johnson, M.T.V., and Ebner, T.J. (2005), 'Position, direction of movement, and speed tuning of cerebellar Purkinje cells during circular manual tracking in monkey', *J. Neurosci.*, 25: 9244–9257.

Rozzi, S.L., Lephart, S.M., Sterner, R., and Kuligowski, L. (1999), 'Balance training for persons with functionally unstable ankles', *J. Orthop. Sports Phys. Ther.*, 29: 478–486.

Runge, C.F., Shupert, C.L., Horak, F.B., and Zajac, F.E. (1999), 'Ankle and hip postural strategies defined by joint torques', *Gait Posture*, 10: 161–170.

Schubert, M., Beck, S., Taube, W., Amtage, F., Faist, M., and Gruber, M. (2008), 'Balance training and ballistic strength training are associated with task-specific corticospinal adaptations', *Eur. J. Neurosci.*, 27: 2007–2018.

Sherrington, C.S. (1910), 'Flexion-reflex of the limb, crossed extension-reflex, and reflex stepping and standing', *J. Physiol.*, 40: 28–121.

Sherrington, C. 1906, *Integrative action of the nervous system*, 1st edn, Constable, London.

Slobounov, S., Wu, T., Hallett, M., Shibasaki, H., Slobounov, E., and Newell, K. (2006), 'Neural underpinning of postural responses to visual field motion', *Biol. Psychol.*, 72: 188–197.

Solopova, I.A., Kazennikov, O.V., Deniskina, N.B., Levik, Y.S., and Ivanenko, Y.P. (2003), 'Postural instability enhances motor responses to transcranial magnetic stimulation in humans', *Neurosci. Lett.*, 337: 25–28.

Song, C.H., Petrofsky, J.S., Lee, S.W., Lee, K.J., and Yim, J.E. (2011), 'Effects of an exercise program on balance and trunk proprioception in older adults with diabetic neuropathies', *Diabetes Technol. Ther.*, 13: 803–811.

Soto, O., Valls-Sole, J., Shanahan, P., and Rothwell, J. (2006), 'Reduction of intracortical inhibition in soleus muscle during postural activity', *J. Neurophysiol.*, 96: 1711–1717.

Stapley, P.J., Beretta, M.V., Dalla, T.E., and Schieppati, M. (2006), 'Neck muscle fatigue and postural control in patients with whiplash injury', *Clin. Neurophysiol.*, 117: 610–622.

Straube, A. and Brandt, T. (1987), 'Importance of the visual and vestibular cortex for self-motion perception in man (circularvection)', *Hum. Neurobiol.*, 6: 211–218.

Taimela, S., Takala, E.P., Asklof, T., Seppala, K., and Parviainen, S. (2000), 'Active treatment of chronic neck pain: a prospective randomized intervention', *Spine* (Phila. Pa 1976), 25: 1021–1027.

Taube, W. (2012), 'Neurophysiological adaptations in response to balance training', *German J. Sports Med.*, 63(9): 163–167.

Taube, W., Bracht, D., Besemer, C., and Gollhofer, A. (2010), 'The effect of inline skating on postural control in elderly people', *Deutsche Zeitschrift für Sportmedizin*, 61: 45–51.

Taube, W., and Gollhofer, A. (2011), 'Control and Training of Postures.' In Komi, P. (Ed.), *Neuromuscular Aspects of Sport Performance*, (p.254–269). Wiley-Blackwell.

Taube, W., Gruber, M., Beck, S., Faist, M., Gollhofer, A., and Schubert, M. (2007a), 'Cortical and spinal adaptations induced by balance training: correlation between stance stability and cortico-spinal activation', *Acta Physiol. (Oxf.)*, 189: 347–358.

Taube, W., Gruber, M., and Gollhofer, A. (2008a), 'Spinal and supraspinal adaptations associated with balance training and their functional relevance', *Acta Physiol. (Oxf.)*, 193: 101–116.

Taube, W., Kullmann, N., Leukel, C., Kurz, O., Amtage, F., and Gollhofer, A. (2007b), 'Differential Reflex Adaptations Following Sensorimotor and Strength Training in Young Elite Athletes', *Int. J. Sports Med.*, 28: 999–1005.

Taube, W., Kullmann, N., Leukel, C., Kurz, O., Amtage, F., and Gollhofer, A. (2007c), 'Differential reflex adaptations following sensorimotor and strength training in young elite athletes', *Int. J. Sports Med.*, 28: 999–1005.

Taube, W., Leukel, C., and Gollhofer, A. (2008b), 'Influence of enhanced visual feedback on postural control and spinal reflex modulation during stance', *Exp. Brain Res.*, 188: 353–361.

Taube, W., Schubert, M., Gruber, M., Beck, S., Faist, M., and Gollhofer, A. (2006), 'Direct cortico-spinal pathways contribute to neuromuscular control of perturbed stance', *J. Appl. Physiol.*, 101: 420–429.

Thacker, S.B., Stroup, D.F., Branche, C.M., Gilchrist, J., Goodman, R.A., and Weitman, E.A. (1999), 'The prevention of ankle sprains in sports – a systematic review of the literature', *Am. J. Sports Med.*, 27: 753–760.

Tjell, C., and Rosenhall, U. (1998). 'Smooth pursuit neck torsion test: a specific test for cervical dizziness'. *Am. J. Otol.*, 19: 76–81.

Tokuno, C.D., Taube, W., and Cresswell, A.G. (2009), 'An enhanced level of motor cortical excitability during the control of human standing', *Acta Physiol. (Oxf.)*, 195: 385–395.

Treleaven, J., Jull, G., and Low Choy, N. (2005), 'Standing balance in persistent whiplash: a comparison between subjects with and without dizziness', *J. Rehabil. Med*, 37: 224–229.

Trimble, M.H. and Koceja, D.M. (1994), 'Modulation of the triceps surae H-reflex with training', *Int. J. Neurosci.*, 76: 293–303.

van Deursen, R.W. and Simoneau, G.G. (1999), 'Foot and ankle sensory neuropathy, proprioception, and postural stability', *J. Orthop. Sports Phys. Ther.*, 29: 718–726.

Van Emmerik, R.E. and Van Wegen, E.E. (2002), 'On the functional aspects of variability in postural control', *Exerc. Sport Sci. Rev.*, 30: 177–183.

Verhagen, E., van der Beek A., Twisk, J., Bouter, L., Bahr, R., and Van Mechelen, W. (2004), 'The effect of a proprioceptive balance board training program for the prevention of ankle sprains: a prospective controlled trial', *Am. J Sports Med.*, 32: 1385–1393.

Vinay, L., Ben Mabrouk, F., Brocard, F., Clarac, F., Jean-Xavier, C., Pearlstein, E., and Pflieger, J.F. (2005), 'Perinatal development of the motor systems involved in postural control', *Neural Plast.*, 12: 131–139.

Visser, J.E. and Bloem, B.R. (2005), 'Role of the basal ganglia in balance control', *Neural Plast.*, 12: 161–174.

Vuillerme, N., Chenu, O., Demongeot, J., and Payan, Y. (2007), 'Controlling posture using a plantar pressure-based, tongue-placed tactile biofeedback system', *Exp. Brain Res.*, 179: 409–414.

Waddington, G.S. and Adams, R.D. (2004), 'The effect of a 5-week wobble-board exercise intervention on ability to discriminate different degrees of ankle inversion, barefoot and wearing shoes: a study in healthy elderly', *J. Am. Geriatr. Soc.*, 52: 573–576.

Wedderkopp, N., Kaltoft, M., Holm, R., and Froberg, K. (2003), 'Comparison of two intervention programmes in young female players in European handball – with and without ankle disc', *Scand. J. Med. Sci. Sports*, 13: 371–375.

Wedderkopp, N., Kaltoft, M., Lundgaard, B., Rosendahl, M., and Froberg, K. (1999), 'Prevention of injuries in young female players in European team handball. A prospective intervention study', *Scand. J. Med. Sci. Sports*, 9: 41–47.

Westlake, K.P., Wu, Y., and Culham, E.G. (2007), 'Sensory-specific balance training in older adults: effect on position, movement, and velocity sense at the ankle', *Phys. Ther.*, 87: 560–568.

Winter, D.A., Patla, A.E., Prince, F., Ishac, M., and Gielo-Perczak, K. (1998), 'Stiffness control of balance in quiet standing', *J. Neurophysiol.*, 80: 1211–1221.

Zia, S., Cody, F., and O'Boyle, D. (2000), 'Joint position sense is impaired by Parkinson's disease', *Ann. Neurol.*, 47: 218–228.

PART IV

Motor control and learning in voluntary actions

14

BODY SCHEMA, ILLUSIONS OF MOVEMENT AND BODY PERCEPTION

Mark Schram Christensen

COPENHAGEN NEURAL CONTROL OF MOVEMENT, DEPARTMENT OF NUTRITION, EXERCISE AND SPORTS &
DEPARTMENT OF NEUROSCIENCE AND PHARMACOLOGY, UNIVERSITY OF COPENHAGEN, DENMARK AND
DANISH RESEARCH CENTRE FOR MAGNETIC RESONANCE, COPENHAGEN UNIVERSITY HOSPITAL,
HVIDOVRE, DENMARK

Introduction

When I grasp the cup in front of me, while I lean back in my chair and rest for a moment when writing these words, my brain is faced with an overwhelming task of integrating motor commands sent to the muscles with the (infinite) amount of sensory information that I receive through my sensory organs for me to fulfil the intention of having a short rest. Besides having to sort out relevant sensory information and keeping track of the ongoing muscle activity, in order to perform these tasks simultaneously, the brain also needs to know where the arm is positioned in space in relation to the keyboard and the cup I reach for, and where my upper body is located relative to the back rest of the chair. How does the brain keep track of this information in order to enable me to perform these multiple actions with ease and grace?

In the first part of the chapter I will try to give an idea of the mechanisms that are involved in keeping track of where the body is located in space, and how the limbs are positioned relative to each other. This, however, requires a through discussion on what type of knowledge is needed in order to give an account of where and how the limbs of the body are located. In psychological and psychiatric literature two different terms are used to describe knowledge about the body: one is 'body schema' and the other is 'body image'. I will give an introduction to this discussion, which is still not settled, and come up with an alternative given through Bayesian statistics, that can account for many of the phenomena that are categorised as related to either body schema or body image. Because of this unsettled debate I will use the term 'body representation' in order to cover the different aspects of processes related to the body, and I will use the term 'body awareness' whenever I speak of the phenomena related to the subjective perception of the body, regardless of whether it is the perception of positions of the limbs or the feeling of 'me' being inside 'my body'. Underneath the general term I will try to distinguish different aspects of body awareness whenever appropriate.

Body awareness and body representations rely on the integration of sensory information from different sensory modalities. We receive information about the body when our skin is touched and when our muscles contract. But when we look down or raise the arm in order to bring it into the visual field, we also receive visual information about the body. Although somatosensory information (the sensation of touch, signalled through receptors in the skin) and proprioception (sensory information from muscles, tendons and joints, signalling the stretch and load of muscles) are extremely important for body representation, it is evident from the scientific literature that body representation and body awareness are influenced by visual information as well. Therefore one cannot speak about one single 'body sense', i.e. one sensory modality that gives rise to the information required for body representations. Body awareness and representation rely on the integration of information from all sensory modalities. Furthermore, a number of experimental manipulations have been carried out, which information about the body has been manipulated. The rubber hand illusion is probably the best-known example. In this chapter I will also give you some examples of these manipulations.

Body awareness is also tightly coupled to movements. We are aware of many of the movements that we carry out, and we feel that we control our own movement, but this awareness of our movements also relies on how we perceive our stationary body. In this chapter I will give some examples of how one can manipulate with the perception of movement, and thereby introduce situations with abnormal movement perception. By the end of the chapter I will have shown how it is possible to combine knowledge about body awareness and movement awareness into a coherent picture of our own moving body.

How has the whole topic of body awareness become interesting for neuroscience? First and foremost, the questions addressed in neuroscientific literature originate from ancient philosophical questions, such as 'Who am I?', 'What is the "self"?' and 'How do we perceive the world?' Because these questions arise in the mind of humans over and over again and have therefore fascinated mankind for centuries, it becomes evident that they should also be addressed in a scientific community dealing with the mind. However, this book is not a philosophical work, nor a book on the mind, but rather a work on motor control and motor learning, so why here? Well, motor control deals with a fundamental question of moving Body X from position A to position B or letting Body X perform action Y. In order to understand how Body X performs these tasks it is important to know how Body X knows the state of itself in order to move from A to B or perform action Y.

Body scheme versus body image

In the scientific literature on how the brain creates a representation of the body, two distinct terms have been used in the past century. These are body schema and body image. Head and Holmes introduced the term 'body schema' a century ago (Head and Holmes, 1911–12) and more recently the term 'body image' has also been introduced as a separate entity (Paillard, 1991).

Body schema is generally considered to be a term that covers a representation in the central nervous system of the body's current position, which is the spatial organisation of the limbs that depends on the biomechanical constraints of the body. This representation is used whenever movements or actions are carried out in order to determine the body's position, that then can be used to plan, control and adjust movements and actions in order to interact with objects and the environment and achieve behavioural goals. The body schema is intimately linked to movements and actions, i.e. a neural representation of the body used for spatial sensorimotor processing (Haggard and Wolpert, 2004).

In contrast body image is generally considered to be a perceptual representation of properties of the body, that are related to how the body appears visually from the outside.

and when our muscles contract. But when we look down or raise the arm in order to bring it into the visual field, we also receive visual information about the body. Although somatosensory information (the sensation of touch, signalled through receptors in the skin) and proprioception (sensory information from muscles, tendons and joints, signalling the stretch and load of muscles) are extremely important for body representation, it is evident from the scientific literature that body representation and body awareness are influenced by visual information as well. Therefore one cannot speak about one single 'body sense', i.e. one sensory modality that gives rise to the information required for body representations. Body awareness and representation rely on the integration of information from all sensory modalities. Furthermore, a number of experimental manipulations have been carried out, which information about the body has been manipulated. The rubber hand illusion is probably the best-known example. In this chapter I will also give you some examples of these manipulations.

Body awareness is also tightly coupled to movements. We are aware of many of the movements that we carry out, and we feel that we control our own movement, but this awareness of our movements also relies on how we perceive our stationary body. In this chapter I will give some examples of how one can manipulate with the perception of movement, and thereby introduce situations with abnormal movement perception. By the end of the chapter I will have shown how it is possible to combine knowledge about body awareness and movement awareness into a coherent picture of our own moving body.

How has the whole topic of body awareness become interesting for neuroscience? First and foremost, the questions addressed in neuroscientific literature originate from ancient philosophical questions, such as 'Who am I?', 'What is the "self"?' and 'How do we perceive the world?' Because these questions arise in the mind of humans over and over again and have therefore fascinated mankind for centuries, it becomes evident that they should also be addressed in a scientific community dealing with the mind. However, this book is not a philosophical work, nor a book on the mind, but rather a work on motor control and motor learning, so why here? Well, motor control deals with a fundamental question of moving Body X from position A to position B or letting Body X perform action Y. In order to understand how Body X performs these tasks it is important to know how Body X knows the state of itself in order to move from A to B or perform action Y.

Body scheme versus body image

In the scientific literature on how the brain creates a representation of the body, two distinct terms have been used in the past century. These are body schema and body image. Head and Holmes introduced the term 'body schema' a century ago (Head and Holmes, 1911–12) and more recently the term 'body image' has also been introduced as a separate entity (Paillard, 1991).

Body schema is generally considered to be a term that covers a representation in the central nervous system of the body's current position, which is the spatial organisation of the limbs that depends on the biomechanical constraints of the body. This representation is used whenever movements or actions are carried out in order to determine the body's position, that then can be used to plan, control and adjust movements and actions in order to interact with objects and the environment and achieve behavioural goals. The body schema is intimately linked to movements and actions, i.e. a neural representation of the body used for spatial sensorimotor processing (Haggard and Wolpert, 2004).

In contrast body image is generally considered to be a perceptual representation of properties of the body, that are related to how the body appears visually from the outside.

These could be properties such as size, shape, or colour. So the bottom line here is that body schema is information related to how the body is located, whereas body image is related to how the body appears perceptually. This suggests that there is a functional distinction between body schema and body image (de Vignemont 2010), whereas the first is body representations that are available for actions and the latter is body representations available for perception. Using this distinction should help us in identifying neural mechanisms that underlie body schema on one side and body image on the other side. Another distinction between body schema and body image is the idea that body image is information that is available for consciousness, i.e. something that you are consciously aware of, whereas body schema is information that you cannot be aware of. In particular the latter distinction has been put forward because of observations made by Mishkin and Ungerleider (1982) and Milner and Goodale (1995). Studies have been carried out in which patients with selective lesions of the brain were able to perform goal direct movements appropriately, but without awareness of their movements and their body, or vice versa. Patients with visual form agnosia cannot report the shape of a visual object, but if they are asked to perform an action directed towards the object, they can perform it correctly. In contrast, patients with optic ataxia can make correct visual judgments, but are unable to perform an action directed towards the object (de Vignemont 2010). A third distinction is a temporal distinction, which proposes that information available for the body schema is adaptable, including information on where the arm is located, is something that can be changed, whereas the information related to the body image, such as the size of the body, is more static.

The body schema, i.e. the neural representation of the body used for spatial sensorimotor processing, has certain properties, which have been nicely summarised by Haggard and Wolpert (2004). First of all, it is spatially oriented, meaning that a stimulus, that is applied to the body surface, is coded in a spatial reference frame that represents the outer world. If something is itching on the foot, you can still scratch the correct foot, even if you have crossed your legs. However, the story is somewhat more detailed. Yamamoto and Kitazawa (2001a) showed that if you ask subjects to judge whether a stimulus is applied first to either one hand or the other, subjects are normally extremely good at this task. If their hands are located in a normal position, they can discriminate the temporal order of two stimuli down to a separation of 70 ms, but if the arms are crossed, subjects require at least an interval of 300 ms between the stimuli to complete the task correctly.

Another property of the body schema is that it is modular. Instead of having a coded representation of all possible body positions, and combinations of body positions, that would be an almost infinitely large database of positions, the body schema is composed of modules, so the foot is part of the leg, the fingers are part of the hand which then again is part of the arm, and these modules can be affected independently, so for instance apraxia patients may have difficulties grasping a spoon, but may be able to reach for the food with the spoon effortlessly (Sirigu et al., 1995).

The body schema is constantly updated according to the movements that the body performs, which is evident when looking at the receptive field of parietal and basal ganglia neurons when the hand is moved (Graziano and Gross, 1993). The body schema is adaptable, which means that it can account for gradual changes of the properties of the body, but it can also adjust to quicker transitions. In an experiment in which Yamamoto and Kitazawa (2001b) used crossed sticks, that the subjects had learned to use, they saw effects similar to those with crossed hands: the temporal order judgments became diminished after they had learned to use the sticks and crossed only the sticks and not the hands.

It is also suggested that the body schema is supramodal, which means that it transforms sensory information from many different sensory modalities into one representation (Lackner, 1988). Thereby the body schema can also be considered to be coherent, which means that it does not give ambiguous representations of the body if sensory information from different modalities is not 100 percent in accordance with each other. In those cases, the body schema tries to make one coherent representation, and it is suggested that the underlying weighting is done according to the amount of information from the different sensory modalities (van Beers and Wolpert, 2002).

But when going through some of the clinical evidence of a clear dissociation between body schema and image, it becomes apparent that because of the way body schema and body image are used differently within scientific sub-disciplines, this distinction vanishes (de Vignemont et al., 2005). Furthermore, if we also consider that part of the body schema is adaptable, for instance to the body size, but that body size is considered to belong to the body image, then the distinction becomes blurred.

A lot of the evidence on body awareness and its neural counterpart comes from a number of neurological or psychiatric cases, that in one way or another have manifested themselves as complications related to the body. Usually, due to lesions in smaller or larger parts of the brain, disorders manifest themselves in various ways, and these have been associated with being either disorders of body image or of body schema. A disorder such as macro/microsomatoagnosia, in which subjects have a distorted awareness of the size and shape of the whole body or body parts only, is usually considered a disorder of body image. Another type of disorder is autotopagnosia in which subjects mislocalise body parts, when they are asked to point to a specific body part. This type of disorder could be considered a disorder of body schema (see for instance Haggard and Wolpert 2004), whereas according to the studies by Sirigu et al. (1991) it is a disorder of body image, but the disorder is not restricted to the subject's own body, but may also affect their ability to point to body parts on another person's body. One of the more peculiar disorders is heterotopagnosia, in which subjects designate parts of their own body as other person's body parts.

The general problem with the distinction between disorders of body schema and of body image is that the same test is used to distinguish between separate disorders that are related to body image and separate disorders related to body schema. The simple test is to ask subjects to point to a specific body part. The test is used, for instance, to distinguish between two disorders, believed to be disorders of body schema, numbsense and deafferentation. The ability to correctly point to a specific body part is believed to be disrupted in deafferentation but not in numbsense. However, the same test is also used to differentiate between disorders of body image, such as apraxia in which the test should not be affected, but it should be affected in autotopagnosia (de Vignemont, 2010).

The case of deafferentation, when patients lose sensation from the neck and below, is considered by some as a deficit of body schema in which they instead use their body image to guide their movement (Gallagher and Cole, 1995), but as de Vignemont (2010) correctly argues, in which the cases where they then have learned to perform movements, do they then still have a deficit of body schema? So depending on which perspective you take on the data that you are looking at, one may argue that a certain deficit is a deficit of body schema and another is one of body image.

An alternative approach, suggested both by Haggard and Wolpert (2004) implicitly and by de Vignemont (2010) more explicitly, is to adopt a Bayesian integration approach to the question of different types of body representations. The Bayesian approach uses the idea that predictions are made by the CNS of the consequences of the movement that the body performs. These predictions can then be tested by comparing the expected sensory consequences of a movement with the actual consequences, and make it possible to incorporate sensory information with motor commands, in order to avoid the strict distinctions between something that is used for movements only, and something that is only related to perception.

A Bayesian approach

When humans interact with the environment, they constantly test their predictions about the properties of the environment and themselves, by comparing the actual consequences of their actions with their predictions of what the action will cause. Whenever predictions and consequences conjure, beliefs are formed, that make up the foundations for future thoughts of how the world behaves. When predictions and consequences do not conjure, two possibilities arise: either the consequences are not perceived as they actually happened or the predictions are wrong. Bayesian statistics (Price 1763) has in addition taught us that an appropriate way to overcome this discrepancy is to update the prediction based on prior belief and the present sensory data. Thereby, a learning algorithm has been established, that seems to use the information available for the central nervous system in order to establish new concepts and ideas of how the world behaves. When it comes to the relation between Bayesian approaches and how we understand the body, a number of elegant studies have been carried out, that seem to suggest that at least some kind of Bayesian approach is used when body awareness is formed. One of the key findings is that the formation of a body representation depends on the integration of information from many sensory channels. As we shall see later, the rubber hand illusion is an example in which visual information is weighted much stronger than proprioceptive information, and therefore gives rise to the illusion of a rubber hand being part of your own body. Another case is in fact the opposite, here somatosensory information is weighted more than visual information. Van Beers and Wolpert (2002) used a classical adaptation paradigm, in which subjects were asked to adapt to manipulations of the position of their fingertip. These manipulations were done either in the depth or azimuth (leftwards). Here they showed that there was selective adaptation of either vision or proprioception depending on whether the manipulation was done in the depth (stronger relative visual adaptation) or in the azimuth (stronger proprioceptive adaptation). This means that, depending on the weighting of either visual information or proprioceptive information, different outcomes of the adaptation process will take place. By generalising this concept of Bayesian integration of different sensory modalities weighted by relevant measures determined from prior experience, it is possible to determine the neural representations of the body that are used to guide our movements, i.e. what would be called the body schema, and give us the perceptual representation of our body, which is considered to be our body image. Thereby one can use one general framework to describe both phenomena.

A specific case of deafferentation

The best-known examples of a disturbance of body awareness are the cases of two patients IW (Cole and Sedgwick, 1992) and GL (Forget and Lamarre, 1987), both of whom suffered from peripheral neuropathy that made them incapable of feeling their body from the neck down. Initially, both patients had great difficulties performing movements, because they lacked sensory feedback, that is constantly used to modify ongoing movement. Nevertheless both have taught themselves to perform a number of daily life activities, and IW has even learned to walk. However, because they lack sensory feedback from the neck down they need to use different strategies when they move. In particular, they rely heavily on vision, and they need to make a strong attentional and intentional effort whenever they have to make a movement. If IW is placed in darkness he cannot stand upright, and if he sneezes while walking he will likely fall, because he then cannot concentrate on the ongoing movements (Cole and Paillard, 1998).

Some interesting observations have been carried out on GL and IW. They are both able to make some types of anticipatory adjustments. If you hold out a heavy object in your hand and someone else removes the object, you will see that your hand moves automatically upwards when the load is removed. If, however, you move the object away yourself, you will make an anticipatory adjustment of the hand so it does not move upwards after the load is removed. Both IW and GL are able to make these anticipatory adjustments. Furthermore, both of them are able to make judgments about weight. In the presence of visual feedback they can make judgments of a 150 g weight with an accuracy of 10 g. Without vision they are impaired in the task, but are still able to judge a 400 g weight with an accuracy of 200 g. This suggests that some sense of effort is preserved even in the absence of sensory feedback, and it may likely be ascribed to their ability to use the information about the predicted sensory consequences that the motor system is producing when a movement is carried out, even in the absence of actual sensory feedback (Cole and Paillard, 1998).

The neural correlates of the body schema

As outlined in the previous section, the concept of body schema is not a unified concept, although it is generally accepted that it is related to the representation of the limbs in a way that can be used for action. This non-unified concept is luckily also reflected in the way the transformation of coordinates and information available for movements about the body is represented in the brain. '*There is no such thing as multipurpose area for space and body schema*' (Rizzolatti, Luppino and Matelli 1998). Recordings from monkeys have given us the most comprehensive picture of how the body is represented in the brain in relation to movements. Penfield's (Penfield and Boldrey, 1937; Rasmussen and Penfield, 1947) now famous drawings of the representation of the human body in the somatosensory cortex (S1) and motor cortex (M1) has told us a lot about the representation of the body in the brain. But besides the direct sensory representations of the body parts in S1 and M1, a number of sub-regions within the posterior parietal cortex, premotor cortex and supplementary motor areas each have representations of the different body parts such as the arm, leg, hand and face, which are used to code specific movements. Both superior parietal lobule (SPL) and inferior parietal lobule (IPL) receive visual and somatosensory information about the body. Sub-regions within SPL and IPL, in particular the posterior parts, process visual information, and the anterior part of SPL predominantly processes somatosensory information, whereas the anterior part of IPL integrates visual and somatosensory information (Rizzolatti et al., 1998). This gives rise to multiple representations of the different body parts in the parietal sub-regions. In area PEci of the SPL there is a full somatosensory map of the body. In other areas, representations of the arm can be found, for instance within medial intraparietal area (MIP), anterior intraparietal area (AIP), PElp (in SPL), PFG (in rostral part of IPL), PF (subregion of the supramarginal gyrus) in the parietal lobe (for anatomical location see Rizzolatti et al., 1998; Pandya and Seltzer, 1982). Similarly, in the frontal motor cortical areas of the monkey, a representation of the arm can be found in area F1 (human homologue: primary motor cortex (M1)), F2 (human homologue: dorsal premotor cortex (PMd)), F3 (human homologue: supplementary motor area (SMA)), F4 (human homologue: ventral premotor cortex (PMv)), F5 (human homologue: Broca's area) and F6 (human homologue: preSMA).

Depending on which sub-area is investigated, different coordinate systems are used to code the location of the arm. Area PE in SPL codes the location of the arm in body centred coordinates (Rizzolatti et al. 1998), and that information is used by M1 to control movements of the arm. However, if one looks at the properties of ventral intraparietal area (VIP) in the intraparietal sulcus, this area has both visual dominant neurons and bimodal, i.e. tactile and visual dominant, neurons. Some of the neurons code in retinal coordinates, i.e. they code differentially depending on where gaze is centred, and others code in egocentrical coordinates, i.e. they can thereby keep track of where the visual information is compared to where the body is. Interestingly, some of the neurons only have receptive fields within the peripersonal space, i.e. close to the body. VIP is anatomically connected to area F4 (human homologue PMv), that also has visually dominant and bimodal dominant neurons.

Illusions of movement

Let us now focus on movements and our phenomenal awareness of movements, i.e. how it feels consciously to perform movements. One faces now a difficult problem, that is to disentangle the relative contributions to awareness of movements of peripheral feedback, performance of the movements and the predictions made by the central nervous system about the consequence of the movements.

From the point of view of phenomenal awareness, illusions of movements are extremely interesting, because they enable the field of cognitive sciences to study situations in which it is possible to dissociate the phenomenal awareness of one's own movements from the actual performance of movements. This can mainly be done by creating situations where subjects perceive that they perform a movement, without actually doing so. In the following we will focus on two aspects: one is the vibration-induced illusion of movement, the other is stimulation of brain areas that induce illusory sensations of movements. Finally, some approaches will be described peripheral feedback and motor commands to sensation and awareness of movements.

Vibration-induced illusion of movement

Goodwin, McCloskey and Matthews (1972) described an illusion in which they vibrated the biceps or triceps muscle tendon of healthy subjects at a frequency of 100 Hz. If subjects had their eyes closed and judged the position of their arm, they made misjudgment of up to 40 degrees. The subjects felt that the position of the arm corresponded to a position where the vibrated muscle was stretched. Later studies by Craske (1977) showed that it is possible to induce illusions of movements that are biomechanically impossible to achieve.

One of the pioneers in the studies of vibration-induced illusions of movements is Jean-Pierre Roll (Albert et al., 2006; Roll et al., 2009; Thyrion and Roll, 2010). The group around him recorded the actual proprioceptive feedback from a series of muscles, while subjects performed movements. Then the firing frequencies of the muscle spindle afferents were translated into a vibration frequency, which then can be applied to subjects. These illusory movements were mimicking writing movements and movements when drawing various geometrical figures. When subjects were asked what type of movements they had perceived when they were stimulated, and when they were asked to perform the movement afterwards, they correctly identified, for instance, the letter that they had experienced being drawn by their hand, and they could reproduce the movement from which the recorded muscle firing patterns that were used for vibration were done. These vibration patterns can be applied to series of muscles and thereby it is possible to induce illusions of both 2D and 3D movements.

What these studies tell us is that our awareness of movements under these circumstances matches quite well the information that is present in the muscle spindles, i.e. the proprioceptive signal that tells us how much our muscles are stretched.

But how can that be related to studies of our body schema and body image? One of the more peculiar phenomena that can be studied with tendon vibration is the so-called Pinocchio illusion (Lackner, 1988). If you ask subjects to pinch their nose with their index finger and thumb, and vibrate the biceps tendon, a mismatch between the senses occurs, because it feels as if the biceps is more stretched then it truly is and at the same time, the fingers are holding the nose. In order to solve this mismatch the illusion of having your nose growing longer starts to develop. One may think why the most appropriate way of solving this mismatch by the brain is to let your nose grow, rather than inhibiting information either from the proprioceptors of the biceps or somatosensation from the fingertips.

One thing is what can be induced peripherally; another thing is how the brain represents these peripheral illusions and copes with the induced mismatches as produced by the Pinocchio illusion. One of the leading researchers in the field of neuroimaging of vibration-induced illusions of movement is Eiichi Naito. He has conducted numerous PET and fMRI studies of the phenomenon. One of the challenges in neuroimaging studies is to find appropriate control situations, in these cases situations which you also vibrate the subjects, so they receive similar sensory input with the small exception that in one situation they should feel an illusion of movement and in the other situation only the vibration. By vibrating nearby skin or vibrating on the bone, without vibrating the muscle and tendon, it is possible to have sufficient control conditions, and an alternative is to use either very high vibration frequency (above 200 Hz) or very low vibration frequency (e.g. 10 Hz).

The first neuroimaging study conducted was a PET study: Naito et al. (1999) showed that regions in the supplementary motor area, caudal cingulate motor area, dorsal premotor cortex and area 4a of M1 displayed increased activation when subjects experienced the illusion of movement. Area 4p, 3b and 1, frontal and parietal operculum were activated in common for the vibration but not related to the illusion of movement. The area that showed strongest activation associated with the illusion was contralateral area 4a in M1, even though subjects did not perform any movements during the task. So not only does the primary motor cortex code activation related to actual performance of a movement (Naito and Ehrsson, 2001), but also activation in relation to illusions of movement.

Using an approach similar to that in the Pinocchio illusion, Naito et al. (2002) conducted studies where the left and right hand were attached to each other while only one hand was vibrated. In this situation, where the two hands were in somatic contact with each other there was a transfer of the illusion from the vibrated to the non-vibrated hand. What the neuroimaging data showed was that only when the subjects felt that the non-vibrated hand was moving, because it was attached to the vibrated hand, was there increased activation of M1. So again the neuroimaging tells us that even in the absence of performing movements, but only in the presence of an illusory movement, M1 codes the sensation of movements.

When inducing illusion of movement of limbs other than the hand, Naito et al. (2007) showed that there are regions such as area 4a, area 6 and the cerebellum, that all show activation for each of the particular limbs that are feeling moved by the vibration in somatotopic specific regions, and in addition hand movements showed activation in the supplementary motor area. Furthermore, it was found that regions in area 44, anterior insula, basal ganglia, cingulate motor area, and in inferior parietal areas, were commonly activated irrespective of which limb the illusion of movement was induced in.

Visually induced illusion of body movement

One of the most compelling ways of manipulating the subject's perception of their own movements is to use distorted visual feedback. Kaneko et al. (2007) used an indirect measure combined with questionnaires to show that subjects had illusory sensations of movement, when a video image of a moving hand was displayed around the same location as their own static hand. Subjects reported that they had felt their hand moving and during this illusory condition excitability of M1 changed, as measured by an increase in the ability to evoked larger muscle twitches with TMS.

Hagura et al. (2007) investigated the relative contributions of visual feedback and proprioceptive feedback to sensation of movement using fMRI. It is generally considered that visual feedback exerts a dominance over proprioceptive feedback in the presence of conflicting sensory information between vision and proprioception, an issue that will be discussed further in relation to the rubber hand illusion. In the experiment Hagura et al. used the combination of a static visual feedback and a vibration induced illusion of movements, affecting proprioception. They found a clear reduction in the sensation of movement when a static visual image was presented. The study further showed that the SPL activity correlated with the amount of visual dominance, i.e. more SPL activity corresponded with less sensation of movement.

Another category of visually induced illusions of movements is found in studies in which visual feedback is manipulated so there is a discrepancy between the performed movement and the visual feedback the movement generates. One of the pioneering studies in this branch of research was the experiment by Nielsen (1963) in which subjects were asked to perform line drawing movements, while they were seated with their hand in a mirror box. Sometimes the subjects did not see their own movement but a line drawing movement performed by the experimenter. If the line drawings by the experimenter deviated from the straight line the subjects were asked to make, the subjects performed compensatory movements in the opposite direction, and at the same time felt that their hand was being drawn passively in the opposite direction. These strange experiences were attributed to 'magnets' or 'unidentified forces', and subjects lost control or their own movements or lost agency. What this line of experiments clearly demonstrates is strong visual dominance, that overrules proprioceptive cues about one's real movements.

In order to separate the conscious subjective reports from the actual motor performance, Fourneret and Jeannerod (1998) asked subject to perform a task very similar to the one presented by Nielsen (1963). In two separate sessions the task was either to identify which trajectory the subjects had made, based on a verbal matching task, or to perform a motor matching task blindfolded. In the verbal task subjects tended to say that they had moved in a straight line even though the bias in either direction had led them to perform a movement in the opposite direction. A small subgroup of participants was able to correctly identify their deviation, but to a lesser extent than had actually taken place. When asked to perform a motor response instead of a verbal matching task, the same pattern was found and most subjects believed they performed a straight movement. Only a small subgroup was able to repeat a slightly deviating movement, but not to the same extent as theoretically would have been expected, if they had completely matched their actual movement.

In order to investigate how the brain solves the conflict between proprioception and vision (without taking into account additional information from motor commands, e.g.. internal models), Balslev et al. (2005) designed an experiment in which subjects were exposed to passive mouse-like movements on a mouse tablet, while they watched a cursor, that either moved congruent with their actual (passive) movement or incongruent with their actual (passive) movement. This task was done during fMRI, and revealed that SMA, precentral gyrus and temporoparietal junction (TPJ) showed increased activation when there was a conflict between vision and proprioception. Farrer et al. (2003) obtained similar results during a PET scan, in which large discrepancies between visual feedback and actual performance of a movement were reflected in the inferior parietal lobe activity. Another study (Shimada et al., 2005) using near infrared spectroscopy (NIRS) showed that superior/medial parts of the parietal lobe were active when subjects experienced synchrony between visual and proprioceptive feedback, whereas asynchronous conditions showed activity in the inferior parietal lobe.

Another type of visual manipulation is to delay the visual feedback and thereby induce absence of the feeling of being the agent of certain behaviours. Leube et al. (2003) asked subjects to judge whether they themselves were the agent of a movement they saw, which was a delayed video feedback signal of their own movements. They found that the longer the delay, the more activity in the superior temporal sulcus. Similarly, Farrer et al. (2007) used delays and found that IPL, in particular increased activity in angular gyrus, reflected increased action discrepancy when a delay was detected although the same delay was used. Similar results where found when subjects were tricked into believing that someone else was causing the delayed action they saw.

Stimulation induced illusion of movement

Another approach to study illusions of movement is stimulation of cortical sites without accompanying movements that nevertheless leads to an illusory sensation of movement. What we need to look for is not the loci for evoking actual movement, but rather areas where stimulation gives rise to an illusory sensation of movement in the absence of any real movements. Going through some of the early neurosurgery literature in which electrical stimulation has been used, Penfield and Boldrey (1937) report in one of their seminal papers on direct cortical stimulation, that stimulation of a postcentral region gave rise to a peripheral sensation of movement. Another paper from Scarff (1940) has a small note from one of the patients, who it is reported that stimulation of the premotor cortex gives rise to a sensation of movement without any accompanying movements, i.e. an illusory sensation of movement. Instead of using direct stimulation of the cortical surface during surgery, Amassian et al. (1989) used the, at that time new, technique of transcranial magnetic stimulation (TMS). With TMS it is possible to induce electrical currents in the nervous tissue underneath the stimulation coil. Thereby it is possible to stimulate brain areas in healthy volunteers without unnecessary invasive procedures.

Amassian et al. (1989) blocked the blood flow to the arm and waited for 25–35 min, resulting in a lack of sensation from and inability to move the blocked arm. In two subjects it was possible to elicit a sensation of movement after stimulation of cortical sites near and around M1. The stimulation intensities used were approximately 1.4 times higher than the intensities needed to evoke a movement during normal sensorimotor conditions without the block. Nevertheless, the results clearly show that even in the absence of any peripheral feedback from the arm, it is possible to induce a central mechanism, that can replace the absent feedback and give rise to an illusory sensation of movement.

Recently I conducted (Christensen et al., 2010) a similar study, but now with the use of brain navigation, in order to carefully determine the strength of the sensation of movement following stimulation of different cortical sites. The strength of the sensation of movement was subjectively evaluated on a scale from 0–5, with 5 as the strongest sensation of movement elicited by a 500 ms train of 20 Hz repetitive TMS during normal sensory conditions. Sensation of movement was then compared with that sensation in order to quantify how strong the sensation was. Not only did we use ischemic nerve block of the arm, but also spinal block of the legs, in order to determine whether the phenomenon was distinct for the arm, and whether the use of other types of paralysis could give rise to the same phenomenon. Using the quantification of the sensation of movement, it was possible to distinguish between the illusory sensation of movement evoked by stimulation over the primary motor cortex and the dorsal premotor cortex. Whereas the sensation of movement was stronger before paralysis when M1 was stimulated compared with premotor cortex, we found that the sensation of movement after PMd stimulation was not reduced even after complete block of the ability to move and sensory feedback. This was the case for both the arm and the leg. On the other hand, sensation of movement after primary motor cortex stimulation was reduced both for the arm and the leg. Other areas such as the supplementary motor area and posterior parietal cortex were also stimulated, but stimulation of these areas did not give rise to an illusory sensation of movement.

However, both the supplementary motor area and parietal cortex, albeit more anterior parietal regions, do indeed have properties that make them interesting in relation to illusory sensations of movement. Fried et al. (1991) had a group of 13 patients undergoing epilepsy surgery, and in order to determine where to perform the operation, part of the medial surface of the brain was stimulated. Some of the stimulations gave rise to overt movement, but in three patients stimulation of six sites within SMA gave rise to illusory sensations of movement, and one patient reported that it felt as if the arm was moving. In four patients another interesting phenomenon occurred, in which stimulation of 15 sites in these four patients gave rise to 'an urge to move', i.e. the feeling that they wanted to move without doing so.

A more recent study (Desmurget et al., 2009) investigated stimulation of parietal regions in three patients, and four patients underwent stimulation of premotor regions, while undergoing tumour surgery. In the three patients who underwent parietal stimulation, it was found that either stimulation gave rise to an 'intention to move', i.e. that they were about to move without doing so, or they felt illusory movements. On the other hand, in the patients who were stimulated in premotor regions, stimulation gave rise to movements that the patients were unaware of.

What do these studies tell us? First of all it is evident that you can experience sensation of movements without overt movement taking place in the body. Furthermore, it seems that several different cortical sites can give rise to illusory sensations of movements, when stimulated under the appropriate conditions, which suggests that it is not one single area of the brain that is responsible for the sensation of movement.

The fact that experiences of movement can take place in the absence of any overt movement suggests that mechanisms other than sensory feedback from muscles, joints and skin contribute to our sensation of movement. One candidate mechanism could be internal models, which are described in Chapter 1 and could contribute to the sensation of movement. The reason for suggesting this mechanism as a possible candidate is that it is a mechanism that makes predictions of the sensory feedback when you perform voluntary movements. This means that it implicitly assumes that some sort of voluntary effort is involved in performing the movement, which also suggests that if the mechanism is involved during voluntary movements, it will not contribute to sensation of movement during a passive movement. This means that in the cases when external electrical or magnetic stimulation is applied to a cortical site, the illusion of movement is induced because the stimulation activates a cortical area, that either directly activates a predictive mechanism, that can influence areas responsible for signalling the sensation of movement, or it means that the stimulated site is directly involved in creating the illusion of movement.

The efferent vs. afferent discussion

One of the central discussions on sensation of movement it to what an extent efferent (Roland, 1978) mechanisms contribute to a sensation of movement, a discussion that goes back to the beginning of the twentieth century, when Sherrington argued for a muscular sense (see Matthews, 1982). It is well known that proprioceptive information (also called kinaesthetic sense) is important for sensation of movement (McCloskey, 1978).

Ellaway et al. (2004) used TMS to answer whether perception of a TMS induced twitch in the muscle is due to sensory feedback or efferent mechanisms that give rise to the sensation. They used a paradigm where subjects had to judge the temporal order of a TMS evoked twitch in one hand and an electrically induced twitch in the other hand. It was argued that because subjects were able correctly to make the judgment of when the actual twitch took place in the left or right hand, they relied on the sensory feedback rather than a central mechanism. A central mechanism evoked by the TMS should presumably be faster than the sensory feedback and hence skew the order of the two events. But because subjects could correctly identify the order, it was argued that a central mechanism was not responsible for the sensation of the movement induced by the muscle twitch, rather it was due to sensory feedback.

Obhi et al. (2009) also tried to disentangle the relative contribution for movement sensation of efferent and afferent mechanisms, in a task in which the subject had to judge the onset of voluntary or passive movements. They came to the conclusion that neither efference nor reafference (the afferent feedback caused by a movement) solely were able to explain the findings of when subjects perceived that their limb moved. In order to clarify this point, Walsh et al. (2010) asked subjects to perform wrist flexion-extension movements while their abilities to sense their hand and move their hand were blocked using ischemic nerve block. They found that the longer subjects had to try to perform the effort-full movement, and the stronger they were asked to perform the movements, the more they perceived their hand to have moved, even though it never moved.

Although the question of contribution of efferent or afferent mechanisms to the sensation of movements has been investigated for decades, studies repeatedly show up suggesting that either one mechanism or the other can contribute to the sensation of movement. So can we draw any conclusions on this matter? One must probably accept that both mechanisms contribute to the sensation of movement, and then use that knowledge in order to accept that it matters for a patient or an athlete if they have reduced sensibility in a limb, when they have to perform a certain task, but otherwise also accept that how you perceive your own movements is not necessarily in agreement with reality, because perception is partly shaped by efferent mechanisms, that may involve predictive mechanisms that may be incorrect.

Where is the sensation of movement located in the brain?

Besides the strong proprioceptive input that the primary somatosensory cortex received in particular in areas 2 and 3a, what we have learned from the studies on illusions of movements, is that areas other than sensory areas seem to play an important role for our sensation of movement. In particular, M1 is involved when we perceive illusions of movement due to vibration of the muscle tendon (Naito, 2004). But using stimulation induced illusions of movement, both Amassian et al. (1989) and Christensen et al. (2010) showed that in the absence of sensory feedback, it was possible to stimulate M1 and premotor cortex and thereby elicit an illusion of movement. These results are somewhat in contrast to Desmurget et al.'s (2009) findings, showing that stimulation of premotor cortex gave rise to overt movements without accompanying sensation of movement, even in the presence of muscle activity and presumably also accompanying sensory reafference. However, stimulation methods were very different in these cases. The SMA is also a likely candidate to give illusions of movement (Fried et al., 1991), and its role may also be to compute mechanisms that can inhibit the reafference from voluntary movements (Haggard and Whitford, 2004).

Finally, the evidence of a role of parietal areas in giving sensations of movements is compelling. In particular, lesion studies have shown that parietal areas (Sirigu et al., 1999) are extremely important when attributing movements to oneself. Furthermore, TMS studies have shown that attributing movements to being performed by oneself can be diminished after TMS of parietal areas (MacDonald and Paus, 2003; Preston and Newport, 2008).

Illusions of body perception

Rubber hand illusions

Since the neural mechanisms of neurological and psychiatric disorders that alter body awareness can be very heterogeneous and the underlying mechanisms are not fully understood, finding a possible model to study body awareness in healthy subjects has been on the wish list for many years. The first systematic study that used a model of alterations of body awareness in healthy subjects was the now famous study of the rubber hand illusion, presented by Botvinick and Cohen (1998), although anecdotal reports date much further back in time (Tastevin, 1937).

The rubber hand illusion (RHI) is a very simple model, that can alter body awareness. A subject sits by a table, one arm is resting on the table. A rubber hand or arm is placed in front of the subject. A screen blocks the vision of the subject's own arm. The subject can now see only the rubber hand in front of them. The experimenter starts to stroke the subject's own hand and the rubber hand simultaneously. In a majority of subjects an illusion occurs after a few minutes (typically as little as 10–30 s) and the subjects start to feel ownership of the rubber hand.

If the touches on the two hands are performed asynchronously the illusion breaks down, so an important factor for the illusion to work is the simultaneous visual input of the skin on the rubber being touched while the subject's own hand is touched. Furthermore, the anatomical posture of the rubber hand should be, if not similar, then at least possible for the subject to achieve. This means that a rubber hand turned 90 of 180 degrees will not elicit the illusion. The distance between the rubber hand and the subject's own hand also seems to play a role. Lloyd (2007) found that a distance between the hands beyond 27.5 cm caused the illusion to decay very rapidly. Interestingly, this distance roughly matches the size of the peripersonal space (Maravita et al., 2003).

One may ask what determines whether a subject feels the RHI or not. One of the obvious factors would be that the rubber hand should resemble the appearance of one's own hand. However, this does not seem to be very important, although objects like wooden sticks do not give rise to the illusion. The size can vary (within reason), and the colour can vary.

Figure 14.1 Illustration of the rubber hand illusion. A rubber hand is placed in front of the subject in the visual field. The subject cannot see her own hand. The experimenter strokes simultaneously the rubber hand and the real hand.

Actually, it seems that subjects judge a rubber hand that they have owned to be more like their own hand, i.e. they change their view of the hand as if they have owned it *'Participants who had experienced the RHI reported significantly greater similarity between their own hand and the rubber hand than participants who had not experienced the RHI'* (Longo et al., 2009).

Besides asking the subjects whether they feel that the rubber hand is their own hand, other less subjective measures are used to determine that the subject really owns the rubber hand. One way is to threaten the rubber hand with knives or needles (Ehrsson et al., 2007). If the subject has incorporated the rubber hand such that it is felt as being a part of the subject's body, it is expected that the threat will lead to some sort of harm avoidance response. One clear measure is to look at the skin conductance response, which increases when you feel threatened. And indeed, this response does change when subject feels ownership of a rubber hand that is being threatened.

The rubber hand illusion has also been investigated using neuroimaging techniques. Ehrsson et al. (2004) showed that when subjects experience ownership of the rubber hand, the intraparietal cortex and ventral premotor cortex show increased activation compared with appropriate control conditions. Furthermore, Ehrsson et al. (2007) also showed that if you threaten the rubber hand, brain activity (as measured with fMRI) increases in areas responsible for pain processing (insula and cingulate cortex).

Another measure that is used to assess the strength of the illusion is a measure of proprioceptive drift, in which participants are asked to point to the location of the felt rubber hand.

The deviation between the actual position of their own hand and the position of their perceived new rubber hand is called the proprioceptive drift. However, it has been questioned in a number of studies whether that measure is a good measure of the perceptual experience of the RHI (Rohde et al., 2011).

Moving rubber hand illusion

A number of studies have used the RHI in a couple of variations: either the rubber hand itself was moving or where the participants had to move their own hand after experiencing the touch variation of the rubber hand illusion.

Tsakiris et al. (2006) performed one of the first examples of a moving rubber hand illusion-like experiment. They used a video projection of the subject's own hand instead of a physical rubber hand. This has the advantage that you can display the moving hand in a different position. This method also gives illusion similar to the stroked RHI. In their experiment they used the proprioceptive drift as the measure of the illusion. In both the tactile version, in which the subjects experienced tactile touch on their own hand and watched it on the projection, and also in an active movement condition and a passive movement condition, the proprioceptive drift was significantly larger than in the conditions when the actual touch or movement was asynchronous with the visual feedback, but there were still proprioceptive drifts taking place in the asynchronous conditions.

Another question that has been addressed is whether a recalibration of the proprioceptive sense takes place, that then gives rise to errors in performing subsequent movements. This was tested in subjects who had to move their own hand after they had experienced the rubber hand illusion.

There is a somewhat mixed outcome of these experiments. Holmes at al. (2006) compared three situations in which subjects either were exposed to a real hand, a rubber hand or a wooden block presented in a mirror. By varying the distance between the real and the apparent (mirror image of a) real hand, rubber hand or wooden block, it was possible to induce a visually caused bias in where the subject's proprioceptive sense would estimate the subject's own hands to be located. Even without the induction of an actual illusion of ownership of the mirror image of the three objects, it was found that a proprioceptive bias was introduced by the presence of the hands. It was largest for the mirror image of the real hand (the subject's own hand) and also present for the rubber hand. However, if the rubber hand was presented in a position that was incongruent with the subject's own hand, the visual bias disappeared.

Kammers et al. (2009), in contrast, could not show a bias even after the experience of the RHI, that had induced a proprioceptive drift. So subjects performed the subsequent reaching task without any difference in kinematical data whether they had experienced the illusion or not. One explanation of the absent finding could be that the kinematical data was low pass filtered to around 10Hz, which means that any corrections taking place faster than 100 ms would not be shown in the data.

Newport et al. (2010) performed a similar set of experiments and showed reaching errors after subjects had experienced a visual image of a hand as their own hand, and subsequently had to reach towards a goal with their real hand. The trick in this setup was that instead of just a single visual hand two visual images of the subject's own hand were displayed for the subjects, one slightly to the left and one to the right of the real hand. When one of the hands was visually stroked in synchrony with the real hand, the subject experienced that particular hand as their own, and conversely when the other was stroked that was felt as their own. Afterwards subjects were asked to make a pointing movement towards a target, and by measuring the end point error it was possible to show that end point error corresponded quite well to whether the left or the right hand was felt as the subject's own hand. Interestingly, by showing touch of both hands, subjects also had an ownership illusion of both hands at the same time, which was first reported by Ehrsson (2009) and later reconfirmed (Guterstam et al., 2011).

The first moving 'RHIs' were made with video feedback, but one study by Dummer et al. (2009) used an actual moving arm, to induce a moving RHI. Subjects were attached to a rubber hand through a wooden rod, allowing them to perform horizontal movements of the whole arm in an active, a passive and a tactile condition. Furthermore, asynchronous conditions were also introduced. The main finding was that in the synchronous conditions, but not in the asynchronous, subjects felt the RHI and there was a statistical trend towards the active condition being stronger than the passive.

When the rubber hand illusion moves from a purely visuo-tactile illusion to become an illusion of movement sensory modalities other than tactile stimuli become important. Not only do the subjects receive cutaneous input, but they also receive input from proprioception, and when voluntary movements are performed during the RHI, motor commands also affect the illusion. In order to disentangle the contribution to the illusion from cutaneous and proprioceptive feedback Walsh et al. (2011) designed a moving RHI experiment by connecting the real index finger with the rubber hand's index finger, in which they used anaesthesia of the digital nerve of the index finger. Walsh et al. (2011) showed that in the absence of cutaneous feedback from the finger it was still possible to induce ownership of the rubber hand, thereby highlighting the importance of proprioceptive feedback to the illusion of ownership. However, contrary to the findings of Tsakiris et al. (2006), Dummer et al. (2009) and Kalckert and Ehrsson (2012), Walsh et al. (2011) found that the passive movements gave stronger illusions than the active.

Recently Kalckert and Ehrsson (2012) conducted a study in which they addressed the difference in feeling of ownership of a moving rubber hand and feeling of agency. They used a moving RHI (mRHI) and dissociated feeling of ownership and agency using two different types of manipulation. In the general setup the subject's real index finger was attached to the rubber hand index finger and when the real hand (which was hidden inside a box) moved, the rubber hand index finger moved. That gave a situation when subjects felt both ownership of and agency over the rubber hand. Applying a situation when the rubber hand was moving asynchronously with the real hand, it was possible to remove both the sensation of ownership and agency. In another condition they rotated the rubber hand 180 degrees which gave rise to a complete loss of ownership, but subjects continued to have a feeling of agency when they actively moved their own hand and thereby also the rubber hand. When the experimenter moved their hand and the rubber hand, so that they experienced a passive movement, the subjects lost agency. This shows that the feeling of agency and ownership can be dissociated.

Out-of-body and full body illusions

If it is possible to feel ownership of a rubber hand, then why not feel ownership over a completely different body? This was probably the question that researchers asked themselves after the rubber hand illusion had been on the stage for some years. In 2007 two research groups independently worked on the idea of inducing an out-of-body illusion. By using stereoscopic video glasses on subjects, Ehrsson (2007) and Lenggenhager et al. (2007) produced an illusion in which the subject's perspective was a couple of metres away from their own body, while they at the same time could see their own body. By gently touching the subject's own body and letting the subject see that they were touched, from a perspective that was some distance away from their present perspective it was possible to induce this illusion.

Petkova and Ehrsson (2008) extended the illusion by letting the subjects believe they were inside a mannequin, so the subjects were placed next to a mannequin and through their stereoscopic glasses they watched recording from a camera placed on the head of the mannequin. When the subject and mannequin were stroked on the body simultaneously, the subject felt that he or she was inside the mannequin. A variation of the study was made, in which the subject and the experimenter were shaking hands, while the subject's vision was from a camera placed on the experimenter's head. That gave the subject the impression that they were standing in front of themselves being in the experimenter's body shaking their own hand.

Body size illusion

Body size illusions have been studied using both video manipulations and vibration manipulations. De Vignemonet et al. (2005) used an approach similar as the Pinocchio illusion (see section on Vibration induced illusion of movement, above) to manipulate the size of one's other finger. So holding one finger while experiencing vibration on the biceps muscle gave rise to an illusion of a longer index finger. Furthermore, Ehrsson et al. (2005) induced a waist-shrinking illusion, here subjects were holding their own waist while vibrators were applied to the wrists. This gave the illusory sensation of the waist being smaller than normal. The latter study was performed during fMRI scanning and revealed that the illusory change in body size gave rise to activation changes in areas where the postcentral gyrus and intraparietal gyrus meet.

Van der Hoort et al. (2011) used an approach similar to the previous explained mannequin illusion, but instead of a real size mannequin they used a 4 m high doll or a 30 cm short Barbie doll. By presenting objects in front of these dolls that were viewed by the subject, they were able to induce an illusion and the subject felt that he/she was either 4 m or 30 cm high. By testing the subject in an imaginary movement task, in which the subject had to imagine performing a certain movement, for instance walk down the corridor and back, it was evident that the time it took for the subjects to imagine this task scaled with their perceived body size. So, the smaller the subject felt, the longer it took to imagine performing the task. In normal situations there is a very high correspondence between the time it takes to imagine performing a task and actually performing it.

Concluding remarks

With the last decade's advancements in neuroscientific methods such as neuroimaging, electrode implants, and non-invasive cortical stimulation and the inclusion of virtual reality techniques in the neuropsychology laboratories around the world, our knowledge of the body, and how important it is for the experience of the self, has grown tremendously. We have moved from a situation in which most cases of abnormal body awareness were considered pathological, and therefore it was difficult to give straight answers to the mechanisms in play due to the possibility of plastic changes accompanying the pathological states. Furthermore, the introduction of the very simple rubber hand illusion paradigm has boosted the number of studies on body awareness, simply because this experimental manipulation is so simple to achieve, it is very accessible and it enables researchers to manipulate with body awareness in healthy volunteers, in a consistent and reliable way. This is a remarkable situation even though an anecdotal description has been available for more than 75 years (Tastevin, 1937).

From the huge quantity of studies on body awareness, a number of conclusions can be derived. First of all, the experience of our body seems to be an amalgamation of information from multiple sensory channels, that are integrated into a singular representation of the body, which is constantly being updated according to the information one receives from the different sensory modalities. Many of the studies highlight the importance of visual information, that in many cases is able to overrule the information from proprioception and somatosensation, when phenomenal awareness of the body is formed. Both the rubber hand illusion experiments and experimentally induced out-of-body experiences are examples of such. Furthermore, when we start to move there is additional information from efferent mechanisms that give rise to the unified integration of information leading to sensations of movement.

References

1. Albert, F., Bergenheim, M., Ribot-Ciscar, E. and Roll, J.-P. (2006) 'The Ia afferent feedback of a given movement evokes the illusion of the same movement when returned to the subject via muscle tendon vibration', *Exp Brain Res* 172 (2): 163–74
2. Amassian, V.E., Cracco, R.Q. and Maccabee, P.J. (1989) 'A sense of movement elicited in paralyzed distal arm by focal magnetic coil stimulation of human motor cortex', *Brain Res* 479 (2): 355–60
3. Balslev, D., Nielsen, F.Å., Paulson, O.B. and Law, I. (2005) 'Right temporoparietal cortex activation during visuo-proprioceptive conflict', *Cereb Cortex* 15 (2): 166–9
4. Botvinick, M., and Cohen, J. (1998) 'Rubber hands "feel" touch that eyes see', *Nature* 391 (6669): 756–756
5. Christensen, M.S., Lundbye-Jensen, J., Grey, M.J., Vejlby, A.D., Belhage, B. and Nielsen, J.B. (2010) 'Illusory sensation of movement induced by repetitive transcranial magnetic stimulation', *PLoS ONE* 5 (10): e13301
6. Cole, J. and Paillard, J. (1998) 'Living without touch and peripheral information about body position and movement: Studies with deafferented subjects', in J. Bermúdez (ed.) *The Body and the Self*, Boston: MIT Press: 1–24
7. Cole, J.D. and Sedgwick, E.M. (1992) 'The perceptions of force and of movement in a man without large myelinated sensory afferents below the neck', *J Physiol (Lond)* 449: 503–15
8. Craske, B. (1977) 'Perception of impossible limb positions induced by tendon vibration', *Science* 196 (4285): 71–3
9. de Vignemont, F., Ehrsson, H.H. and Haggard, P. (2005) 'Bodily illusions modulate tactile perception', *Curr Biol* 15 (14): 1286–90
10. de Vignemont, F. (2010) 'Body schema and body image – pros and cons', *Neuropsychologia* 48 (3): 669–80
11. Desmurget, M., Reilly, K.T., Richard, N., Szathmari, A, Mottolese, C. and Sirigu, A. (2009) 'Movement intention after parietal cortex stimulation in humans', *Science* 324 (5928): 811–13
12. Dummer, T., Piot-Annand, A., Neal, T. and Moore, C. (2009) 'Movement and the rubber hand illusion', *Perception* 38 (2): 271–80
13. Ehrsson, H.H., Kito, T., Sadato, N., Passingham, R.E. and Naito, E. (2005) 'Neural substrate of body size: illusory feeling of shrinking of the waist', *PLoS Biology* 3 (12): e412
14. Ehrsson, H.H., Spence, C. and Passingham, R.E. (2004) 'That's my hand! Activity in premotor cortex reflects feeling of ownership of a limb', *Science* 305 (5685): 875–7
15. Ehrsson, H.H., Wiech, K., Weiskopf, N., Dolan, R.J. and Passingham, R.E. (2007) 'Threatening a rubber hand that you feel is yours elicits a cortical anxiety response', *Proc Natl Acad Sci USA* 104 (23): 9828–33
16. Ehrsson, H.H. (2009) 'How many arms make a pair? Perceptual illusion of having an additional limb', *Perception* 38 (2): 310–12
17. Ehrsson, H.H. (2007) 'The experimental induction of out-of-body experiences', *Science* 317 (5841): 1048

18. Ellaway, P.H., Prochazka, A., Chen, M. and Gauthier, M.J. (2004) 'The sense of movement elicited by transcranial magnetic stimulation in humans is due to sensory feedback', *J Physiol (Lond)* 556 (Pt 2): 651–60

19. Farrer, C., Franck, N., Georgieff, N., Frith, C.D., Decety, J. and Jeannerod, M. (2003) 'Modulating the experience of agency: a positron emission tomography study', *Neuroimage* 18 (2): 324–33

20. Farrer, C., Frey, S.H., Van Horn, J.D., Tunik, E., Turk, D., Inati, S. and Grafton, S.T. (2007) 'The angular gyrus computes action awareness representations', *Cereb Cortex* 18 (2): 254–61

21. Forget, R. and Lamarre, Y. (1987) 'Rapid elbow flexion in the absence of proprioceptive and cutaneous feedback', *Hum Neurobiol* 6 (1): 27–37

22. Fourneret, P. and Jeannerod, M. (1998) 'Limited conscious monitoring of motor performance in normal subjects', *Neuropsychologia* 36 (11): 1133–40

23. Fried, I., Katz, A., McCarthy, G., Sass, K.J., Williamson, P., Spencer, S.S. and Spencer, D.D. (1991) 'Functional organization of human supplementary motor cortex studied by electrical stimulation', *J Neurosci* 11 (11): 3656–66

24. Gallagher, S. and Cole, J. (1995) 'Body image and body schema in a deafferented subject', *J Mind Behavior* 16: 369–90

25. Goodwin, G.M., McCloskey, D.I. and Matthews, P.B.C. (1972) 'Proprioceptive illusions induced by muscle vibration: Contribution by muscle spindles to perception?', *Science* 175 (4028): 1382–4

26. Graziano, M.S. and Gross, C.G. (1993) 'A bimodal map of space: somatosensory receptive fields in the macaque putamen with corresponding visual receptive fields', *Exp Brain Res* 97 (1): 96–109

27. Guterstam, A., Petkova, V. and Ehrsson, H.H. (2011) 'The illusion of owning a third arm', *PLoS One* 6 (2): e17208

28. Haggard, P. and Whitford, B. (2004) 'Supplementary motor area provides an efferent signal for sensory suppression', *Brain Res Cog Brain Res* 19 (1): 52–8

29. Haggard, P. and Wolpert, D.M. (2004) 'Disorders of Body Scheme'. In H.-J. Freund, M. Jeannerod, M. Hallett and Leiguarda (eds.) *Higher-Order Motor Disorders*, Oxford: Oxford University Press

30. Hagura, N., Takei, T., Hirose, S., Aramaki Y., Matsumura, M., Sadato, N. and Naito, E. (2007) 'Activity in the posterior parietal cortex mediates visual dominance over kinesthesia', *J Neurosci* 27 (26): 7047–53

31. Head, H. and Holmes, H.G. (1911–1912) 'Sensory disturbances from cerebral lesions', *Brain*, 34, 102–254

32. Holmes, N.P., Snijders, H.J. and Spence, C. (2006) 'Reaching with alien limbs: visual exposure to prosthetic hands in a mirror biases proprioception without accompanying illusions of ownership', *Percept Psychophys* 68 (4): 685–701

33. Kalckert, A. and Ehrsson, H.H. (2012) 'Moving a rubber hand that feels like your own: a dissociation of ownership and agency', *Front. Hum. Neurosci.* 6: 1–14

34. Kammers, M.P.M., de Vignemont, F., Verhagen, L. and Dijkerman, H.C. (2009) 'The rubber hand illusion in action', *Neuropsychologia* 47 (1): 204–1

35. Kaneko, F., Yasojima, T and Kizuka, T. (2007) 'Kinesthetic illusory feeling induced by a finger movement movie effects on corticomotor excitability', *Neuroscience* 149 (4): 976–84

36. Lackner J.R. (1988) 'Some proprioceptive influences on the perceptual representation of body shape and orientation', *Brain* 111 (Pt 2): 281–97

37. Lenggenhager, B., Tadi, T., Metzinger, T. and Blanke, O. (2007) 'Video ergo sum: manipulating bodily self-consciousness', *Science* 317 (5841): 1096–9

38. Leube, D.T., Knoblich, G., Erb, M., Grodd, W., Bartels, M. and Kircher, T.T.J. (2003) 'The neural correlates of perceiving one's own movements', *Neuroimage* 20 (4): 2084–90

39. Lloyd, D.M. (2007) 'Spatial limits on referred touch to an alien limb may reflect boundaries of visuo-tactile peripersonal space surrounding the hand', *Brain Cogn* 64 (1): 104–9

40. Longo, M.R., Schüur, F., Kammers, M.J.M., Tsakiris, M. and Haggard P. (2009) 'Self awareness and the body image', *Acta Psychol (Amst)* 132 (2): 166–72

41. MacDonald, P.A. and Paus, T. (2003) 'The role of parietal cortex in awareness of self-generated movements: a transcranial magnetic stimulation study', *Cereb Cortex* (2003) 13 (9): 962–7

42. Maravita, A., Spence, C. and Driver, J. (2003) 'Multisensory integration and the body schema: close to hand and within reach', *Curr Biol* 13 (13): R531–9

43. Matthews, P.B. (1982) 'Where does Sherrington's "muscular sense" originate? Muscles, joints, corollary discharges?', *Annu Rev Neurosci* 5: 189–218

44. McCloskey, D.I. (1978) 'Kinesthetic sensibility', *Physiol Rev* 58 (4): 763–820

45. Milner, A. D. and Goodale, M. A. (1995) *The Visual Brain in Action*. Oxford: Oxford University Press.

46. Mishkin, M. and Ungerleider, L.G. (1982) 'Contribution of striate inputs to the visuospatial functions of parieto-preoccipital cortex in monkeys', *Behav Brain Res* 6 (1): 57–77

47. Naito, E. and Ehrsson, H.H. (2001) 'Kinesthetic illusion of wrist movement activates motor-related areas', *Neuroreport* 12 (17): 3805–9

48. Naito, E., Nakashima, T., Kito, T., Aramaki, Y, Okada, T. and Sadato, N. (2007) 'Human limb-specific and non-limb-specific brain representations during kinesthetic illusory movements of the upper and lower extremities', *Eur J Neurosci* 25 (11): 3476–87

49. Naito, E., Roland, P.E. and Ehrsson, H.H. (2002) 'I feel my hand moving: a new role of the primary motor cortex in somatic perception of limb movement', *Neuron* 36 (5): 979–88

50. Naito, E., Ehrsson, H.H., Geyer, S., Zilles, K. and Roland, P.E. (1999) 'Illusory arm movements activate cortical motor areas: a positron emission tomography study', *J Neurosci* 19 (14): 6134–44

51. Naito, E. (2004) 'Sensing limb movements in the motor cortex: how humans sense limb movement', *The Neuroscientist* 10 (1): 73–82

52. Newport, R., Pearce, R. and Preston, C. (2010) 'Fake hands in action: embodiment and control of supernumerary limbs', *Exp Brain Res* 204 (3): 385–95

53. Nielsen, T.I. (1963) 'Volition: a new experimental approach', *Scand J Psychol* 4 (1): 225–30

54. Obhi, S.S., Planetta, P.J. and Scantlebury, J. (2009) 'On the signals underlying conscious awareness of action', *Cognition* 110 (1): 65–73

55. Paillard, J. (ed) (1991) 'Knowing where and knowing how to get there', in J. Paillard (ed.) *Brain and Space*, Oxford: Oxford University Press.

56. Pandya, D.N. and Seltzer, B. (1982) 'Intrinsic connections and architectonics of posterior parietal cortex in the rhesus monkey', *J Comp Neurol* 204(2): 196–210

57. Penfield, W. and Boldrey, E. (1937) 'Somatic motor and sensory representation in the cerebral cortex of man as studied by electrical stimulation', *Brain* 60: 389–443

58. Petkova, V. and Ehrsson, H.H. (2008) 'If I were you: perceptual illusion of body swapping', *PLoS One* 3 (12): e3832

59. Preston, C. and Newport, R. (2008) 'Misattribution of movement agency following right parietal TMS', *Soc Cogn Affect Neurosci* 3 (1): 26–32

60. Price, R. (1763) 'An essay towards solving a problem in the doctrine of chances. By the late Rev. Mr. Bayes', *Phil Trans Roy Soc Lond* 53: 370–418

61. Rasmussen, T. and Penfield, W. (1947) 'The human sensorimotor cortex as studied by electrical stimulation', *Fed Proc* 6 (1): 184–5

62. Rizzolatti, G., Luppino, G. and Matelli, M. (1998) 'The organization of the cortical motor system: new concepts', *Electroencephalogr Clin Neurophysiol* 106 (4): 283–96

63. Rohde, M., De Luca, M. and Ernst, M.O. (2011) 'The rubber hand illusion: Feeling of ownership and proprioceptive drift do not go hand in hand', *PLoS One* 6 (6): e21659

64. Roland, P.E. (1978) 'Sensory feedback to the cerebral cortex during voluntary movement in man', *Behav Brain Sci* 1: 129–71

65. Roll, J.-P., Albert, F, Thyrion, C., Ribot-Cisar, E, Bergenheim, M. and Mattei, B. (2009) 'Inducing any virtual two-dimensional movement in humans by applying muscle tendon vibration', *J Neurophysiol* 101 (2): 816–23

66. Scarff, J.E. (1940) 'Primary cortical centers for movements of upper and lower limbs in man: Observations based on electrical stimulation', *Arch Neuro Psychiatr* 44 (2): 243–99

67. Shimada, S., Hiraki, K. and Oda, I. (2005) 'The parietal role in the sense of self-ownership with temporal discrepancy between visual and proprioceptive feedbacks', *Neuroimage* 24 (4): 1225–32

68. Sirigu, A., Grafman, J., Bressler, K. and Sunderland, T. (1991) 'Multiple representations contribute to body knowledge processing: Evidence from a case of autotopagnosia', *Brain* 114 (Pt 1B): 629–42

69. Sirigu, A., Cohen, L., Duhamel, J.R., Pillon, B., Dubois, B. and Agid, Y. (1995) 'A selective impairment of hand posture for object utilization in apraxia', *Cortex* 31 (1): 41–55

70. Sirigu, A., Daprati, E., Pradat-Diehl, P., Franck, N. and Jeannerod, M. (1999) 'Perception of self-generated movement following left parietal lesion', *Brain* 122 (Pt 10): 1867–74

71. Tastevin, J. (1937) 'En partant de l'expérience d'Aristote les déplacements artificiels des parties du corps ne sont pas suivis par le sentiment de ces parties ni par les sensations qu'on peut y produire', *L'encéphale* 32, 57–84.

72. Thyrion, C. and Roll, J.-P. (2010) 'Predicting any arm movement feedback to induce three-dimensional illusory movements in humans', *J Neurophysiol* 104 (2): 949–59

73. Tsakiris, M., Prabhu, G. and Haggard, P. (2006) 'Having a body versus moving your body: how agency structures body-ownership', *Cons Cogn* 15 (2): 423–32

74. Van Beers, R.J., Wolpert, D.M. (2002) 'When feeling is more important than seeing in sensorimotor adaptation', *Curr Biol* 12 (10): 834–7

75. Van der Hoort B., Guterstam, A. and Ehrsson, H.H. (2011) 'Being Barbie: the size of one's own body determines the perceived size of the world', *PLoS One* 6 (5): e20195

76. Walsh, L.D., Gandevia, SC., and Taylor, J.L. (2010) 'Illusory movements of a phantom hand grade with the duration and magnitude of motor commands', *J Physiol (Lond)* 588 (Pt 8): 1269–80

77. Walsh, L.D., Moseley, G.L., Taylor, J.T. and Gandevia, S.C. (2011) 'Proprioceptive signals contribute to the sense of body ownership', *J Physiol (Lond)* 589 (Pt 12): 3009–21

78. Yamamoto, S. and Kitazawa, S. (2001a) 'Reversal of subjective temporal order due to arm crossing', *Nat Neurosci* 4 (7): 759–65

79. Yamamoto, S. and Kitazawa, S. (2001b) 'Sensation at the tips of invisible tools'. *Nat Neurosci* 4 (10): 979–80

15

VOLUNTARY MOVEMENT

Limitations and consequences of the anatomy and physiology of motor pathways

John C. Rothwell[1] and Jens Bo Nielsen[2]

[1] UCL INSTITUTE OF NEUROLOGY, LONDON
[2] DEPARTMENT OF NEUROSCIENCE AND PHARMACOLOGY, UNIVERSITY OF COPENHAGEN

Introduction

Voluntary movements arise from activity in the cerebral cortex, and the instructions are then carried to the spinal cord in order to activate motoneurones and produce movement. Voluntary inputs from the cortex do not have private access to spinal motoneurones, separated from any other systems. In the vast majority of cases, cortical inputs first contact interneurones which then relay the commands to motoneurones. Since the same interneurones also receive continuous input from sensory receptors (and hence might be thought to participate in spinal reflexes) as well as from interneurones from other parts of the spinal cord, this means that by the time cortical input reaches motoneurones it has been filtered by multiple lower level systems. In higher primates and in man, cortical input can access some motoneurones via a special direct pathway (the corticomotoneuronal pathway), which is often supposed to play a critical role in volitional movement. However, even if this input is strong (and there is little comparative evidence on this) the excitability of motoneurones will have been biased by the multiple other inputs that each one receives. Thus, even this connection does not guarantee the brain a straightforward control of muscle.

In this chapter, we will summarize the basic organization of voluntary motor control. We describe how the anatomy of the body determines some of the basic features of spinal cord organization. Within this organization a number of specialized circuits that prioritize particular patterns of connections have developed. The excitability of all of these circuits can be modulated on relatively fast time scales, such as that of the walking cycle or in the tenths of a second prior to onset of a movement. However, much longer term changes can occur in the 'set points' about which the circuits operate, increasing or decreasing their importance relative to other pathways. These latter processes are usually termed 'plasticity' and contribute, for example, to the increase in stretch reflexes after spinal injury or physical training.

Cortical inputs are superimposed on and filtered by this organization. However, the basic cortical circuits differ from those in the spinal cord in two important anatomical and physiological respects. Anatomically, the representations of individual muscles, particularly those

304

controlling the hand and fingers, are highly intermixed compared with the much more segregated anatomy of muscle representations in the cord. We argue below that distributed cortical projections allow for flexibility of connections between muscle representations, and therefore are critical to the flexibility of movements unrestricted by postural demands. Physiologically, the cortex is the main gateway for visual inputs to enter and influence motor control. This is particularly relevant during reaching with the arm and during the swing phase of gait for the leg. In both cases, the limbs are relatively free from feedback control from gravitational and contact force sensors, and can therefore be driven to a large extent by visual inputs.

Although we have the sensation that we control our muscles directly during a voluntary movement, the reality is that a rough concept at the cortex is transformed into the reality of movement by interactions at levels of the central nervous systems (CNS) of which we are never aware. The anatomy and physiology of the connections mean that if a volitional command were formulated in some hypothetical centre, it would be extremely difficult to predict the consequences with any certainty unless the state of every interposed connection were known in advance.

The chapter begins with a description of the organization of the corticospinal tract (CST) and the special features that distinguish it from other descending pathways. Then we consider how the CST input interacts with spinal circuits and their innate reflexes. Finally, we consider to what extent the CST tract really is the source of volitional input to cord and consider to what extent it participates in reflex functions like the spinal cord.

The motor cortex, corticospinal tract and voluntary movements

Summary of organization in non-human primates

The motor areas of cortex are defined by the fact that (1) all of them send axons directly to the spinal cord in the corticospinal tract, and (2) all are densely interconnected with each other. These regions are the primary motor area (Brodmann's area 4, mainly located in the anterior bank of the central sulcus), premotor cortex (the lateral part of area 6 of Brodmann), which is usually divided into dorsal and ventral in human studies, the supplementary motor area (the medial part of Brodmann's area 6) and (3) regions of the cingulate cortex ventral to the supplementary motor area (Dum & Strick, 2002). About 40 percent of the corticospinal fibers come from the primary motor cortex, whereas the cingulate and supplementary areas supply only about 20 percent each and the premotor areas some 10 percent (Lemon, 2008). All of these areas of cortex also project to brainstem areas that give rise to reticulospinal tracts, giving them an indirect, cortico-reticulo-spinal route to the spinal cord as well as the direct corticospinal route. The primary motor cortex is thought to have fewer of these indirect connections than other motor areas, suggesting that its output represents the most 'favored' cortical access to spinal cord circuits. There are also projections to the lateral vestibulospinal system and, in monkeys and cats, to the rubrospinal system. The status of the latter is debated in humans (Onodera & Hicks, 2009; Hicks & Onodera 2012). It has been claimed on the basis of cell counts in the magnocellular portion of the red nucleus that the rubrospinal tract is vestigial in humans and that its role has been taken over by an expanded corticospinal tract. However, the data is scanty and further investigation is required (Onodera & Hicks, 2009; Hicks & Onodera 2012).

The corticospinal tract travels in the lateral column of the spinal cord and has large numbers of terminations in segments controlling distal muscles of the arm and leg, but also innervates all sections of the cord. About 90 percent of the projections are crossed. A smaller proportion project ipsilaterally and lies in the ventral funiculus of the spinal cord. In primates most of the terminations of the corticospinal tract are on interneurones in the spinal grey matter with a smaller number of direct monosynaptic inputs to motoneurones, particularly those inner-vating the distal muscles of the extremities. These connections represent the only way the cortex can interact directly with the motor apparatus (corticomotoneuronal connections). In other vertebrates the cortex only has indirect access to the motoneurones through the spinal interneurones or brain stem nuclei. This is also true in rodents despite some early claims to the contrary (Alstermark et al., 2004).

The reticulospinal tracts arise from areas of the pontine and medullary reticular formation and the majority project bilaterally to innervate more proximal and axial muscles, although recent electrophysiological work suggests that they may even have some connections to the most distal muscles of the hand (Baker, 2011). In the cat, the pontine portion of the tract is in the ventral portion of the cord, whereas the medullary portion is in the lateral columns. Virtually, all reticulospinal connections synapse onto interneurones of the spinal grey matter, with few monosynaptic connections to motoneurones.

Consistent with this general anatomical picture of the distal bias of the corticospinal system, lesion studies in primates show that the main permanent deficit of corticospinal section (pyramidotomy) is a disorder of distal muscle control. Although animals could grasp objects, like the bars of the cage, with some force, and hence could activate intrinsic hand muscles, they had difficulty in manipulating objects with a precision grip and in performing isolated finger movements (fractionation). They also had great difficulty in releasing food from the hand and often retrieved it by rooting into the hand with the snout. This does not mean to say that the corticospinal input is not involved in control of more proximal muscles. Within 6 to 10 hours after the pyramidotomy, the animals could sit unsupported, but their arms hung loosely from the shoulder. After 24 hours, they could stand but could only take a few steps. It was only after several weeks that they recovered sufficient shoulder muscle control to grasp objects directly without circumduction of the shoulder. From this it would seem that the contribution of corticospinal input to more proximal muscles could eventually be compensated, perhaps by indirect cortico-reticulospinal inputs, whereas there was a permanent deficit in manipulative control. Finally, it is worth noting that all movements remained slower and fatigued more rapidly than in the intact animal.

The majority of cortically initiated movements that remain in the pyramidotomized animals are presumably made using rubrospinal and reticulospinal systems, and give some measure of the importance of these connections in the intact state. Isolated lesions of the reticulospinal system are impossible to achieve because the fibers are intermixed with other tracts along most of their path, in the lateral and ventromedial columns to the spinal cord. Experiments by Lawrence and Kuypers (1968a; b) suggest that lesion of these tracts produces much more profound deficits in overall motor control than a pyramidotomy. Lesions to the lateral pathways impair hand movements as well as reducing the capacity of monkeys to flex the elbow. Lesions of the ventromedial systems produce a loss of more automatic functions, leading to a flexion bias of the trunk and limbs as well as severe impairment of axial and proximal movement. The conclusion is that the ventromedial systems are involved in main-taining a stable base upon which movements of the distal extremities can be superimposed.

Descending motor tracts in human

Transcranial magnetic stimulation (TMS) of the motor cortex can be used to activate the output of the motor cortex (Rothwell, 1997). This produces short latency responses, mainly in contralateral muscles, that are largest and of lowest threshold in the hand and forearm (Brouwer & Ashby, 1990; Rothwell, 1997). Several lines of evidence suggest that the TMS pulse leads to activity in the fast conducting component of the corticospinal tract. Thus, the size of the response correlates directly with the anatomical integrity of the corticospinal tract in patients with capsular lesions after stroke (Perez & Cohen, 2009). Similar recordings from the spinal epidural space show that the descending activity, evoked by cortical stimulation, travels at a velocity of around 60 to 70 m/s consistent with conduction in large diameter myelinated axons (Burke et al., 1990). The very short onset latency of the EMG response and the extremely brief period of increased firing probability that it causes in individual motorneurones is strong evidence that at least the onset of the EMG response is due to activity in the monosynaptic corticomotoneuronal component of the corticospinal tract (Palmer & Ashby, 1992; Brouwer & Ashby, 1992).

Such studies have suggested that the corticomotoneuronal projection is more widespread in humans than in primates. Presumed monosynaptic responses can be obtained in most, if not all, muscles of the arm and leg (Rothwell, 1997). Motor cortical TMS can even activate trunk muscles at short latency, again consistent with a corticomotoneuronal connection. Although it is difficult to estimate the strength of the connection in terms of mV depolarisation of a spinal motoneurone, rough calculations give values of 3 to 5 mV in total. Studies with TMS indicate that the amplitude of the input is greatest for intrinsic hand muscles, lower for forearm muscles and biceps, and very small in triceps and deltoid (Brouwer & Ashby, 1990; Rothwell, 1997). In contrast, stimulation with transcranial high voltage electrical stimulation, which tends to activate corticospinal axons directly rather than transynaptically, shows very large monosynaptic EPSPs in motoneurones of deltoid and even pectoralis major, that are equal in size to those in hand muscles (Rothwell, 1997). This discrepancy has never been resolved. In the leg, similar work suggests that there are also monosynaptic inputs to all leg muscles, that have been investigated, but without any clear proximal to distal gradient (Brouwer & Ashby, 1992; Petersen et al., 2003). There are, however, clear differences for muscles working at the same joint. Thus, tibialis anterior appears to have monosynaptic inputs which are of at least the same size as intrinsic hand muscles, whereas soleus motoneurones generally show some of the smallest monosynaptic inputs that have been recorded (Nielsen et al. 1993a, b; 1995). There are very few cases of isolated descending tract lesions in human patients that can be compared with the animal data. The lesion of the internal capsule, that is common in subcortical stroke, disrupts both the direct and indirect cortical connections to the spinal cord. Thus, we see to a mixed degree the combined effect of a corticospinal and reticulospinal lesion. Rare surgical cases show that the effect is quite different from a pure corticospinal lesion. Bucy and colleagues (1964) sectioned the middle third of the cerebral peduncle, which is mainly devoted to the corticospinal tract, to control abnormal involuntary movements (hemiballismus) in a patient who had suffered a natural unilateral lesion of the subthalamic nucleus. They reported that the lesion abolished the involuntary movement, but that there was no spasticity and movement control remained good. However, the lesion spared some corticospinal fibers which may have contributed to functional recovery.

Other descending pathways are more difficult to study in humans. The reticulospinal system can be assessed using the startle reflex (Rothwell, 2006). There is excellent evidence that in rats the startle response is conducted in reticulospinal tracts, and in humans the same is also likely to be true since (1) startle can be observed in anencephalic children born with no cerebral cortex, and can even be exaggerated in some patients with dense motor stroke; and (2) EMG studies of the spread of the startle reflex in healthy humans show that the response begins in muscles innervated by caudal cranial nerves (sternocleidomastoid, XI nerve), spreading up the brainstem to innervate mentalis (VII nerve) and then masseter muscles (V nerve). Some patients with excessive startle have lesions located in brainstem areas. Consistent with a reticulospinal projection, the startle reflex primarily affects proximal muscles and is rarely seen, for example, in the hand (Rothwell, 2006).

Motor circuits in the spinal cord: organization, control and modulation

Recent work on development of spinal cord in mice models has suggested that the location and types of neurone within the grey matter as well as the connections that each receives depend to a large extent on the dorso-ventral and medio-lateral position of the progenitor cells of each subpopulation (Jessell et al., 2011). In the mouse, the motoneurones that will innervate distal leg muscles are located dorsally in the ventral grey matter whereas the proximal moto-neurones lie ventrally. The fact that motoneurones innervating the same muscle lie close together increases the likelihood that they will receive common inputs. In addition, they tend to be located near other synergist muscles that will also share more of the input than motoneu-rones pools that are located at a distance. Motoneurone pools usually occupy several adjacent spinal segments, forming columns within a spatially segregated part of the ventral grey matter. Motor columns from synergist muscles likewise group together over several segments.

The terminations of sensory afferents to the spinal cord also seem to produce terminals and synapses that are position dependent. Thus, inputs from the proximal muscles terminate more ventrally in the cord than inputs from distal muscles, so that they tend to synapse among homonymous motoneurones. This organization is complemented by a similar organization of spinal interneurones, which also appear to develop, at least partially, according to their initial position in the cord. For example, Renshaw interneurones are located only in ventral parts of the spinal grey. Here they tend to receive inputs from and project to motoneurones that innervate proximal muscles with far fewer connections to distal motoneurones. The organi-zation of other groups of interneurones is less well studied, but work suggests that there may be over 100 different types of neurone, each clustered within specific areas of cord and projecting to specific populations of motoneurones (Jessell et al., 2011). Their position, as for motoneurones, may be an important factor in determining the inputs they receive as well as the output connections that they make.

It is likely that similar rules govern the organization of the human spinal cord. In this case, however, motoneurones that innervate distal motoneurones lie laterally whereas those innervating proximal muscles are medial. Flexor motoneurones are dorsal to extensor motoneurones. This organization then accounts for the termination pattern of the different descending motor systems. Those in the lateral columns, and in particular the corticospinal tract, tend to innervate distal muscles whereas those in the ventral columns innervate proximal motoneurones (Lemon, 2008).

Connectivity in the spinal cord

The circuitry of the spinal cord has presumably evolved to solve the problem of how best to arrange connections between the muscles that it innervates and the limited numbers and types of sensory receptor that innervate the body. We do not know the rules that specify spinal organization in any detail. However, one striking observation is that most of the connections between sensory input and motoneurone output are indirect, going via interneurones rather than direct sensory-motor pathways (Jankowska, 2001; 2008). An obvious exception to this is the monosynaptic connection between primary muscle spindle afferents and their homony-mous motoneurones. However, this seems very much to be a special case rather than the rule.

One advantage of having interposed interneurones is that they are an effective way of allowing the spinal circuitry to switch between different states. For example, in the two fundamental states of stance and gait, connectivity during posture should be arranged in order to resist perturbations of the body whereas during gait postural control must be released and movement allowed. Going from posture to movement means turning off the connections that assure stability and turning on those that allow movement. If different sets of interneurones are involved in these states, it would not be too difficult to suppress one set and excite the other to switch between states. An alternative arrangement could be for all inputs to have direct contacts with all motoneurones. A system of presynaptic inhibition could then select which inputs would be active in any particular state. However, such an organization would be more complex to implement in terms of accuracy of spatial connections. Nevertheless, the fact that presynaptic inhibition exists suggests that this strategy may sometimes be appropriate (Rudomin, 2002).

A second advantage of interneurones is that they can specialize in producing different patterns or rhythms of activity. This could be a special property of individual neurones or a property of an interconnected network of neurones, such as envisaged for the locomotor pattern generator.

In the absence of any formal theoretical justification, spinal interneurones are at present classified according to their involvement in different reflexes and their sensory input modality. In effect, the connectivity of the interneurones is viewed as allowing reflexes to associate particular patterns of input with particular patterns of output. Furthermore, since the excitability of the interneurones can be controlled, the importance of the connection can be modulated according to circumstance. This was pointed out very clearly in the early work by John Eccles and his co-workers in the 1950s and the subsequent work by Anders Lundberg and his co-workers, who emphasized the integrative properties of the interneurones (Alstermark et al., 2010; Hultborn, 2006). It is a general finding that every single interneurone receives input not only from the sensory modality which is the basis for its classification (e.g. as a 'Ia inhibitory interneurone', 'Ib inhibitory interneurone', 'gr. II interneurone' or 'flexor reflex afferent interneurone'), but also from a number of other sensory afferent modalities, other interneurones and a number of descending pathways (e.g. corticospinal, vestibulospinal, reticulospinal). Such a convergence of inputs means that it can be something of an arbitrary decision whether a neurone that receives, for example, Ia and Ib input is a Ia- or a Ib- interneurone or an interneurone that integrates Ia and Ib. In fact, the current classification of interneurones and the emphasis on reflex function of the spinal cord is much more determined by the available techniques for investigation and the ease with which peripheral nerves may be activated, than any real physiological raison d'être. It is not an unrealistic possibility that spinal interneurones with a slight turn of events could have been classified based on their supraspinal input as taking part in different voluntary movements rather than the current classification based on afferent input as taking part in different reflex actions. This was realized already by Sherrington (1906) more than 100 years ago when he wrote:

> A simple reflex is probably a purely abstract conception, because all parts of the nervous system are connected together and no part of it is probably ever capable of reaction without affecting and being affected by various other parts, and it is a system certainly never absolutely at rest. But the simple reflex is a convenient, if not a probable, fiction.

It may be time to give up the rather old-fashioned idea of separate reflexes and voluntary movements and instead focus on the sensory contribution to voluntary movement (Figure 15.1).

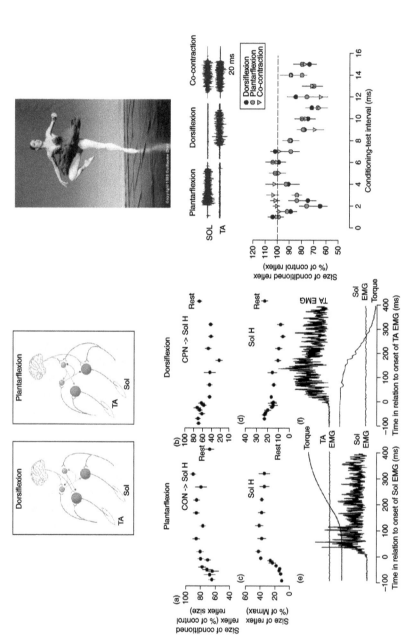

Figure 15.1 Modulation of reciprocal inhibition during voluntary movement. Ia inhibitory interneurones are characterized by receiving input from Ia afferents from one muscle and inhibiting the motoneurones of the antagonists. However, the same interneurones also receive input from descending motor fibres including the corticospinal tract. At the onset of dorsiflexion corticomotoneuronal cells projecting to tibialis anterior motoneurones thus inhibit soleus motoneurones through collaterals to Ia inhibitory interneurones (in the left diagram). At the onset of plantarflexion another set of Ia inhibitory interneurones, which inhibit tibialis anterior motoneurones, are activated by collaterals from corticomotoneuronal cells projecting to the soleus motoneurones (diagram to the right). When evaluating the excitability of soleus motoneurones by monosynaptic (H−)reflex testing a large suppression of the H-reflex is thus seen at short intervals when the common peroneal nerve to the tibialis anterior muscle is stimulated during dorsiflexion (black circles in graph to the right). During plantarflexion this inhibition is smaller due to mutual inhibition of the interneurones projecting to the tibialis anterior motoneurones (grey circles). When both muscles need to be activated in a stabilizing co-contraction around the ankle joint, the Ia inhibitory interneurones are decoupled from the normal parallel activation with their corresponding motoneurones and reciprocal inhibition is efficiently suppressed (triangles). Adapted from Nielsen & Kagamihara (1992).

Making spinal motoneurones discharge

As Sherrington pointed out, the spinal motoneurones are the final common path, which integrates information from the rest of the central nervous system and sensory feedback information from the body. The discharge of every single motoneurone and thus the activation of every single muscle fiber is determined by the integrated depolarization from on average 10,000 synaptic inputs arising from a number of different sensory modalities, spinal interneurones and supraspinal control centres. A precise estimate of the relative contribution of each of these inputs is difficult to obtain, but Powers and Binder (2001) have calculated, based on the synaptic activation of spinal motoneurones in the cat by electrical stimulation of different sensory afferents and descending fibers, that none of the main sources of motoneuronal activation provide sufficient drive on their own to depolarize the motoneurones sufficiently to reach the discharge frequencies observed under normal physiological conditions. In fact, the total drive from all these sources is insufficient to drive the motoneurones to their maximal firing frequencies observed during dynamic conditions without internal amplification mechanisms in the motoneurones, but this is another story. All muscle activity is thus the result of summation of the graded input from a large number of input sources (voluntary, automatic, conscious, non-conscious, reflex, sensory etc.), although some sources naturally do contribute more than others depending on the specific circumstances.

An estimate of the importance of some sensory inputs in supplementing/maintaining muscle activity during voluntary contraction comes from examining what happens when the input is suddenly removed. If an active muscle is suddenly shortened by an external perturbation, length and load sensitive afferents stop discharging momentarily. When such an unloading of the ankle plantarflexors is performed, in midstance a pronounced drop in soleus EMG activity is observed (Sinkjaer et al., 2000; af Klint et al., 2009). This indicates that under normal circumstances, activity in length and load sensitive afferents from the muscles contributes to the activation of the muscle. The short latency of the drop in EMG can only be explained by cessation of activity in a spinal pathway, and the inference is therefore that feedback mechanisms from the active muscle–tendon complex helps to maintain activity in the muscle. It has similarly been found in cats that flexion is initiated immediately, if activity in load-sensitive afferents from extensors is suddenly removed (for instance when the cat steps into a hole in the ground), but prolonged if activity in the load-sensitive afferents is maintained (Pearson, 2008).

Interaction of supraspinal inputs from cortex with the circuitry of the spinal cord

Segmental interactions

Given this summary of the anatomy, it is clear that any command elaborated at the cortex can be influenced at a number of different levels by interactions with other circuits, both in the brainstem (in the case of cortico-reticulospinal pathways) and in the spinal cord. Corticospinal pathways mostly connect to spinal interneurones where they share inputs with other descending pathways as well as with sensory afferents from the periphery and other spinal interneurones. Even the most direct, corticomotoneuronal pathway, which appears to give the cortex a direct access to motor outputs, will be influenced by the level of excitation provided to motorneurones from interneurones and sensory afferents. Thus, the idea that the cortex can have some preferential control over motoneuronal output during volitional movement is misleading. Instead, the cortical input to the spinal cord should be viewed as using or modulating the output of the spinal circuitry itself. There is no separation between 'spinal reflexes' and 'cortical voluntary movement'. Instead, it is important to focus on how the neuronal machinery in the spinal cord may provide an extremely flexible tool for the execution of voluntary movements.

A clear example of the interdependence of spinal and supraspinal inputs comes from considering the activity in Ia inhibitory interneurones (Crone & Nielsen, 1994). These interneurones receive Ia afferent input from an agonist muscle and inhibit the motoneurones of the antagonist. However, they are also activated by collaterals from the descending motor tracts to the agonist motoneurones (Figure 15.2). It is these connections that are responsible for increased activity in the interneurones prior to and at the onset of contraction of the agonist muscle, thereby preventing elicitation of a stretch reflex and unwanted activation of the antagonist when the antagonist muscle is suddenly stretched (Crone & Nielsen, 1994). Later during the static phase of the contraction, the activity of the interneurones declines despite continued activation of the agonist motoneurones. At this time pronounced inhibition of the antagonist motoneurones is no longer required since the risk of eliciting a stretch reflex in the antagonist is greatly reduced. During one and the same movement, the activation of the Ia inhibitory interneurones is thus highly flexible and adjusted according to the functional requirements at any given time. They are not controlled rigidly in parallel with the agonist motoneurones, but may be decoupled from this activation when required. This is most clearly seen during tasks where simultaneous activation of pairs of antagonist muscles is required (Nielsen & Kagamihara, 1992). During balancing, co-contraction of antagonist dorsi- and plantarflexors is often required to stabilize the ankle joint, and in this situation the activity of the Ia inhibitory interneurones is greatly reduced in order to facilitate the simultaneous activation of the muscles (Figure 15.2). This flexibility is also seen as long-term adjustments

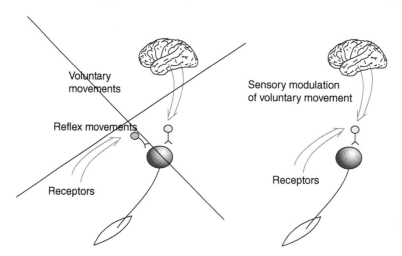

Figure 15.2 Sensory afferents and descending motor commands converge on common interneurones: reflexes are a part of voluntary movement. A distinction between reflex movements and voluntary movements is often made. However, since descending fibres and sensory afferents converge on the same interneurones in the spinal cord it is not reasonable to make this distinction in relation to the normal human movement repertoire. It seems more correct to simply refer to the descending and sensory contributions to the activation of the muscles during movement.

of the activity of the interneurones according to changes in the environment or the functional requirements during task performance. Following lesions of the descending motor tracts, transmission in the Ia afferent inhibitory pathway between ankle flexors and extensors is often reduced as an adaptation to the lack of activation from central pathways (Crone et al., 1994). It is also small in ballet dancers, possibly as an adaptation in relation to the functional requirement of stabilizing co-contraction around the ankle joint, and it may be up- or down-regulated in relation to training of specific motor tasks (Nielsen et al., 1993b).

Propriospinal interactions

In addition to being filtered by segmental circuitry, supraspinal inputs also interact with a special category of interneurones known as propriospinal interneurones (Alstermark et al., 2007; Pierrot-Deseilligny and Marchand-Pauvert 2002). These cells have axons that span several segments of cord and are well positioned to participate in movements that involve coordinated action of both proximal and distal muscles of a limb. The classic example for the arm is a reach-to-grasp movement in which the shoulder and elbow propel the hand towards a target. In the cat, propriospinal neurones in the C3-C4 spinal segments have been convincingly shown to mediate part of the supraspinal motor command to hand and arm motoneurones during reaching movements (Alstermark et al., 2007). Similar propriospinal neurones have been shown to exist also in monkey and human with an organization roughly similar to that in the cat (Alstermark et al., 2007; Pierrot-Deseilligny and Marchand-Pauvert, 2002). It should be noted that some early papers disputed whether this class of interneurones existed in primates, and it was argued that they may not contribute significantly to muscle activation during functional conditions. However, it appears that in those experiments, the interneurones may have been well suppressed by feedforward inhibition that was exacerbated by anaesthesia. At the present time, the evidence indicates that a functional propriospinal pathway is used in arm reaching and that it exists in primates and humans.

What is the functional role of the corticomotoneronal (CM) system?

Lawrence and Kuypers (1968a; b) thought that the most important new contribution of the corticomotoneuronal system was to allow individuation of finger movement, which is a prerequisite for precision grip and dexterous manipulation of objects with the hand. This line of thought has been followed up by a number of elegant experiments in monkeys and humans by Roger Lemon and co-workers, which have demonstrated the importance of the CM projections to hand and finger motoneurones for dexterity (Lemon, 1999; Lemon, 2008). Thus, there is no doubt of the general truth in this basic concept, but it should be pointed out that subcortical systems are already quite capable of highly specific and carefully timed control of individual muscles. Swallowing is an excellent example of exquisite control of more than 40 individual muscles, in which the cortex has no role other than to initiate the whole sequence. Locomotion is another example where co-ordinated activity in more than 50 muscles is generated by networks in the spinal cord without any need of supraspinal control at least in vertebrates other than humans. Precision of movement is also no prerogative of the hand. In fact we can make (and sense) movements of the shoulder with much greater precision than the hand. So why do spinal circuits not control the hand in the same way and achieve similar independence and precision as can be achieved with cortical input? We hypothesize that there are at least two advantages of cortical control. The first is adaptability which emerges as a consequence of the anatomy of the cortical motor representation. The second is integration of visual input which is particularly important in shaping the hand to manipulate objects. Individuation and precision are secondary consequences of this organization.

Flexibility of control

The spinal cord appears to have a highly patterned spatial organization in which groups of motoneurones and interneurones with common projection targets are located in close proximity. This means that the inputs that project to the spinal cord, which also terminate in particular spatial locations, will recruit outputs in a relatively restricted pattern. Effectively, this means that a limited number of output patterns can be 'wired into' spinal cord organization. This anatomy may well account for the patterned outputs that are often reported after intraspinal stimulation.

The organization of the motor cortex is much more of a mosaic than that of the spinal cord (Dum & Strick, 2002). Although groups of neurones that project to distal and proximal limb muscles are located in common areas of the cortex, within those areas the outputs intermingle to a far greater degree than occurs in the spinal grey matter. For example, in the forelimb region of cortex, outputs to hand muscles can be spread throughout the area and can intermingle with flexor, extensor and synergist muscles in seemingly random patterns. Work suggests that although the groups of neurones with common output targets are more intimately connected than those that have different targets (Capaday et al., 2009), this distributed organization may provide the cortex with more flexibility than can occur in the spinal cord. The corticomotoneuronal connection allows these flexible patterns of activity to go directly to spinal motorneurones without being filtered through the spinal networks.

A secondary consequence is that a direct connection would be able to conduct highly synchronized input commands and produce precisely timed phasic motor output. Several lines of evidence show that the corticospinal inputs can be synchronized in order to increase the effectiveness of the efferent volley. Synchronized CM activity prior to and at the onset of movement may be especially important to recruit motoneurones quickly and synchronously in ballistic motor tasks (Fetz & Cheney, 1987; Nielsen et al., 1995). Indeed, EMG recordings show that very short synchronous bursts of activity are characteristic of many fractionated finger movements, such as writing and tool use as well as some of the muscle activity during gait (e.g. the two synchronized bursts of TA EMG activity in the swing phase and the short bursts of knee flexor and extensor EMG activity around heel strike, all lasting 100 to 200 ms). Throwing and jumping also involve very short lasting and highly synchronized activation of the involved muscles. Interposing interneuronal synapses in these connections would tend to remove synchrony and smooth out the command. This is indeed what is seen following corticospinal lesion (Farmer et al., 1993). The CM system may thus also be at the heart of human evolution in view of the evolutionary advantage of being able to throw something at an animal in order to kill and eat it. This requires the ability to judge movement of the animal based on visual information, to predict the outcome of the throw in terms of direction and speed based on prior knowledge and to transform this information into appropriately timed powerful and synchronized activation of a large number of muscles throughout the body. The CM system appears to be suitably positioned for this.

Many experiments on animals and on humans have shown that learning new tasks is accompanied by synaptic plasticity in both motor cortex and spinal cord. Interestingly, plasticity is less easy to demonstrate in spinal than in cortical networks. Whether this is a true difference or a limitation of the present methodology is unclear. However, there may be an advantage in leaving spinal circuits relatively stable, since if the lower level wiring changes, all motor programs sent down from a cortical level will be affected. It is probably better to maintain the lower level 'hardware' relatively constant and change movement by changing the upper level 'software' command structure. Plasticity in the highly interconnected network of the motor cortex could be used to switch patterns of motor commands and adapt quickly to new tasks. The one exception to this general rule is consistent with this interpretation: plasticity of presynaptic inhibition is relatively easy to demonstrate in the spinal cord. Presynaptic inhibition is ideally suited to fine-tune spinal networks without disrupting the overall function.

Visual control of movement

Visual input to the motor cortex may well make use of the flexibility inherent in corticomotoneuronal output. One-third of the cortex, particularly in the parietal and premotor areas, is devoted to visual processing. A considerable part of this is used for shaping/orienting our hand in preparation for grasping and manipulating objects or for moving the legs to avoid or to target visually identified objects. Sending this information to the motor cortex gives the visual input access to a large variety of patterns of muscular output that can be adapted to suit the characteristics of individual objects (e.g. size, weight, slipperiness, position of centre of gravity, etc.). The corticomotoneuronal pathway ensures that these patterns reach motorneurones directly, without being influenced too strongly by local somatosensory inputs. As noted above, the special status of the arm as a non-load bearing appendage frees its control systems from many of the constraints induced by gravity, allowing visual input to control movement directly.

The same type of argument applies for the corticomotoneuronal projection to leg muscles. Visual control is important when one of our legs is in the air. Such control can steer the foot clear of obstacles or make sure that we do not step on things that we would rather not have on our shoes. Once the leg is on the ground, however, visual information is of little use. It seems reasonable to assume that this explains the much more pronounced corticomotoneuronal control of the TA muscle, which is mainly active when the leg is in the air, than of any other leg muscle.

The reflex function of the motor cortex

So far we have imagined that a volitional movement is formulated within the cerebral cortex and that information flows in one direction to motor cortex and spinal cord. Movement occurs either via supraspinal control of spinal networks or via the direct corticomotoneuronal pathway that is particularly important for movement of the hand. However, as in all other parts of the nervous system, information flows bidirectionally: in this case, there are relatively direct and strong inputs from sensory receptors in the limbs to motor cortical, and particularly corticomotoneuronal, outputs in the brain.

Activation of muscle and cutaneous sensory afferents from the fingers tends to evoke rather small muscle responses at short latencies compatible with transmission in spinal (reflex) pathways (Macefield et al., 1996). The largest responses are seen at somewhat longer latencies that are consistent with ascending transmission to the thalamus onwards to the cortex and then down again to the spinal cord via the corticospinal tract. Convincing evidence from both animals and humans has shown that these 'late' reflexes are indeed mediated by transcortical reflex pathways. To what extent similar transcortical reflexes also contribute to activation of more proximal arm muscles is still a matter of debate (Macefield et al., 1996; Fellows et al., 1996). As for the leg, convincing evidence has also been presented for a contribution by transcortical reflex pathways to late muscular responses in the TA muscle evoked by muscle and cutaneous sensory afferents. Similar strong evidence is not present for any other leg muscles and it is reasonable to doubt whether transcortical reflex pathways contribute to late reflex responses in leg muscles other than the TA muscle (Christensen et al., 2000). This is consistent with the pronounced corticospinal control of the TA muscle as compared to most other leg muscles.

Muscles that are best equipped with direct corticomotoneuronal projections thus appear to be equipped with the most potent transcortical reflexes. For these muscles, sensory afferents have little strength at spinal level, but much more so at the cortical level. Why is this? We have postulated that one of the functions of the corticomotoneuronal system is to provide flexible visual control of distal arm (and leg) muscles which bypasses much of the intrinsic spinal circuits. If so, then there is clearly a need to incorporate somatosensory information into this command as soon as the hand contacts an object that may be manipulated. Cutaneous and muscle spindle input can give information about the slipperiness and compressibility of a surface while spindles can signal unexpected contacts. Consistent with this idea, transcortical stretch reflexes are largest in both antagonist ankle muscles when the leg is on the ground rather than in the air (Christensen et al., 2000). Perhaps this is one function of the transcortical reflex loop. Effectively, CM cells may thus to some extent be seen as spinal neurones (propriospinal neurones?) which have been displaced through evolution to the cortex.

A second possible function of the transcortical loops is in coordinating reflex responses to disturbances of movement during free arm movement. Control of reaching is complex because of interaction torques between elbow and shoulder joints, which mean that sometimes patterns of muscle activity are more related to preventing unwanted interactions rather than producing active movement at a joint. Disturbances to one part of the limb therefore have implications for the control of all muscles in the limb, meaning that simple reflexes limited to the homonymous muscle are insufficient to compensate. Instead, what is required is a coordinated contraction of all the muscles in the limb. It appears that this is in fact what happens when perturbations are applied to one joint in a free arm pointing task. It seems as if reflexes recruit responses in many muscles and that these are modulated according to the state of the limb when the perturbation is applied.

Although there is good evidence that such corrections occur via a cortical pathway, it is not clear why this should be superior to multi-segmental interactions at a spinal level. One possibility is that control of interaction torques is solved by cortical mechanisms, so that if reflex corrections are to be organized appropriately, the simplest solution is to use the same mechanisms.

The unconscious, automatic nature of voluntary movements

It is possible to elicit transcortical reflexes in subjects without them noticing it or being aware of it. Obviously, activation of the primary motor cortex (at least by sensory input) is insufficient to elicit any conscious awareness of movement – not to speak of any experience of any intention linked to that movement. This is also confirmed by the multitude of observations of subjects who are able to react and even make rather elaborate choices in choice–reaction tasks without being aware of it (Christensen et al., 2008). Clearly, volition is not involved in any of these cases. So why is it that the motor cortex is always linked to voluntary movements? As we have pointed out here, the motor cortex may differ from the spinal cord in degree of flexibility and a larger possibility of integrating visual input, but otherwise there are no differences between the motor cortex and the spinal cord circuitries (after all CM cells project to motoneurones and receive sensory input much like any good old-fashioned interneurone) that could warrant a significantly different role in our conscious experience of control of the movements that we perform. What we are proposing is that it is not the degree of perceived volition which determines to what extent the motor cortex is involved in a given task, but rather the need for flexible visual control. Although locomotion is often said to be automatic (meaning not involving volition) whereas reaching is thought to be voluntary and at least much less automatic, the evidence suggests that it is only in our perception that this is the case. Reaching can be done without conscious awareness and without volition, and locomotion definitely involves activity in the motor cortex and does require volition for initiation, just like any other (voluntary) movement. In studies on gait, comparison is often made to tasks in which subjects perform static contraction of a muscle, and this is then called a voluntary contraction to emphasize that gait does not involve volition and so also likely not motor cortical involvement. We hope that we have made it clear that there is little to support this distinction between automatic and voluntary tasks. We need to consider the integration between supraspinal and spinal control centers for any specific task in order to understand how that task is controlled by the nervous system, and try to avoid putting into it volition and voluntary which in any case are terms that belong in philosophy or in specific inquiries aimed at unravelling the mechanisms of our cognitive abilities and our conscious experiences.

References

Af Klint R, Nielsen JB, Sinkjaer T, Grey MJ. (2009). Sudden drop in ground support produces force-related unload response in human overground walking. *J Neurophysiol.* 101(4):1705–12.

Alstermark B, Isa T, Pettersson LG, Sasaki S. (2007). The C3-C4 propriospinal system in the cat and monkey: a spinal pre-motoneuronal centre for voluntary motor control. *Acta Physiol (Oxf).* 189(2):123–40

Alstermark B, Ogawa J, Isa T. (2004). Lack of monosynaptic corticomotoneuronal EPSPs in rats: disynaptic EPSPs mediated via reticulospinal neurones and polysynaptic EPSPs via segmental interneurones. *J Neurophysiol.* 91(4):1832–9.

Alstermark B, Hultborn H, Jankowska E, Pettersson LG. (2010). Anders Lundberg (1920–2009). *Exp Brain Res.* 200(3–4):193–5.

Baker SN. (2011). The primate reticulospinal tract, hand function and functional recovery. *J Physiol.* 589: 5603–5312.

Brouwer B, Ashby P. (1990). Corticospinal projections to upper and lower limb spinal motoneurones in man. *Electroencephalogr Clin Neurophysiol.* 76(6):509–19.

Brouwer B, Ashby P. (1992). Corticospinal projections to lower limb motoneurones in man. *Exp Brain Res.* 89(3):649–54

Bucy PC, Keplinger JE, Siqueira EB. (1964). Destruction of the 'pyramidal tract' in man. *J Neurosurg.* 21:285–298.

Burke D, Hicks RG, Stephen JP. (1990). Corticospinal volleys evoked by anodal and cathodal stimulation of the human motor cortex. *J Physiol.* 425:283–99.

Capaday C, Ethier C, Brizzi L, Sik A, van Vreeswijk, C., Gingras D. (2009). On the nature of the intrinsic connectivity of the cat motor cortex: evidence for a recurrent neural network topology. *J Neurophysiol.* 102:2131–2141.

Christensen MS, Kristiansen L, Rowe JB, Nielsen JB. (2008). Action-blindsight in healthy subjects after transcranial magnetic stimulation. *Proc Natl Acad Sci U S A.* 105(4):1353–7

Christensen LO, Petersen N, Andersen JB, Sinkjaer T, Nielsen JB. (2000). Evidence for transcortical reflex pathways in the lower limb of man. *Prog Neurobiol.* 62(3):251–72

Crone C, Nielsen J. (1994). Central control of disynaptic reciprocal inhibition in humans. *Acta Physiol Scand.* 152(4):351–63

Crone C, Nielsen J, Petersen N, Ballegaard M, Hultborn H. (1994). Disynaptic reciprocal inhibition of ankle extensors in spastic patients. *Brain.* 117 (Pt 5):1161–8.

Dum RP, Strick PL. (2002). Motor areas in the frontal lobe of the primate. *Physiol Behav.* 77:677–682.

Farmer SF, Swash M, Ingram DA, Stephens JA. (1993). Changes in motor unit synchronization following central nervous lesions in man. *J Physiol.* 463:83–105.

Fellows SJ, Töpper R, Schwarz M, Thilmann AF, Noth J. (1996). Stretch reflexes of the proximal arm in a patient with mirror movements: absence of bilateral long-latency components. *Electroencephalogr Clin Neurophysiol.* 101(2):79–83.

Fetz EE, Cheney PD. (1987). Functional relations between primate motor cortex cells and muscles: fixed and flexible. *Ciba Found Symp.* 132:98–117.

Hicks TP, Onodera S. (2012). The mammalian red nucleus and its role in motor systems, including the emergence of bipedalism and language. *Prog Neurobiol.* 96(2):165–75.

Hultborn H. (2006). Spinal reflexes, mechanisms and concepts: from Eccles to Lundberg and beyond. *Prog Neurobiol.* 78(3–5):215–32.

Jankowska E. (2001). Spinal interneuronal systems: identification, multifunctional character and reconfigurations in mammals. *J Physiol.* 533(Pt 1):31–40

Jankowska E. (2008). Spinal interneuronal networks in the cat: elementary components. *Brain Res Rev.* 57(1):46–55

Jessell TM, Surmeli G, Kelly JS. (2011). Motor neurones and the sense of place. *Neuron* 72:419–424.

Lawrence DG, Kuypers HG. (1968a). The functional organization of the motor system in the monkey. I. The effects of bilateral pyramidal lesions. *Brain* 91:1–14.

Lawrence DG, Kuypers HG. (1968b). The functional organization of the motor system in the monkey. II. The effects of lesions of the descending brain-stem pathways. *Brain* 91:15–36.

Lemon RN. (2008). Descending pathways in motor control. *Annu Rev Neurosci.* 31:195–218.

Lemon RN. (1999). Neural control of dexterity: what has been achieved? *Exp Brain Res.* 128(1–2):6–12.

Macefield VG, Rothwell JC, Day BL. (1996). The contribution of transcortical pathways to long-latency stretch and tactile reflexes in human hand muscles. *Exp Brain Res.* 108(1):147–54.

Nielsen J, Kagamihara Y. (1992). The regulation of disynaptic reciprocal Ia inhibition during co-contraction of antagonistic muscles in man. *J Physiol.* 456:373–91.

Nielsen J, Petersen N. (1995). Changes in the effect of magnetic brain stimulation accompanying voluntary dynamic contraction in man. *J Physiol.* 484 (Pt 3):777–89.

Nielsen J, Petersen N, Deuschl G, Ballegaard M. (1993a). Task-related changes in the effect of magnetic brain stimulation on spinal neurones in man. *J Physiol.* 471:223–43.

Nielsen J, Crone C, Hultborn H. (1993b). H-reflexes are smaller in dancers from The Royal Danish Ballet than in well-trained athletes. *Eur J Appl Physiol Occup Physiol.* 66(2):116–21.

Nielsen J, Petersen N, Ballegaard M. (1995). Latency of effects evoked by electrical and magnetic brain stimulation in lower limb motoneurones in man. *J Physiol.* 484 (Pt 3):791–802.

Onodera S, Hicks TP. (2009). A comparative neuroanatomical study of the red nucleus of the cat, macaque and human. *PLoS One.* 4(8):e6623.

Palmer E, Ashby P. (1992). Corticospinal projections to upper limb motoneurones in humans. *J Physiol.* 448:397–412.

Pearson KG. (2008). Role of sensory feedback in the control of stance duration in walking cats. *Brain Res Rev.* 57(1):222–7

Perez MA, Cohen LG. (2009). The corticospinal system and transcranial magnetic stimulation in stroke. *Top Stroke Rehabil.* 16(4):254–69

Petersen NT, Pyndt HS, Nielsen JB. (2003). Investigating human motor control by transcranial magnetic stimulation. *Exp Brain Res.* 152(1):1–16.

Pierrot-Deseilligny E, Marchand-Pauvert V. (2002). A cervical propriospinal system in man. *Adv Exp Med Biol.* 508:273–279.

Powers RK, Binder MD. (2001). Input-output functions of mammalian motoneurones. *Rev Physiol Biochem Pharmacol.* 143:137–263.

Rothwell JC. (1997). Techniques and mechanisms of action of transcranial stimulation of the human motor cortex. *J Neurosci Methods* 74:113–122.

Rothwell JC. (2006). The startle reflex, voluntary movement, and the reticulospinal tract. *Suppl Clin Neurophysiol.* 58:223–231.

Rudomin P. (2002). Selectivity of the central control of sensory information in the mammalian spinal cord. *Adv Exp Med Biol.* 508:157–70

Sherrington CS. (1906). The Integrative Action of the Nervous System. New Haven: Yale University Press.

Sinkjaer T, Andersen JB, Ladouceur M, Christensen LO, Nielsen JB. (2000). Major role for sensory feedback in soleus EMG activity in the stance phase of walking in man. *J Physiol.* 523 Pt 3: 817–27.

16

ACUTE AND LONG-TERM NEURAL ADAPTATIONS TO TRAINING

Jacques Duchateau,[1] *Tibor Hortobágyi*[2] *and Roger M. Enoka*[3]

[1] LABORATORY OF APPLIED BIOLOGY, UNIVERSITÉ LIBRE DE BRUXELLES, BELGIUM
[2] CENTER FOR HUMAN MOVEMENT SCIENCES, UNIVERSITY MEDICAL CENTER, GRONINGEN, THE NETHERLANDS
[3] DEPARTMENT OF INTEGRATIVE PHYSIOLOGY, UNIVERSITY OF COLORADO, BOULDER, USA

Introduction

The purpose of this chapter is to discuss the contribution of neural changes to the adaptations of the neuromuscular system in response to practice and training. The main goal is to emphasize that these adaptations can be elicited with a variety of interventions, but that a more focused drive is necessary to optimize the improvements to a given task. To that end, the chapter addresses three topics: (1) the task specificity of muscle activation, (2) the acute adjustments and long-term neural adaptations to ballistic (fast) contractions, and (3) these same adjustments and adaptations to unilateral exercise.

Task specificity of muscle activation

The purposeful movements we perform are controlled by activating thousands of motor units (an ensemble that comprises a motor neuron, its axon and the muscle fibers that the axon innervates) located in the multiple muscles selected for the action. Because the activation characteristics vary with the requirements of a task, the adaptations elicited by practicing an action tend to be relevant only for that action (Jensen et al., 2005; Schubert et al., 2008; Vila-Chã et al., 2010). A strength-training intervention that increases the size and strength of a muscle, for example, will not improve performance on every task in which the muscle is engaged (Aagaard et al., 1996; Caserotti et al., 2008; Rutherford & Jones, 1986; Tracy and Enoka, 2006). Furthermore, physical-training interventions can increase muscle strength in the absence of an increase in muscle size (Bezerra et al., 2009; Maffiuletti & Martin, 2001; Oliveira et al., 2010) and the increases in muscle strength and motor function are typically much greater than the relative increase in muscle size (Häkkinen et al., 2000; Suetta et al., 2004).

These observations indicate that training-induced gains in performance can only be achieved with appropriate adaptations in the properties and activation of muscle (Häkkinen et al., 2003; Hortobágyi et al., 2001a; Jones & Rutherford, 1987; Maffiuletti & Martin, 2001; Norrbrand et al., 2008). The purpose of the first section of this chapter is to discuss three lesser-known examples of task conditions that modulate muscle activation during relatively modest levels of physical activity and thereby influence the outcomes that can be realized with a training intervention.

Light-load training

Although the classic approach to strength training is to prescribe a program that involves lifting heavy loads (Aagaard et al., 2002; Häkkinen et al., 2000), there is some concern that clinical populations may not tolerate such a vigorous regimen (Andersen et al., 2009; Petterson et al., 2009; Rahmann et al., 2009). As an alternative approach to strength training, some work has examined the effectiveness of training with light loads (<50 percent of maximum), especially as a countermeasure for the declines in muscle strength and function that accompany advancing age.

In contrast to the results achieved with the more traditional approaches to strength training (Bottaro et al., 2007; Caserotti et al., 2008; Holsgaard-Larsen et al., 2011; Miszko et al., 2003; Mueller et al., 2009), the outcomes from light-load training by older adults have been more equivocal for gains on functional tasks (Earles et al., 2001; Fatouros et al., 2005; Vincent et al., 2002). A period of several weeks of training limb muscles with a light load, for example, elicits modest increases in strength and either minimal or no improvement in performance on functional tasks (de Vreede et al., 2005; Manini et al., 2007; Seynnes et al., 2004; Tracy and Enoka, 2006). Furthermore, the practice of functional tasks has also produced mixed outcomes on motor function in older adults. For example, 20 weeks of Tai Chi exercises by older adults improved the strength of the knee extensors and force steadiness during submaximal isometric contractions (Christou et al., 2003), six practice sessions of a pegboard test improved force steadiness during submaximal isometric contractions involving index finger abduction but not the pinch grip (Marmon et al., 2011), and eight weeks of practicing a ball-rolling exercise with the fingers of one hand improved both pinch force steadiness and time to complete a pegboard test but not to influence grip strength (Ranganathan et al., 2001).

Taken together, these observations suggest that some features of the tasks being performed during both light-load training and training with functional tasks can elicit a broader spectrum of adaptations that depend less on the specific tasks being performed compared with the outcomes achieved with traditional strength training. A training method developed by Yasushi Koyama called 'Beginning Movement Load (BML) Training' provides a convincing example of this interaction. The training is performed on machines designed by Koyama to involve actions about multiple degrees of freedom, a load that is modulated by a patented gear system, the activation of muscles in a sequence that progresses from proximal to distal, muscle actions that comprise a passive lengthening phase followed immediately by an active shortening contraction, minimal coactivation of antagonist muscles and the involvement of trunk and proximal (shoulder and hip) muscles to establish a stable base of support (Figure 16.1; Koyama et al., 2010). In a randomized trial involving 24 older adults, Kobayashi et al. (in press) found that eight weeks of BML training with light loads (30 percent of the one-repetition maximum) on seven different BML machines that did not specifically target either the knee extensor or elbow muscle produced significant increases in the strength of the knee extensor muscles (32 percent) but not the elbow flexors (10 percent), improved the steadiness of submaximal isometric contractions with both the knee extensors and elbow flexors, and reduced performance times for ascending and descending stairs and one-legged balance, but not chair-rise time.

The improved steadiness during submaximal, isometric contractions (10 percent, 30 percent, and 65 percent of maximal voluntary contraction [MVC] force) for both the knee extensor and elbow flexor muscle groups suggests the training elicited adaptations that influenced the motor output from the spinal cord (Farina et al., 2009; Negro et al., 2009). Other studies have also found that functional training can improve force steadiness during isometric contractions performed by older adults (Christou et al., 2003; Marmon et al., 2011; Ranganathan et al., 2001). The results of a computational study suggest that changes in steadiness (force fluctuations) during isometric contractions are attributable to modulation of common low-frequency oscillations in motor neuron discharge rates (Dideriksen et al., 2012 press). In contrast to the results achieved with light-load training, strength training of the knee extensors does not improve force steadiness during isometric contractions (Bellew, 2002; Hortobágyi et al., 2001b; Manini et al., 2005; Tracy et al., 2004; Tracy & Enoka, 2006), which is consistent with the assertion that traditional strength-training approaches confer more specificity in the adaptations they elicit.

The results obtained by Kobayashi et al. (in press) also suggest that the light-load training on the BML machines improved the ability of proximal muscles to provide postural support. Although four of the machines exercised the upper body and three the lower body, six of the machines required the participant to maintain a stable posture while seated and one required this while standing. The decreases in performance times for the functional tests were only moderately associated with the various gains in MVC force and force steadiness, which suggests that other adaptations were responsible (Le Bozec & Bouisset, 2004; Rutherford & Jones, 1986).

Given that traditional strength training exhibited by older adults has no more than a modest influence on the postural control (Granacher et al., 2009) but balance training can elicit significant increases in the rate of force development (Gruber et al., 2007; Schubert et al., 2008), it might be speculated that BML training can improve postural control in older adults but this remains to be determined.

These studies indicate that light-load training involving non-planar actions about multiple degrees of freedom seems to be capable of eliciting functionally meaningful adaptations in older adults, perhaps by modulating muscle activation at a systemic level. These findings contrast with the more task-specific adaptations associated with traditional strength training.

Load compliance

The adaptations that can be elicited by physical training depend on the mechanical properties of the load encountered by the body parts involved. Loads can possess varying levels of compliance, elasticity, and viscosity that will each require unique activation strategies during the performance of an action (Darainy et al., 2004; Debicki & Gribble, 2004; Gomi & Osu, 1998; Krutky et al., 2010). For example, both the distribution of activity among synergist muscles and the responsiveness of stretch-reflex pathways differ with load compliance (Akazawa et al., 1983; Baudry & Enoka, 2009; Buchanan & Lloyd, 1995; Doemges & Rack, 1992a, 1992b; Perreault et al., 2008). As a consequence of these adjustments, motor unit activity varies with load compliance when an isometric contraction is sustained with the same net muscle torque (Enoka et al., 2011).

The influence of load compliance on muscle activation has been studied most thoroughly during the performance of fatiguing contractions. The typical approach has been to compare muscle activation when participants were required to sustain a submaximal (<30 percent MVC force) contraction with the same net muscle torque either by pushing against a rigid restraint (low compliance) or by supporting an inertial load (high compliance). The subject is required to maintain a constant force (force control) during the low-compliance task and to hold the limb in the same position (position control) during the high-compliance task.

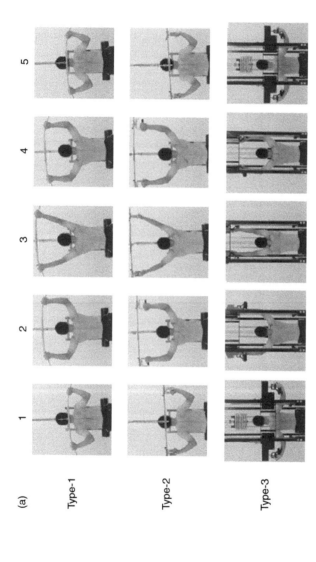

(a)

Type-1

Type-2

Type-3

1   2   3   4   5

Figure 16.1 The lat. pull-down exercise as performed with a load that was 30 percent of maximum on a conventional machine (Type 1) and one designed for Beginning Movement Load (BML) training (Type 3).

(a) Video images of five positions during the exercise as performed on three types of machines: 1 = beginning of shortening and beginning of shortening contraction; 5 = end of shortening contraction. The Type-1 machine confines angular displacements about the shoulder and elbow to a frontal place. The Type-2 machine allows some rotation about the pronation-supination of the forearm (dodge movement) in addition to the frontal plane motion. The Type-3 machine (BML) combines the frontal-plane and longitudinal rotation with rotation in a transverse plane (internal-external rotation) about the shoulder joint. A more detailed view of the BML machine can be found in Figure 1B in Koyama et al. (2010).

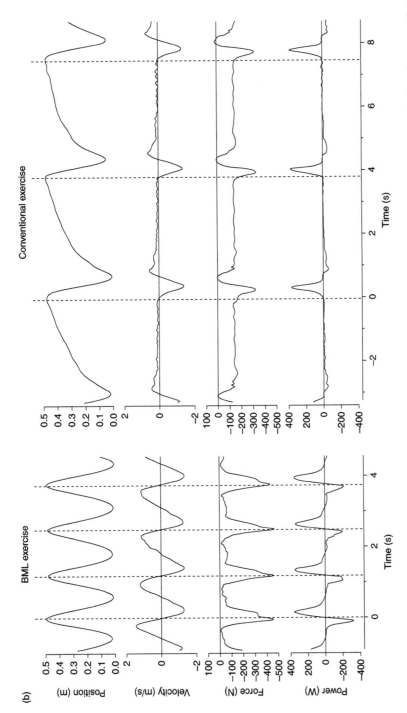

(b) Kinematics, force, and power for the bar during several repetitions of the lat. pull-down exercise as performed on the BML (left) and convential (right) machines. A zero position (top trace) indicates the minimal position of the bar (maximal position of the weight stack), a negative velocity denotes a downward displacement, a negative force corresponds to the subject pulling down on the weight stack, and positive and negative powers represents production (passive lengthening) and absorption (shortening contraction), respectively. Note that the force applied to the bar is close to zero for most of each repetition on the BML machine, but not the conventional machine. (Reproduced from Koyama et al., 2010.)

Position control is more challenging than force control when the target force is less than ~45 percent MVC force (Maluf et al., 2005; Rudroff et al., 2011), which results in a lesser endurance time for fatiguing contractions that require position control (Baudry et al., 2009b; Hunter et al., 2005, 2008; Klass et al., 2008b; Maluf et al., 2005; Rudroff et al., 2011).

EMG amplitude for the agonist muscles is identical at the onset of the two fatiguing contractions due to the muscles exerting the same net muscle torque, which means the greater difficulty associated with position control arises during the contractions. The difference is not attributable to antagonist coactivation as the relative increase in EMG amplitude for the agonist and antagonist muscles is similar during the two tasks (Enoka et al., 2011). Rather, position control is characterized by more rapid changes in motor unit activity in agonist muscles, including greater decreases in discharge rate, a more pronounced increase in the variability of discharge times, and a faster recruitment of additional motor units (Baudry et al., 2009b; Mottram et al., 2005; Rudroff et al., 2010). Furthermore, the modulation of motor neuron activity during position control appears to be more focused on spinal networks than on adjustments that influence the responsiveness of descending pathways (Baudry et al., 2009a; Klass et al., 2008b). It has been hypothesized that the major difference between the two conditions is attributable to a greater stretch-reflex responsiveness required to maintain a constant limb position while supporting a compliant load (Enoka et al., 2011).

The different demands of force and position control appear to be analogous to those associated with closed- and open-chain kinetic exercises used in rehabilitation. The classic distinction between these two types of exercises is whether (closed chain) or not (open chain) the feet or hands exert a force directly against the ground during the action (Fitzgerald, 1997). Closed-chain kinetic exercises are considered more functional because they tend to involve displacements about multiple degrees of freedom in contrast to an isolated action about a single degree of freedom (open-chain kinetic exercises). Much of the current interest in these two types of exercises focuses on which of the two approaches is more effective for rehabilitation after reconstruction of an anterior cruciate ligament (Fleming et al., 2005; Grodski & Marks, 2008; Tagesson et al., 2008). Anecdotal reports by patients, however, indicate that closed-chain kinetic exercises are more challenging, presumably because these exercises require the explicit control of body position rather than simply exerting a force against an external load. Nonetheless, the field has not yet examined the two types of exercises within the framework of force and position control.

Limb orientation

It is not possible to produce even an isolated action, such as flexing the elbow joint or abducting the index finger, by activating only the muscles that span the involved joint. In addition to generating the intended displacement, the nervous system must also activate muscles both to provide the appropriate postural support and to counteract unwanted torques. These requirements can limit the amount of activity that can be generated in a target muscle (Barry et al., 2008; Riley et al., 2008; Rudroff et al., 2007).

In the initial studies on the influence of load compliance on endurance time, Hunter et al. (2002) found that subjects could sustain an isometric contraction with the elbow flexors (forearm horizontal and upper arm vertical) at a target force of 15 percent MVC with force control for approximately twice as long as with position control (mean ± SD; 1402 ± 728 s vs. 702 ± 582 s). When the position of the arm was changed so that the upper arm was horizontal and the two tasks were compared, the difference in endurance time was less: time to failure during position control (high compliance) was 51 ± 26 percent of that for force control when the upper arm was vertical (Hunter et al., 2002), and 77 ± 21 percent when it was horizontal (Rudroff et al., 2005).

The reason for the difference in relative endurance times for the two arm postures is attributable to the greater postural demand on the shoulder muscles when the upper arm is in a vertical position. In this position, the upper arm was abducted from the trunk around 15 degrees and the mass supported from the wrist caused an extensor torque about the elbow joint and external-rotation torque about the shoulder joint. To keep the elbow joint at 90 degrees and the arm in the same position, subjects also had to activate rotator cuff muscles (supraspinatus, infraspinatus, teres minor) to counteract the secondary load (external rotation). Furthermore, the rate of increase in EMG activity in these muscles was greater during position control for this position (Rudroff et al., 2007). Consequently, the postural demands on the shoulder muscles exacerbated the influence of load compliance on endurance time of the elbow flexors.

In examining the interaction between load compliance and limb posture, it was also noted that the relative endurance time for the two types of fatiguing contractions (force and position control) could be influenced by the orientation of the forearm. When the fatiguing contractions were performed at a target force of 20 percent MVC with the upper arm horizontal, endurance time was briefer for position control (477 ± 276 s) than for force control (609 ± 250 s) when the forearm was supinated, but not when it was in a neutral position (478 ± 260 s and 476 ± 246 s, respectively) (Rudroff et al., 2005; 2007). Furthermore, the rate of increase in EMG amplitude for the elbow flexors (biceps brachii, brachialis, brachioradialis) was greater during position control than during force control when the forearm was supinated, whereas the rates of increase did not differ for the two tasks when the forearm was in a neutral position.

Although biceps brachii, brachialis and brachioradialis all contribute to an elbow flexor torque, they can produce opposing actions about the pronation–supination axis of the forearm (Zhang et al., 1998). Due to these interactions, changes in forearm orientation can engage inhibitory reflex pathways and thereby modulate the relative activation of these muscles (Naito et al., 1996; 1998). For example, spike-triggered stimulation of the radial nerve innervating brachioradialis reduced the discharge rate of motor units in biceps brachii, but this effect was greatest when the forearm was supinated and least when it was pronated (Barry et al., 2008). Thus, the modulation of biceps brachii EMG activity by feedback from brachioradialis varied with forearm orientation.

The functional significance of the inhibitory projection from brachioradialis to the motor units of biceps brachii was underscored by demonstrating an association between the strength of the pathway and the endurance time for a fatiguing contraction. Riley et al. (2008) found that three practice sessions of a fatiguing contraction requiring force control at a target force of 20 percent MVC with the elbow flexors increased endurance time (57.5 ± 67.5 percent), reduced the rate of increase in EMG amplitude for biceps brachii only, and decreased the strength of the inhibitory pathway from brachioradialis to biceps brachii. The decrease in reflex inhibition was the main predictor of the increase in endurance time ($r^2 = 0.60$). However, three practice sessions of a comparable fatiguing contraction requiring position control do not influence endurance time (Hunter et al., 2003), and no study has yet measured the strength of the inhibitory pathway during position control.

Nonetheless, some evidence suggests that manipulating the orientation of a limb about its longitudinal axis may influence the outcomes that can be achieved with training. BML exercises, for example, include a rotation about the longitudinal axis of the limb that is known as a dodge movement (Koyama et al., 2010) as it is shown for the 'lat. pull-down' exercise in

Figure 16.1. The forearm is rotating from a supinated position at the beginning of the short-ening contraction (Position 3 in Figure 16.1A) to a pronated position at the end of the lift (Position 5 in Figure 16.1A). Koyama (1999) asserts that the dodge movement facilitates non-planar motion that augments the range of motion achieved during the exercise and enables a sequence of muscle activity that is similar to unconstrained throwing-striking actions. Certainly, the EMG activity for shoulder, elbow, and forearm muscles during an elbow flexion action differs when the forearm is pronated or supinated and whether it rotates into supination or pronation (Tsuchiya & Suzuki, 2011), as would be expected from the reflex pathways between these muscles. However, the functional significance of the dodge move-ment remains to be demonstrated.

In contrast to the better-known attributes of task specificity in training adaptations, the three examples discussed in the first section of this chapter (light-load training, load compli-ance, and limb orientation) demonstrate how the requirements of a task can modulate muscle activation during relatively modest levels of physical activity. It is likely that these features of training tasks can be exploited to augment the transfer of performance gains to activities of daily living.

Adaptations underlying performance gains of ballistic contractions

This section examines the acute and long-term neural adaptations in response to repeated performances of a specific action. The task selected to illustrate the types of adaptations that can occur is a muscle contraction performed as fast as possible, classically referred to as 'ballistic contraction'. A ballistic contraction is usually defined as a contraction performed as fast as possible that is followed immediately by relaxation. It is characterized by a high rate of force development and a brief time to reach peak force (~100–200 ms; Desmedt & Godaux, 1977; Van Cutsem et al., 1998). Because it takes longer (>300 ms) to reach the maximal force a muscle can produce, the peak force achieved during a ballistic contraction is less than the maximal muscle force (Aagaard et al., 2002; Duchateau & Baudry, 2011). Ballistic actions are often encountered in activities of daily life and in various sport movements. For example, to catch a falling object or respond to a sudden perturbation with a limb or the whole body, a rapid reaction is necessary and movements used in such conditions require a rapid contraction of the involved muscles. In sporting activities, the rate of force development more accurately predicts performance in explosive movement (for example, a smash in volleyball or in tennis) than maximal force.

Ballistic contraction performed under different initial conditions

Muscle activation during a ballistic contraction, as assessed by surface EMG, is characterized by a brief and highly synchronized burst of activity. The magnitude of the activation and hence the force produced by a muscle depend on the number of motor units activated (motor unit recruitment) and the rates at which motor neurons discharge action potentials (rate coding). Despite some initial suggestions to the contrary, the recruitment order of motor units (size principle) during rapid (ballistic) contractions is similar to that during slower contractions (Duchateau & Enoka, 2011). However, the relative contributions of recruitment and rate coding to the force exerted by a muscle vary with contraction speed. Slow contractions can be characterized by the progressive activation of motor units to an upper limit of recruitment that is about 80-90 percent of the maximum force in most limb muscles (De Luca et al., 1982; Kukulka & Clamann, 1981; Van Cutsem et al., 1997). However, motor units are recruited at much lower forces (recruitment thresholds) during ballistic contractions. For example, most motor units in the tibialis anterior are recruited during a ballistic contraction when the force is only one-third of maximum (Desmedt & Godaux, 1977). Furthermore, the reduction in recruitment threshold of motor unit with an increase in contraction speed is greater for slow-contracting muscles (e.g., soleus) than for fast-contracting muscles (e.g., masseter; Desmedt & Godaux, 1978).

The increase in muscle force beyond the upper limit of motor unit recruitment is entirely attributable to rate coding. In most muscles, the maximal rate at which motor units discharge action potentials during strong isometric contractions is 30-60 Hz (Duchateau & Enoka, 2011). In contrast, the instantaneous discharge rates at the initiation of a rapid contraction often reach values of 60-120 Hz in untrained subjects with few action potentials discharged (~1–6) by each motor unit (Desmedt & Godaux, 1977; Van Cutsem & Duchateau, 2005).

The discharge characteristics of single motor units during ballistic contractions can be modulated by the initial conditions under which the task is performed. For example, when a ballistic contraction is superimposed on a submaximal isometric contraction (~25 percent MVC), the maximal rate of force development is 16 percent less than that achieved during a ballistic contraction performed without prior activation (Van Cutsem & Duchateau, 2005). Although the electromechanical delay is shortened, the EMG activity during the super-imposed ballistic contraction is less synchronized due to a significant decline in the average discharge rate (−22 percent) and in the percentage of motor units (6.2 vs. 15.5 percent) that exhibit double discharges (doublet) at brief intervals (≤ 5 ms between successive action potentials). However, the reduction in the rate of force development during the superimposed ballistic contraction is abolished with the addition of a brief silent period (classically called premotor silent period) at the transition between the sustained and ballistic actions. The silent period presumably enables motor neurons to achieve a non-refractory state leading to a more synchronous recruitment and a greater discharge rate of motor units during the subsequent ballistic contraction (Tsukahara et al., 1995; Van Cutsem & Duchateau, 2005).

These studies demonstrate that ballistic contractions require a specific activation of units compared with the recruitment and rate coding characteristics underlying slow (e.g. maximal isometric) contractions. The work further indicates that the initial conditions can modulate the capacity for rate coding and hence the performance of a fast voluntary contraction.

Acute effects of practicing ballistic contractions

The acquisition of a new motor skill is a gradual process that requires many repetitions over a long period of time. Although simple movements can be learned quickly, more specialized skills typically require years of intensive practice. Skill acquisition involves modifications in the central nervous system, and the need for prolonged practice presumably reflects the importance of the learning processes that produce these modifications (Adkins et al., 2006; Jensen et al., 2005; Perez et al., 2004; Wolpaw, 1997).

Few studies have investigated the influence of a single practice session on the ballistic contractions performed with muscles that span a single joint. Among them, Lee and colleagues (2010) studied the influence of practicing ballistic contractions that involved abduction of the index finger as fast as possible. Subjects performed 300 movements in sets of ten repetitions. Practice of the ballistic task significantly improved peak abduction acceleration by 64 percent and 93 percent after 150 and 300 repetitions, respectively. The increase in acceleration was accompanied by an increase in excitability of the corticospinal track as assessed by the size of the motor evoked potential (MEP) induced by transcranial magnetic stimulation (TMS). The MEP amplitude in the first dorsal interosseous of the trained hand increased significantly by 43 percent and 63 percent after 150 and 300 repetitions, respectively.

Similar observations were reported for abductor pollicis brevis (Rogasch et al., 2009) and for first dorsal interosseous (Hinder et al., 2011) of both young and elderly subjects. However, Hinder et al. (2011) reported increases in excitability, assessed by MEP changes in response to TMS, together with decreases in short-interval intracortical inhibition tested by paired TMS pulses in both young and elderly adults, whereas Rogasch et al. (2009) only observed an increase in excitability in young subjects and no corticospinal changes in elderly adults despite significant improvements in kinematics after a single session of practice. These results underscore the use-dependent corticomotor plasticity of the CNS network controlling muscle activity during ballistic contractions performed by healthy young subjects. However, the contrasting results on the influence of a single practice session on corticospinal excitability in elderly adults, despite significant improvement in task performance, suggest either a reduced capacity to modulate cortical activity or the involvement of different loci in the adaptation.

In agreement with previous studies (Carroll et al., 2008; Jensen et al., 2005; Karni et al., 1995; Muellbacher et al., 2001; 2002; Pascual-Leone et al., 1994; Rogasch et al., 2009), the observation reported by Lee and colleagues (2010) indicates that the motor cortex is implicated in human motor learning and that improvement in ballistic actions depends on adaptations in the motor cortex. To assess the stability of the improvements in the ballistic contractions after a single practice session, low-frequency repetitive TMS (rTMS) was delivered for 15 min. over the motor cortex hot-spot for first dorsal interosseous to cause transient depression (~30 min.) in cortical excitability (Muellbacher et al., 2000). The practice-induced increases in peak finger acceleration and MEP amplitude were reduced (13 percent and 28 percent, respectively) following rTMS of the motor cortex, which indicates that rTMS disrupts the performance gains in ballistic actions (see also Baraduc et al., 2004; Muellbacher et al., 2002). This suggestion is consistent with the observation of an increased MEP in response to TMS without change by more direct stimulation of the descending tract (brainstem and cervical stimulations) after a session of ballistic actions (Muellbacher et al., 2001). Together, these studies emphasize that the primary motor cortex is involved in encoding elementary aspects of motor learning and that several practice sessions are needed to secure the acquisition of a new motor task.

Training effects of ballistic contractions

Several studies have shown that different types of training can increase the rate of force development during a ballistic contraction. For example, a few weeks of strength training with heavy loads (Aagaard et al., 2002; Andersen et al., 2009) or light loads (Gruber et al., 2007; Häkkinen et al., 1985; Van Cutsem et al., 1998) and sensorimotor training (Gruber et al., 2007) are effective in improving the performance of ballistic actions. Although such training methods can enhance the contractile kinetics of the trained muscle (Duchateau & Hainaut, 1984), most of the adaptations occur within the nervous system (Aagaard, 2003; Carroll et al., 2011; Duchateau et al., 2006; Van Cutsem et al., 1998). For example, the surface EMG activity of the agonist muscles at the onset of the ballistic contraction is increased and more synchronized after training (Aagaard et al., 2002; Gruber & Gollhofer, 2004; Gruber et al., 2007; Van Cutsem et al., 1998). It has also been reported that the intent to contract a muscle as fast as possible, rather than the actual movement velocity itself, is a major determinant of the velocity-dependent training response (Behm & Sale, 1993).

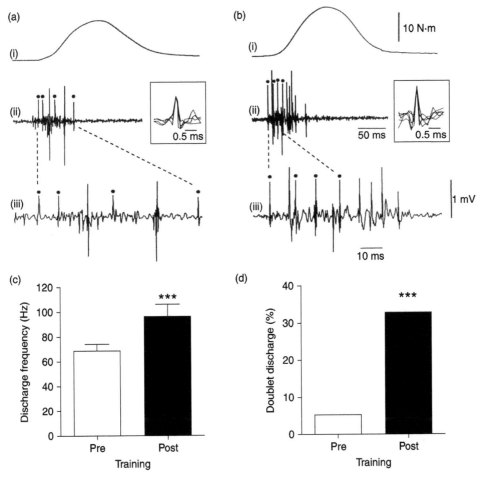

Figure 16.2 Comparison of motor units behavior from the tibialis anterior during ballistic contractions of similar force levels (~40 percent MVC), before (a) and after (b) 3 months of dynamic training. Traces correspond to the ankle dorsiflexors force (i) and intramuscular EMG plotted at slow (ii) and fast (iii) speeds. Panel (a) shows the typical discharge pattern of a motor unit in untrained muscle, consisting in a brief time elapse between the first two spikes followed by longer interspike periods. Panel (b) illustrates the usual motor unit behavior in trained muscle showing that the high onset of instantaneous discharge frequency is maintained during the subsequent spikes. The asterisks indicate the discharge frequency of the same motor unit and their traces are superimposed with an extended display in the insert. In (c) the histograms illustrate the mean maximal discharge frequency, averaged for the first three interspike intervals, during ballistic contractions for the whole motor units population analyzed before and after training. In (d) the histograms illustrate the percentage of motor units from the total population analyzed that displayed doublet discharges of ≤ 5 ms, before and after training. ★★★Denote significant difference (P<0.001) before and after training. Data are from Van Cutsem et al. (1998).

Van Cutsem and colleagues (1998) found that dynamic training can change motor unit behavior during ballistic contractions. In this study, initial discharge rates of motor units in the tibialis anterior were compared before and after three months of training with ballistic contractions of the dorsiflexors muscles against a moderate load (30–40 percent MVC).

Although no change was observed in the recruitment order of motor units, the average instantaneous discharge rate of the first four action potentials increased from 69 Hz before training to 96 Hz after training. In addition to the increase in discharge rate for the first interspike interval, each motor unit was able to maintain a greater rate during successive discharges and training increased the number of motor units (5 to 33 percent) that discharged with doublets ($\leq$ 5 ms). Thus, the increase in the rate of force development during the ballistic contraction was mainly achieved by an adaptation in motor unit discharge rate.

The specificity of the adaptations achieved with ballistic actions has been investigated in a few studies. For example, Gruber and colleagues (2007) compared two modalities of training the triceps surae muscles: ballistic contractions with loads of 30–40 percent MVC and sensorimotor (balance) training that comprised postural tasks on various devices (wobbling board, spinning top, soft mat and cushion). Four weeks of training led to a greater gain in maximal rate of force development after ballistic training (48 percent) than sensorimotor training (14 percent), whereas MVC force remained unchanged after both types of training. Furthermore, there was an increase in EMG activity during the first 100 ms of activation in both the soleus and gastrocnemius muscles after ballistic training, but not after sensorimotor training. These observations suggest that the two training modalities induced different neural adaptations and that the gains were greater for the type of training that was more similar to the test task.

Despite evidence suggesting a major role for neural mechanisms in the adaptations elicited by ballistic training, the loci of the adaptations along the neuraxis are unknown (Carroll et al., 2011; Duchateau et al., 2006; Duchateau & Baudry, 2011). The changes that can be evoked in the neural circuits with training appear to be produced by adaptations located at both supraspinal and spinal levels.

Supraspinal level

Despite the observation that several weeks of skill training can increase corticospinal excitability (Jensen et al., 2005; Pascual-Leone et al., 1995; Perez et al., 2004), strength training does not seem to be accompanied by similar adaptations (Carroll et al., 2001; Jensen et al., 2005). In contrast, Beck and colleagues (2007) reported that MEP size increased in the tibialis anterior after four weeks of ballistic training of the dorsiflexor muscles against a load of 30–40 percent of 1-repetition maximum (RM). This adaptation was not associated with changes in either short-interval intracortical inhibition or intracortical facilitation. However, a subsequent study by the same group on the triceps surae muscles (Schubert et al., 2008) used a collision technique of subthreshold TMS and Hoffmann reflex (H-reflex – an equivalent of the stretch reflex evoked by electrical stimulation of the muscle Ia afferents) to demonstrate a training-related modulation of the fast corticospinal projections for the trained task. Because spinal excitability tested by the H-reflex did not change in either study, the interpretation was that the changes most likely occurred at a supraspinal level. Although there is no clear explanation for the increase in MEP size after ballistic training, the results further demonstrated that the adaptations were specific to the task because no changes were observed at rest or in non-trained movements (Schubert et al., 2008).

The task specificity of these adaptations likely explains the discrepant results from different studies. Indeed, a training effect may only be observable when the training and testing conditions are similar. Moreover, the absence of an increase in corticospinal excitability after training (see Hortobágyi et al., 2009) does not mean that no adaptation occurred at the supraspinal level. As shown from functional imaging (fMRI) studies in humans, an enlarged region of activation in the sensorimotor cortex has been observed after strength training

(Farthing et al., 2007) and adjustments in cortical activity have been reported during motor learning (Doyon & Benali, 2005). However, these changes subsequently shifted to subcortical structures, such as the basal ganglia, the cerebellum and the limbic structures, for building and consolidation of the motor memory trace (Doyon & Benali, 2005). Functionally, such a shift in the activated regions of the brain from cortical to subcortical system will render the task more automatic (Floyer-Lea & Matthews, 2004; Puttemans et al., 2005). Consistent with the observations on humans, animal studies have also demonstrated that motor training can induce structural and functional adaptations within several motor areas, including basal ganglia, cerebellum, and red nucleus (Adkins et al, 2006).

Spinal level

In contrast to the few studies that have reported an increased amplitude of the H-reflex by strength training when tested during muscle activation (Aagaard et al., 2002; Duclay et al., 2008; Holtermann et al., 2007; Lagerquist et al., 2006; Scaglioni et al., 2002; Taube et al., 2007), a four-week program of ballistic and sensorimotor training did not change the amplitude of the H-reflex (Schubert et al., 2008; Taube et al., 2007). This observation suggests that spinal reflex circuits can adapt specifically to the particular requirements of the task. However, it can be challenging to determine the functional significance of adaptations in spinal reflex pathways. For example, training with rapid isometric contractions increases reciprocal inhibition at the onset of a fast contraction. However, the magnitude of the increase in disynaptic reciprocal inhibition was not correlated with the increase in the rate of force development (Geertsen et al., 2008). Nonetheless, the results were interpreted to indicate that the adaptation depressed activation of antagonist muscles to facilitate faster movements. The absence of a change in disynaptic reciprocal inhibition at rest suggests that training did not induce any plastic changes in the pathway itself. Rather, the larger facilitation of the pathway at the onset of the ballistic contraction after training was explained by a parallel increase in supraspinal drive to the motor neurons of the ankle dorsiflexors and the inhibitory interneurons responsible for reciprocal inhibition of the antagonist motor neurons (Geertsen et al., 2008).

In addition to a possible role for changes in synaptic connectivity in the spinal network, other evidence also indicates that the properties of the motor neurons themselves can be altered by long-term physical activity (Gardiner, 2006). In this context, Beaumont and Gardiner (2003) reported that endurance training in rats changed the biophysical properties of motor neurons, which resulted in a more hyperpolarized resting membrane potential, increased threshold for spike initiation, and faster rise times for antidromic spikes. These adaptations, which likely reflect alterations in ionic conductance of the membrane of the motor neuron, can modify the recruitment thresholds and discharge patterns of the neurons. Indirect evidence suggests that long-term ballistic training may cause adaptations at a motor neuron level that may explain the greater occurrence of doublet discharges after three months of ballistic training (Van Cutsem et al., 1998). Doublet discharges, which are often observed at the onset of motor neuron discharge in response to fast current injection (Baldissera et al., 1987), may be related to changes in the intrinsic properties of the motor neuron (Granit et al., 1963). In human studies, doublet discharges of short intervals (3.2 – 4 ms) in response to TMS have also been reported (Bawa & Lemon, 1993; Gandevia & Rothwell, 1987). A possible explanation for doublets is that motor neurons capable of producing a double discharge may undergo a state of increased depolarization (delayed depolarization) that occurs during the falling phase of the action potential (Garland & Griffin, 1999). The motor neuron remains responsive to strong synaptic input during this period, and thereby susceptible to reach voltage threshold for the generation of a second action potential. This discussion is consistent with the reduced discharge rate and the number of doublet discharges during ballistic contraction observed during ageing (Klass et al., 2008a), as it is associated with a prolongation of the after-hyperpolarization phase of motor neurons in human (Piotrkiewicz et al., 2007).

In summary, the neural adaptations that accompany acute and long-term changes in the rate of force development during a ballistic contraction are diverse and appear to be distributed along the entire neuraxis. Furthermore, the reported experimental data indicate that when the practice task corresponds to a ballistic contraction, the improvement in motor performance and the magnitude of neural adaptation are superior to unspecific training modality.

Adaptations in the nervous system to unilateral exercise

This section will examine the acute activation patterns and chronic adaptations in the nervous system produced by forceful unilateral muscle contractions in intact humans. A brief section will review the clinical relevance of these adaptations in the context of cross-education. Several reviews have discussed critical aspects of unilateral exercise not presented here, including laterality, skill transfer, and bimanual control (Carroll et al., 2006; Carson, 2005; Farthing, 2009; Hortobágyi, 2005; Swinnen & Wenderoth, 2004; Teixeira, 2000; Zhou, 2000).

Acute cortical and spinal responses to forceful unilateral muscle contractions

According to classical physiology texts, an effortful voluntary contraction of a muscle group on one side of the body is the result of the contralateral activation of the primary motor cortex (M1) and other neural structures involved in movement generation. This view holds that the motor command that generates unilateral movement is confined to the target muscles on one side of the body. Recent data modify this view. Behavioral observations show that during a *forceful* muscle contraction it is nearly impossible to contract only the target muscle on one side because EMG electrodes placed on adjacent muscles and on the homologous muscle pair in the opposite limb reveal as much as about 20 percent associated activity (Hortobágyi et al., 1997; Zijdewind et al., 2006). When healthy volunteers performed forceful unilateral muscle contractions repeatedly, coactivation of the homologous muscle pair increased and it became increasingly difficult to willfully suppress this activity (Hortobágyi et al., 2011). As muscular fatigue sets in, activity in non-target muscles also tends to increase (Zijdewind & Kernell, 2001). These observations suggest that the unilaterally intended motor command activates the ipsilateral M1 through some paths in the brain, spine, or both or, more likely, there is actually a concurrent bilateral activation of the motor areas in the two hemispheres as the effort is planned and executed and the intra- and interhemispheric effects act on the already ongoing activity (Figure 16.3).

Changes in brain activation

TMS and imaging studies revealed that brain activation is not confined to the hemisphere generating the command for the unilateral effort. A standard experimental arrangement is that volunteers contract a muscle on one side of the body and during these efforts TMS pulse is delivered to M1 that is ipsilateral to the contraction and the MEPs recorded from the resting contralateral muscle. Following the publication of the first TMS data in the ipsilateral M1 during unilateral muscle contractions (Hess et al., 1986), there were inconsistent findings concerning the activation of the ipsilateral M1 (Chiappa et al., 1991; Samii et al, 1997). When

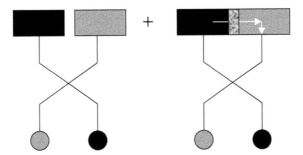

Figure 16.3 A mechanism of cross education-evoked adaptations. The left panel shows, in a posterior view, that activation of the left motor areas (black rectangle) produces contraction of a muscle on the right side (black circle). Imaging studies revealed that unilateral motor activity simultaneously also activates, to a lesser extent, most motor areas in the right hemisphere, illustrated by the gray rectangle, producing in some cases unintended associated activity in the left hand (gray circle). This repeated bilateral activation of the right, 'non-involved' motor areas in a unilateral exercise program lowers the activation threshold of the affected cells, as shown by twitch-interpolation studies, using TMS and peripheral nerve stimulation. The right panel shows that concurrently and in combination with the bilateral activation (denoted by the + sign), activity-dependent changes modulate excitability of interhemispheric paths, including interhemispheric inhibition, through the corpus callosum (shaded box in middle), allowing the 'untrained' motor areas to increase neural drive to the untrained muscle when called for during a muscle contraction. This model does not show and postulate a role for ipsilateral paths to contribute to cross-education.

healthy adults contracted the right abductor pollicis brevis, there was a monotonic increase in MEP size recorded from the resting left abductor pollicis brevis as contraction intensity increased from 40 to 100 percent of maximum (Muellbacher et al., 2000). However, there also was a doubling in F wave amplitude, a measure of spinal excitability. These data suggested the involvement of spinal and cortical mechanisms for the increase in the activation of the left abductor pollicis brevis (Muellbacher et al., 2000).

However, the absence of changes in MEPs recorded in resting muscles produced by transelectrical stimulation (TES), a type of stimulation which activates the corticospinal axons directly without affecting the corticomotor cells, and changes in MEPs evoked by TMS led to the conclusion that the facilitation seen in the resting hand muscle was due to cortical mechanisms (Tinazzi & Zanette, 1998). Since these early papers, several studies have confirmed the involvement and activation of the ipsilateral M1 during strong unilateral muscle contractions and found this activation to increase with increasing contraction intensity without increases in measures of spinal excitability (Howatson et al., 2011; Perez & Cohen, 2008; Stinear et al., 2001). Although cross-education intervention studies seem to produce spatially specific force transfer to the homologous muscle pair after unilateral training (Hortobágyi et al., 2011; but see Lee et al., 2009; Sariyildiz et al., 2011), MEPs in the right, antagonist wrist extensors also increased substantially during left wrist flexion. This lack of spatial specificity is probably related to the propinquity of the cortico-corticoneurons driving wrist flexors and extensors in primates (Cheney et al., 1985) and humans (Graziano et al., 2004).

These increases in activation follow a task-specific pattern. When healthy humans contract their arm or foot muscles corticospinal excitability in the contracting muscle is significantly lower during lengthening than during shortening contractions (Abbruzzese et al., 1994; Gruber et al., 2009; Nordlund et al., 2002; Sekiguchi et al, 2001; Sekiguchi et al., 2003; Uematsu et al., 2010; Duclay et al., 2011), presumably due to differences in excitability of the motor neuron pool (Abbruzzese, et al., 1994; Duclay et al., 2011) and reductions in maximum excitation level and in the gain of the corticospinal tract during lengthening compared to shortening contractions (Sekiguchi et al., 2001; Sekiguchi et al., 2003). Such task-specific differences in excitability in the contracting muscle predict task-specific activation of the ipsilateral M1. The recruitment curve in the resting vs. active muscle was indeed task-specific but opposite in direction: MEPs were about 20 percent greater in the resting right flexor carpi radialis while the left wrist flexors performed forceful lengthening compared with shortening contractions (Howatson et al., 2011) even though the right hand was motionless and the associated EMG activity (Zijdewind et al., 2006) in the right flexor carpi radialis was minimal and similar when the left wrist flexors lengthened and shortened. Thus, afferent feedback from the right wrist muscles played a minimal role in modulating MEP size. In addition, experiments using strong lengthening and shortening contractions in combination with increasing TMS intensities from 1.0 to 1.8 of resting motor threshold revealed that the ipsilateral M1 had a greater capacity to respond to TMS during shortening than during lengthening contractions of the left wrist flexors. The data suggest that interhemispheric effects are capable to task specifically modulate the corticospinal responses to TMS (Avanzino et al., 2007; Lee et al., 2007; Perez & Cohen, 2009).

The nature of the ipsilateral M1's involvement has been also probed using paired-pulse TMS experiments. Strikingly, these studies revealed a decrease and even a near abolishment of short-interval intracortical inhibition (Howatson et al., 2011; Muellbacher et al., 2000; Perez & Cohen, 2008), being a measure of the activity in the intracortical GABAergic inhibitory interneurons (Kujirai et al., 1993) and an increase in intracortical facilitation (Howatson et al., 2011). Lessened intracortical inhibition and heightened intracortical facilitation suggest an activity-dependent modulation of short-interval intracortical inhibition and intracortical facilitation. These changes might contribute to facilitate corticospinal excitability in the ipsilateral M1 during forceful unilateral contractions (Ziemann et al., 1996), a mechanism that is supported by the strong correlation between the short-interval intracortical inhibition and the size of the maximal MEP (Perez & Cohen, 2008). Because muscle contraction at increasing intensities reduces and relaxation increases short-interval intracortical inhibition in the ipsilateral M1 (Buccolieri et al., 2004; Perez & Cohen, 2008), the modulation of activation must arise within the ipsilateral M1. There is not only less and less intracortical inhibition in the ipsilateral M1 as contraction intensity increases, there is also unique task specificity in the activation pattern: the diminishment is substantially greater during lengthening than during shortening contractions (Howatson et al., 2011). In combination with other data, the possibility exists that instead of the increased afferent input, differences in the descending motor command and activation of brain areas that link function of the motor cortices during muscle lengthening vs. shortening may cause the contraction-specific modulation of ipsilateral motor cortical output.

In addition to intracortical processes, the state of connections linking the two motor hemispheres also changes during a forceful unilateral muscle contraction. Fibers traversing in the corpus callosum between the two M1s are sparse and exclusively excitatory glutamatergic (Salerno & Georgesco, 1996; Wahl & Ziemann, 2008). These neurons have excitatory effects on cortical cells and also on inhibitory interneurons, producing a net inhibitory effect (Baumer et al., 2006; Di Lazzaro et al., 1999; Hanajima et al., 2001; Sohn et al., 2003). The excitatory effect precedes the inhibitory effect during a unilateral muscle contraction

(Innocenti, 1986; Naito et al., 1970; Salerno & Georgesco, 1996). In contrast, the sensory areas have virtually no connections with the contralateral M1 or sensory areas as shown in temporal lobe epilepsy patients who underwent invasive EEG monitoring with bilaterally implanted subdural electrodes (Terada et al., 2012).

Forceful unilateral muscle contractions produce activity-dependent changes in interhemispheric inhibition, albeit these TMS experiments are methodologically complex. In one experimental condition in which the size of the conditioning pulse was adjusted, the magnitude of interhemispheric inhibition from the active to the 'non-active' hemisphere decreased and these reductions did not correlate with the MEP facilitation measured by recruitment curves (Perez & Cohen, 2008). This lack of correlation suggests that interhemispheric inhibition has less effect than short-interval intracortical inhibition on MEP facilitation in the 'non-active' M1 during forceful unilateral muscle contraction. In other words, the interaction of inter- and intracortical processes seems to play an important role in the tuning of the corticospinal motor output arising from the 'non-active', ipsilateral M1 during strong, unilateral contractions (Figure 16.3).

Changes in interhemispheric inhibition are not only activity- but also task-dependent. Interhemispheric inhibition in M1 ipsilateral to the contraction diminished during shortening and further diminished during forceful lengthening contractions (Howatson et al., 2011). The cause of the task-dependent variation in interhemispheric inhibition is unknown. EEG, EMG, and TMS studies showed that the descending command is different between lengthening and shortening muscle contractions and perhaps this difference can modulate interhemispheric inhibition in a task-specific fashion (Fang et al., 2001; Grabiner & Owings, 2002; Gruber et al., 2009). As an extension of these studies, when subjects were not reminded to keep the 'resting' arm free of the associated EMG activity during a strong unilateral muscle contraction, the source of the associated activity was M1 ipsilateral to contraction. This observation was based on the similar inhibition in the silent period measured in the associated and matched voluntary activity, suggesting that the two M1s are probably activated concurrently during a forceful unilateral contraction (Zijdewind et al., 2006) (Figure 16.3).

The interhemispheric effects would require 10-12 ms to exert their influence from the active to the 'non-active' M1. Thus, instead of an overflow from the active hemisphere to the 'non-active' hemisphere to generate the activity in M1 ipsilateral to the forceful contraction, a model is more likely in which the two hemispheres are actually activated simultaneously (Mayston et al., 1999) as the unilateral effort is planned and executed and the intra- and interhemispheric effects would act on the already ongoing activity (Figure 16.3). EEG (Crone et al., 1998; Urbano et al., 1998), fMRI (Cramer et al., 1999; Dettmers et al., 1996; Kim et al., 1993; Rao et al., 1993; Sehm et al., 2009; Shibasaki et al., 1993), MEG (Cheyne & Weinberg, 1989; Cheyne et al., 1995; Salmelin et al., 1995), MRCP (Mayer et al., 1999), and positron emission tomography (PET) (Kawashima et al., 1998) data strongly suggest simultaneous excitation of both M1s during unilateral finger movements. Interhemispheric excitation and inhibition, while acting on short-interval intracortical inhibition and intracortical facilitation (Perez & Cohen, 2009), would then shape the ongoing activity in M1 ipsilateral to the effort.

Changes at the spinal level

There are no direct pathways interconnecting motor neurons on the left and the right side but there are rich and complex circuits that comprise multiple interneurons at the segmental level and reflex actions that act on the descending command. When intact humans perform a strong unilateral muscle contraction, instead of an observable facilitation akin to the Jendrassik maneuver, there is a large (Uematsu et al., 2010) and long-lasting depression of the H-reflex (Bikmullina et al., 2006; Hortobágyi et al., 2003; Howatson et al., 2011; Slivko & Bogutskaya, 2008). The depression is intensity-dependent and in several subjects the strong unilateral efforts virtually abolish the H-reflex. While MEPs, short-interval intracortical inhibition and interhemispheric inhibition exhibit specificity according to muscle lengthening and short-ening, the H-reflex does not. The depression in the H-reflex probably does not reflect a decrease in excitability of the motor neuron pool because these strong unilateral contractions did not affect cervicomedullary MEPs which activate motor neurons through the cortico-spinal tract (Hortobágyi et al., 2003; Taylor & Gandevia, 2004). Thus, there was no inhibi-tion or disfacilitation of the motor neurons. In addition, the H-reflex was still depressed despite some background EMG activity that would be expected to facilitate the reflex, suggesting that H-reflex depression associated with contralateral voluntary contraction occurs via a presynaptic mechanism. Curiously, the depression reverts to about 40 percent facilita-tion in percutaneous electrical stimulation-evoked muscle contractions. Coupled with the observation that stimulation of the median nerve at the wrist or of a purely cutaneous nerve in the forearm, at non-painful levels, weakly facilitates the H-reflex, the results suggest that the descending activity associated with a contralateral voluntary contraction produces a strong presynaptic inhibition of Ia afferents. Results from recent studies have confirmed the task-specific facilitation of MEPs and reductions in intracortical inhibition, and not task-specific depression of the H-reflex (Howatson et al., 2011; Uematsu et al., 2010). The H-reflex depression, a unique and prolonged form of presynaptic inhibition, has also been observed during unilateral rhythmic upper and lower limb movements (Carson et al., 2004; McIlroy et al., 1992), suggesting a role for afferent inputs to the motor neurons in the resting limb. The functional significance of the H-reflex depression is unclear, and perhaps it is a mechanism to shape the input to the 'inactive' muscle.

Chronic adaptations to unilateral exercise and its clinical relevance

It is somewhat paradoxical that a poorly controlled single-subject motor learning experiment described in the psychomotor literature more than 100 years ago (Scripture et al., 1894) set the stage for what is now a mainstream effort to decipher how the nervous system controls voluntary movement generated on one side of the body, and how such unilateral voluntary movements bring about functionally meaningful changes in the inactive, contralateral homologous muscles. Here we focus on only one particular experimental arrangement in which healthy young subjects exercise one limb muscle with near maximal voluntary effort and electrical stimulation-evoked muscle contractions (Bezerra et al., 2009; Hortobágyi et al., 1999). Meta analyses of the randomized clinical trials included in reviews showed that the magnitude of cross-education is about 8 percent (Carroll et al., 2006; Munn et al., 2004). However, the mechanisms mediating these small, statistically, and clinically significant effects remain elusive because the repeated use of TMS and imaging in training studies is methodologically challenging, complex, and cost-prohibitive.

Cortical mechanisms

There are only a handful of studies that have examined the mechanisms of how repeated forceful muscle contractions produce the cross-education effects. The magnitude of cross-education seems to be direction-specific (Farthing, 2009; Farthing et al., 2005), corroborating data in the motor learning literature (Latash, 1999; Teixeira, 2000). Curiously, cross-education occurred only after training the dominant hand in right-handed individuals but there was no cross-education when right-handed individuals trained the left, non-dominant wrist (Farthing et al., 2005). Dozens of studies reported that the magnitude of cross-education is proportional to strength gains of the trained arm (Zhou, 2000). However, the normally coupled strength gains became dissociated when right-handed individuals exercised the non-dominant left wrist, its strength improved 42 percent but produced non-significant 9 percent cross-education (Farthing et al., 2005). Transfer of the ability to produce maximal voluntary force may occur in both directions if the task is unfamiliar to both limbs and transfer may be asymmetrical if the task is less familiar to the non-dominant limb; however, this interpretation of the data has not been tested (Farthing et al., 2005). Neuroanatomical differences between the two hemispheres could in part underlie the greater cross-education from right to left arms because in right-handed individuals the left central sulcus lies deeper, the volume of left M1 globus pallidus is larger (Amunts et al., 1996; Kooistra & Heilman, 1988) and the whole-brain functional connectivity with left M1 is more extensive than with right M1 (Guye et al., 2003).

So far, there exists only one fMRI study that examined changes in brain activation after resistance training in the trained and untrained brain (Farthing et al., 2007). After 24 sessions of maximal isometric ulnar deviation in the right arm of right-handed subjects maximal strength increased 45 percent. The unusual movement used in the training also produced a large cross-education effect of 47 percent. As if a forceful practice of ulnar deviation represented an elementary form of motor learning, fMRI analyses revealed an enlarged region of activation in contralateral sensorimotor cortex, left temporal lobe, and the cerebellum during muscle contractions with the untrained left arm. These data are consistent with a model of adaptations in which the unilateral ulnar deviation practice concurrently activated both hemispheres (Figure 16.3). Such activation normally produces associated (EMG) activity in the homologous muscle pair without overt movement (Zijdewind et al., 2006). A speculative conclusion is that when subjects performed ulnar deviation test contractions with the untrained arm after training, the brain areas that became repeatedly activated by the six-week-long practice were available and the motor programs within these brain areas were accessible to the untrained wrist muscles.

Another and perhaps concurrent mechanism could be that the unilateral training modified the inhibition–excitation balance in the interhemispheric connections. Evidence for this mechanism comes from a TMS study in which healthy volunteers exercised first dorsal interosseous of the dominant right hand and performed 1000 voluntary index finger abduction at 80 percent of the maximum force over 20 sessions (Hortobágyi et al., 2011). Training increased maximal strength 50 percent in the trained and 28 percent in the untrained, left first dorsal interosseus. Corticospinal excitability in the untrained M1, measured with TMS before and after every fifth session, increased 6 percent at rest. These changes, like those in short-interval intracortical inhibition and intracortical facilitation, did not correlate with cross-education. When weak and strong TMS pulses were delivered to the untrained M1 on a background of a weak and strong muscle contraction of the trained right first dorsal interosseous muscle, excitability of the untrained M1 increased significantly. Interhemispheric inhibition decreased 9 percent acutely within sessions and 31 percent chronically over 20 sessions and these chronic reductions progressively became more strongly associated with cross-education. Furthermore, these TMS data correlate with imaging data (Farthing et al., 2007): in both cases there was an increase in excitability of the untrained M1 measured not at rest but during muscle contraction. The TMS data seems to suggest that both the capacity and sensitivity of the untrained M1 increased to drive the untrained muscle during the test contractions after training. The strong correlations between reductions in interhemispheric inhibition and the magnitude of cross-education suggest a functional role for interhemispheric inhibition mediating cross-education but it is unclear how these reductions would affect mirror activity and bilateral coordination. Data from acute limb immobilization and conditioning of the opposite hemisphere with paired-associative stimulation seem to confirm the use-dependent modulation of hemispheric balance in humans (Avanzino et al., 2011; Shin & Sohn, 2011).

Other approaches to invoke a role for cortical mechanisms in cross-education provide a somewhat conflicting picture with the data presented above. For example, mental imagery produced cross-education inconsistently (Farthing et al., 2007; Yue & Cole, 1992), most likely because mental imagery does not always produce strength increases in the target muscle itself (Herbert et al., 1998; Ranganathan et al., 2004; Zijdewind et al., 2003). An increase in voluntary activation of the muscles exhibiting cross-education would suggest the involvement of a cortical mechanism. Although TMS and electrical twitch interpolation techniques, which represent the completeness of motor unit activation in the forcefully contracted muscle, revealed increases in muscle activation in the cross-educated muscle, the magnitude of changes in muscle activation and cross-education were not correlated or did not correlate (Lee et al., 2009; Shima et al., 2002), making it difficult to assign functional significance to changes in muscle activation.

Spinal mechanisms

Even though some initial randomized trials suggested that spinal mechanisms mediate cross-education brought about by electromyostimulation (Cabric & Appell, 1987) it is now clear that such a procedure has profound effects on multiple brain structures in addition to spinal circuits (Hortobágyi & Maffiuletti, 2011). Recent studies nonetheless found no conclusive evidence for adaptations in the H-reflex at rest or during muscle contraction (Fimland et al., 2009; Lagerquist et al., 2006). The results from these studies are also difficult to interpret because the changes in H-reflex amplitude were dissociated in the trained and the cross-educated muscle: one study showed an increased presynaptic inhibition in the trained muscle but no changes in the cross-educated plantarflexors, whereas another study showed no changes in H-reflex amplitude in either muscle (Fimland et al., 2009; Lagerquist et al., 2006). Although there was a significant increase in the amplitude of the volition or V-wave in the cross-educated soleus muscle, this measure does not provide conclusive evidence whether the changes in reflex excitability were the result of increased descending motor drive, presynaptic inhibition, or both (Fimland, et al., 2009). Unlike the data obtained in the plantarflexors, subtle changes in specific H-reflex parameters revealed adaptations in the Ia afferent system in the cross-educated dorsiflexors following five weeks of high-intensity isometric dorsiflexion training but these cross-education effects and changes in reflex properties were not correlated (Dragert & Zehr, 2011). Together, these data suggest that there are probably segmental adaptations, but current experimental approaches are insensitive to detect these changes in these segmental mechanisms associated with cross-education. Future studies should perhaps examine the most profound effect noted in acute, cross-sectional studies, i.e. the changes in the amount of depression of the H-reflex (Hortobágyi et al., 2003) and reciprocal inhibition in the resting limb conditioned with parametric increases in unilateral muscle of contractions of the opposite limb.

Clinical implications

Clinical benefits of cross-education are often discussed, doubted (Carroll et al., 2006), and remain virtually unexplored and unproven. Early clinical trials, without proper experimental controls, reported reductions in functional losses in orthopedic patients' arm function (Stromberg, 1988). Ulnar deviation strength training of the dominant right arm for three weeks attenuated strength loss induced by immobilization of the non-dominant left arm in healthy, injury-free patients (Farthing et al., 2009). A group of subjects trained the right arm and wore a cast on the left arm (Cast-Train). A second group wore a cast and did not train, and the control group received no treatment. Remarkably, the immobilized arm of the cast-train group did not lose muscle strength or size. However, the muscle strength and size in the immobilized arm of the casted group decreased. Thus, cross-education had a protective effect on strength and muscle loss (Farthing et al., 2009). In a follow-up study unilateral handgrip training of the free limb maintained strength (1 percent) in the opposite immobilized limb and this maintenance effect was associated with an increase in the volume of activation in the contralateral M1 as shown by fMRI (Farthing et al., 2011). In contrast, for the non-training control group, there was a significant decrease in strength of the immobilized limb (11 percent) without changes in the contralateral M1 activation. It is unclear if the fMRI-detected increase in M1 activation could be causally linked to the increase in motor unit activation measured in other studies by TMS- and peripheral nerve stimulation-evoked twitch interpolation (Lee et al., 2009; Shima et al., 2002). It is also unknown if there is a difference in muscle and brain activation between subjects who had a cast and trained vs. those who did not train. Previous immobilization studies showed that, at least in the leg muscles, there could be as much as about 20 percent of maximum activation of muscles inside the cast (Hortobágyi et al., 2000) and it is known that even such low-intensity strength training can improve maximal voluntary strength (Laidlaw et al., 1999).

It remains to be seen how resting and immobilized muscles could avoid strength-loss and even muscle atrophy. The untested possibility exists that forceful unilateral muscle contractions produce simultaneous bilateral activation of the two motor cortices, as shown by numerous imaging studies, and this repeated activation of the 'uninvolved' motor areas lowers the activation threshold of the affected cells, as shown by twitch-interpolation studies (Figure 16.3). Concurrently and in combination with the bilateral activation, activity-dependent changes modulate excitability of interhemispheric paths, including interhemispheric inhibition, through the corpus callosum, allowing the 'untrained' motor areas to increase neural drive to the untrained muscle.

In total, although much remains unknown about the mechanism mediating strength-maintaining clinical effects produced by cross-education, there are some encouraging data suggesting that such effects exist in certain conditions. Because recovery from an actual injury or surgery is multifactorial, future studies will have to replicate these recent clinical findings not in healthy individuals but in *patients* who suffered a fracture, have weakness due to an ACL-reconstruction surgery, are recovering from a rotator cuff tear surgery or a stroke.

General conclusion

The examples reviewed in this chapter describe how the activation signal sent to the muscles depends on the requirements of the task being performed. Small changes in the biomechanical constraints of the task can have a substantial influence on the performance of the task. In contrast to the well-known concept of task specificity in training adaptations, it is interesting to note that training with a light load is capable of eliciting neural adaptations, at least in untrained, sedentary young and elderly adults, and these adaptations can improve the performance both in the task that was practiced and in other functional tasks, such as activities of daily living. Even unilateral muscle contractions involve multiple elements in the central nervous system, and the performance adaptations are not confined to the muscle targeted by the intervention and can include adaptations in the homologous muscle pair in the opposite limb. Nonetheless, the second section of the chapter underscores that a more focused drive from the nervous system to muscle is necessary to optimize performance gain of a given task. With such interventions, neural adaptations begin after a single practice session of repeated actions but several sessions are needed to stabilize the performance gains.

Adaptations at a number of sites within the nervous system appear capable of contributing to training-induced changes. Despite some progress in recent years, the loci and the precise nature of the adaptations to practice and training, the principles that govern the specificity of the adaptations, and the mechanisms that underlie the transfer to other movements are still largely unknown. Such knowledge has major implications for clinical practice and sports performance.

References

Aagaard, P. (2003) 'Training-induced changes in neural function', *Exercise and Sport Sciences Reviews*, 31: 61–7.

Aagaard, P., Simonsen, E.B., Andersen, J.L., Magnusson, P. and Dyhre-Poulsen, P. (2002) 'Increased rate of force development and neural drive of human skeletal muscle following resistance training', *Journal of Applied Physiology*, 93: 1318–26.

Aagaard, P., Simonsen, E.B., Trolle, M., Bangsbo, J. and Klausen, K. (1996) 'Specificity of training velocity and training load on gains in isokinetic knee joint strength', *Acta Physiologica Scandinavica*, 156: 123–9.

Abbruzzese, G., Morena, M., Spadavecchia, L. and Schieppati, M. (1994) 'Response of arm flexor muscles to magnetic and electrical brain stimulation during shortening and lengthening tasks in man', *Journal of Physiology*, 481: 499–507.

Adkins, D.L., Boychuk, J., Remple, M.S. and Kleim, J.A. (2006) 'Motor training induces experience-specific patterns of plasticity across motor cortex and spinal cord', *Journal of Applied Physiology*, 101: 1776–82.

Akazawa, K., Milner, T.E. and Stein, R.B. (1983) 'Modulation of reflex EMG and stiffness in response to stretch of a human finger muscle', *Journal of Neurophysiology*, 49: 16–27.

Amunts, K., Schlaug, G., Schleicher, A., Steinmetz, H., Dabringhaus, A., Roland, P.E., et al. (1996) 'Asymmetry in the human motor cortex and handedness', *Neuroimage*, 4: 216–22.

Andersen, L.L., Andersen, J.L., Suetta, C., Kjaer, M., Søgaard, K. and Sjøgaard, G. (2009) 'Effect of contrasting physical exercise interventions on rapid force capactiy of chronically painful muscles', *Journal of Applied Physiology*, 107: 1413–19.

Avanzino, L., Bassolino, M., Pozzo, T. and Bove, M. (2011) 'Use-dependent hemispheric balance', *Journal of Neurosciences*, 31: 3423–8.

Avanzino, L., Teo, J.T. and Rothwell, J.C. (2007) 'Intracortical circuits modulate transcallosal inhibition in humans', *Journal of Physiology*, 583: 99–114.

Baldissera, F., Campadelli, P., and Piccinelli, L. (1987) 'The dynamic response of cat gastrocnemius motor units investigated by ramp-current injection into their motoneurones', *Journal of Physiology*, 387: 317–30.

Baraduc, P., Lang, N., Rothwell, J.C. and Wolpert, D.M. (2004) 'Consolidation of dynamic motor learning is not disrupted by rTMS of primary motor cortex', *Current Biology*, 14: 252–6.

Barry, B.K., Riley, Z.A., Pascoe, M.A. and Enoka, R.M. (2008) 'A spinal pathway between synergists can modulate activity in human elbow flexor muscles', *Experimental Brain Research*, 190: 347–59.

Baudry, S. and Enoka, R.M. (2009) 'Influence of load type on presynaptic inhibition of Ia afferent input onto two synergist muscles', *Experimental Brain Research*, 199: 83–8.

Baudry, S., Jordan, K. and Enoka, R.M. (2009a) 'Heteronymous reflex responses in a hand muscle when maintaining constant finger force or position at different contraction intensities', *Clinical Neurophysiology*, 120: 210–17.

Baudry, S., Rudroff, T., Pierpoint, L. and Enoka, R.M. (2009b) 'Load type influences motor unit recruitment in biceps brachii during a sustained contraction', *Journal of Neurophysiology*, 102: 1725–35.

Baumer, T., Bock, F., Koch, G., Lange, R., Rothwell, J.C., Siebner, H.R., et al. (2006) 'Magnetic stimulation of human premotor or motor cortex produces interhemispheric facilitation through distinct pathways', *Journal of Physiology*, 572: 857–68.

Bawa, P. and Lemon, R.N. (1993) 'Recruitment of motor units in response to transcranial magnetic stimulation in man', *Journal of Physiology*, 471: 445–64.

Beaumont, E. and Gardiner, P.F. (2003) 'Endurance training alters the biophysical properties of hind-limb motoneurons in rats', *Muscle & Nerve*, 27: 228–36.

Beck, S. Taube W., Gruber M., Amtage F., Gollhofer A. and Schubert M. (2007) 'Task-specific changes in motor evoked potentials of lower limb muscles after different training interventions', *Brain Research*, 1179: 51–60.

Behm, D.G. and Sale, D.G. (1993) 'Intended rather than actual movement velocity determines velocity-specific training response', *Journal of Applied Physiology*, 74: 359–68.

Bellew, J.W. (2002) 'The effect of strength training on control of force in older men and women', *Aging Clinical and Experimental Research*, 14: 35–41.

Bezerra, P., Zhou, S., Crowley, Z., Brooks, L. and Hooper, A. (2009) 'Effects of unilateral electromy-ostimulation superimposed on voluntary training on strength and cross-sectional area', *Muscle & Nerve* 40: 430–7.

Bikmullina, R. Kh., Rozental, A.N. and Pleshchinskii, I.N. (2006) 'Modulation of soleus H-reflex during dorsal and plantar flexions in the human ankle joint', *Human Physiology*, 32: 593–8.

Bottaro, M., Machado, S.N., Nogueira, W., Scales, R. and Veloso, J. (2007) 'Effect of high versus low-velocity resistance training on muscular fitness and functional performance in older men', *European Journal of Applied Physiology*, 99: 257–64.

Buccolieri, A., Abbruzzese, G. and Rothwell, J.C. (2004) 'Relaxation from a voluntary contraction is preceded by increased excitability of motor cortical inhibitory circuits', *Journal of Physiology*, 558: 685–95.

Buchanan, T.S. and Lloyd, D.G. (1995) 'Muscle activity is different for humans performing static tasks which require force control and position control', *Neuroscience Letters*, 128: 61–4.

Cabric, M. and Appell, H.J. (1987) 'Effect of electrical stimulation of high and low frequency on maximum isometric force and some morphological characteristics in men', *International Journal of Sports Medicine*, 8: 256–60.

Carroll, T.J., Riek, S. and Carson, R.G. (2001) 'Corticospinal responses to motor training revealed by transcranial magnetic stimulation', *Exercise and Sport Sciences Reviews*, 29: 54–9.

Carroll, T.J., Lee, M., Hsu, M. and Sayde, J. (2008) 'Unilateral practice of a ballistic movement causes bilateral increases in performance and corticospinal excitability', *Journal of Applied Physiology*, 104: 1656–64.

Carroll, T.J., Selvanayagam, V.S., Riek, S. and Semmler, J.G. (2011) 'Neural adaptations to strength training: moving beyond transcranial magnetic stimulation and reflex studies', *Acta Physiologica*, 202: 119–40.

Carroll, T.J., Herbert, R.D., Munn, J., Lee, M. and Gandevia, S.C. (2006) 'Contralateral effects of unilateral strength training: evidence and possible mechanisms', *Journal of Applied Physiology*, 101: 1514–22.

Carson, R.G. (2005) 'Neural pathways mediating bilateral interactions between the upper limbs', *Brain Research Review*, 49: 641–62.

Carson, R.G., Riek, S., Mackey, D.C., Meichenbaum, D.P., Willms, K., Forner, M., et al. (2004) 'Excitability changes in human forearm corticospinal projections and spinal reflex pathways during rhythmic voluntary movement of the opposite limb', *Journal of Physiology*, 560: 929–40.

Caserotti, P., Aagaard, P., Larsen, J.B. and Puggaard, L. (2008) 'Explosive heavy-resistance training in old and very old adults: changes in rapid muscle force, strength and power', *Scandinavian Journal of Medicine and Science in Sports*, 18: 773–82.

Cheney, P.D., Fetz, E.E. and Palmer, S.S. (1985) 'Patterns of facilitation and suppression of antagonist forelimb muscles from motor cortex sites in the awake monkey', *Journal of Neurophysiology*, 53: 805–20.

Cheyne, D. and Weinberg, H. (1989) 'Neuromagnetic fields accompanying unilateral finger movements: pre-movement and movement-evoked fields', *Experimental Brain Research*, 78: 604–12.

Cheyne, D., Weinberg, H., Gaetz, W. and Jantzen, K.J. (1995) 'Motor cortex activity and predicting side of movement: neural network and dipole analysis of pre-movement magnetic fields', *Neuroscienci Letters*, 188: 81–4.

Chiappa, K.H., Cros, D., Day, B., Fang, J.J., Macdonell, R. and Mavroudakis, N. (1991) 'Magnetic stimulation of the human motor cortex: ipsilateral and contralateral facilitation effects', *Electroencephalography and Clinical Neurophysiology*, Suppl, 43: 186–201.

Christou, E.A., Yang, Y. and Rosengren, K.S. (2003) 'Taiji training improves knee extensor strength and force control in older adults', *Journal of Gerontology*, 58A: 763–6.

Cramer, S.C., Finklestein, S.P., Schaechter, J.D., Bush, G. and Rosen, B.R. (1999) 'Activation of distinct motor cortex regions during ipsilateral and contralateral finger movements', *Journal of Neurophysiology*, 81: 383–7.

Crone, N.E., Miglioretti, D.L., Gordon, B., Sieracki, J.M., Wilson, M.T., Uematsu, S., et al. (1998) 'Functional mapping of human sensorimotor cortex with electrocorticographic spectral analysis. I. Alpha and beta event-related desynchronization', *Brain*, 121: 2271–99.

Darainy, M., Malfait, N., Gribble, P.L., Towhidkhah, F. and Ostry, D.J. (2004) 'Learning to control arm stiffness under static conditions', *Journal of Neurophysiology*, 92: 3344–50.

Debicki, D.B. and Gribble, P.L. (2004) 'Inter-joint coupling strategy during adaptation to novel viscous loads in human arm movement', *Journal of Neurophysiology*, 92: 754–65.

De Luca, C.J., LeFever, R.S., McCue, M.P. and Xenakis, A.P. (1982) 'Behaviour of human motor units in different muscles during linearly varying contractions', *Journal of Physiology*, 329: 113–28.

Desmedt, J.E. and Godaux, E. (1977) 'Ballistic contractions in man: characteristic recruitment pattern of single motor units of the tibialis anterior muscle', *Journal of Physiology*, 264: 673–93.

Desmedt, J.E. and Godaux, E. (1978) 'Ballistic contractions in fast or slow human muscles: discharge patterns of single motor units', *Journal of Physiology*, 285: 185–96.

Dettmers, C., Ridding, M.C., Stephan, K.M., Lemon, R.N., Rothwell, J.C. and Frackowiak, R.S. (1996) 'Comparison of regional cerebral blood flow with transcranial magnetic stimulation at different forces', *Journal of Applied Physiology*, 81: 596–603.

de Vreede, P.L., Samson, M.M., Van Meeteren, N.L.U., Duursma, S.A. and Verhaar, H.J.J. (2005) 'Functional-task exercise versus resistance strength exercise to improve daily function in older women: a randomized, controlled trial', *Journal of the American Geriatric Society*, 53: 2–10.

Dideriksen, J.L., Negro, F., Enoka, R.M. and Farina, D. (2012) 'Motor unit recruitment strategies and muscle properties determine the influence of synaptic noise on force steadiness', *Journal of Neurophysiology*, 107: 3357–69.

Di Lazzaro, V., Oliviero, A., Profice, P., Insola, A., Mazzone, P., Tonali, P., et al. (1999) 'Direct demonstration of interhemispheric inhibition of the human motor cortex produced by transcranial magnetic stimulation', *Experimental Brain Research*, 124: 520–4.

Doemges, F. and Rack, P.M. (1992a) 'Changes in the stretch reflex of the human first dorsal interosseus muscle during different tasks', *Journal of Physiology*, 447: 563–73.

Doemges, F. and Rack, P.M. (1992b) 'Task-dependent changes in the response of human wrist joints to mechanical disturbance', *Journal of Physiology*, 447: 575–85.

Doyon, J. and Benali, H. (2005) 'Reorganization and plasticity in the adult brain during learning of motor skills', *Current Opinion in Neurobiology*, 15: 161–7.

Dragert, K. and Zehr, E.P. (2011) 'Bilateral neuromuscular plasticity from unilateral training of the ankle dorsiflexors', *Experimental Brain Research*, 208: 217–27.

Duchateau, J. and Baudry, S. (2011) 'Training adaptation of the neuromuscular system', in P.V. Komi (ed.) *Neuromuscular Aspects of Sport Performance* (pp. 216–53), Oxford: Wiley-Blackwell.

Duchateau, J. and Enoka, R.M. (2002) 'Neural adaptations with chronic activity patterns in able-bodied humans', *American Journal of Physical Medicine & Rehabilitation*, 81: S17–27.

Duchateau, J. and Enoka, R.M. (2011) 'Human motor unit recordings: origins and insight into the integrated motor system', *Brain Research*, 1409: 42–61.

Duchateau, J. and Hainaut, K. (1984) 'Isometric or dynamic training: differential effects on mechanical properties of a human muscle', *Journal of Applied Physiology*, 56: 296–301.

Duchateau, J., Semmler, J.G. and Enoka, R.M. (2006) 'Training adaptations in the behavior of human motor units', *Journal of Applied Physiology*, 101: 1766–75.

Duclay, J., Martin, A., Robbe, A. and Pousson, M. (2008) 'Spinal reflex plasticity during maximal dynamic contractions after eccentric training', *Medicine and Science in Sports and Exercise*, 40: 722–34.

Duclay, J., Pasquet B, Martin A and Duchateau J. (2011) 'Specific modulation of corticospinal and spinal excitabilities during maximal voluntary isometric, shortening and lengthening contractions in synergist muscles', *Journal of Physiology*, 589: 2901–6.

Earles, D.R., Judge, J.O. and Gunnarsson, O.T. (2001) 'Velocity training induces power-specific adaptations in highly functioning older adults', *Archives of Physical Medicine and Rehabilitation*, 82: 872–8.

Enoka, R.M., Baudry, S., Rudroff, T., Farina, D., Klass, M. and Duchateau, J. (2011) 'Unraveling the neurophysiology of muscle fatigue', *Journal of Electromyography and Kinesiology*, 21: 208–19.

Fang, Y., Siemionow, V., Sahgal, V., Xiong, F. and Yue, G.H. (2001) 'Greater movement-related cortical potential during human eccentric versus concentric muscle contractions', *Journal of Neurophysiology*, 86: 1764–72.

Farina, D., Holobar, A., Merletti, R., Enoka, R.M. (2009). Decoding the neural drive to muscles from the surface electromyogram. *Clinical Neurophysiology* 121, 1616–23.

Farthing, J.P. (2009) 'Cross-education of strength depends on limb dominance: implications for theory and application', *Exercise and Sport Sciences Reviews*, 37: 179–87.

Farthing, J.P., Borowsky, R., Chilibeck, P.D., Binsted, G. and Sarty, G.E. (2007) 'Neuro-physiological adaptations associated with cross-education of strength', *Brain Topography*, 20: 77–88.

Farthing, J.P., Chilibeck, P.D. and Binsted, G. (2005) 'Cross-education of arm muscular strength is unidirectional in right-handed individuals', *Medicine and Science in Sports and Exercise*, 37: 1594–1600.

Farthing, J.P., Krentz, J.R. and Magnus, C.R. (2009) 'Strength training the free limb attenuates strength loss during unilateral immobilization', *Journal of Applied Physiology*, 106: 830–6.

Farthing, J.P., Krentz, J.R., Magnus, C.R., Barss, T.S., Lanovaz, J.L., Cummine, J., et al. (2011) 'Changes in functional magnetic resonance imaging cortical activation with cross education to an immobilized limb', *Medicine and Science in Sports and Exercise*, 43: 1394–1405.

Fatouros, I.G., Kambas, A., Katrabasas, I., Nikolaidis, K., Chatzinikolaou, A., Leontsini, D., et al. (2005) 'Strength training and detraining effects on muscular strength, anaerobic power, and mobility of inactive older men are intensity dependent', *British Journal of Sports Medicine*, 39: 776–80.

Fimland, M.S., Helgerud, J., Solstad, G.M., Iversen, V.M., Leivseth, G. and Hoff, J. (2009) 'Neural adaptations underlying cross-education after unilateral strength training', *European Journal of Applied Physiology*, 107: 723–30.

Fitzgerald, G.K. (1997) 'Open versus closed kinetic chain exercise: issues in rehabilitation after anterior cruciate ligament reconstructive surgery', *Physical Therapy*, 77: 1747–54.

Fleming, B.C., Okensdahl, K. and Beynnon, B.D. (2005) 'Open- or closed-chain kinetic chain exercises after anterior cruciate ligament reconstruction?', *Exercise and Sport Sciences Reviews*, 33: 134–40.

Floyer-Lea, A. and Matthews, P.M. (2004) 'Changing brain networks for visuomotor control with increased movement automaticity', *Journal of Neurophysiology*, 92: 2405–12.

Gandevia, S.C. and Rothwell, J.C. (1987) 'Knowledge of motor commands and the recruitment of human motoneurons', *Brain*, 110: 1117–30.

Gardiner, P.F. (2006) 'Changes in alpha-motoneuron properties with altered physical activity levels', *Exercise and Sport Sciences Reviews*, 34: 54–8.

Garland, S.J. and Griffin, L. (1999) 'Motor unit double discharges: statistical anomaly or functional entity?', *Canadian Journal of Applied Physiology*, 24: 113–30.

Geertsen, S.S., Lundbye-Jensen, J. and Nielsen, J.B. (2008) 'Increased central facilitation of antagonist reciprocal inhibition at the onset of dorsiflexion following explosive strength training', *Journal of Applied Physiology*, 105: 915–22.

Gomi, H. and Osu, R. (1998) 'Task-dependent viscoelasticity of human multijoint arm and its spatial characteristics for interaction with environments', *Journal of Neuroscience*, 18: 8965–78.

Grabiner, M.D. and Owings, T.M. (2002) 'EMG differences between concentric and eccentric maximum voluntary contractions are evident prior to movement onset', *Experimental Brain Research*, 145: 505–11.

Granacher, U., Gruber, M. and Gollhofer, A. (2009) 'Resistance training and neuromuscular performance in seniors', *International Journal of Sports Medicine*, 30: 652–7.

Granit, R., Kernell, D. and Shortess, G.K. (1963) 'Quantitative aspects of repetitive firing of mammalian motoneurones, caused by injected currents', *Journal of Physiology*, 168: 911–31.

Graziano, M.S., Patel, K.T. and Taylor, C.S. (2004) 'Mapping from motor cortex to biceps and triceps altered by elbow angle', *Journal of Neurophysiology*, 92: 395–407.

Grodski, M. and Marks, R. (2008) 'Exercises following anterior cruciate ligament reconstructive surgery: biomechanical considerations and efficacy of current approaches', *Research in Sports Medicine*, 16: 75–96.

Gruber, M. and Gollhofer, A. (2004) 'Impact of sensorimotor training on the rate of force development and neural activation', *European Journal of Applied Physiology*, 92: 98–105.

Gruber, M., Gruber S.B., Taube W., Schubert M., Beck S.C. and Gollhofer A. (2007) 'Differential effects of ballistic versus sensorimotor training on rate of force development and neural activation in humans', *Journal of Strength and Conditioning Research*, 21: 274–82.

Gruber, M., Linnamo, V., Strojnik, V., Rantalainen, T. and Avela, J. (2009) 'Excitability at the motoneuron pool and motor cortex is specifically modulated in lengthening compared to isometric contractions', *Journal of Neurophysiology*, 101: 2030–40.

Guye, M., Parker, G.J., Symms, M., Boulby, P., Wheeler-Kingshott, C.A., Salek-Haddadi, A., et al. (2003) 'Combined functional MRI and tractography to demonstrate the connectivity of the human primary motor cortex in vivo', *Neuroimage*, 19: 1349–60.

Häkkinen, K., Alen, M., Kallinen, M., Newton, R.U. and Kraemer, W.J. (2000) 'Neuromuscular adaptation during prolonged strength training, detraining and re-strength-training in middle-aged and elderly people', *European Journal of Applied Physiology*, 83: 51–62.

Häkkinen, K., Alen, M., Kraemer, W.J., Gorostiaga, E., Izquierdo, M., Rusko, H., et al. (2003) 'Neuromuscular adaptations during concurrent strength and endurance training versus strength training', *European Journal of Applied Physiology*, 89: 42–52.

Häkkinen, K., Komi, P.V. and Alén, M. (1985) 'Effect of explosive type strength training on isometric force- and relaxation-time, electromyographic and muscle fibre characteristics of leg extensor muscles', *Acta Physiologica Scandinavica*, 125: 587–600.

Hanajima, R., Ugawa, Y., Machii, K., Mochizuki, H., Terao, Y., Enomoto, H., et al. (2001) 'Interhemispheric facilitation of the hand motor area in humans', *Journal of Physiology*, 531: 849–59.

Herbert, R.D., Dean, C. and Gandevia, S.C. (1998) 'Effects of real and imagined training on voluntary muscle activation during maximal isometric contractions', *Acta Physiologica Scandinavica*, 163: 361–8.

Hess, C.W., Mills, K.R. and Murray, N.M. (1986) 'Magnetic stimulation of the human brain: facilitation of motor responses by voluntary contraction of ipsilateral and contralateral muscles with additional observations on an amputee', *Neuroscience Letters*, 71: 235–40.

Hinder, M.R., Schmidt, M.W., Garry, M.I., Carroll, T.J. and Summers, J.J. (2011) 'Absence of cross-limb transfer of performance gains following ballistic motor practice in older adults', *Journal of Applied Physiology*, 110: 166–75.

Holsgaard-Larsen, A., Caserotti, P., Puggard, L. and Aagaard, P. (2011) 'Stair-ascent performance in elderly women: effect of explosive strength training', *Journal of Aging and Physical Activity*, 19: 117–36.

Holtermann, A., Roeleveld, K., Engström, M. and Sand, T. (2007) 'Enhanced H-reflex with resistance training is related to increased rate of force development', *European Journal of Applied Physiology*, 101: 301–12.

Hortobágyi, T. (2005) 'Mechanisms of unilateral interventions producing contralateral adaptations', *IEEE Engineering and Medicine and Biology Magazine*, 24: 22–8.

Hortobágyi, T. and Maffiuletti, N.A. (2011) 'Neural adaptations to electrical stimulation strength training', *European Journal of Applied Physiology*, 111: 2439–49.

Hortobágyi, T., DeVita, P., Money, J. and Barrier, J. (2001a) 'Effects of standard and eccentric overload strength training in young women', *Medicine and Science in Sports and Exercise*, 33: 1206–12.

Hortobágyi, T., Dempsey, L., Fraser, D., Zheng, D., Hamilton, G., Lambert, J., et al. (2000) 'Changes in muscle strength, muscle fibre size and myofibrillar gene expression after immobilization and retraining in humans', *Journal of Physiology*, 524: 293–304.

Hortobágyi, T., Hill, J.P. and Lambert, N.J. (1997) 'Greater cross education following training with muscle lengthening than shortening', *Medicine and Science in Sports and Exercise*, 29: 107–12.

Hortobágyi, T., Richardson, S.P., Lomarev, M., Shamim, E., Meunier, S., Russman, H., et al. (2011) 'Interhemispheric plasticity in humans', *Medicine and Science in Sports and Exercise*, 43: 1188–99.

Hortobágyi, T., Scott, K., Lambert, J., Hamilton, G. and Tracy, J. (1999) 'Cross-education of muscle strength is greater with stimulated than voluntary contractions', *Motor Control*, 3: 205–19.

Hortobágyi, T., Taylor, J.L., Russell, G., Petersen, N. and Gandevia, S.C. (2003) 'Changes in segmental and motor cortical output with contralateral muscle contractions and altered sensory inputs in humans', *Journal of Neurophysiology*, 90: 2451–9.

Hortobágyi, T., Tunnel, D., Moody, J., Beam, S. and DeVita, P. (2001b) 'Low- or high-intensity strength training partially restores impaired quadriceps force accuracy and steadiness in aged adults', *Journal of Gerontology*, 56: B38–47.

Howatson, G., Taylor, M.B., Rider, P., Motawar, B.R., McNally, M.P., Solnik, S., et al. (2011) 'Ipsilateral motor cortical responses to TMS during lengthening and shortening of the contralateral wrist flexors', *European Journal of Neuroscience*, 33: 978–90.

Hunter, S.K., Lepers, R., MacGillis, C.J. and Enoka, R.M. (2003) 'Activation among elbow flexor muscles differs when maintaining arm position during a fatiguing contraction', *Journal of Applied Physiology*, 94: 2439–47.

Hunter, S.K., Rochette, L., Critchlow, A. and Enoka, R.M. (2005) 'Time to task failure differs with load type when old adults perform a submaximal fatiguing contraction', *Muscle & Nerve* 31: 730–40.

Hunter, S.K., Ryan, D.R., Ortega, J.D. and Enoka, R.M. (2002) 'Task differences with the same torque load alter the endurance time of submaximal fatiguing contractions in humans', *Journal of Neurophysiology*, 88: 3087–96.

Hunter, S.K., Yoon, T., Farinella, J., Griffith, E.E. and Ng, A.V. (2008) 'Time to task failure and muscle activation vary with load type for submaximal fatiguing contraction with the lower leg', *Journal of Applied Physiology*, 105: 463–72.

Innocenti, G.M. (1986) 'General organisation of callosal connections in the cerebral cortex', in E.G. Jones & A. Peters (eds.), *Cerebral Cortex* (pp. 291–353), New York: Plenum Press.

Jensen, J.L., Marstrand, P.C.D. and Nielsen, J.B. (2005) 'Motor skill training and strength training are associated with different plastic changes in the central nervous system', *Journal of Applied Physiology*, 99: 1558–68.

Jones, D.A. and Rutherford, O.M. (1987) 'Human muscle strength training: the effects of three different regimens and the nature of the resultant changes', *Journal of Physiology*, 391: 1–11.

Karni A., Meyer G., Jezzard P., Adams M.M., Turner R. and Ungerleider L.G. (1995) 'Functional MRI evidence for adult motor cortex plasticity during motor skill learning', *Nature*, 377: 155–8.

Kawashima, R., Matsumura, M., Sadato, N., Naito, E., Waki, A., Nakamura, S., et al. (1998) 'Regional cerebral blood flow changes in human brain related to ipsilateral and contralateral complex hand movements – a PET study', *European Journal of Neuroscience*, 10: 2254–60.

Kim, S.G., Ashe, J., Georgopoulos, A.P., Merkle, H., Ellermann, J.M., Menon, R.S., et al. (1993) 'Functional imaging of human motor cortex at high magnetic field', *Journal of Neurophysiology*, 69: 297–302.

Klass, M., Baudry, S. and Duchateau, J. (2008a) 'Age-related decline in rate of torque development is accompanied by lower maximal motor unit discharge frequency during fast contractions', *Journal of Applied Physiology*, 104: 739–46.

Klass, M., Lévénez, M., Enoka, R.M. and Duchateau, J. (2008b) 'Spinal mechanisms contribute to differences in time to failure of submaximal fatiguing contractions performed with different tasks', *Journal of Neurophysiology*, 99: 1096–1104.

Kobayashi, H., Koyama, Y., Enoka, R.M. and Suzuki, S. (2012) 'A unique form of light-load training improves steadiness and performance on some functional tasks in older adults', *Scandinavian Journal of Medicine and Science in Sports*, Apr 11. doi: 10.1111/j.1600–0838. 2012. 01460x. [Epub ahead of print].

Kooistra, C.A. and Heilman, K.M. (1988) 'Motor dominance and lateral asymmetry of the globus pallidus', *Neurology*, 38: 388–90.

Koyama, Y. (1999) 'The baseball training revolution by BML theory'. (In Japanese). *Baseball Magazine*, Tokyo.

Koyama, Y., Kobayashi, H., Suzuki, S. and Enoka, R.M. (2010) 'Enhancing the weight training experience: a comparison of limb kinematics and EMG activity on three machines', *European Journal of Applied Physiology*, 109: 789–801.

Krutky, M.A., Ravichandran, V.J., Trumbower, R.D. and Perreault, E.J. (2010) 'Interactions between limb and environmental mechanics influence stretch reflex sensitivity in the human arm', *Journal of Neurophysiology*, 103: 429–40.

Kujirai, T., Caramia, M.D., Rothwell, J.C., Day, B.L., Thompson, P.D., Ferbert, A., et al. (1993) 'Corticocortical inhibition in human motor cortex', *Journal of Physiology*, 471: 501–19.

Kukulka, C.G. and Clamann, H.P. (1981) 'Comparison of the recruitment and discharge properties of motor units in human brachial biceps and adductor pollicis during isometric contractions', *Brain Research* 219: 45–55.

Lagerquist, O., Zehr, E.P. and Docherty, D. (2006) 'Increased spinal reflex excitability is not associated with neural plasticity underlying the cross-education effect', *Journal of Applied Physiology*, 100: 83–90.

Laidlaw, D.H., Kornatz, K.W., Keen, D.A., Suzuki, S. and Enoka, R.M. (1999) 'Strength training improves the steadiness of slow lengthening contractions performed by old adults', *Journal of Applied Physiology*, 87: 1786–95.

Latash, M. L. (1999) 'Mirror writing: learning, transfer, and implications for internal inverse models', *Journal of Motor Behavior*, 31: 107–11.

Le Bozec, S. and Bouisset, S. (2004) 'Does postural chain mobility influence muscular control in sitting ramp pushes?', *Experimental Brain Research*, 158: 427–37.

Lee, M., Hinder, M.R., Gandevia, S.C. and Carroll, T.J. (2010) 'The ipsilateral motor cortex contributes to cross-limb transfer of performance gains after ballistic motor practice', *Journal of Physiology*, 588: 201–12.

Lee, M., Gandevia, S.C. and Carroll, T.J. (2009) 'Unilateral strength training increases voluntary activation of the opposite untrained limb', *Clinical Neurophysiology*, 120: 802–8.

Lee, H., Gunraj, C. and Chen, R. (2007) 'The effects of inhibitory and facilitatory intracortical circuits on interhemispheric inhibition in the human motor cortex', *Journal of Physiology*, 580: 1021–32.

Maffiuletti, N.A. and Martin, A. (2001) 'Progressive versus rapid rate of contraction during 7 wk of isometric resistance training', *Medicine and Science in Sports and Exercise*, 33: 1220–7.

Maluf, K.S., Shinohara, M., Stephenson, J.L. and Enoka, R.M. (2005) 'Muscle activation and time to task failure differ with load type and contraction intensity for a human hand muscle', *Experimental Brain Research*, 167: 165–77.

Manini, T.M., Clark, B.C., Tracy, B.L., Burke, J. and Ploutz-Snyder, L. (2005) 'Resistance and functional training reduces knee extensor position fluctuations in functionally limited older adults', *European Journal of Applied Physiology*, 95: 436–46.

Manini, T., Marko, M., VanArnam, T., Cook, S., Fernhall, B., Burke, J., et al. (2007) 'Efficacy of resistance and task-specific exercise in older adults who modify tasks of everyday life', *Journal of Gerontology*, 62A: 616–23.

Marmon, A.R., Gould, J.R. and Enoka, R.M. (2011) 'Practicing a functional task improves steadiness with hand muscles in older adults', *Medicine and Science in Sports and Exercise*, 43: 1531–7.

Mayer, M., Schulze, S., Danek, A. and Botzel, K. (1999) 'Dipole source analysis in persistent mirror movements', *Brain Topography*, 12: 49–60.

Mayston, M.J., Harrison, L.M. and Stephens, J.A. (1999) 'A neurophysiological study of mirror movements in adults and children', *Annals Neurology*, 45: 583–94.

McIlroy, W.E., Collins, D.F. and Brooke, J.D. (1992) 'Movement features and H-reflex modulation. II. Passive rotation, movement velocity and single leg movement', *Brain Research*, 582: 85–93.

Miszko, T.A., Cress, M.E., Slade, J.M., Covey, C.J., Agrawal, S.K. and Doerr, C.E. (2003) 'Effects of strength and power training on physical function in community-dwelling older adults', *Journal of Gerontology*, 58A: 171–5.

Mottram, C.J., Jakobi, J.M., Semmler, J.G. and Enoka, R.M. (2005) 'Motor-unit activity differs with load type during a fatiguing contraction', *Journal of Neurophysiology*, 93: 1381–92.

Muellbacher, W., Facchini, S., Boroojerdi, B. and Hallett, M. (2000) 'Changes in motor cortex excitability during ipsilateral hand muscle activation in humans', *Clinical Neurophysiology*, 111: 344–9.

Muellbacher, W., Ziemann, U., Boroojerdi, B., Cohen, L. and Hallett, M. (2001) 'Role of the human motor cortex in rapid motor learning', *Experimental Brain Research*, 136: 431–8.

Muellbacher W., Ziemann U., Wissel J., Dang N., Kofler M., Facchini S., et al. (2002) 'Early consolidation in human primary motor cortex', *Nature*, 415: 640–4.

Mueller, M., Breil, F. A., Vogt, M., Steiner, R., Lippuner, K., Popp, A., et al. (2009) 'Different response to eccentric and concentric training in older men and women', *European Journal of Applied Physiology*, 107: 145–53.

Munn, J., Herbert, R.D. and Gandevia, S.C. (2004) 'Contralateral effects of unilateral resistance training: a meta-analysis', *Journal of Applied Physiology*, 96: 1861–6.

Naito, H., Nakamura, H., Kurosaki, T. and Tamura, Y. (1970) 'Transcallosal excitatory postsynaptic potentials of fast and slow pyramidal tract cells in cat sensorimotor cortex', *Brain Research*, 19: 299–301.

Naito, A, Shindo, M., Miyasaka, T., Sun, Y.I., Momoi, H. and Chishima, M. (1998) 'Inhibitory projection from pronator teres to biceps brachii motoneurones in human', *Experimental Brain Research*, 121: 99–102.

Naito, A., Shindo, M., Miyasaka, T., Sun, Y.I. and Morita, H. (1996) 'Inhibitory projection from brachioradialis to biceps brachii motoneurones in human', *Experimental Brain Research*, 111: 483–6.

Negro, F., Holobar, A., Farina, D. (2009). Fluctuations in isometric muscle force can be described by one linear projection of low-frequency components of motor unit discharge rates. *Journal of Physiology*, 587, 5925–5938.

Nordlund, M.M., Thorstensson, A. and Cresswell, A.G. (2002) 'Variations in the soleus H-reflex as a function of activation during controlled lengthening and shortening actions', *Brain Research*, 952: 301–7.

Norrbrand, L., Fluckey, J.D., Pozzo, M. and Tesch, P.A. (2008) 'Resistance training using eccentric overload induces early adaptations in skeletal muscle size', *European Journal of Applied Physiology*, 102: 271–81.

Oliveira, A.S., Corvino, R.B., Goncalves, M., Caputo, F. and Denadal, B.S. (2010) 'Effects of a single habituation session on neuromuscular isokinetic profile at different movement velocities', *European Journal of Applied Physiology*, 110: 1127–33.

Pascual-Leone, A., Grafman, J. and Hallett, M. (1994) 'Modulation of cortical motor output maps during development of implicit and explicit knowledge', *Science*, 263: 1287–9.

Perez, M.A. and Cohen, L.G. (2008) 'Mechanisms underlying functional changes in the primary motor cortex ipsilateral to an active hand', *Journal of Neuroscience*, 28: 5631–40.

Perez, M.A. and Cohen, L.G. (2009) 'Scaling of motor cortical excitability during unimanual force generation', *Cortex*, 45: 1065–71.

Perez, M.A., Lungholt, B.K.S., Nyborg, K. and Nielsen, J.B. (2004) 'Motor skill training induces changes in the excitability of the leg cortical area in healthy humans', *Experimental Brain Research*, 159: 197–205.

Perreault, E.J., Chen, K., Trumbower, R.D. and Lewis, G.N. (2008) 'Interactions with compliant loads alter stretch reflex gains but not intermuscular coordination', *Journal of Neurophysiology*, 99: 2101–13.

Petterson, S.C., Mizner, R.L., Stevens, J.E., Raisis, L., Bodenstab, A., Newcomb, et al. (2009) 'Improved function from progressive strengthening interventions after total knee arthroplasty: a randomized clinical trial with an imbedded prospective cohort', *Arthritis and Rheumatism*, 61: 174–83.

Piotrkiewicz, M., Kudina, L., Mierzejewska, J., Jakubiec, M. and Hausmanowa-Petrusewicz, I. (2007) 'Age-related change in duration of afterhyperpolarization of human motoneurones', *Journal of Physiology*, 585: 483–90.

Puttemans, V., Wenderoth, N. and Swinnen, S.P. (2005) 'Changes in brain activation during the acquisition of a multifrequency bimanual coordination task: from the cognitive stage to advanced levels of automaticity', *Journal of Neuroscience* 25: 4270–8.

Rahmann, A.E., Brauer, S.G. and Nitz, J.C. (2009) 'A specific inpatient aquatic physiotherapy program improves strength after total hip or knee replacement surgery: a randomized controlled trial', *Archives of Physical Medicine and Rehabilitation*, 90: 745–55.

Ranganathan, V.K., Siemionow, V., Sahgal, V., Liu, J.Z. and Yue, G.H. (2001) 'Skilled finger movement exercise improves hand function', *Journal of Gerontology*, 56A: M518–22.

Ranganathan, V.K., Siemionow, V., Liu, J.Z., Sahgal, V. and Yue, G.H. (2004) 'From mental power to muscle power – gaining strength by using the mind', *Neuropsychologia*, 42: 944–56.

Rao, S.M., Binder, J.R., Bandettini, P.A., Hammeke, T.A., Yetkin, F.Z., Jesmanowicz, A., et al. (1993) 'Functional magnetic resonance imaging of complex human movements', *Neurology*, 43: 2311–8.

Riley, Z.A., Baudry, S. and Enoka, R.M. (2008) 'Reflex inhibition in human biceps brachii decreases with practice of a fatiguing contraction', *Journal of Neurophysiology*, 100: 2843–51.

Rogasch, N.C., Dartnall, T.J., Cirillo, J., Nordstrom, M.A. and Semmler, J.G. (2009) 'Corticomotor plasticity and learning of a ballistic thumb training task are diminished in older adults', *Journal of Applied Physiology*, 107: 1874–83.

Rudroff, T., Barry, B.K., Stone, A.L., Barry, C.J. and Enoka, R.M. (2007) 'Accessory muscle activity contributes to the variation in time to task failure for different arm postures and loads', *Journal of Applied Physiology*, 102: 1000–6.

Rudroff, T., Jordan, K., Enoka, J.A., Matthews, S.D. and Enoka, R.M. (2010) 'Discharge of biceps brachii motor units is modulated by load compliance and forearm posture', *Experimental Brain Research*, 202: 111–20.

Rudroff, T., Justice, J.N., Holmes, M.R., Matthews, S.D. and Enoka, R.M. (2011) 'Muscle activity and time to task failure differ with load compliance and target force for elbow flexor muscles', *Journal of Applied Physiology*, 110: 125–36.

Rudroff, T., Poston, B., Shin, I-S., Bojsen-Møller, J. and Enoka, R.M. (2005) 'Net excitation of the motor unit pool varies with load type during fatiguing contractions.', *Muscle & Nerve*, 31: 78–87.

Rutherford, O.M. and Jones, D.A. (1986) 'The role of learning and coordination in strength training', *European Journal of Applied Physiology*, 55: 100–5.

Salerno, A. and Georgesco, M. (1996) 'Interhemispheric facilitation and inhibition studied in man with double magnetic stimulation', *Electroencephalography and Clinical Neurophysiology*, 101: 395–403.

Salmelin, R., Forss, N., Knuutila, J. and Hari, R. (1995) 'Bilateral activation of the human somato-motor cortex by distal hand movements', *Electroencephalography and Clinical Neurophysiology*, 95: 444–52.

Samii, A., Canos, M., Ikoma, K., Wassermann, E.M. and Hallett, M. (1997) 'Absence of facilitation or depression of motor evoked potentials after contralateral homologous muscle activation', *Electroencephalography and Clinical Neurophysiology*, 105: 241–5.

Sariyildiz, M., Karacan, I., Rezvani, A., Ergin, O. and Cidem, M. (2011) 'Cross-education of muscle strength: cross-training effects are not confined to untrained contralateral homologous muscle', *Scandinavian Journal of Medicine and Science in Sports*, 21: e359–64.

Scaglioni G., Ferri A., Minetti A.E., Martin A., Van Hoecke J., Capodaglio P., et al. (2002) 'Plantar flexor activation capacity and H reflex in older adults: adaptations to strength training', *Journal of Applied Physiology*, 92: 2292–302.

Schubert, M., Beck S., Taube W., Amtage F., Faist M. and Gruber M. (2008) 'Balance training and ballistic strength training are associated with task-specific corticospinal adaptations', *European Journal of Neuroscience*, 27: 2007–18.

Scripture, E.W., Smith, T.L. and Brown, E.M. (1894) 'On education of muscular control and power', *Studies Yale Psychological Laboratory*, 2: 114–9.

Sehm, B., Perez, M. A., Xu, B., Hidler, J. and Cohen, L.G. (2009) 'Functional neuroanatomy of mirroring during a unimanual force generation task', *Cerebral Cortex*, 20: 34–45.

Sekiguchi, H., Kimura, T., Yamanaka, K. and Nakazawa, K. (2001) 'Lower excitability of the corticospinal tract to transcranial magnetic stimulation during lengthening contractions in human elbow flexors', *Neuroscience Letters*, 312: 83–86.

Sekiguchi, H., Nakazawa, K. and Suzuki, S. (2003) 'Differences in recruitment properties of the corticospinal pathway between lengthening and shortening contractions in human soleus muscle', *Brain Research*, 977: 169–79.

Seynnes, O., Singh, M.A.F., Hue, O., Pras, P., Legros, P. and Bernard, P.L. (2004) 'Physiological and functional responses to low-moderate versus high-intensity progressive resistance training in frail elders', *Journal of Gerontology*, 59A: 503–9.

Shibasaki, H., Sadato, N., Lyshkow, H., Yonekura, Y., Honda, M., Nagamine, T., et al. (1993) 'Both primary motor cortex and supplementary motor area play an important role in complex finger movement', *Brain*, 116: 1387–98.

Shima, N., Ishida, K., Katayama, K., Morotome, Y., Sato, Y. and Miyamura, M. (2002) 'Cross education of muscular strength during unilateral resistance training and detraining', *European Journal of Applied Physiology*, 86: 287–94.

Shin, H.W. and Sohn, Y.H. (2011) 'Interhemispheric transfer of paired associative stimulation-induced plasticity in the human motor cortex', *Neuroreport*, 22: 166–70.

Slivko, É.I. and Bogutskaya, G.A. (2008) 'Long-lasting inhibition of the soleus muscle H reflex related to realization of voluntary arm movements in humans', *Neurophysiology*, 40: 221–7.

Sohn, Y.H., Jung, H.Y., Kaelin-Lang, A. and Hallett, M. (2003) 'Excitability of the ipsilateral motor cortex during phasic voluntary hand movement', *Experimental Brain Research*, 148: 176–85.

Stinear, C.M., Walker, K.S. and Byblow, W.D. (2001) 'Symmetric facilitation between motor cortices during contraction of ipsilateral hand muscles', *Experimental Brain Research*, 139: 101–5.

Stromberg, B.V. (1988) 'Contralateral therapy in upper extremity rehabilitation', *American Journal of Physical Medicine*, 65: 135–43.

Suetta, C., Magnusson, S.P., Rosted, A., Aagaard, P., Jakobsen, A.K., Larsen, L.H., et al. (2004) 'Resistance training in the early postoperative phase reduces hospitalization and leads to muscle hypertrophy in elderly hip surgery patients – a controlled, randomized study', *Journal of the American Geriatric Society*, 52: 2016–22.

Swinnen, S.P. and Wenderoth, N. (2004) 'Two hands, one brain: cognitive neuroscience of bimanual skill', *Trends in Cognitive Sciences*, 8: 18–25.

Tagesson, S., Oberg, B., Good, L. and Kvist, J. (2008) 'A comprehensive rehabilitation program with quadriceps strengthening in closed versus open kinetic chain exercise in patients with anterior cruciate ligament deficiency: a randomized clinical trial evaluating dynamic tibial translation and muscle function', *American Journal of Sports Medicine*, 36: 298–307.

Taube W., Kullmann N., Leukel C., Kurz O., Amtage F. and Gollhofer A. (2007) 'Cortical and spinal adaptations induced by balance training: correlation between stance stability and corticospinal activation', *Acta Physiologica Scandinavica*, 189: 347–58.

Taylor, J.L. and Gandevia, S.C. (2004) 'Noninvasive stimulation of the human corticospinal tract', *Journal of Applied Physiology*, 96: 1496–1503.

Teixeira, L.A. (2000) 'Timing and force components in bilateral transfer of learning', *Brain and Cognition*, 44: 455–69.

Terada, K., Umeoka, S., Usui, N., Baba, K., Usui, K., Fujitani, S., et al. (2012) 'Uneven interhemispheric connections between left and right primary sensori-motor areas', *Human Brain Mapping*, 33: 14–26.

Tinazzi, M. and Zanette, G. (1998) 'Modulation of ipsilateral motor cortex in man during unimanual finger movements of different complexities', *Neuroscience Letters*, 244: 121–4.

Tracy, B.L. and Enoka, R.M. (2006) 'Steadiness training with light loads in the knee extensors of elderly adults', *Medicine and Science in Sports and Exercise*, 38: 735–45.

Tracy, B.L., Byrnes, W.C. and Enoka, R.M. (2004) 'Strength training reduces force fluctuations during anisometric contractions of the quadriceps femoris muscles in old adults', *Journal of Applied Physiology*, 96: 1530–40.

Tsuchiya, K. and Suzuki, S. (2011) 'Influence of forearm orientation on modulation of electromyographic activity of shoulder muscles during elbow flexion tasks in healthy adults', *Proceedings of the XXIIIrd Congress of the International Society of Biomechanics*, Brussels, Belgium, July 3–7, p. 172.

Tsukahara R., Aoki H., Yabe K. and Mano T. (1995) 'Effects of premotion silent period on single motor unit firing at initiation of a rapid contraction', *Electroencephalography and Clinical Neurophysiology*, 97: 223–30.

Uematsu, A., Obata, H., Endoh, T., Kitamura, T., Hortobagyi, T., Nakazawa, K., et al. (2010) 'Asymmetrical modulation of corticospinal excitability in the contracting and resting contralateral wrist flexors during unilateral shortening, lengthening and isometric contractions', *Experimental Brain Research*, 206: 59–69.

Urbano, A., Babiloni, C., Onorati, P., Carducci, F., Ambrosini, A., Fattorini, L., et al. (1998) 'Responses of human primary sensorimotor and supplementary motor areas to internally triggered unilateral and simultaneous bilateral one-digit movements. A high-resolution EEG study', *European Journal of Neuroscience*, 10: 765–70.

Van Cutsem, M. and Duchateau, J. (2005) 'Preceding muscle activity influences motor unit discharge and rate of torque development during ballistic contractions in humans', *Journal of Physiology*, 562: 635–44.

Van Cutsem, M., Duchateau, J. and Hainaut, K. (1998) 'Changes in single motor unit behaviour contribute to the increase in contraction speed after dynamic training in humans', *Journal of Physiology*, 513: 295–305.

Van Cutsem M., Feiereisen P., Duchateau J. and Hainaut K. (1997) 'Mechanical properties and behaviour of motor units in the tibialis anterior during voluntary contractions', *Canadian Journal of Applied Physiology*, 22: 585–97.

Vila-Chã, E., Falla, D. and Farina, D. (2010) 'Motor unit behavior during submaximal contractions following six weeks of either endurance or strength training', *Journal of Applied Physiology*, 109: 1455–66.

Vincent, K.R., Braith, R.W., Feldman, R.A., Magyari, P.M., Cutler, R.B., Persin, S.A., et al. (2002) 'Resistance exercise and physical performance in adults aged 60 to 83', *Journal of the American Geriatric Society*, 50: 1100–7.

Wahl, M. and Ziemann, U. (2008) 'The human motor corpus callosum', *Nature Reviews Neuroscience*, 19: 451–66.

Wolpaw, J.R. (1997) 'The complex structure of a simple memory', *Trends in Neurosciences*, 20: 588–94.

Yue, G. and Cole, K.J. (1992) 'Strength increases from the motor program: comparison of training with maximal voluntary and imagined muscle contractions', *Journal of Neurophysiology*, 67: 1114–23.

Zhang, L., Butler, J., Nishida, T., Nuber, G., Huang, H. and Rymer, W.Z. (1998) 'In vivo determination of the direction of rotation and moment-arm relationship of individual elbow muscles', *Journal of Biomechanical Engineering*, 120: 625–33.

Zhou, S. (2000) 'Chronic neural adaptations to unilateral exercise: Mechanisms of cross education', *Exercise and Sport Sciences Reviews*, 28: 177–84.

Ziemann, U., Bruns, D. and Paulus, W. (1996) 'Enhancement of human motor cortex inhibition by the dopamine receptor agonist pergolide: evidence from transcranial magnetic stimulation', *Neuroscience Letters*, 208: 187–90.

Zijdewind, I., Butler, J. E., Gandevia, S.C. and Taylor, J.L. (2006) 'The origin of activity in the biceps brachii muscle during voluntary contractions of the contralateral elbow flexor muscles', *Experimental Brain Research*, 175: 526–35.

Zijdewind, I. and Kernell, D. (2001) 'Bilateral interactions during contractions of intrinsic hand muscles', *Journal of Neurophysiology*, 85: 1907–13.

Zijdewind, I., Toering, S.T., Bessem, B., Van Der Laan, O. and Diercks, R.L. (2003) 'Effects of imagery motor training on torque production of ankle plantar flexor muscles', *Muscle Nerve*, 28: 168–73.

PART V

Challenges in motor control and learning

17

MOTOR CONTROL AND MOTOR LEARNING UNDER FATIGUE CONDITIONS

Janet L. Taylor

NEUROSCIENCE RESEARCH AUSTRALIA AND THE UNIVERSITY OF NEW SOUTH WALES

Introduction

Fatigue is a term that has a wide range of meanings. It is a common symptom in many illnesses where fatigue is used to describe feelings of sleepiness, of general lack of energy, and of the need for abnormally high effort to perform tasks, whether physical or mental (e.g., Dinges, 1995; Hossain et al., 2005; Johansson et al., 2009; Smets et al., 1995; Torres-Harding & Jason, 2005; Zwarts et al., 2008). It may be associated with physical weakness or difficulty in performing cognitive tasks. In health, similar feelings of fatigue can occur after physical or mental exertion. Fatigue is also used more specifically in relation to an episode of physical or mental work but even then it has both a perceptual and a performance aspect. That is, during physical or mental exercise people have both sensations of fatigue and a decrement in maximal performance, which can also be described as fatigue (e.g., Boksem & Tops, 2008; Enoka et al., 2011; Gandevia, 2001; Knicker et al., 2011). Finally, fatigue is used to refer to activity-dependent changes of the muscles or nervous system when these lead to a decrease in motor output (e.g., Enoka & Duchateau, 2008; Gandevia, 2001; Taylor & Gandevia, 2008). Ash (1914) described fatigue as 'a comprehensive term which in its widest application embraces all those immediate and temporary changes, whether of a functional or organic character, which take place within an organism or any of its constituent parts as a direct result of its own exertions, and which tend to interfere with or inhibit the organism's further activities.'

The current chapter will not encompass all aspects of fatigue but will concentrate on fatigue brought about by physical exercise and its effects on motor performance and motor learning. The chapter will use the terms muscle fatigue, peripheral fatigue, central fatigue, supraspinal fatigue, mental fatigue and sensations of fatigue to refer to various aspects of fatigue brought about by exercise (see Table 17.1). The chapter will describe how the muscle and nervous system change during repeated or sustained motor activity before going on to discuss how these changes may impact on the performance of motor tasks and on the learning of new motor skills.

Table 17.1 Aspects of fatigue

Muscle fatigue	• any exercise-induced reduction in the ability of a muscle to generate force or power
Peripheral fatigue	• fatigue produced by changes at or distal to the neuromuscular junction; a component of muscle fatigue
Central fatigue	• a progressive exercise-induced reduction in maximal voluntary activation; a component of muscle fatigue
Supraspinal fatigue	• fatigue produced by suboptimal output from the motor cortex; a component of central fatigue
Sensations of muscle fatigue	• perception of muscle weakness/increased effort to perform a motor task
	• exercise-related muscle discomfort/pain/burning
Mental fatigue	• 'tiredness or exhaustion' experienced after demanding cognitive activity

Fatigue caused by exercise

Muscle fatigue can be defined as an exercise-related decrease in the maximal force or power produced by a muscle or muscle group whether or not the current task can be performed (Bigland-Ritchie & Woods, 1984; Gandevia, 2001). In humans, the maximal force or power of a muscle is generally measured by performance of a maximal voluntary contraction (Figure 17.1). In a voluntary contraction the nervous system drives the muscle fibers, therefore muscle fatigue can occur through changes in either the muscle or the nervous system or both. Often muscle fatigue can occur without an overt impairment of motor performance because most tasks do not require maximal output from the muscles (e.g., Gates & Dingwell, 2008; Morin et al., 2012). However, neuromuscular activity must be altered to compensate for the muscle fatigue and sustain the performance (e.g., Bigland-Ritchie et al., 1986a, 1986b; Gates & Dingwell, 2008; Hunter et al., 2004; Singh et al., 2010; Taylor & Gandevia, 2008).

To produce a voluntary movement or apply a force requires a chain of events starting in the brain and ending in the muscle. At its simplest the chain comprises motor planning in premotor areas of cortex, production of output from the motor cortex via corticospinal neurons, activation of motoneurons in the spinal cord, and activation of muscle fibers with the generation of action potentials along the sarcolemma, excitation-contraction coupling and the production of mechanical output through cross-bridge mechanisms. Changes that contribute to muscle fatigue occur at multiple sites throughout this chain (Gandevia, 2001). Muscle fatigue is often divided into two: peripheral fatigue, which refers to changes at the neuromuscular junction or in the muscle itself, and central fatigue, which refers to changes within the central nervous system (Bigland-Ritchie et al., 1978, 1986b; Gandevia, 1988, 2001; Gandevia et al., 1995) (Figure 17.2). The level of this division is chosen for convenience. Stimulation of the peripheral motor nerve can activate the muscle without the involvement of the central nervous system (e.g., Merton, 1954; Millet et al., 2011). Hence, any exercise-related decrease in the muscle force evoked by peripheral nerve stimulation is called peripheral fatigue.

Peripheral fatigue: Mechanisms and consequences

For stimulation of a peripheral motor axon to evoke force from a muscle, there must be conduction of the action potential to the terminals of the motor axon, synaptic activation of the muscle fiber across the neuromuscular junction, action potential generation and propagation

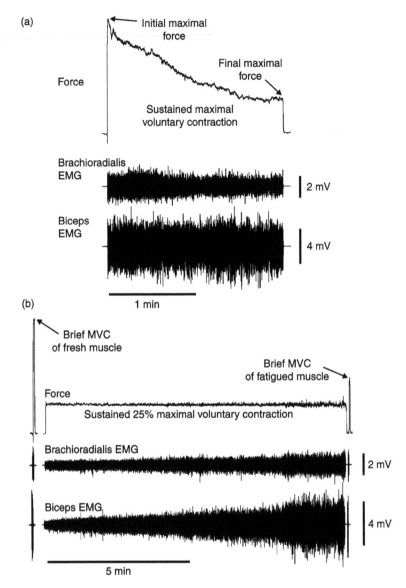

Figure 17.1 Force and electromyographic activity (EMG) in sustained maximal and submaximal isometric elbow flexion contractions. (a) During the sustained maximal voluntary contraction (MVC) force starts to fall immediately. Initially, it falls quickly and then it continues to fall at a much slower rate. The fall in maximal voluntary force indicates fatigue. Force is unsteady throughout the contraction. Changes in surface EMG reflect changes in the muscle fiber action potentials and motor unit firing rates. Surface EMG often decreases with slowing of muscle fiber conduction velocity and motor unit firing. (b) A brief MVC performed before a sustained 25 percent maximum contraction produced more force than a brief MVC performed after the sustained contraction. This fall indicates that the submaximal contraction produced fatigue. During a sustained submaximal voluntary contraction, force can initially be maintained. EMG increases, reflecting the recruitment of motor units to compensate for peripheral fatigue. As fatigue progresses, the EMG becomes more variable with more bursts and force also becomes variable. See Hunter and Enoka (2003).

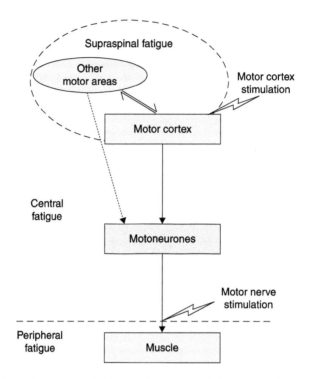

Figure 17.2 Peripheral fatigue, central fatigue and supraspinal fatigue. In people, peripheral fatigue can be identified by a reduction in the force produced by stimulation of the motor nerve (or the muscle). This stimulation bypasses the nervous system and any change in the force produced by it must occur through processes at or distal to the point of stimulation. Central fatigue can be identified by an increase in the force increment evoked by motor nerve stimulation during a maximal voluntary effort. The increment (superimposed twitch) reflects any failure of the nervous system to drive the muscle maximally. Any processes proximal to the site of motor nerve stimulation can contribute to central fatigue. Supraspinal fatigue is a part of central fatigue. It can be identified by an increase in the superimposed twitch evoked by motor cortical stimulation during a maximal effort. Processes at or above the motor cortex can contribute to supraspinal fatigue.

along the sarcolemma of the muscle fiber and down the t tubules, excitation–contraction coupling to release calcium from the sarcoplasmic reticulum into the cytoplasm of the muscle fiber, and cross-bridge formation and cycling. All of these processes can be altered by repetitive activation of the nerve and muscle, and could contribute to a decrease in force output from the muscle fiber (Allen et al., 2008; Bigland-Ritchie et al., 1979; Lamb, 2002; Place et al., 2010; Sieck & Prakash, 1995; Vucic et al., 2010).

In practice, there is little evidence for failure in the nerve fiber or at the neuromuscular junction in voluntary contractions in healthy humans. Conduction in the motor axons is very slightly slowed with repetitive activation but this does not lead to conduction block, so action potentials continue to be conveyed to the nerve terminals (Kiernan et al., 2004; Vagg et al., 1998). If axons are electrically stimulated at high rates, then transmission across the neuromuscular junction can fail so that the muscle fibers no longer respond to the nerve action potential with one-to-one firing (Bigland-Ritchie et al., 1979; Trontelj et al., 2002). However, this does not occur in *voluntary* contractions as the motoneurons only produce high firing rates for short periods of time (Fuglevand & Keen, 2003). Block at the neuromuscular junction can be a problem in disease. It is the mechanism of weakness and fatigue in myeasthenia gravis (e.g., Cantor, 2010).

Failure of muscle fiber excitation also seems to have a limited role in human voluntary contractions. From its initiation at the neuromuscular junction, the muscle fiber action potential is conducted along the sarcolemma to reach all parts of the muscle fiber and also propagates down the t-tubular system. With each muscle fiber action potential there is an efflux of K+ from the muscle cell. With repeated activity, K+ tends to accumulate extracellularly and to depolarize the cell membrane. High frequency activation of muscle fibers *in vitro* and stimulated contractions *in vivo* show that changes in muscle fiber membrane excitability can have large effects on muscle force (Allen et al., 2008; Fuglevand, 1995; Jones et al., 1979; Pagala et al., 1994). In voluntary contractions, the conduction of the muscle fiber action potential slows and potential amplitude can change (Balog et al., 1994; Lannergren & Westerblad, 1986), but such changes seem not to have a large role in fatigue. Among other factors, physiological firing rates mitigate against loss of force through failure of excitation (for review Allen et al., 2008; Fuglevand & Keen, 2003; Thomas et al., 2006).

Thus, the major contributors to peripheral fatigue during voluntary exercise are likely to be within the muscle fibers. They include mechanisms which decrease release of calcium ions from the sarcoplasmic reticulum, which decrease the sensitivity of myofibrillary proteins to calcium ions, which alter the kinetics of cross-bridge cycling, and which impair the pumping of calcium ions back into the sarcoplasmic reticulum (for reviews, Allen et al., 2008; Lamb, 2002; Place et al., 2010) (Figure 17.3). It is likely that some of the products of metabolism mediate some of these effects (Allen et al., 2008). However, in the context of the neural control of voluntary movement and motor learning, what is important is not the precise intracellular mechanisms of muscle fatigue but their consequences for the contractile properties of the muscle fibers. With fatigue, muscle fibers produce less force, their speed of shortening is reduced, and the force–velocity relationship is altered (e.g., Burke et al., 1973; de Haan et al., 1989; De Ruiter et al., 1999; Jones, 2010; Jones et al., 2006). Hence, power output is reduced more than the force. In addition, the relaxation rate of the fatigued muscle fibers is also slowed (e.g., de Haan et al., 1989; Edwards et al., 1975; Fitts et al., 1982; Jones et al., 2006). However, muscle fibers do not all fatigue at the same rate (Westerblad et al., 2010). The smaller, slower muscle fibers that depend most on aerobic metabolism, fatigue much more slowly than the larger, faster fibers (Burke et al., 1973; Karatzaferi et al., 2001; Sargeant, 2007) although the association between the size, speed and fatiguability of motor units in humans is not completely clear (Enoka, 2011; Fuglevand et al., 1999). In voluntary contractions, the small, fatigue-resistant fibers are recruited at lower forces than the larger, fatigable fibers (Henneman et al., 1974; Milner-Brown et al., 1973; Monster & Chan, 1977) (Figure 17.4). Thus, the muscle fibers that are engaged in sustained postural or locomotor tasks fatigue slowly whereas fibers that are recruited for high force or power output fatigue much more quickly (Sargeant, 2007).

Central fatigue

There are many changes within the nervous system during a fatiguing motor task. Some changes may compensate for the altered contractile properties of the muscle so that the task can still be performed despite peripheral fatigue, but other changes may contribute to the loss of maximal force or power (Enoka et al., 2011; Gandevia, 2001; Taylor & Gandevia, 2008). That is, changes in neural drive can contribute to muscle fatigue.

Voluntary activation of a muscle refers to the level of neural drive to a muscle and is expressed as a function of the muscle's force-producing capability. To measure voluntary

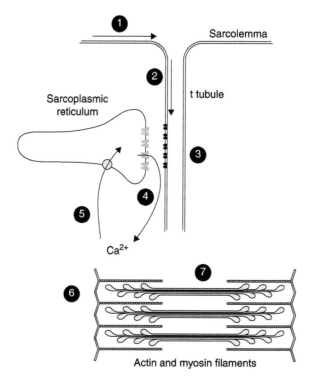

Figure 17.3   Mechanisms of peripheral fatigue. The production of force by a muscle fiber requires the generation of an action potential and its propagation along the sarcolemma and down the t tubules. Here, there are voltage sensors which interact with the sarcoplasmic reticulum to cause release of Ca^{2+} through ryanodine receptors into the cytoplasm. Ca^{2+} interacts with the myosin filaments to allow cross-bridge cycling. With repetitive activation, changes can occur at all these sites and may contribute to fatigue (see Allen et al., 2008).

1. Transmission of the surface membrane action potential impaired by external accumulation of K^+ and internal increase in Na^+.
2. Conduction of the t-tubule action potential impaired by external accumulation of K^+ and internal increase in Na^+.
3. Activation of voltage sensors reduced by (i) reduced action potential amplitude, (ii) voltage sensor inactivation.
4. Sarcoplasmic reticulum Ca^{2+} release reduced (i) by precipitation of Ca^{2+} phosphate (ii) multiple mechanisms act on the ryanodine receptors through which Ca^{2+} is released e.g. increased Mg^{2+}, decreased ATP, phosphate.
5. Ca^{2+} reuptake into the sarcoplasmic reticulum inhibited by increased phosphate, ADP and reactive oxygen species and also by decreased ATP.
6. Ca^{2+} sensitivity of myofibrillar proteins reduced by increased phosphate, decreased pH and increased reactive oxygen species.
7. Maximum Ca^{2+} activated force reduced by increased phosphate and shortening velocity reduced by increased ADP.

activation, the peripheral motor nerve to the muscle can be stimulated during a voluntary contraction so that a twitch-like increment in force is evoked (e.g., Babault et al., 2001; Belanger & McComas, 1981; Merton, 1954; Shield & Zhou, 2004; Taylor, 2009). The size of the superimposed twitch is inversely related to the force produced by voluntary contraction (e.g., Allen et al., 1998; Behm et al., 1996; Merton, 1954) (Figure 17.5). It reflects the muscle

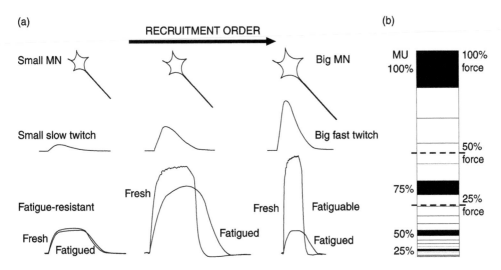

Figure 17.4 Recruitment of motor units. (a) The biophysical properties of the motoneurons lead to a reproducible recruitment order in which smaller motoneurons are recruited before larger ones. The properties of the muscle fibers motoneurons supply tend to match the motoneurons. Each of the small motoneuron supplies a small number of slow muscle fibers. These motor units provide low forces but are very slow to fatigue as illustrated by the lack of change in tetanic force after repeated tetani. In contrast each of the largest motoneurons supplies a larger number of fast muscle fibers. These motor units provide large forces but quickly lose force and slow in their rate of contraction with repeated tetani. (b) The amount of force produced by the largest motor unit in a muscle can be up to one hundred times that produced by the smallest. In a model of human adductor pollicis muscle, Herbert and Gandevia (Herbert & Gandevia, 1999) used a 50 fold range of forces. This diagram illustrates the forces produced by motor units in their model. Each block represents 6 motor units out of a total of 120 motor units. The diagram shows that the smallest 25 percent of motor units (MU) contribute less than 5 percent of the muscle's force whereas the largest 25 percent contribute around two-thirds of the force.

fibers that have not been recruited or are firing too slowly to produce a fused contraction at the moment of stimulation. The superimposed twitch is compared with a twitch evoked when the muscle is at rest to quantify how much of the muscle's force has been recruited voluntarily. During maximal voluntary contractions, force usually approaches the muscle's maximum but rarely reaches it (Allen et al., 1998; Gandevia, 2001; Shield & Zhou, 2004). That is, voluntary activation is usually less than 100 percent. During exercise, subjects' ability to activate muscle maximally falls further (Belanger & McComas, 1981; Gandevia, 2001; Merton, 1954) (Figure 17.6a). This represents central fatigue, which can be defined as a progressive exercise-related failure of voluntary activation of a muscle (Gandevia, 2001).

The activation of muscle fibers depends on the firing of motoneurons. Each motoneuron provides the sole innervation to a number of muscle fibers (between 10 and 1500 in different muscles) and together these form a motor unit (Buchthal & Schmalbruch, 1980; Duchateau & Enoka, 2011; Liddell & Sherrington, 1925). In a motor unit, each action potential in the motoneuron produces an action potential in each of the muscle fibers. Hence, with the development of central fatigue, if muscle fibers are not recruited or are not firing fast enough to

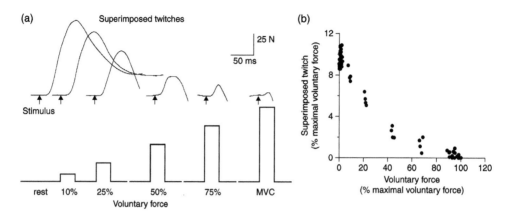

Figure 17.5 Superimposed twitches and voluntary force. (a) Twitches were evoked in human trapezius muscle by electrical stimulation (arrows) of the accessory nerve in the neck. (Taylor et al., 2009). The force of shoulder elevation was measured. The largest twitch (on the left) was evoked with the muscle relaxed, whereas the smallest twitch (on the right) was evoked during a maximal voluntary contraction (MVC). The traces have been offset from their baseline force for the illustration and each trace is an average of several trials. (b) The amplitude of the superimposed twitch plotted against voluntary force for trapezius muscle for one subject. The data are those illustrated by the twitches in (a) The relationship between the superimposed twitch and voluntary force is near linear for most muscles. This relationship allows voluntary activation to be calculated by comparing the superimposed twitch to the twitch evoked from the resting muscle.

produce a fused contraction, this is because motoneurons are not recruited or are firing too slowly (Merton, 1954). Therefore, the possible mechanisms of central fatigue can be considered in terms of factors that can affect motoneuron firing. These include changes in properties of the motoneurons, and changes in the excitatory or inhibitory inputs to motoneurons (Gandevia, 2001; Taylor & Gandevia, 2008). Such inputs include sensory feedback from the periphery and descending input from the brain. Although it remains uncertain precisely which mechanisms contribute to central fatigue and to what extent, some possibilities will be discussed below.

Motoneurons during fatigue

It is likely that motoneurons become less responsive to excitatory input when they are activated repetitively such as during a fatiguing contraction (Figure 17.7). This could be due to a change in the properties of the motoneurons or to inhibition of the motoneurons by synaptic input. In animal studies, motoneurons show a change in their properties with repetitive activation. They initially fire at a high rate in response to a constant injected current but this slows quickly and then more slowly (Button et al., 2007; Kernell & Monster, 1982; Sawczuk et al., 1997). Some findings in people are consistent with such a change. When subjects are given visual or auditory feedback, they can often keep a single motor unit firing at a steady rate for several minutes. However, over time, the motoneuron requires more and more voluntary descending excitatory input to generate the same firing rate (Johnson et al., 2004). Further evidence comes from a reduction in the size of responses to corticospinal tract

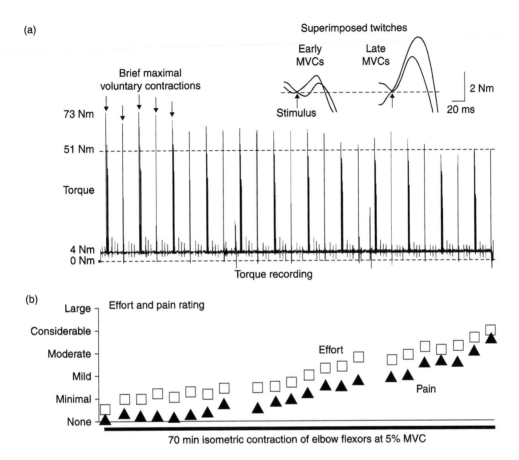

Figure 17.6 Weak contractions lead to a slow progression of fatigue. Subjects performed a very weak (5 percent maximum) isometric contraction of the elbow flexor muscles for 70 minutes (Smith et al., 2007). Occasionally they performed a brief (2–3 s) maximal voluntary contraction (MVC) and stimuli were delivered to evoke superimposed twitches. Subjects regularly reported how much effort it took to maintain the weak contraction and the level of pain in the elbow flexor muscles. (a) Continuous force trace for a single subject shows that the weak contraction of 4 Nm was maintained for 70 min. The torque produced in MVCs decreases slowly indicating progressive fatigue. At the top right are superimposed twitches evoked during 2 MVCs near the beginning and 2 MVCs near the end of the prolonged weak contraction. Over the course of the sustained contraction, the superimposed twitches became bigger. This increase is a marker of central fatigue. (b) Mean data showing the progressive increase in reported effort (squares) to hold the weak contraction. Although the sustained contraction remains <10 percent of maximal, subjects report that it takes considerable effort (rating ~4 out of 10). Muscle pain also increases in the same way.

stimulation during fatiguing contractions (Butler et al., 2003; McNeil et al., 2009, 2011). An increase in inhibitory input may also play a part in decreasing the responsiveness of some motoneurons during fatigue. Small-diameter muscle afferents increase firing during exercise and have been shown to inhibit motoneurons innervating the elbow extensors but facilitate motoneurons innervating the elbow flexors (Martin et al., 2006). Their effects on other motoneuron pools are less clear. Spinal reflex responses in hand muscles and plantar flexors

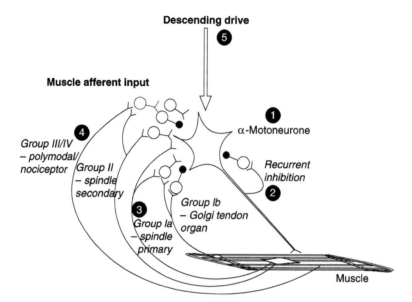

Figure 17.7 Influences on the motoneuron pool during fatigue. Some of the loss of force of fatigue occurs because the motoneurons no longer fire fast enough to drive the muscle fibers to produce their maximal force. This diagram is a simplified representation of muscle afferent input to the motoneurons (see Gandevia, 1998). The numbers indicate some sites relevant to fatigue.

1. Repetitive firing of the motoneuron makes it less excitable. Spike frequency adaptation is one possible mechanism.
2. Recurrent inhibition. Influence in fatigue remains uncertain.
3. Group Ia afferents (muscle spindle primary endings) (i) ongoing firing probably decreases reducing excitation to the motoneurons, (ii) stretch sensitivity may increase.
4. Group III/IV afferents (small-diameter myelinated and non-myelinated afferents). Firing increases with (i) facilitation of some motoneuron pools and inhibition of others, (ii) presynaptic inhibition of Ia afferents, (iii) effects on gamma-motoneurons which alter the sensitivity of muscle spindle afferents.
5. Descending drive (i) becomes suboptimal during maximal efforts (supraspinal fatigue), (ii) can increase during submaximal efforts to overcome decreases in motoneuron excitability and to recruit extra motor units to compensate for impairment of force production by the muscle.

are reduced (Duchateau et al., 2002; Duchateau & Hainaut, 1993; Garland, 1991) but this could be the result of presynaptic inhibition of the Ia afferents (Pettorossi et al., 1999; Rossi et al., 1999a, 1999b) rather than inhibition of the motoneurons themselves.

Apart from a reduction in their responsiveness, motoneuron recruitment and firing may also be reduced by a decrease in the excitatory input to the motoneurons. Some excitatory input during contractions is provided by afferent feedback. In particular, muscle spindle afferents fire during voluntary contractions and facilitate the motoneurons of the muscles in which they are located (Bongiovanni & Hagbarth, 1990; Griffin et al., 2001). There is evidence that this facilitation may decline during fatigue as muscle spindle firing rates decrease during sustained voluntary contractions (Macefield et al., 1991). In addition, short-latency stretch reflexes and H-reflexes are reduced (Duchateau et al., 2002; Duchateau & Hainaut, 1993). Regardless of the mechanism of these reflex changes they suggest a reduction in the Ia afferent input to the motoneurons.

Motor cortex during fatigue

Descending drive to the motoneurons becomes suboptimal during fatigue. Transcranial magnetic stimulation over the motor cortex can be used to evoke superimposed twitches in some muscle groups and can provide another measure of voluntary activation (Gandevia et al., 1996; Lee et al., 2008; Sidhu et al., 2009a; Todd et al., 2003). If cortical stimulation evokes a superimposed twitch during a subject's maximal effort, it indicates that not only were the motoneurons not firing sufficiently to activate the muscle maximally at the moment of stimulation but also some output from the motor cortex remained untapped by the voluntary effort. During fatiguing contractions, the superimposed twitch evoked by cortical stimulation increases in size (e.g., Gandevia et al., 1996; Goodall et al., 2010; Sidhu et al., 2009b; Søgaard et al., 2006; Taylor et al., 2000, 2006). This indicates that there is a failure of the motor cortex to produce sufficient drive to the motoneurons. This failure is known as supraspinal fatigue. It should be noted that the occurrence of suboptimal drive from the motor cortex does not necessarily mean a decrease in descending drive as the decreased responsiveness of the motoneurons suggests that they require more drive to produce the same output.

It is not clear whether events in the motor cortex itself contribute directly to supraspinal fatigue or whether the failure occurs because of input to the motor cortex. Indeed, the behavior of cells in the motor cortex during fatiguing exercise is not yet clear. Responses to transcranial magnetic stimulation suggest both increased excitation and increased inhibition *during* fatiguing voluntary contraction, but indicate the opposite changes, decreased excitation and decreased inhibition, when the muscle is at rest between or after contractions. More specifically, short-latency excitatory responses (motor evoked potentials, MEPs) grow in size during both sustained maximal and submaximal efforts, but at the same time, the silent period, which is an inhibitory response following the MEP, increases in duration (e.g., Ljubisavljevic et al., 1996; McKay et al., 1996; Taylor et al., 1996; Taylor & Gandevia, 2001). Furthermore, inhibition caused by subthreshold magnetic stimuli increases in depth (Seifert & Petersen, 2010). The net effect of these changes on corticospinal output is not known, and their contribution to supraspinal fatigue is doubtful because, at least under some circumstances, supraspinal fatigue can be seen when these changes are not evident (Gandevia et al., 1996). Other changes are seen when the muscle is at rest after the exercise. The MEP is reduced for many minutes (e.g., Samii et al., 1996a, 1996b, 1998; Taylor & Gandevia, 2001) but again, any effect on voluntary output is doubtful as depression of the MEP is not seen during voluntary activity. The depression is not due to increased intracortical inhibition, as paired-pulse stimulation shows that intracortical inhibition is reduced after exercise (Benwell et al., 2006; Maruyama et al., 2006; Takahashi et al., 2009). Thus, although responses to TMS show that various fatigue-related changes occur in the motor cortex, neither the mechanisms of these changes nor their effects on motor performance are known. Functional magnetic resonance imaging (fMRI) has also been used to examine cortical behavior during fatigue (Liu et al., 2003, 2005). Fatiguing maximal or submaximal abduction contractions of the index finger lead to increases in activity in a network of motor areas including the sensorimotor cortex, supplementary motor area, premotor and cingulate cortex (Post et al., 2009; van Duinen et al., 2007). However, some of this activity may represent drive to other muscles that are increasingly recruited with fatigue.

Muscle afferents in fatigue

While muscle fatigue is defined in terms of the output of the motor pathway, fatiguing exercise also results in changes in sensory feedback from the muscle. This feedback influences central fatigue through interactions with motor output at the spinal and cortical levels. In addition, feedback from the muscle subserves the sensations of muscle fatigue such as discomfort or weakness. Muscles are well supplied with sensory endings which give feedback to the nervous system on mechanical and chemical changes in the muscle. Both the large- and small-diameter muscle afferents interact with the motor pathway at multiple levels and also contribute to perception. Muscle spindles and Golgi tendon organs are supplied by large-diameter myelinated axons and contribute to proprioception (Gandevia, 1996; Liu et al., 2005; Windhorst, 2007, 2008). Muscle spindle endings provide feedback on muscle length and changes in muscle length (Matthews, 1981; Proske & Gregory, 2002). Their signals contribute to the perceptions of joint position and movement. Golgi tendon organs give a dynamic signal of the force produced by the muscle and contribute to the perception of muscle forces (Horcholle-Bossavit et al., 1990; Proske & Gregory, 2002). The muscle is also supplied by the smaller myelinated group III and unmyelinated group IV afferents, which are mechanically and chemically sensitive (Graven-Nielsen & Mense, 2001; Kaufman et al., 1983; Kaufman & Rybicki, 1987; Kaufman et al., 1984; Mense, 2009; Mense & Stahnke, 1983). These afferents subserve the sensations of muscle fatigue and pain.

The effects of fatigue on the firing of large-diameter muscle afferents are not completely clear. Muscle spindle firing rates measured through microneurography decrease during a sustained submaximal contraction but this is a very restricted example of exercise (Macefield et al., 1991). The decrease in spindle firing is supported by indirect evidence which shows a decrease in reflex responses to muscle stretch, although reflexes rely on the excitability of the motoneurons as well as on the afferent volley evoked from the muscle spindles (Avela et al., 1999; Duchateau & Hainaut, 1993). Finally, subjects' perceptions of limb movement and position can be impaired during fatigue, which also suggests some impairment of spindle firing (e.g., Carpenter et al., 1998; Forestier et al., 2002; Hiemstra et al., 2001; Voight et al., 1996). In contrast, studies in animals have suggested that muscle spindles may have increased sensitivity to dynamic movements in fatigue (Fischer & Schafer, 2005; Hill, 2001). Differences may be due to the influence of fusimotor drive, which is the motor supply to the specialized intrafusal muscle fibers on which the muscle spindle sensory endings are located (Birznieks et al., 2008; Pedersen et al., 1998).

The firing of Golgi tendon organs reflects the force produced by the muscle fibers that insert through the tendon organs. Thus, if these fibers produce less force the tendon organ's firing is reduced, but there is little evidence that fatigue has other independent effects on tendon organ firing. With very strong contractions, the tendon organs may show some adaptation (Thompson et al., 1990) but in general the ensemble discharge of the tendon organs from a muscle should continue to signal the force output of the muscle during fatigue. Tendon organ firing could contribute to sensations of weakness as their input will be lower than expected for a given motor command.

During fatiguing contractions, the firing of group III and IV afferents increases. Some of the afferents tend to be more mechanically sensitive and fire throughout contractions whereas others fire as metabolites accumulate in the muscle (Kaufman et al., 1983; Kaufman & Rybicki, 1987; Rotto & Kaufman, 1988). Ischemia also increases group III and IV afferent firing in strong contractions when intramuscular pressure is high and prevents blood flow to the muscle (Hayes et al., 2006; Kaufman et al., 1984). These afferents have diverse effects during exercise. They contribute to reflex increases in blood pressure, heart rate and ventilation (e.g., Mitchell & Johnson, 2003; Mitchell et al., 1983; Secher & Amann, 2012). At a segmental level, they act to inhibit the motoneurons of some muscles and facilitate the motoneurons of other muscles (Kniffki et al., 1979, 1981; Martin et al., 2006). They also decrease excitatory input to the motoneurons by presynaptic inhibition of the muscle spindle primary (Ia) afferents (Pettorossi et al., 1999; Rossi et al., 1999b), and may mediate other changes in muscle spindle firing through effects on the fusimotor neurons (Ljubisavljevic & Anastasijevic, 1994; Pedersen et al., 1998). At a cortical level, they contribute to supraspinal fatigue although this may not be through a direct effect on the motor cortex (Gandevia et al., 1996). Finally, they convey those sensations of fatigue that are described as muscle pain or burning (O'Connor & Cook, 1999).

Mental fatigue

Mental fatigue refers to feelings of 'tiredness or exhaustion' that may be experienced after periods of demanding cognitive activity (Boksem & Tops, 2008). It does not directly fit into the scheme of muscle fatigue or fatigue produced by exercise, in that it has not been shown to reduce the force or power of maximal voluntary contractions, nor are the feelings of 'tiredness' generated through feedback from muscle afferents. However, there may well be interactions between fatigue caused by physical exertion and fatigue caused by mental exertion. Mental fatigue produced by prolonged cognitive effort has recently been show to influence exercise endurance (Marcora et al., 2009). Where mental fatigue plays a part in the decision to stop exercise, it may be a contributor to central fatigue. In addition, the cognitive effects of mental fatigue such as decreased attention, increased distractibility, reduced monitoring for correct performance of the task and poor selection of actions (e.g., Boksem et al., 2005; Boksem et al., 2006; van der Linden et al., 2003) are likely to be important for the performance of many complex motor tasks. Reaction times can be prolonged and sensory gating is impaired (van der Linden & Eling, 2006; van der Linden et al., 2006). With motor tasks like driving a car, performance deteriorates with time, not because of muscle fatigue, but because of mental fatigue (Campagne et al., 2004; Philip et al., 2003).

It is not clear whether exercise itself can lead to mental fatigue. Occasional studies indicate impairment of some cognitive functions immediately after high intensity or prolonged exercise (e.g., Coco et al., 2009; Eich & Metcalfe, 2009; Lo Bue-Estes et al., 2008; see Tomporowski, 2003). However, acute bouts of aerobic exercise are usually followed by improvements in cognitive performance (for review Lambourne & Tomporowski, 2010). Even when a cognitive task is performed at the same time as a motor task, cognitive functioning is often improved although tasks that involve the prefrontal cortex may be impaired (Brisswalter et al., 2002; Dietrich & Sparling, 2004; Lambourne & Tomporowski, 2010; Tomporowski, 2003). Dietrich and Audiffren (2011) propose that whole-body (locomotor) exercise improves cognitive functions that employ implicit processing and impairs those that require explicit processing. In contrast, performance of a fatiguing motor task with the arm or hand can impair simultaneous cognitive performance as tested with reaction time tasks (Lorist et al., 2002; Terrier & Forestier, 2009; Zijdewind et al., 2006). Presumably, motor tasks that require a high level of accuracy and attention will cause mental fatigue if prolonged, and in turn, the performance of such tasks may well be susceptible to impairment by mental fatigue.

Fatigue and motor performance

In general, muscle fatigue caused by exercise leads to reduced force output from the muscle fibers and slowing of their contraction and relaxation. Motoneurons require more descending drive to maintain their firing rates. Subjects report that more effort is needed to perform the same actions, and the firing of small-diameter fatigue-sensitive muscle afferents is perceived as discomfort and fatigue of the muscles (Figure 17.6B). Other effects include more variability in force output, clumsiness and a reduction in accuracy of skilled movements.

Variation in fatigue across motor units in a muscle

Fatigue does not affect all motor units in a similar way, even when only a single muscle performs a task. When a high force or high velocity muscle contraction is performed, then most of the motor units in the muscle are working but the large, fast-twitch muscle fibers are most easily fatigued. Thus, most of the loss of force comes from these fibers, whereas the smaller, slower fibers are little fatigued (Carpentier et al., 2001; Gollnick et al., 1974a, 1974b; Karatzaferi et al., 2001; Sargeant, 2007). In contrast, with a lower force muscle contraction, only a proportion of the motor units in the muscle are engaged in the task. The recruitment of motoneurons in a voluntary contraction follows the size principle (Henneman et al., 1974; Milner-Brown et al., 1973), with smaller, low-threshold motoneurons recruited before larger ones. Furthermore, low-threshold motoneurons innervate fatigue-resistant muscle fibers and produce low forces (Burke et al., 1973; Cope & Clark, 1991; Gordon et al., 2004). Hence, fatigue progresses slowly in a low force task (e.g., Kouzaki & Shinohara, 2006; Smith et al., 2007) (Figure 17.6) but the motor units that are first affected must be relatively fatigue-resistant because the more fatigable motor units are not recruited (Beltman et al., 2004). Both larger (high-threshold) and smaller motoneurons become less excitable with repeated firing (McNeil et al., 2009, 2011) but larger motoneurons show more spike frequency adaptation (Button et al., 2007; Kernell & Monster, 1982; Spielmann et al., 1993), which suggests that high threshold motor units may be more susceptible to fatigue at the motoneuron level as well as in the muscle fibers.

The pattern of motor unit recruitment and the tendency of later-recruited motor units to fatigue more quickly mean that the effects of muscle fatigue on motor performance depend both on characteristics of the fatiguing exercise and also on the task to be performed with the fatigued muscle. For example, if fatigue has been brought about by exercise which requires high force (or power), then performance of a task that requires only low-threshold motor units may be little affected (Nessler et al., 2011). Fatigue caused by high power sprints has little effect on the kinetics or the muscles engaged in low power cycling or running (Morin et al., 2012). However, if the task to be performed involves the generation of maximal force or power, then its performance must be impaired. Contractions will be weaker and slower (e.g., Bigland-Ritchie et al., 1983; De Ruiter et al., 1999; James et al., 1995; Jones et al., 2006; Sargeant, 2007).

In the contrasting condition, when fatigue starts to develop during a submaximal task, additional motor units can be recruited to compensate for the decrease in the force generated by those motor units which were initially engaged in the task (e.g., Adam & De Luca, 2003, 2005; Bigland-Ritchie et al., 1986a, 1986b; Kuchinad et al., 2004). This recruitment requires additional descending drive and can usually be seen as an increase in the electrical activity recorded from the muscle by using surface electromyography (e.g., Bigland-Ritchie et al., 1986a; Fuglevand et al., 1993; Löscher et al., 1996; Petrofsky & Phillips, 1985; Schmitt et al., 2009; Søgaard et al., 2006) (see Figure 17.1b). The additional recruitment is also associated with an increase in the effort required to perform the task (Hasson et al., 1989; Kilbom et al., 1983; Søgaard et al., 2006) (Figure 17.6b). There is also increased force variability as larger motor units provide less fine grading of force output. As fatigue is progressive, the recruitment of new units is also progressive if the task is maintained (Figure 17.8). In addition, the newly-recruited higher-threshold motor units tend to be more fatigable. They lose force more quickly than those originally employed in the task, so that further units need

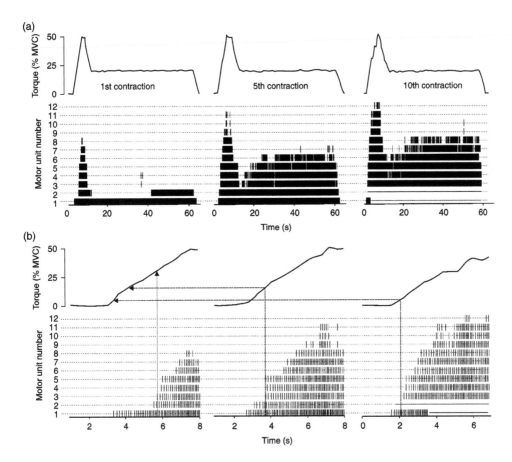

Figure 17.8 Recruitment of motor units during fatigue. Subjects made repeated knee extensor contractions. In each contraction they increased torque to 50 percent maximal voluntary contraction (MVC), then decreased to 20 percent MVC which they held for ~1 min. Multiple motor units were recorded during the contractions. (a) Three contractions (1st, 5th and 10th) in the series performed by one subject are shown. The torque profiles are very similar for the three contractions. Each firing of a motor unit is represented by a vertical line. Motor unit 1 is first recruited and is on the bottom row, motor unit 2 on the next row and so on. In the 1st contraction, 8 motor units were recruited during the ramp up to 50 percent MVC but only one of these initially fired during the maintained 20 percent MVC. As the muscle became progressively fatigued both within contractions and with successive contractions, more and more MUs were recruited. Solid lines for motor units 1 and 2 in the final contraction indicate that they could no longer be identified. (b) The starts of the same three contractions are shown on an expanded scale. Motor units maintain their order of recruitment but are recruited at lower forces in the successive contractions. This suggests that each motor unit produces less force with fatigue. Recruitment thresholds for motor unit 3 are shown. Data from Adam and DeLuca (2003).

to be recruited more and more quickly. Ultimately, the task can no longer be performed (Hunter et al., 2004; Maluf & Enoka, 2005). An example of such a submaximal task is holding a book at arm's length. Initially, it is easy. It gradually requires more effort and eventually can no longer be performed.

If a submaximal task is completed before all motor units are recruited (i.e. someone takes the book from your hand before you can no longer hold it), then fatigue will be concentrated in the mid-range motor units. The largest motor units will have performed no work and will not be fatigued, whereas the smallest units will have performed work but they are the least fatigable. Hence, for the whole muscle, maximal force will be reduced because the mid-range units have lost some ability to produce force, but the speed of contraction may be preserved because the biggest units are still fresh. In this situation, performance of a submaximal task that uses the fatigued muscle fibers will require more motor units and so more descending drive than if the muscle were fresh. The recruitment of the larger non-fatigued motor units is likely to add variability to force and movement.

Variability in muscle output

Force output is more variable and movements are less smooth during and after fatiguing contractions. There are long-lasting increases in tremor across all frequencies up to 15 Hz after strong efforts (Furness et al., 1977). In surgeons, physiological tremor increases with the duration of operations (Slack et al., 2009). During sustained submaximal contractions, increasing bursts are seen in the EMG (Hunter & Enoka, 2003) (Figure 17.1b). It is likely that multiple mechanisms contribute to increased force fluctuations during fatigue and may depend on the level of muscle activation. Some of the increased variability may be the recruitment of larger motor units, which will create larger force variations if they are firing at low rates and are not fused. Changes in muscle afferent feedback and the stretch reflex loop may contribute to increases in physiological tremor with a frequency of 8-12 Hz (Cresswell & Löscher, 2000; Sanes, 1985) but central factors may also contribute as an effort performed during nerve block can increase subsequent tremor (Furness et al., 1977). Furthermore, fatigue of one arm can increase tremor in the other arm (Morrison & Sosnoff, 2010) and fatigue of one finger can increase tremor in all the fingers (Hwang et al., 2009). In contrast, in stronger sustained submaximal contractions, increases in force variability are not due to physiological tremor but occur at other frequencies (Ebenbichler et al., 2000). Cross-correlation between motor units suggests an increase in common drive to the motoneurons, which may be related to altered muscle spindle input (Contessa et al., 2009).

Co-ordination of muscles

Generally, tasks are not performed by one muscle or even by synergists acting about one joint. They require multiple muscles acting about multiple joints and this provides many more degrees of freedom for the nervous system to control, and also for the nervous system to use to compensate for fatigue of particular muscles (e.g., Danion et al., 2001; Gates & Dingwell, 2008; Gorelick et al., 2003; Huffenus et al., 2006; Rodacki et al., 2002; Singh et al., 2010; van Dieen et al., 1996). For example, in tasks like throwing a ball or hammering or sawing, altered patterns of muscle recruitment can compensate for fatigue of specific muscle groups so that the endpoint of the movement (e.g. the ball, hammer tip or saw) can be maintained (Cote et al., 2008; Cote et al., 2002; Forestier & Nougier, 1998; Huffenus et al., 2006). However, with higher levels of fatigue compensation is no longer possible (Bonnard et al., 1994). Furthermore, as in single joint tasks, if the performance requires near maximal force or power, it will be impaired by fatigue. For example, the velocity of the ball decreases as the arm muscles of baseball pitchers, or the leg muscles of footballers, become fatigued (Apriantono et al., 2006; Escamilla et al., 2007). Moreover, some of the ability to compensate for fatigue in complex tasks depends on expertise and so is presumably the result of learning (Aune et al., 2008; Kiryu et al., 2011).

Motor learning with fatigued muscles

It is difficult to encapsulate the effects of muscle fatigue on motor learning because both fatigue and motor learning cover a wide range of mechanisms and behaviors. However, there are two main questions to ask about the effects of fatigue on motor learning. First, does fatigue impair the process of learning so that it takes longer or is less effective? Second, is there some specificity to learning a task with muscle fatigue so that it is then performed best under the same conditions of muscle fatigue, and is that performance better than if the task was learnt with fresh muscles? Various combinations of learning task and fatigue-producing exercise have been used to attempt to answer these questions. Both positive and negative answers have resulted with no clear consensus. A number of possible factors that may influence whether studies demonstrate impaired motor learning are discussed below.

Does fatigue impair the process of learning?

Complexity of task

While some studies report impaired motor learning after fatiguing exercise, others do not. One study in mice has suggested that learning complex tasks may be more impaired than simple tasks (Mizunoya et al., 2004). Preceding fatiguing exercise did not affect the performance of mice on a simple variant of a water maze, in which the mice must locate an underwater platform, but performance on a more complex variant was reduced. However, in humans, studies which suggest that the process of motor learning is not impaired by fatigue have included a range of tasks from some that largely use the arms, such as adaptation to a force-field during reaching (Takahashi et al., 2006), visuomotor tracking (Mierau et al., 2009), and bouncing a volleyball off a wall to hit a target (Cotten et al., 1974), through to tasks that use the whole body, such as tracking with a large stylus that was held in two hands and required sway of the body (Williams & Cooper, 1976) or tasks with multiple components (run and balance a ball, kick a ball, bounce and catch a ball etc. Stockard, 1974). Similarly, studies which suggest that the process of learning is impaired by exercise-related fatigue have also ranged from targeted single arm movements (Kerr, 1993), through to whole body tasks such as ladder climbing (Arnett et al., 2000a, 2000b; Carron, 1972), and complex sequences of hopping and jumping (Berger & Smith-Hale, 1991). Thus, the particular muscle groups used and the task complexity do not appear to be critical factors in the impairment of motor learning by fatigue.

Level of peripheral fatigue of muscles employed by the task

Under some circumstances one would expect that the level of peripheral fatigue should impair motor learning. One theoretical limit is when muscles that are critical to the task have lost so much force or power that the task is impossible. For example, we can think of learning to perform a task that requires accurate submaximal force output from a muscle group. With fatigue of the muscle, then, as discussed above, recruitment of more motor units will be required to produce the target force than if the muscle was fresh. To recruit the 'extra' motor units will require more descending drive, and it is likely that sensory feedback from the fatigued muscle will be altered. However, if fatigue is severe enough, then recruitment of all the motor units may not be sufficient to compensate for the fatigue-induced weakness of the muscle fibers. It will not be possible to produce the target force and it would be surprising if practice could improve subjects' ability to produce the target accurately. The same argument should hold for more complex tasks although it may be difficult to identify whether fatigue of a specific muscle or muscle group has pushed the task beyond the capacity of the musculoskeletal system. Furthermore, the critical level of fatigue must depend on the task. If the task involves high forces or fast movements, say a gymnastic manoeuvre, then relatively little loss of power may put it beyond capacity, whereas if it requires low forces and slow movements, say writing with a pen, then only extreme peripheral fatigue would make it impossible.

If peripheral fatigue limits performance, it may impair motor learning, even when the task is not impossible for the muscles to carry out. For some tasks, the aim is to perform as quickly or as strongly as possible. When the muscle is weaker and slower through peripheral fatigue, then the best possible performance of the task will be reduced from that possible with fresh muscle. If performance after practice with fatigued muscles is not as good as that after practice with fresh muscles, then learning of the skill may be considered reduced. The need to use as much power as possible is an overt requirement in sporting skills like shot put or high-jump. It is less obvious – but the requirement for speed and muscle strength is also likely to be a factor in tasks such as ladder climbing, in which as many rungs as possible must be climbed before overbalancing. It may contribute to reported impairment in learning this task with fatigue (e.g., Arnett et al., 2000a, 2000b; Carron, 1972).

A third aspect to consider is whether it is slower or more difficult to learn tasks when they require recruitment of a larger part of the muscle or muscle group. That is, they require more motor output and more effort. Although evidence is limited, it seems that tasks which require the use of bigger forces (and more effort) are not necessarily slower or more difficult to learn than those which require low forces (Enebo & Sherwood, 2005). Furthermore, when motor adaptation was studied by subjects reaching toward a target within an applied force-field, the time course of adaptation to the force-field was not altered after the muscles used in reaching were fatigued (Takahashi et al., 2006). This lack of effect on learning occurred despite changes in movement quality, which were probably due to the increase in the proportion of muscle recruited. Although a high force or power task may not be innately difficult to learn, practice of it will quickly start to fatigue the muscle which may then limit possible performance or even make the task impossible.

Changes in sensory feedback

Sensory feedback from the muscle has a role in motor learning (e.g., Avanzino et al., 2012; Ghez et al., 1990; Lundbye-Jensen et al., 2011; Rothwell et al., 1982) and, as noted above, muscle afferent feedback is altered after fatiguing exercise. There are changes in the firing of proprioceptive, large-diameter muscle afferents with fatigue and also in the small-diameter afferents that convey the feelings of muscle pain and fatigue. Both large-diameter and small-diameter muscle afferents can also influence firing of neurons at multiple sites in the motor pathway. However, it is not known whether these changes affect motor learning.

While proprioceptive feedback is changed when tasks are performed with fatigued muscle, it is not clear whether the signals are degraded, or altered because different motor output is required to complete the task, or both. However, learning of various motor tasks can take place even with deliberate interference with proprioceptive input through muscle vibration and can occur at the same rate as when proprioception is intact even if performance is impaired (Radhakrishnan et al., 2008; Vidoni & Boyd, 2008). Indeed, visuomotor adaptation is possible even with no large-diameter input from the body (Bernier et al., 2006), although other studies show conflicting results (Bard et al., 1995; Guedon et al., 1998). One possibility is that altered somatosensory feedback impairs adaptation to mechanical perturbations but not to visual distortions (Pipereit et al., 2006). Thus, changes in proprioceptive feedback with fatigue are unlikely to impair learning of tasks which rely on visual feedback (Williams & Cooper, 1976). There may well be an impact on tasks where there is no visual guidance but as yet this is uncertain.

The effect of small-diameter muscle afferent firing on motor learning is also unclear. Experimental muscle pain in humans causes changes in motor output (e.g., Ervilha et al., 2005; Farina et al., 2005; Graven-Nielsen et al., 1997) but in one simple motor learning task, it does not change increases in finger acceleration with practice, nor the associated alteration in finger movements evoked by cortical stimulation (Ingham et al., 2011). However, one study of novel tongue movements has shown that learning can be impaired both by anesthesia of the tip of the tongue and by intra-oral pain induced through surface application of capsaicin (Boudreau et al., 2010).

Changes in the central nervous system

In general, central fatigue is usually seen with respect to activation of the muscles that have taken part in the exercise rather than being a general effect, so that when a task is learnt by muscles that did not take part in the fatiguing exercise neither performance nor learning is impaired (Apriantono et al., 2006; Dickinson et al., 1979). In addition, central fatigue is often short-lived after exercise. This means that peripheral fatigue is almost always present whenever there is central fatigue and this makes it difficult to identify any specific effects of central fatigue on motor learning. One feature of central fatigue that may be important is the mismatch between effort and motoneuron output that occurs in some low-force fatiguing contractions (e.g., Smith et al., 2007; Søgaard et al., 2006). This may be problematic if motor output to perform a task is learnt in terms of effort as suggested by Takahashi et al. (2006). Motor cortical stimulation after fatiguing exercise has shown a decrease in cortical excitability for tens of minutes, although this change is seen only when the muscles are at rest and does not appear to be directly linked to performance (e.g., Avanzino et al., 2011; Brasil-Neto et al., 1993; Samii et al., 1996b). Recently, it has been suggested that it represents a type of cortical plasticity and that it can interact with artificially induced plastic changes in the motor cortex (Milanovic et al., 2011). This suggests that it may impact on motor learning but as yet there is no evidence.

Very long-term exercise is likely to impact on learning if there is associated sleep deprivation. When mice were kept for several days in a cage in which the floor was water-covered their exercise endurance and spatial memory were impaired and this was associated with a reduction in dendritic spines on corticospinal neurons (Chen et al., 2009). This is likely to be associated with reduced plasticity in the motor cortex. In humans, sleep deprivation and fatigue lead to slower learning of motor skills, as seen in medical residents on night shift who learnt endovascular techniques more slowly than their day-shift colleagues (Naughton et al., 2011). However, these effects are unlikely to result from muscle fatigue and any associated central nervous system changes but are linked to disturbed sleep.

Is there specificity in learning during fatigue?

During muscle fatigue, as discussed, muscles are generally weaker and slower so that different motor commands are needed to perform a task with fatigued muscles compared with non-fatigued muscles. Furthermore, proprioceptive feedback is not the same in the fatigued and non-fatigued states and these changes may be compounded by altered responses of motoneurons and motor cortical cells. These differences suggest that learning a skill during muscle fatigue is not the same as learning the skill without fatigue. When subjects learnt to reach in a force-field with fatigued muscles, retention was tested with muscles which were partially recovered from fatigue (Takahashi et al., 2006). Performance suggested that subjects retained an internal model consistent with the muscle being fatigued and produced motor commands that recruited more muscle than if they had learnt the task in an unfatigued state. However, like the evidence for an effect of fatigue on motor learning, the evidence for specificity in learning more complex tasks during fatigue is mixed. Some studies have suggested an advantage in learning under fatigued conditions when retention of learning is assessed under the same conditions (Kerr, 1993; Williams & Cooper, 1976) but other studies have not found an advantage (Williams et al., 1976b) or find that it is partial or only occurs with a specific level of fatigue or with men but not women (Anshel & Novak, 1989; Arnett et al., 2000a, 2000b; Stockard, 1974; Williams et al., 1976a). These mixed results suggest that any advantage of learning during fatigue may be small or inconsistent, or may only be demonstrated under specific conditions. Factors that may be important include the level of fatigue of muscles involved in the task, whether task performance is impaired by the fatigue and whether the task is under visual guidance or relies on proprioceptive feedback.

For example, Kerr (1993) describes a task in which participants learnt a complex arm trajectory. Subjects learnt the movement with fresh or fatigued muscles and then, on a second day, some performed the movement with fresh muscles and some with fatigued muscles. Those subjects who acquired the movement and then performed it under the same conditions (either both with fatigued or both with fresh muscles) performed better in terms of timing and overall error than those who swapped conditions. In this study, the muscles which were fatigued were involved in the task. Thus, it is likely that different neural output was required in the fatigued and non-fatigued conditions. However, the task did not rely on high forces or fast movements so that adequate performance was possible despite fatigue. Finally, subjects did not receive on-line visual feedback of their performance but only of each completed trajectory so that the primary feedback during the movement was proprioceptive.

In contrast, Anshel and Novak (1989) examined shot put, in which high explosive forces must be engaged with skilled timing. Here, when the best performance relies on maximally recruiting appropriate muscles, the neural output with and without fatigue is not likely to be very different so that any advantage of learning the task with fatigued muscles is likely to be small. Similarly, in a ladder climbing task where the aim was to climb as many rungs as possible in 20 s, performance was likely limited by the muscles' output so that better control of that output had little scope to improve performance (Williams et al., 1976a). A ladder climbing task, in which the aim was to climb as high as possible without overbalancing, may move the balance of requirements from muscle output towards accurate control. Here, performance when acquiring the skill was less impaired by fatigue for men than for women, and when retention of learning was tested during fatigue, there was an advantage in learning during fatigue for men but not women (Arnett et al., 2000b). This suggests that such an advantage may only be fully realised if the combination of fatigue and task is such that altered neural command can largely compensate for muscle fatigue so that performance can be similar in the non-fatigued and fatigued conditions.

It has been suggested that specificity for learning under fatigued conditions does not occur for tasks that rely highly on visual or other exteroceptive feedback (Williams et al., 1976b; Williams & Cooper, 1976). However, in visuomotor adaptation, aftereffects, which imply sensorimotor recalibration, occur when visual feedback is given during movements but not when it is given after movements (Hinder et al., 2010). Furthermore, adaptation to the fatigued state of the muscle occurs during visual guidance and results in motor commands that are appropriate for the fatigued muscle (Takahashi et al., 2006). On the other hand, it is unknown whether adapting to a force-field with fatigued muscles improves retest performance with fatigued muscles. Presumably, adjustment of motor command to the present state of the muscle occurs all the time and so may be retained only in the short term (Demougeot & Papaxanthis, 2011; Kording et al., 2007). Alternatively, if a visually guided task requires relatively slow movements which allow for multiple error corrections, then the motor system may adapt to a new muscle state very quickly.

Summary and conclusions

During muscle fatigue, the muscle becomes weaker and slower, the motoneurons become harder to drive, cortical activity changes and motor tasks become more effortful. Proprioceptive sensory input from the muscles is altered and the firing of small-diameter muscle afferents gives the sensations of muscle fatigue and/or pain. Force output becomes more variable. The progress of these processes and their relative importance is task-dependent. Only muscle fibers that are active become fatigued and it is likely that the active motoneurons are also most affected. High levels of motor output quickly lead to impaired performance, whereas with low levels of activity, the progress of fatigue is slow and, in addition, fatigue of the active motor units can be compensated by recruitment of more units. Multiple muscles are involved in most tasks and this allows the nervous system to maintain performance of a task by varying drive to different muscles. However, when fatigue is great enough performance will be impaired. The level of fatigue that impairs performance is also task-dependent. Strong or fast muscle contractions will be impaired with a relatively small loss of force or power. Weak, slow muscle contractions can be performed despite relatively large drops in force and power, but may be impaired by tremor or other mechanisms of increased force variability.

While sleep deprivation seems likely to impair the process of learning, it remains unclear whether muscle fatigue does so because it is difficult to disentangle learning from the impairment of performance. Often the expert performance of a motor task relies on strength or speed, so that when less skill is demonstrated with fatigued than non-fatigued muscle after the same practice it could be a deficit in either, or both, performance and learning. In reaching in a force field, the time course of adaptation is the same with and without fatigue, but this is unlikely to be the same process as learning a skill.

If a motor task is performed using fatigued motor units, then motor commands must differ compared to those when there is no fatigue. Proprioceptive feedback will also differ. However, the advantage of learning a task during fatigue to its later performance under the same fatigue conditions seems slight. An advantage is likely to be apparent only under specific circumstances when the fatigue and task are such that the motor commands differ appreciably, yet the task can be performed just as well with the fatigued and non-fatigued muscle. Furthermore, this advantage may be quite short-lasting as the adaptation of motor commands to altered muscle states occurs continuously.

References

Adam, A., and De Luca, C. J. (2003) 'Recruitment order of motor units in human vastus lateralis muscle is maintained during fatiguing contractions', *J. Neurophysiol.*, 90: 2919–2927.

Adam, A., and De Luca, C. J. (2005) 'Firing rates of motor units in human vastus lateralis muscle during fatiguing isometric contractions', *J. Appl. Physiol.*, 99: 268–280.

Allen, D. G., Lamb, G. D., and Westerblad, H. (2008) 'Skeletal muscle fatigue: cellular mechanisms', *Physiol. Rev.*, 88: 287–332.

Allen, G. M., McKenzie, D. K., and Gandevia, S. C. (1998) 'Twitch interpolation of the elbow flexor muscles at high forces', *Muscle Nerve*, 21: 318–328.

Anshel, M. H., and Novak, J. (1989) 'Effects of different intensities of fatigue on performing a sport skill requiring explosive muscular effort: a test of the specificity of practice principle', *Percept. Mot. Skills*, 69: 1379–1389.

Apriantono, T., Nunome, H., Ikegami, Y., and Sano, S. (2006) 'The effect of muscle fatigue on instep kicking kinetics and kinematics in association football', *J. Sports Sci.*, 24: 951–960.

Arnett, M. G., Bennett, C., Gilmartin, K., and DeLuccia, D. (2000a) 'The effect of specific anaerobically induced fatigue on skill acquisition and transfer performance', *Physic. Educat.*, 57: 216–221.

Arnett, M. G., DeLuccia, D., and Gilmartin, K. (2000b) 'Male and female differences and the specificity of fatigue on skill acquisition and transfer performance', *Res. Q. Exerc. Sport*, 71: 201–205.

Ash, I. E. (1914) 'Fatigue and its effects upon control', *Arch. Psychol.*, 31: 1–61.

Aune, T. K., Ingvaldsen, R. P., and Ettema, G. J. (2008) 'Effect of physical fatigue on motor control at different skill levels', *Percept. Mot. Skills*, 106: 371–386.

Avanzino, L., Giannini, A., Tacchino, A., Abbruzzese, G., and Bove, M. (2012) 'The role of proprioception in the consolidation of ipsilateral 1Hz-rTMS effects on motor performance', *Clin. Neurophysiol.*, 123: 577–581.

Avanzino, L., Tacchino, A., Abbruzzese, G., Quartarone, A., Ghilardi, M. F., Bonzano, L., Ruggeri, P., and Bove, M. (2011) 'Recovery of motor performance deterioration induced by a demanding finger motor task does not follow cortical excitability dynamics', *Neuroscience*, 174: 84–90.

Avela, J., Kyrolainen, H., Komi, P. V., and Rama, D. (1999) 'Reduced reflex sensitivity persists several days after long-lasting stretch-shortening cycle exercise', *J. Appl. Physiol.*, 86: 1292–1300.

Babault, N., Pousson, M., Ballay, Y., and Van Hoecke, J. (2001) 'Activation of human quadriceps femoris during isometric, concentric, and eccentric contractions', *J. Appl. Physiol.*, 91: 2628–2634.

Balog, E. M., Thompson, L. V., and Fitts, R. H. (1994) 'Role of sarcolemma action potentials and excitability in muscle fatigue', *J. Appl. Physiol.*, 76: 2157–2162.

Bard, C., Fleury, M., Teasdale, N., Paillard, J., and Nougier, V. (1995) 'Contribution of proprioception for calibrating and updating the motor space', *Can. J. Physiol. Pharmacol.*, 73: 246–254.

Behm, D. G., St-Pierre, D. M., and Perez, D. (1996) 'Muscle inactivation: assessment of interpolated twitch technique', *J. Appl. Physiol.*, 81: 2267–2273.

Belanger, A. Y., and McComas, A. J. (1981) 'Extent of motor unit activation during effort', *J. Appl. Physiol.*, 51: 1131–1135.

Beltman, J. G., de Haan, A., Haan, H., Gerrits, H. L., van Mechelen, W., and Sargeant, A. J. (2004) 'Metabolically assessed muscle fibre recruitment in brief isometric contractions at different intensities', *Eur. J. Appl. Physiol.*, 92: 485–492.

Benwell, N. M., Sacco, P., Hammond, G. R., Byrnes, M. L., Mastaglia, F. L., and Thickbroom, G. W. (2006) 'Short-interval cortical inhibition and corticomotor excitability with fatiguing hand exercise: a central adaptation to fatigue?' *Exp. Brain Res.*, 170: 191–198.

Berger, R. A., and Smith-Hale, L. A. (1991) 'Effects of fatigue on performance and learning of a gross motor task', *J. Appl. Sport. Sci. Res.*, 5: 155–161.

Bernier, P. M., Chua, R., Bard, C., and Franks, I. M. (2006) 'Updating of an internal model without proprioception: a deafferentation study', *Neuroreport*, 17: 1421–1425.

Bigland-Ritchie, B., Cafarelli, E., and Vollestad, N. K. (1986a) 'Fatigue of submaximal static contractions', *Acta Physiol. Scand. Suppl.*, 556 137–148.

Bigland-Ritchie, B., Furbush, F., and Woods, J. J. (1986b) 'Fatigue of intermittent submaximal voluntary contractions: central and peripheral factors', *J. Appl. Physiol.*, 61: 421–429.

Bigland-Ritchie, B., Johansson, R., Lippold, O. C., and Woods, J. J. (1983) 'Contractile speed and EMG changes during fatigue of sustained maximal voluntary contractions', *J. Neurophysiol.*, 50: 313–324.

Bigland-Ritchie, B., Jones, D. A., Hosking, G. P., and Edwards, R. H. (1978) 'Central and peripheral fatigue in sustained maximum voluntary contractions of human quadriceps muscle', *Clin. Sci. Mol. Med.*, 54: 609–614.

Bigland-Ritchie, B., Jones, D. A., and Woods, J. J. (1979) 'Excitation frequency and muscle fatigue: electrical responses during human voluntary and stimulated contractions', *Exp. Neurol.*, 64: 414–427.

Bigland-Ritchie, B., and Woods, J. J. (1984) 'Changes in muscle contractile properties and neural control during human muscular fatigue', *Muscle Nerve*, 7: 691–699.

Birznieks, I., Burton, A. R., and Macefield, V. G. (2008) 'The effects of experimental muscle and skin pain on the static stretch sensitivity of human muscle spindles in relaxed leg muscles', *J. Physiol.*, 586: 2713–2723.

Boksem, M. A., Meijman, T. F., and Lorist, M. M. (2005) 'Effects of mental fatigue on attention: an ERP study', *Brain Res. Cogn. Brain Res.*, 25: 107–116.

Boksem, M. A., Meijman, T. F., and Lorist, M. M. (2006) 'Mental fatigue, motivation and action monitoring', *Biol. Psychol.*, 72: 123–132.

Boksem, M. A., and Tops, M. (2008) 'Mental fatigue: costs and benefits', *Brain Res. Rev.*, 59: 125–139.

Bongiovanni, L. G., and Hagbarth, K. E. (1990) 'Tonic vibration reflexes elicited during fatigue from maximal voluntary contractions in man', *J. Physiol.*, 423: 1–14.

Bonnard, M., Sirin, A. V., Oddsson, L., and Thorstensson, A. (1994) 'Different strategies to compensate for the effects of fatigue revealed by neuromuscular adaptation processes in humans', *Neurosci. Lett.*, 166: 101–105.

Boudreau, S. A., Hennings, K., Svensson, P., Sessle, B. J., and Arendt-Nielsen, L. (2010) 'The effects of training time, sensory loss and pain on human motor learning', *J. Oral Rehabil.*, 37: 704–718.

Brasil-Neto, J. P., Pascual-Leone, A., Valls-Sole, J., Cammarota, A., Cohen, L. G., and Hallett, M. (1993) 'Postexercise depression of motor evoked potentials: a measure of central nervous system fatigue', *Exp. Brain Res.*, 93: 181–184.

Brisswalter, J., Collardeau, M., and Rene, A. (2002) 'Effects of acute physical exercise characteristics on cognitive performance', *Sports Med.*, 32: 555–566.

Buchthal, F., and Schmalbruch, H. (1980) 'Motor unit of mammalian muscle', *Physiol. Rev.*, 60: 90–142.

Burke, R. E., Levine, D. N., Tsairis, P., and Zajac, F. E., 3rd. (1973) 'Physiological types and histochemical profiles in motor units of the cat gastrocnemius', *J. Physiol.*, 234: 723–748.

Butler, J. E., Taylor, J. L., and Gandevia, S. C. (2003) 'Responses of human motoneurons to corticospinal stimulation during maximal voluntary contractions and ischemia', *J. Neurosci.*, 23: 10224–10230.

Button, D. C., Kalmar, J. M., Gardiner, K., Cahill, F., and Gardiner, P. F. (2007) 'Spike frequency adaptation of rat hindlimb motoneurons', *J. Appl. Physiol.*, 102: 1041–1050.

Campagne, A., Pebayle, T., and Muzet, A. (2004) 'Correlation between driving errors and vigilance level: influence of the driver's age', *Physiol. Behav.*, 80: 515–524.

Cantor, F. (2010) 'Central and peripheral fatigue: exemplified by multiple sclerosis and myasthenia gravis', *PM&R*, 2: 399–405.

Carpenter, J. E., Blasier, R. B., and Pellizzon, G. G. (1998) 'The effects of muscle fatigue on shoulder joint position sense', *Am. J. Sports Med.*, 26: 262–265.

Carpentier, A., Duchateau, J., and Hainaut, K. (2001) 'Motor unit behaviour and contractile changes during fatigue in the human first dorsal interosseus', *J. Physiol.*, 534: 903–912.

Carron, A. V. (1972) 'Motor performance and learning under physical fatigue', *Med. Sci. Sports Exerc.*, 4: 101–106.

Chen, J. R., Wang, T. J., Huang, H. Y., Chen, L. J., Huang, Y. S., Wang, Y. J., and Tseng, G. F. (2009) 'Fatigue reversibly reduced cortical and hippocampal dendritic spines concurrent with compromise of motor endurance and spatial memory', *Neuroscience*, 161: 1104–1113.

Coco, M., Di Corrado, D., Calogero, R. A., Perciavalle, V., Maci, T., and Perciavalle, V. (2009) 'Attentional processes and blood lactate levels', *Brain Res.*, 1302: 205–211.

Contessa, P., Adam, A., and De Luca, C. J. (2009) 'Motor unit control and force fluctuation during fatigue', *J. Appl. Physiol.*, 107: 235–243.

Cope, T. C., and Clark, B. D. (1991) 'Motor-unit recruitment in the decerebrate cat: several unit properties are equally good predictors of order', *J. Neurophysiol.*, 66: 1127–1138.

Cote, J. N., Feldman, A. G., Mathieu, P. A., and Levin, M. F. (2008) 'Effects of fatigue on intermuscular coordination during repetitive hammering', *Motor Control*, 12: 79–92.

Cote, J. N., Mathieu, P. A., Levin, M. F., and Feldman, A. G. (2002) 'Movement reorganization to compensate for fatigue during sawing', *Exp. Brain Res.*, 146: 394–398.

Cotten, D. J., Spieth, W. R., Thomas, J. R., and Biasiotto, J. L. (1974) 'Local and total body fatigue effects on learning and performance of a gross motor skill', *Med. Sci. Sports*, 6: 151–153.

Cresswell, A. G., and Löscher, W. N. (2000) 'Significance of peripheral afferent input to the alpha-motoneurone pool for enhancement of tremor during an isometric fatiguing contraction', *Eur. J. Appl. Physiol.*, 82: 129–136.

Danion, F., Latash, M. L., Li, Z. M., and Zatsiorsky, V. M. (2001) 'The effect of a fatiguing exercise by the index finger on single- and multi-finger force production tasks', *Exp. Brain Res.*, 138: 322–329.

de Haan, A., Jones, D. A., and Sargeant, A. J. (1989) 'Changes in velocity of shortening, power output and relaxation rate during fatigue of rat medial gastrocnemius muscle', *Pflugers Arch.*, 413: 422–428.

De Ruiter, C. J., Jones, D. A., Sargeant, A. J., and De Haan, A. (1999) 'The measurement of force/velocity relationships of fresh and fatigued human adductor pollicis muscle', *Eur. J. Appl. Physiol. Occup. Physiol.*, 80: 386–393.

Demougeot, L., and Papaxanthis, C. (2011) 'Muscle fatigue affects mental simulation of action', *J. Neurosci.*, 31: 10712–10720.

Dickinson, J., Medhurst, C., and Whittingham, N. (1979) 'Warm-up and fatigue in skill acquisition and performance', *J. Mot. Behav.*, 11: 81–86.

Dietrich, A., and Audiffren, M. (2011) 'The reticular-activating hypofrontality (RAH) model of acute exercise', *Neurosci. Biobehav. Rev.*, 35: 1305–1325.

Dietrich, A., and Sparling, P. B. (2004) 'Endurance exercise selectively impairs prefrontal-dependent cognition', *Brain Cogn.*, 55: 516–524.

Dinges, D. F. (1995) 'An overview of sleepiness and accidents', *J. Sleep Res.*, 4: 4–14.

Duchateau, J., Balestra, C., Carpentier, A., and Hainaut, K. (2002) 'Reflex regulation during sustained and intermittent submaximal contractions in humans', *J. Physiol.*, 541: 959–967.

Duchateau, J., and Enoka, R. M. (2011) 'Human motor unit recordings: origins and insight into the integrated motor system', *Brain Res.*, 1409: 42–61.

Duchateau, J., and Hainaut, K. (1993) 'Behaviour of short and long latency reflexes in fatigued human muscles', *J. Physiol.*, 471: 787–799.

Ebenbichler, G. R., Kollmitzer, J., Erim, Z., Löscher, W. N., Kerschan, K., Posch, M., Nowotny, T., Kranzl, A., Wober, C., and Bochdansky, T. (2000) 'Load-dependence of fatigue related changes in tremor around 10 Hz', *Clin. Neurophysiol.*, 111: 106–111.

Edwards, R. H., Hill, D. K., and Jones, D. A. (1975) 'Metabolic changes associated with the slowing of relaxation in fatigued mouse muscle', *J. Physiol.*, 251: 287–301.

Eich, T. S., and Metcalfe, J. (2009) 'Effects of the stress of marathon running on implicit and explicit memory', *Psychon. Bull. Rev.*, 16: 475–479.

Enebo, B., and Sherwood, D. (2005) 'Experience and practice organization in learning a simulated high-velocity low-amplitude task', *J. Manipulative Physiol. Ther.*, 28: 33–43.

Enoka, R. M. (2011) 'Muscle fatigue – from motor units to clinical symptoms', *J. Biomech.*, 45: 427–433.

Enoka, R. M., Baudry, S., Rudroff, T., Farina, D., Klass, M., and Duchateau, J. (2011) 'Unraveling the neurophysiology of muscle fatigue', *J. Electromyogr. Kinesiol.*, 21: 208–219.

Enoka, R. M., and Duchateau, J. (2008) 'Muscle fatigue: what, why and how it influences muscle function', *J. Physiol.*, 586: 11–23.

Ervilha, U. F., Farina, D., Arendt-Nielsen, L., and Graven-Nielsen, T. (2005) 'Experimental muscle pain changes motor control strategies in dynamic contractions', *Exp. Brain Res.*, 164: 215–224.

Escamilla, R. F., Barrentine, S. W., Fleisig, G. S., Zheng, N., Takada, Y., Kingsley, D., and Andrews, J. R. (2007) 'Pitching biomechanics as a pitcher approaches muscular fatigue during a simulated baseball game', *Am. J. Sports Med.*, 35: 23–33.

Farina, D., Arendt-Nielsen, L., and Graven-Nielsen, T. (2005) 'Experimental muscle pain reduces initial motor unit discharge rates during sustained submaximal contractions', *J. Appl. Physiol.*, 98: 999–1005.

Fischer, M., and Schafer, S. S. (2005) 'Effects of changes in pH on the afferent impulse activity of isolated cat muscle spindles', *Brain Res.*, 1043: 163–178.

Fitts, R. H., Courtright, J. B., Kim, D. H., and Witzmann, F. A. (1982) 'Muscle fatigue with prolonged exercise: contractile and biochemical alterations', *Am. J. Physiol.*, 242: C65–73.

Forestier, N., and Nougier, V. (1998) 'The effects of muscular fatigue on the coordination of a multi-joint movement in human', *Neurosci. Lett.*, 252: 187–190.

Forestier, N., Teasdale, N., and Nougier, V. (2002) 'Alteration of the position sense at the ankle induced by muscular fatigue in humans', *Med. Sci. Sports Exerc.*, 34: 117–122.

Fuglevand, A. J. (1995) 'The role of the sarcolemma action potential in fatigue', *Adv. Exp. Med. Biol.*, 384: 101–108.

Fuglevand, A. J., and Keen, D. A. (2003) 'Re-evaluation of muscle wisdom in the human adductor pollicis using physiological rates of stimulation', *J. Physiol.*, 549: 865–875.

Fuglevand, A. J., Macefield, V. G., and Bigland-Ritchie, B. (1999) 'Force-frequency and fatigue properties of motor units in muscles that control digits of the human hand', *J. Neurophysiol.*, 81: 1718–1729.

Fuglevand, A. J., Zackowski, K. M., Huey, K. A., and Enoka, R. M. (1993) 'Impairment of neuromuscular propagation during human fatiguing contractions at submaximal forces', *J. Physiol.*, 460: 549–572.

Furness, P., Jessop, J., and Lippold, O. C. (1977) 'Long-lasting increases in the tremor of human hand muscles following brief, strong effort', *J. Physiol.*, 265: 821–831.

Gandevia, S. C. (1988) 'Physiological accompaniments of human muscle fatigue', *Aust. Paediatr. J.*, 24 Suppl 1: 104–108.

Gandevia, S. C. (1996) 'Kinaesthesia: roles for afferent signals and motor commands', in *Handbook of Physiology, section 12, Exercise: regulation and integration of multiple systems.*, L. B. Rowell and T. J. Sheperd, eds, Oxford University Press, New York, pp. 128–172.

Gandevia, S. C. (1998) 'Neural control in human muscle fatigue: changes in muscle afferents, motoneurones and motor cortical drive', *Acta Physiol. Scand.*, 162: 275–283.

Gandevia, S. C. (2001) 'Spinal and supraspinal factors in human muscle fatigue', *Physiol. Rev.*, 81: 1725–1789.

Gandevia, S. C., Allen, G. M., Butler, J. E., and Taylor, J. L. (1996) 'Supraspinal factors in human muscle fatigue: evidence for suboptimal output from the motor cortex', *J. Physiol.*, 490: 529–536.

Gandevia, S. C., Allen, G. M., and McKenzie, D. K. (1995) 'Central fatigue. Critical issues, quantification and practical implications', *Adv. Exp. Med. Biol.*, 384: 281–294.

Garland, S. J. (1991) 'Role of small diameter afferents in reflex inhibition during human muscle fatigue', *J. Physiol.*, 435: 547–558.

Gates, D. H., and Dingwell, J. B. (2008) 'The effects of neuromuscular fatigue on task performance during repetitive goal-directed movements', *Exp. Brain Res.*, 187: 573–585.

Ghez, C., Gordon, J., Ghilardi, M. F., Christakos, C. N., and Cooper, S. E. (1990) 'Roles of proprioceptive input in the programming of arm trajectories', *Cold Spring Harb. Symp. Quant. Biol.*, 55: 837–847.

Gollnick, P. D., Karlsson, J., Piehl, K., and Saltin, B. (1974a) 'Selective glycogen depletion in skeletal muscle fibres of man following sustained contractions', *J. Physiol.*, 241: 59–67.

Gollnick, P. D., Piehl, K., and Saltin, B. (1974b) 'Selective glycogen depletion pattern in human muscle fibres after exercise of varying intensity and at varying pedalling rates', *J. Physiol.*, 241: 45–57.

Goodall, S., Ross, E. Z., and Romer, L. M. (2010) 'Effect of graded hypoxia on supraspinal contributions to fatigue with unilateral knee-extensor contractions', *J. Appl. Physiol.*, 109: 1842–1851.

Gordon, T., Thomas, C. K., Munson, J. B., and Stein, R. B. (2004) 'The resilience of the size principle in the organization of motor unit properties in normal and reinnervated adult skeletal muscles', *Can. J. Physiol. Pharmacol.*, 82: 645–661.

Gorelick, M., Brown, J. M., and Groeller, H. (2003) 'Short-duration fatigue alters neuromuscular coordination of trunk musculature: implications for injury', *Appl. Ergon.*, 34: 317–325.

Graven-Nielsen, T., and Mense, S. (2001) 'The peripheral apparatus of muscle pain: evidence from animal and human studies', *Clin. J. Pain*, 17: 2–10.

Graven-Nielsen, T., Svensson, P., and Arendt-Nielsen, L. (1997) 'Effects of experimental muscle pain on muscle activity and co-ordination during static and dynamic motor function', *Electroencephalogr. Clin. Neurophysiol.*, 105: 156–164.

Griffin, L., Garland, S. J., Ivanova, T., and Gossen, E. R. (2001) 'Muscle vibration sustains motor unit firing rate during submaximal isometric fatigue in humans', *J. Physiol.*, 535: 929–936.

Guedon, O., Gauthier, G., Cole, J., Vercher, J. L., and Blouin, J. (1998) 'Adaptation in visuomanual tracking depends on intact proprioception', *J. Mot. Behav.*, 30: 234–248.

Hasson, S. M., Williams, J. H., and Signorile, J. F. (1989) 'Fatigue-induced changes in myoelectric signal characteristics and perceived exertion', *Can. J. Sport Sci.*, 14: 99–102.

Hayes, S. G., Kindig, A. E., and Kaufman, M. P. (2006) 'Cyclooxygenase blockade attenuates responses of group III and IV muscle afferents to dynamic exercise in cats', *Am. J. Physiol. Heart Circ. Physiol.*, 290: H2239–2246.

Henneman, E., Clamann, H. P., Gillies, J. D., and Skinner, R. D. (1974) 'Rank order of motoneurons within a pool: law of combination', *J. Neurophysiol.*, 37: 1338–1349.

Herbert, R. D., and Gandevia, S. C. (1999) 'Twitch interpolation in human muscles: mechanisms and implications for measurement of voluntary activation', *J. Neurophysiol.*, 82: 2271–2283.

Hiemstra, L. A., Lo, I. K., and Fowler, P. J. (2001) 'Effect of fatigue on knee proprioception: implications for dynamic stabilization', *J. Orthop. Sports Phys. Ther.*, 31: 598–605.

Hill, J. M. (2001) 'Increase in the discharge of muscle spindles during diaphragm fatigue', *Brain Res.*, 918: 166–170.

Hinder, M. R., Riek, S., Tresilian, J. R., de Rugy, A., and Carson, R. G. (2010) 'Real-time error detection but not error correction drives automatic visuomotor adaptation', *Exp. Brain Res.*, 201: 191–207.

Horcholle-Bossavit, G., Jami, L., Petit, J., Vejsada, R., and Zytnicki, D. (1990) 'Ensemble discharge from Golgi tendon organs of cat peroneus tertius muscle', *J. Neurophysiol.*, 64: 813–821.

Hossain, J. L., Ahmad, P., Reinish, L. W., Kayumov, L., Hossain, N. K., and Shapiro, C. M. (2005) 'Subjective fatigue and subjective sleepiness: two independent consequences of sleep disorders?' *J. Sleep Res.*, 14: 245–253.

Huffenus, A. F., Amarantini, D., and Forestier, N. (2006) 'Effects of distal and proximal arm muscles fatigue on multi-joint movement organization', *Exp. Brain Res.*, 170: 438–447.

Hunter, S. K., Duchateau, J., and Enoka, R. M. (2004) 'Muscle fatigue and the mechanisms of task failure', *Exerc. Sport Sci. Rev.*, 32: 44–49.

Hunter, S. K., and Enoka, R. M. (2003) 'Changes in muscle activation can prolong the endurance time of a submaximal isometric contraction in humans', *J. Appl. Physiol.*, 94: 108–118.

Hwang, I. S., Yang, Z. R., Huang, C. T., and Guo, M. C. (2009) 'Reorganization of multidigit physiological tremors after repetitive contractions of a single finger', *J. Appl. Physiol.*, 106: 966–974.

Ingham, D., Tucker, K. J., Tsao, H., and Hodges, P. W. (2011) 'The effect of pain on training-induced plasticity of the corticomotor system', *Eur. J. Pain*, 15: 1028–1034.

James, C., Sacco, P., and Jones, D. A. (1995) 'Loss of power during fatigue of human leg muscles', *J. Physiol.*, 484 (Pt 1): 237–246.

Johansson, B., Berglund, P., and Ronnback, L. (2009) 'Mental fatigue and impaired information processing after mild and moderate traumatic brain injury', *Brain Inj.*, 23: 1027–1040.

Johnson, K. V., Edwards, S. C., Van Tongeren, C., and Bawa, P. (2004) 'Properties of human motor units after prolonged activity at a constant firing rate', *Exp. Brain Res.*, 154: 479–487.

Jones, D. A. (2010) 'Changes in the force-velocity relationship of fatigued muscle: implications for power production and possible causes', *J. Physiol.*, 588: 2977–2986.

Jones, D. A., Bigland-Ritchie, B., and Edwards, R. H. (1979) 'Excitation frequency and muscle fatigue: mechanical responses during voluntary and stimulated contractions', *Exp. Neurol.*, 64: 401–413.

Jones, D. A., de Ruiter, C. J., and de Haan, A. (2006) 'Change in contractile properties of human muscle in relationship to the loss of power and slowing of relaxation seen with fatigue', *J. Physiol.*, 576: 913–922.

Karatzaferi, C., de Haan, A., van Mechelen, W., and Sargeant, A. J. (2001) 'Metabolism changes in single human fibres during brief maximal exercise', *Exp. Physiol.*, 86: 411–415.

Kaufman, M. P., Longhurst, J. C., Rybicki, K. J., Wallach, J. H., and Mitchell, J. H. (1983) 'Effects of static muscular contraction on impulse activity of groups III and IV afferents in cats', *J. Appl. Physiol.*, 55: 105–112.

Kaufman, M. P., and Rybicki, K. J. (1987) 'Discharge properties of group III and IV muscle afferents: their responses to mechanical and metabolic stimuli', *Circ. Res.*, 61: 160–65.

Kaufman, M. P., Rybicki, K. J., Waldrop, T. G., and Ordway, G. A. (1984) 'Effect of ischemia on responses of group III and IV afferents to contraction', *J. Appl. Physiol.*, 57: 644–650.

Kernell, D., and Monster, A. W. (1982) 'Time course and properties of late adaptation in spinal motoneurones of the cat', *Exp. Brain Res.*, 46: 191–196.

Kerr, T. L. (1993) *An analysis of the effects of fatigue and specificity on motor learning.* Masters thesis. Simon Fraser University, Vancouver, BC.

Kiernan, M. C., Lin, C. S., and Burke, D. (2004) 'Differences in activity-dependent hyperpolarization in human sensory and motor axons', *J. Physiol.*, 558: 341–349.

Kilbom, A., Gamberale, F., Persson, J., and Annwall, G. (1983) 'Physiological and psychological indices of fatigue during static contractions', *Eur. J. Appl. Physiol. Occup. Physiol.*, 50: 179–193.

Kiryu, T., Murayama, T., and Ushiyama, Y. (2011) 'Influence of muscular fatigue on skiing performance during parallel turns', *Conf. Proc. IEEE Eng. Med. Biol. Soc.*, 2011: 8187–8190.

Knicker, A. J., Renshaw, I., Oldham, A. R., and Cairns, S. P. (2011) 'Interactive processes link the multiple symptoms of fatigue in sport competition', *Sports Med.*, 41: 307–328.

Kniffki, K. D., Schomburg, E. D., and Steffens, H. (1979) 'Synaptic responses of lumbar alpha-motoneurones to chemical algesic stimulation of skeletal muscle in spinal cats', *Brain Res.*, 160: 549–552.

Kniffki, K. D., Schomburg, E. D., and Steffens, H. (1981) 'Synaptic effects from chemically activated fine muscle afferents upon alpha-motoneurones in decerebrate and spinal cats', *Brain Res.*, 206: 361–370.

Kording, K. P., Tenenbaum, J. B., and Shadmehr, R. (2007) 'The dynamics of memory as a consequence of optimal adaptation to a changing body', *Nat. Neurosci.*, 10: 779–786.

Kouzaki, M., and Shinohara, M. (2006) 'The frequency of alternate muscle activity is associated with the attenuation in muscle fatigue', *J. Appl. Physiol.*, 101: 715–720.

Kuchinad, R. A., Ivanova, T. D., and Garland, S. J. (2004) 'Modulation of motor unit discharge rate and H-reflex amplitude during submaximal fatigue of the human soleus muscle', *Exp. Brain Res.*, 158: 345–355.

Lamb, G. D. (2002) 'Excitation-contraction coupling and fatigue mechanisms in skeletal muscle: studies with mechanically skinned fibres', *J. Muscle Res. Cell Motil.*, 23: 81–91.

Lambourne, K., and Tomporowski, P. (2010) 'The effect of exercise-induced arousal on cognitive task performance: a meta-regression analysis', *Brain Res.*, 1341: 12–24.

Lannergren, J., and Westerblad, H. (1986) 'Force and membrane potential during and after fatiguing, continuous high-frequency stimulation of single *Xenopus* muscle fibres', *Acta Physiol. Scand.*, 128: 359–368.

Lee, M., Gandevia, S. C., and Carroll, T. J. (2008) 'Cortical voluntary activation can be reliably measured in human wrist extensors using transcranial magnetic stimulation', *Clin. Neurophysiol.*, 119: 1130–1138.

Liddell, E. G. T., and Sherrington, C. S. (1925) 'Recruitment and some other factors of reflex inhibition', *Proc. R. Soc. Lond. B. Biol. Sci.*, 97: 488–518.

Liu, J. Z., Shan, Z. Y., Zhang, L. D., Sahgal, V., Brown, R. W., and Yue, G. H. (2003) 'Human brain activation during sustained and intermittent submaximal fatigue muscle contractions: an FMRI study', *J. Neurophysiol.*, 90: 300–312.

Liu, J. Z., Zhang, L., Yao, B., Sahgal, V., and Yue, G. H. (2005) 'Fatigue induced by intermittent maximal voluntary contractions is associated with significant losses in muscle output but limited reductions in functional MRI-measured brain activation level', *Brain Res.*, 1040: 44–54.

Ljubisavljevic, M., and Anastasijevic, R. (1994) 'Fusimotor-induced changes in muscle spindle outflow and responsiveness in muscle fatigue in decerebrate cats', *Neuroscience*, 63: 339–348.

Ljubisavljevic, M., Milanovic, S., Radovanovic, S., Vukcevic, I., Kostic, V., and Anastasijevic, R. (1996) 'Central changes in muscle fatigue during sustained submaximal isometric voluntary contraction as revealed by transcranial magnetic stimulation', *Electroencephalogr. Clin. Neurophysiol.*, 101: 281–288.

Lo Bue-Estes, C., Willer, B., Burton, H., Leddy, J. J., Wilding, G. E., and Horvath, P. J. (2008) 'Short-term exercise to exhaustion and its effects on cognitive function in young women', *Percept. Mot. Skills*, 107: 933–945.

Lorist, M. M., Kernell, D., Meijman, T. F., and Zijdewind, I. (2002) 'Motor fatigue and cognitive task performance in humans', *J. Physiol.*, 545: 313–319.

Löscher, W. N., Cresswell, A. G., and Thorstensson, A. (1996) 'Excitatory drive to the alpha-motoneuron pool during a fatiguing submaximal contraction in man', *J. Physiol.*, 491: 271–280.

Lundbye-Jensen, J., Petersen, T. H., Rothwell, J. C., and Nielsen, J. B. (2011) 'Interference in ballistic motor learning: specificity and role of sensory error signals', *PloS One*, 6: e17451.

Macefield, G., Hagbarth, K. E., Gorman, R., Gandevia, S. C., and Burke, D. (1991) 'Decline in spindle support to alpha-motoneurones during sustained voluntary contractions', *J. Physiol.*, 440: 497–512.

Maluf, K. S., and Enoka, R. M. (2005) 'Task failure during fatiguing contractions performed by humans', *J. Appl. Physiol.*, 99: 389–396.

Marcora, S. M., Staiano, W., and Manning, V. (2009) 'Mental fatigue impairs physical performance in humans', *J. Appl. Physiol.*, 106: 857–864.

Martin, P. G., Smith, J. L., Butler, J. E., Gandevia, S. C., and Taylor, J. L. (2006) 'Fatigue-sensitive afferents inhibit extensor but not flexor motoneurons in humans', *J. Neurosci.*, 26: 4796–4802.

Maruyama, A., Matsunaga, K., Tanaka, N., and Rothwell, J. C. (2006) 'Muscle fatigue decreases short-interval intracortical inhibition after exhaustive intermittent tasks', *Clin. Neurophysiol.*, 117: 864–870.

Matthews, P. B. (1981) 'Evolving views on the internal operation and functional role of the muscle spindle', *J. Physiol.*, 320: 1–30.

McKay, W. B., Stokic, D. S., Sherwood, A. M., Vrbova, G., and Dimitrijevic, M. R. (1996) 'Effect of fatiguing maximal voluntary contraction on excitatory and inhibitory responses elicited by transcranial magnetic motor cortex stimulation', *Muscle Nerve*, 19: 1017–1024.

McNeil, C. J., Giesebrecht, S., Gandevia, S. C., and Taylor, J. L. (2011) 'Behaviour of the motoneurone pool in a fatiguing submaximal contraction', *J. Physiol.*, 589: 3533–3544.

McNeil, C. J., Martin, P. G., Gandevia, S. C., and Taylor, J. L. (2009) 'The response to paired motor cortical stimuli is abolished at a spinal level during human muscle fatigue', *J. Physiol.*, 587: 5601–5612.

Mense, S. (2009) 'Algesic agents exciting muscle nociceptors', *Exp. Brain Res.*, 196: 89–100.

Mense, S., and Stahnke, M. (1983) 'Responses in muscle afferent fibres of slow conduction velocity to contractions and ischaemia in the cat', *J. Physiol.*, 342: 383–397.

Merton, P. A. (1954) 'Voluntary strength and fatigue', *J. Physiol.*, 123: 553–564.

Mierau, A., Schneider, S., Abel, T., Askew, C., Werner, S., and Struder, H. K. (2009) 'Improved sensorimotor adaptation after exhaustive exercise is accompanied by altered brain activity', *Physiol. Behav.*, 96: 115–121.

Milanovic, S., Filipovic, S. R., Blesic, S., Ilic, T. V., Dhanasekaran, S., and Ljubisavljevic, M. (2011) 'Paired-associative stimulation can modulate muscle fatigue induced motor cortex excitability changes', *Behav. Brain Res.*, 223: 30–35.

Millet, G. Y., Martin, V., Martin, A., and Verges, S. (2011) 'Electrical stimulation for testing neuromuscular function: from sport to pathology', *Eur. J. Appl. Physiol.*, 111: 2489–2500.

Milner-Brown, H. S., Stein, R. B., and Yemm, R. (1973) 'The orderly recruitment of human motor units during voluntary isometric contractions', *J. Physiol.*, 230: 359–370.

Mitchell, G. S., and Johnson, S. M. (2003) 'Neuroplasticity in respiratory motor control', *J. Appl. Physiol.*, 94: 358–374.

Mitchell, J. H., Kaufman, M. P., and Iwamoto, G. A. (1983) 'The exercise pressor reflex: its cardiovascular effects, afferent mechanisms, and central pathways', *Annu. Rev. Physiol.*, 45: 229–242.

Mizunoya, W., Oyaizu, S., Hirayama, A., and Fushiki, T. (2004) 'Effects of physical fatigue in mice on learning performance in a water maze', *Biosci. Biotechnol. Biochem.*, 68: 827–834.

Monster, A. W., and Chan, H. (1977) 'Isometric force production by motor units of extensor digitorum communis muscle in man', *J. Neurophysiol.*, 40: 1432–1443.

Morin, J. B., Tomazin, K., Samozino, P., Edouard, P., and Millet, G. Y. (2012) 'High-intensity sprint fatigue does not alter constant-submaximal velocity running mechanics and spring-mass behavior', *Eur. J. Appl. Physiol.* 112 (4): 1419–1428.

Morrison, S., and Sosnoff, J. J. (2010) 'The impact of localized fatigue on contralateral tremor and muscle activity is exacerbated by standing posture', *J. Electromyogr. Kinesiol.*, 20: 1211–1218.

Naughton, P. A., Aggarwal, R., Wang, T. T., Van Herzeele, I., Keeling, A. N., Darzi, A. W., and Cheshire, N. J. (2011) 'Skills training after night shift work enables acquisition of endovascular technical skills on a virtual reality simulator', *J. Vasc. Surg.*, 53: 858–866.

Nessler, J. A., Huynh, H., and McDougal, M. (2011) 'A single bout of resistance exercise does not affect nonlinear dynamics of lower extremity kinematics during treadmill walking', *Gait Posture*, 34: 285–287.

O'Connor, P. J., and Cook, D. B. (1999) 'Exercise and pain: the neurobiology, measurement, and laboratory study of pain in relation to exercise in humans', *Exerc. Sport Sci. Rev.*, 27: 119–166.

Pagala, M., Ravindran, K., Amaladevi, B., Namba, T., and Grob, D. (1994) 'Potassium and caffeine contractures of mouse muscles before and after fatiguing stimulation', *Muscle Nerve*, 17: 852–859.

Pedersen, J., Ljubisavljevic, M., Bergenheim, M., and Johansson, H. (1998) 'Alterations in information transmission in ensembles of primary muscle spindle afferents after muscle fatigue in heteronymous muscle', *Neuroscience*, 84: 953–959.

Petrofsky, J. S., and Phillips, C. A. (1985) 'Discharge characteristics of motor units and the surface EMG during fatiguing isometric contractions at submaximal tensions', *Aviat. Space Environ. Med.*, 56: 581–586.

Pettorossi, V. E., Della Torre, G., Bortolami, R., and Brunetti, O. (1999) 'The role of capsaicin-sensitive muscle afferents in fatigue-induced modulation of the monosynaptic reflex in the rat', *J. Physiol.*, 515: 599–607.

Philip, P., Taillard, J., Klein, E., Sagaspe, P., Charles, A., Davies, W. L., Guilleminault, C., and Bioulac, B. (2003) 'Effect of fatigue on performance measured by a driving simulator in automobile drivers', *J. Psychosom. Res.*, 55: 197–200.

Pipereit, K., Bock, O., and Vercher, J. L. (2006) 'The contribution of proprioceptive feedback to sensorimotor adaptation', *Exp. Brain Res.*, 174: 45–52.

Place, N., Yamada, T., Bruton, J. D., and Westerblad, H. (2010) 'Muscle fatigue: from observations in humans to underlying mechanisms studied in intact single muscle fibres', *Eur. J. Appl. Physiol.*, 110: 1–15.

Post, M., Steens, A., Renken, R., Maurits, N. M., and Zijdewind, I. (2009) 'Voluntary activation and cortical activity during a sustained maximal contraction: an fMRI study', *Hum. Brain Mapp.*, 30: 1014–1027.

Proske, U., and Gregory, J. E. (2002) 'Signalling properties of muscle spindles and tendon organs', *Adv. Exp. Med. Biol.*, 508: 5–12.

Radhakrishnan, S. M., Baker, S. N., and Jackson, A. (2008) 'Learning a novel myoelectric-controlled interface task', *J. Neurophysiol.*, 100: 2397–2408.

Rodacki, A. L., Fowler, N. E., and Bennett, S. J. (2002) 'Vertical jump coordination: fatigue effects', *Med. Sci. Sports Exerc.*, 34: 105–116.

Rossi, A., Decchi, B., Dami, S., Della Volpe, R., and Groccia, V. (1999a) 'On the effect of chemically activated fine muscle afferents on interneurones mediating group I non-reciprocal inhibition of extensor ankle and knee muscles in humans', *Brain Res.*, 815: 106–110.

Rossi, A., Decchi, B., and Ginanneschi, F. (1999b) 'Presynaptic excitability changes of group Ia fibres to muscle nociceptive stimulation in humans', *Brain Res.*, 818: 12–22.

Rothwell, J. C., Traub, M. M., Day, B. L., Obeso, J. A., Thomas, P. K., and Marsden, C. D. (1982) 'Manual motor performance in a deafferented man', *Brain*, 105: 515–542.

Rotto, D. M., and Kaufman, M. P. (1988) 'Effect of metabolic products of muscular contraction on discharge of group III and IV afferents', *J. Appl. Physiol.*, 64: 2306–2313.

Samii, A., Lopez-Devine, J., Wasserman, E. M., Dalakas, M. C., Clark, K., Grafman, J., and Hallett, M. (1998) 'Normal postexercise facilitation and depression of motor evoked potentials in postpolio patients', *Muscle Nerve*, 21: 948–950.

Samii, A., Wassermann, E. M., Ikoma, K., Mercuri, B., George, M. S., O'Fallon, A., Dale, J. K., Straus, S. E., and Hallett, M. (1996a) 'Decreased postexercise facilitation of motor evoked potentials in patients with chronic fatigue syndrome or depression', *Neurology*, 47: 1410–1414.

Samii, A., Wassermann, E. M., Ikoma, K., Mercuri, B., and Hallett, M. (1996b) 'Characterization of postexercise facilitation and depression of motor evoked potentials to transcranial magnetic stimulation', *Neurology*, 46: 1376–1382.

Sanes, J. N. (1985) 'Absence of enhanced physiological tremor in patients without muscle or cutaneous afferents', *J. Neurol. Neurosurg. Psychiatry*, 48: 645–649.

Sargeant, A. J. (2007) 'Structural and functional determinants of human muscle power', *Exp. Physiol.*, 92: 323–331.

Sawczuk, A., Powers, R. K., and Binder, M. D. (1997) 'Contribution of outward currents to spike-frequency adaptation in hypoglossal motoneurons of the rat', *J. Neurophysiol.*, 78: 2246–2253.

Schmitt, K., DelloRusso, C., and Fregosi, R. F. (2009) 'Force-EMG changes during sustained contractions of a human upper airway muscle', *J. Neurophysiol.*, 101: 558–568.

Secher, N. H., and Amann, M. (2012) 'Human investigations into the exercise pressor reflex', *Exp. Physiol.*, 97: 59–69.

Seifert, T., and Petersen, N. C. (2010) 'Changes in presumed motor cortical activity during fatiguing muscle contraction in humans', *Acta Physiol. (Oxf.)*, 199: 317–326.

Shield, A., and Zhou, S. (2004) 'Assessing voluntary muscle activation with the twitch interpolation technique', *Sports Med.*, 34: 253–267.

Sidhu, S. K., Bentley, D. J., and Carroll, T. J. (2009a) 'Cortical voluntary activation of the human knee extensors can be reliably estimated using transcranial magnetic stimulation', *Muscle Nerve*, 39: 186–196.

Sidhu, S. K., Bentley, D. J., and Carroll, T. J. (2009b) 'Locomotor exercise induces long-lasting impairments in the capacity of the human motor cortex to voluntarily activate knee extensor muscles', *J. Appl. Physiol.*, 106: 556–565.

Sieck, G. C., and Prakash, Y. S. (1995) 'Fatigue at the neuromuscular junction. Branch point vs. presynaptic vs. postsynaptic mechanisms', *Adv. Exp. Med. Biol.*, 384: 83–100.

Singh, T., Varadhan, S. K., Zatsiorsky, V. M., and Latash, M. L. (2010) 'Fatigue and motor redundancy: adaptive increase in finger force variance in multi-finger tasks', *J. Neurophysiol.*, 103: 2990–3000.

Slack, P. S., Coulson, C. J., Ma, X., Pracy, P., Parmar, S., and Webster, K. (2009) 'The effect of operating time on surgeon's hand tremor', *Eur. Arch. Otorhinolaryngol.*, 266: 137–141.

Smets, E. M., Garssen, B., Bonke, B., and De Haes, J. C. (1995) 'The Multidimensional Fatigue Inventory (MFI) psychometric qualities of an instrument to assess fatigue', *J. Psychosom. Res.*, 39: 315–325.

Smith, J. L., Martin, P. G., Gandevia, S. C., and Taylor, J. L. (2007) 'Sustained contraction at very low forces produces prominent supraspinal fatigue in human elbow flexor muscles', *J. Appl. Physiol.*, 103: 560–568.

Søgaard, K., Gandevia, S. C., Todd, G., Petersen, N. T., and Taylor, J. L. (2006) 'The effect of sustained low-intensity contractions on supraspinal fatigue in human elbow flexor muscles', *J. Physiol.*, 573: 511–523.

Spielmann, J. M., Laouris, Y., Nordstrom, M. A., Robinson, G. A., Reinking, R. M., and Stuart, D. G. (1993) 'Adaptation of cat motoneurons to sustained and intermittent extracellular activation', *J. Physiol.*, 464: 75–120.

Stockard, J. R. (1974) 'Prior physical exertion in learning a novel gross-motor task', *Percept. Mot. Skills*, 38: 146.

Takahashi, C. D., Nemet, D., Rose-Gottron, C. M., Larson, J. K., Cooper, D. M., and Reinkensmeyer, D. J. (2006) 'Effect of muscle fatigue on internal model formation and retention during reaching with the arm', *J. Appl. Physiol.*, 100: 695–706.

Takahashi, K., Maruyama, A., Maeda, M., Etoh, S., Hirakoba, K., Kawahira, K., and Rothwell, J. C. (2009) 'Unilateral grip fatigue reduces short interval intracortical inhibition in ipsilateral primary motor cortex', *Clin. Neurophysiol.*, 120: 198–203.

Taylor, J. L. (2009) 'Point: the interpolated twitch does/does not provide a valid measure of the voluntary activation of muscle', *J. Appl. Physiol.*, 107: 354–355.

Taylor, J. L., Allen, G. M., Butler, J. E., and Gandevia, S. C. (2000) 'Supraspinal fatigue during intermittent maximal voluntary contractions of the human elbow flexors', *J. Appl. Physiol.*, 89: 305–313.

Taylor, J. L., Butler, J. E., Allen, G. M., and Gandevia, S. C. (1996) 'Changes in motor cortical excitability during human muscle fatigue', *J. Physiol.*, 490: 519–528.

Taylor, J. L., and Gandevia, S. C. (2001) 'Transcranial magnetic stimulation and human muscle fatigue', *Muscle Nerve*, 24: 18–29.

Taylor, J. L., and Gandevia, S. C. (2008) 'A comparison of central aspects of fatigue in submaximal and maximal voluntary contractions', *J. Appl. Physiol.*, 104: 542–550.

Taylor, J. L., Olsen, H. B., Sjogaard, G., and Sogaard, K. (2009) 'Voluntary activation of trapezius measured with twitch interpolation', *J. Electromyogr. Kinesiol.* 19 (4): 584–590.

Taylor, J. L., Todd, G., and Gandevia, S. C. (2006) 'Evidence for a supraspinal contribution to human muscle fatigue', *Clin. Exp. Pharmacol. Physiol.*, 33: 400–405.

Terrier, R., and Forestier, N. (2009) 'Cognitive cost of motor reorganizations associated with muscular fatigue during a repetitive pointing task', *J. Electromyogr. Kinesiol.*, 19: e487–493.

Thomas, C. K., Johansson, R. S., and Bigland-Ritchie, B. (2006) 'EMG changes in human thenar motor units with force potentiation and fatigue', *J. Neurophysiol.*, 95: 1518–1526.

Thompson, S., Gregory, J. E., and Proske, U. (1990) 'Errors in force estimation can be explained by tendon organ desensitization', *Exp. Brain Res.*, 79: 365–372.

Todd, G., Taylor, J. L., and Gandevia, S. C. (2003) 'Measurement of voluntary activation of fresh and fatigued human muscles using transcranial magnetic stimulation', *J. Physiol.*, 551: 661–671.

Tomporowski, P. D. (2003) 'Effects of acute bouts of exercise on cognition', *Acta Psychol. (Amst.)*, 112: 297–324.

Torres-Harding, S., and Jason, L. A. (2005) 'What is fatigue? History and epidemiology', in *Fatigue as a Window to the Brain*, J. DeLuca, ed., MIT Press, Cambridge, MA, pp. 3–17.

Trontelj, J. V., Mihelin, M., and Khuraibet, A. (2002) 'Safety margin at single neuromuscular junctions', *Muscle Nerve*, Suppl 11: S21–27.

Vagg, R., Mogyoros, I., Kiernan, M. C., and Burke, D. (1998) 'Activity-dependent hyperpolarization of human motor axons produced by natural activity', *J. Physiol.*, 507: 919–925.

van der Linden, D., and Eling, P. (2006) 'Mental fatigue disturbs local processing more than global processing', *Psychol. Res.*, 70: 395–402.

van der Linden, D., Frese, M., and Meijman, T. F. (2003) 'Mental fatigue and the control of cognitive processes: effects on perseveration and planning', *Acta Psychol. (Amst.)*, 113: 45–65.

van der Linden, D., Massar, S. A., Schellekens, A. F., Ellenbroek, B. A., and Verkes, R. J. (2006) 'Disrupted sensorimotor gating due to mental fatigue: preliminary evidence', *Int. J. Psychophysiol.*, 62: 168–174.

van Dieen, J. H., Toussaint, H. M., Maurice, C., and Mientjes, M. (1996) 'Fatigue-related changes in the coordination of lifting and their effect on low back load', *J. Mot. Behav.*, 28: 304–314.

van Duinen, H., Renken, R., Maurits, N., and Zijdewind, I. (2007) 'Effects of motor fatigue on human brain activity, an fMRI study', *Neuroimage*, 35: 1438–1449.

Vidoni, E. D., and Boyd, L. A. (2008) 'Motor sequence learning occurs despite disrupted visual and proprioceptive feedback', *Behav Brain Funct.*, 4: 32.

Voight, M. L., Hardin, J. A., Blackburn, T. A., Tippett, S., and Canner, G. C. (1996) 'The effects of muscle fatigue on and the relationship of arm dominance to shoulder proprioception', *J. Orthop. Sports Phys. Ther.*, 23: 348–352.

Vucic, S., Burke, D., and Kiernan, M. C. (2010) 'Fatigue in multiple sclerosis: mechanisms and management', *Clin. Neurophysiol.*, 121: 809–817.

Westerblad, H., Bruton, J. D., and Katz, A. (2010) 'Skeletal muscle: energy metabolism, fiber types, fatigue and adaptability', *Exp. Cell Res.*, 316: 3093–3099.

Williams, L. R., Daniell-Smith, J. H., and Gunson, L. K. (1976a) 'Specificity of training for motor skill under physical fatigue', *Med. Sci. Sports*, 8: 162–167.

Williams, L. R. T., Cooper, C. M., and Gilkison, B. K. (1976b) 'Specificity principle and performance and learning a gross tracking task under interpolated heavy physical exercise', *Percept. Mot. Skills*, 42: 1339–1346.

Williams, L. R. T., and Cooper, E. L. (1976) 'Fatigue in learning and performance of a gross tracking task', *Percept. Mot. Skills*, 42: 1287–1294.

Windhorst, U. (2007) 'Muscle proprioceptive feedback and spinal networks', *Brain Res. Bull.*, 73: 155–202.

Windhorst, U. (2008) 'Muscle spindles are multi-functional', *Brain Res. Bull.*, 75: 507–508.

Zijdewind, I., van Duinen, H., Zielman, R., and Lorist, M. M. (2006) 'Interaction between force production and cognitive performance in humans', *Clin. Neurophysiol.*, 117: 660–667.

Zwarts, M. J., Bleijenberg, G., and van Engelen, B. G. (2008) 'Clinical neurophysiology of fatigue', *Clin. Neurophysiol.*, 119: 2–10.

18

MOVEMENT DISORDERS

Implications for the understanding of motor control

Michèle Hubli and Volker Dietz

SPINAL CORD INJURY CENTER, BALGRIST UNIVERSITY HOSPITAL,
UNIVERSITY OF ZURICH, SWITZERLAND

Introduction

Damage of the CNS after an SCI (spinal cord injury), stroke or in PD leads to primary deficits and secondary adaptations of the neuronal control of complex, natural movements in movement disorders. The knowledge of the pathophysiology underlying movement disorders can lead to a more precise definition of the specific requirements for the rehabilitation of the affected patients. This is an important aspect since people live longer nowadays and movement disorders have become one of the most rapidly expanding fields in medicine, leading to increasing costs for treatment and rehabilitation. Conversely, the advances in studies of patients with movement disorders allow us a better understanding of movement control in healthy states.

Diverse changes in the neuronal and/or biomechanical mechanisms can contribute to a movement disorder. It is possible to assess the neuronal and biomechanical contributions to a movement and consequently to their changes in movement disorders by recordings of electromyographic (EMG) activity from several leg muscles and the resulting biomechanical parameters such as joint movements and forces. Furthermore, the actual spinal neuronal function underlying movements can be tested by peripheral nerve stimulation during functional movements such as walking and stepping over obstacles (Hubli et al., 2011, Kloter et al., 2011, Michel et al., 2008). However, the two main sources of leg muscle activation during locomotion, i.e., neuronal central programming of a movement and selective processing of afferent input as well as the interaction of proprioceptive reflex systems can only be separated to a limited degree as yet.

The term 'movement disorders' used in this chapter focuses on the impaired locomotion of patients with SCI, stroke or PD. For the understanding of the pathophysiology, the basic neuronal mechanisms of human locomotion are summarized.

Neuronal control of human locomotion

Locomotion is a cyclic subconscious day-by-day movement. The question how the CNS is able to coordinate limb movements in such a seemingly 'simple' and automated manner challenges modern neuroscience. Locomotion is an interaction of programmed innate pattern of leg muscle activation (see newborn stepping) and an appropriate modulation of muscle activation which has to adapt continuously to the actual requirements, i.e., to ground conditions. The central program selects the optimal context-specific information from a large flow of peripheral inputs and incorporates it into a pattern which leads to a smooth performance of movement.

Locomotion in mammals, including humans, is largely shaped by spinal neuronal networks within the spinal cord, i.e., the central pattern generator (CPG), comprising a set of neurons responsible for creating an autonomous motor pattern (Grillner, 1985, Grillner & Wallen, 1985). For the control of locomotion, however, afferent information from a variety of sources, e.g., visual, vestibular and proprioceptive systems is integrated and interpreted by the CPGs. How important, for example, the proprioceptive input is for movement generation, becomes clear when studying deafferented patients who lack proprioceptive feedback information during movements (Cole & Sedgwick, 1992, Sanes et al., 1985). These patients suffer a permanent loss of large sensory myelinated fibers which results in adaptive strategies to achieve a secure gait. For example, they preserve the body balance by enlarging the base of support by leaning forward, e.g., inclined body, and by a dominant visual control of their steps (Lajoie et al., 1996).

Spinal reflexes (SR) and descending pathways such as the reticulospinal, vestibulospinal, rubrospinal and corticospinal tracts converge on common spinal interneurons, the CPGs (Dietz, 2002b, Edgerton et al., 2004). Each of these tracts incorporates peripheral sensory modalities as well as information from the cerebellum and other supraspinal structures for a context-specific locomotor output (Edgerton et al., 2004). The corticospinal control locomotion in human is phase-dependent (Schubert et al., 1997). This enables the subjects to circumvent surface irregularities or obstacles without losing postural stability.

Most of the basic mechanisms underlying locomotion do not fundamentally differ between bipeds and quadrupeds (Duysens, 2002). Quadrupedal locomotion is characterized by the coordination of forelimb and hindlimb rhythmic activities generated by common spinal neuronal control mechanisms, i.e., long propriospinal neurons coupling the cervical and lumbar segments (Cazalets & Bertrand, 2000, Nathan et al., 1996). The neuronal coupling and coordination of upper and lower limb in quadrupedal locomotion is basically preserved in bipedal gait. Only during locomotion, but not during sitting or standing, unilateral tibial nerve stimulation leads to reflex responses in the proximal muscles of both arms (Dietz et al., 2001, Michel et al., 2008). Such a task-dependent coupling of thoraco-lumbar and cervical locomotor centers is flexible and allows humans to use the upper limb for fine, skilled movements or alternatively, for locomotor tasks, e.g., arm swing as a residual function of quadrupedal locomotion (Dietz, 2002a). The quadrupedal coordination of bipedal gait is altered in a specific way after a stroke and in PD.

Spastic movement disorder

Clinical presentation

From a clinical point of view, the spastic movement disorder is reflected in, e.g., exaggerated reflexes leading to an increased muscle tone as the basis for the movement disorder. Spasticity develops after an upper motoneuron lesion, which can be located in the cerebrum or the spinal cord. It can be caused by stroke, multiple sclerosis, SCI, brain injury, cerebral paresis, or other neurological conditions. The clinical diagnosis of spasticity is based on physical signs under passive conditions, i.e., exaggerated tendon reflexes, clonus and muscle hypertonia reflected as a velocity-dependent resistance of a muscle to stretching. In recent years, studies on functional movements in spastic patients indicate that the clinical signs of spasticity are poorly related to the spastic movement disorder. The characteristic spastic movement disorder is mainly due to secondary changes of muscle properties which require a specific treatment, not focusing on isolated clinical parameters, but on the pathophysiological mechanisms underlying the disorder of functional movements which impairs the patient.

Spinal reflex behavior in the clinical condition

A severe CNS lesion leads to alterations in SR function. During the first weeks after a lesion, called the spinal shock, flaccid paralysis is accompanied with a loss of tendon tap reflexes, while the H-reflex, an electrophysiological correlate to the tendon tap reflex without the contribution of muscle spindles, can be elicited (Hiersemenzel et al., 2000). The loss of tendon tap reflexes is attributed to a general decrease of excitability of α- and γ-motoneurons as well as interneurons due to the sudden loss of supraspinal drive (Elbasiouny et al., 2010). After four to six weeks, the clinical signs of spasticity, i.e., exaggerated reflexes and increased muscle tone, evolve. In patients with fully developed hemiparetic spasticity, the threshold of the soleus stretch reflex is decreased (Levin & Feldman, 1994, Nielsen & Sinkjaer, 1996), possibly due to increased motoneuronal excitability (Powers et al., 1988). Many studies have investigated the neuronal mechanisms underlying exaggerated tendon reflexes in patients with spasticity. The short-latency reflex is mediated by fast conducting Ia nerve fibers from the muscle spindles to the spinal cord. After a severe acute central lesion the short-latency reflexes are abolished followed by a hyper-reflexia due to neuronal reorganization that occurs in both cats (Mendell, 1984) and humans (Carr et al., 1993). Evidence from experiments in humans and animal models demonstrated that alterations in the excitability of excitatory and inhibitory spinal pathways occur after central lesions. Novel connections by axonal sprouting, synaptic strengthening, removed depression of previously inactive connections may cause changes in the strength of inhibition among neuronal circuits. A reduction of presynaptic inhibition of group Ia nerve fibers occurs (Burke & Ashby, 1972) which coincides with the hyper-excitability of tendon tap reflexes (short-latency reflexes). As well as the reduced presynaptic inhibition of Ia afferents, several other mechanisms may also contribute to spasticity: increased reciprocal Ia inhibition (Boorman et al., 1991, Knutsson et al., 1997), alterations in postactivation depression (Nielsen et al., 1993, Thompson et al., 1998), deficient disynaptic reciprocal inhibition (Crone et al., 1994), disinhibition of group II pathways (Nardone & Schieppati, 2005, Remy-Neris et al., 2003) or hyperactivity of fusimotoneurons (γ-motoneurons) (Dietrichson, 1973, Rushworth, 1960). The reduction of presynaptic inhibition of Ia afferents is stronger in the legs of para- compared to hemiplegic patients (Faist et al., 1994). However, the degree of muscle hypertonia assessed by the clinical Ashworth scale is not related to the decreased presynaptic inhibition of Ia afferents (Faist et al., 1994).

Another SR frequently investigated in patients with spasticity is the flexor reflex, a poly-synaptic SR evoked by noxious electrical stimulation. Flexor reflexes are believed to be exaggerated after a central nervous lesion and to cause muscle spasms after severe SCI (Ditunno et al., 2004). However, muscle spasms might be also evoked by spontaneous firing of moto-neurons during rest (Gorassini et al., 2004) due to a receptor upregulation and/or neuronal sprouting (Goldberger & Murray, 1988, Little et al., 1999). After acute, complete SCI, flexor reflex excitability and spastic muscle tone develop in parallel (Hiersemenzel et al., 2000). However, a few months later, the severity/frequency of spasms increases while flexor reflex amplitude decreases (Hiersemenzel et al., 2000). This observation suggests a poor relationship between the activity flexor reflexes and muscle spasms.

The treatment of spasticity is directed towards reducing hyper-reflexia as it was thought that exaggerated reflexes are responsible for the increased muscle tone and therefore also the movement disorder. However, several studies revealed that increased muscle tone is associated with muscle fiber contractures, loss of sarcomeres, rather than with reflex hyperexcitability (O'Dwyer & Ada, 1996, O'Dwyer et al., 1996). In particular, torque motor experiments in leg extensors and arm flexors (antigravity muscles) indicate a major, non-reflex contribution to the spastic muscle tone (Hufschmidt & Mauritz, 1985, Sinkjaer et al., 1993).

Muscular changes

Exaggerated stretch or flexor reflexes elicited under passive conditions, usually done in the clinic bedside examinations, can hardly be solely responsible for the increased muscle tone in spastic patients. Although spasticity is neural in origin, there is evidence for the development of abnormalities in mechanical characteristics of spastic muscles after a CNS lesion. Secondary changes in intrinsic and extrinsic muscle fiber properties contribute to spastic muscle tone (for review see Dietz & Sinkjaer, 2007). Muscle hypertonia is defined as a velocity-dependent resistance of the muscle to stretching and is clinically assessed by the Ashworth scale. In patients with chronic stroke, spastic muscle hypertonia is associated with increased muscle EMG activity exceeding that measured in healthy subjects (Dietz et al., 1991, Hufschmidt & Mauritz, 1985). Despite this extra EMG activity, passive stiffness (e.g. muscle contracture) at the ankle joint is also increased and contributes to the clinically defined spastic muscle hypertonia after stroke. In addition, muscle tone of the non-affected side of stroke patients is not completely normal, e.g., it shows some increase in tone compared to the muscles of healthy subjects (Thilmann et al., 1990).

There is evidence that abnormal stretch reflex activity is insufficient to explain the increased muscle tone in subjects with stroke or multiple sclerosis even in the clinical (passive) conditions (Galiana et al., 2005, Hufschmidt & Mauritz, 1985, O'Dwyer & Ada, 1996, Sinkjaer et al., 1993). Reflex-mediated stiffness in ankle extensors (Sinkjaer et al., 1993) and elbow flexors (Dietz et al., 1991, Ibrahim et al., 1993, Powers et al., 1988) in patients with spasticity (stroke or multiple sclerosis) is within the range of healthy subjects and is just slightly increased in subjects with SCI (Mirbagheri et al., 2001). Therefore, other factors must contribute to spastic muscle tone, such as loss of sarcomeres in series along the myofibrils leading to subclinical contractures (O'Dwyer et al., 1996) and changes in collagen tissue and tendons (Hufschmidt & Mauritz, 1985, Sinkjaer & Magnussen, 1994, Sinkjaer et al., 1993). In addition, changes of biomechanical conditions of a spastic muscle might also have effects on the stretch reflex behavior (via group III/IV muscle afferents) in stroke patients (Kamper et al., 2001, Schmit et al., 2002).

Functional movement

Patients with upper motoneuron lesions typically show movement disorders like gait impairments. They can be evaluated by recordings of electrophysiological and biomechanical parameters. An adequate treatment should address the mechanisms underlying the impaired function and not be directed to isolated clinical symptoms. The physical signs, e.g., exaggerated reflexes assessed under passive conditions, bear little relationship to the patient's movement disorder. This observation has been made by extensive investigations on functional movements of the leg (Berger et al., 1984, Dietz & Berger, 1983) and the arm (Dietz et al., 1991, Ibrahim et al., 1993, Powers et al., 1989).

Locomotor EMG pattern

In patients with spastic hemi- or paraparesis a pattern of leg muscle activation during locomotion can be assessed by EMG recordings. The timing of the pattern, i.e., the reciprocal activation of leg flexors and extensors, is comparable to that in healthy subjects (Dietz, 2003b, Kautz et al., 2006, Maegele et al., 2002). In contrast, leg muscle activity is, depending on the severity of paresis, reduced and has a lower amplitude compared with that in the non-affected side of hemiparetic patients or in healthy subjects (Berger et al., 1984, Dietz & Berger, 1983) (Figure 18.1). The spastic-paretic foot with its plantar-flexed position leads to a premature leg extensor activation during the stance phase of gait (Knutsson & Richards, 1979). Also in healthy subjects a premature leg extensor activation occurs, i.e., before ground contact when they are voluntarily tip-toeing. The contribution of afferent feedback to soleus activity is low in patients with spasticity (Mazzaro et al., 2007). Correspondingly, the leg extensor amplitude modulation during stance is diminished or lacking (Dietz, 2002b).

In general, studies on functional movement show that assessment of clinical spasticity does not relate to gait disorders after stroke or SCI (Ada et al., 1998, Dietz & Sinkjaer, 2007). The reduction of long-latency reflex components rather than exaggerated short-latency (tendon tap) reflexes plays an important role in the impaired movement performance and equilibrium control during upright standing (Nardone et al., 2001) and gait in stoke subjects (Berger et al., 1984).

Propriospinal reflexes

Changing external demands and ground irregularities during gait require an adaptive locomotor pattern. The adaptation in healthy subjects is achieved by proprioceptive input that continuously

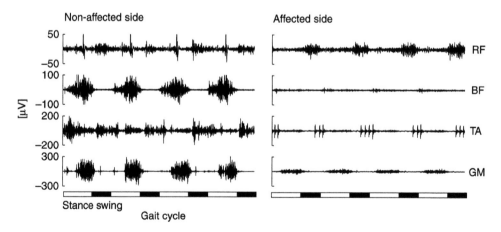

Figure 18.1 Locomotor EMG pattern in a hemiparetic stroke patient. Leg muscle activity (RF rectus femoris, BF biceps femoris, TA tibialis anterior, GM gastrocnemius medialis) during treadmill walking (2 km/h) of the non-affected and affected leg of a stroke subject (21 months after stroke) (from Hubli et al. (2012a)).

modulates the programmed pattern according to information from peripheral sensors, e.g., receptors in muscles, tendons, joints and the skin. One of the essential functions of proprioceptors is to detect unexpected events and to initiate rapid compensatory muscle responses. Three main reflex systems are involved in locomotion: monosynaptic (short-latency) reflexes mediated via Ia afferents, polysynaptic cutaneous reflexes mediated by skin afferents and polysynaptic reflexes integrating afferent inputs from different sources (Dietz, 2002b). While short-latency reflex activity is inhibited (Yang & Stein, 1990), polysynaptic (or long-latency) reflex contribution is essential for leg extensor activity during gait.

In healthy subjects, Ia afferent input to the spinal cord (H-reflex and short-latency stretch reflex) becomes suppressed by presynaptic inhibition during the swing phase (Capaday & Stein, 1987). In patients with spastic paresis, however, this physiological reflex modulation is impaired (Jones & Yang, 1994, Sinkjaer et al., 1996, Sinkjaer et al., 1995). Presynaptic inhibition of Ia afferents is reduced and short-latency stretch reflexes commonly appear in the leg extensor muscles during the transition from swing to stance phase. The inability to suppress stretch reflex excitability in the extensors during the swing phase of gait might contribute to impaired gait in spastic patients (Dietz, 2002b, Faist et al., 1996, Faist et al., 1999, Jones & Yang, 1994, Sinkjaer et al., 1996, Sinkjaer et al., 1995). Also cutaneous reflex modulation is reduced in the leg extensors during gait (Jones & Yang, 1994). Correspondingly, the fast regulation of motoneuron discharges, which characterizes functional muscle activation, is absent in spastic patients (Dietz et al., 1986, Rosenfalck & Andreassen, 1980). This could correspond to a loss of EMG modulation during gait.

Studies on the compensatory reactions to gait perturbations (e.g. short acceleration of a treadmill during stance phase of gait) show an appearance of small short-latency stretch reflexes followed by strong compensatory long-latency (polysynaptic) reflexes in extensor (Dietz, 2002b, Dietz, 2003a) and dorsiflexor (Christensen et al., 2001) muscles in healthy legs. In contrast, in spastic legs, short-latency reflexes appear without significant long-latency reflex activity (Berger et al., 1984, Sinkjaer et al., 1999). In addition, following translational stance displacements in hemiparetic patients, the amplitudes of the compensatory leg muscle EMG response is smaller on the spastic compared to the non-affected side (Dietz & Berger, 1984). Obviously exaggerated short-latency reflexes in spastic patients are associated with an absence or reduction of functionally essential polysynaptic (long-latency) reflexes. The loss of long-latency reflexes leads to a reduced proprioceptive contribution to the leg muscle activation during gait and concurrently to less well-modulated leg muscle activity (Dietz, 2001). These findings might result from an impaired gating of peripheral input by spinal neuronal circuits after a central lesion (Dietz & Sinkjaer, 2007) leading to a lower capacity of reflex modulation and consequently to lower adaptation of muscle activity to ground conditions (Mazzaro et al., 2007). This might contribute to the spastic movement disorder (Burne et al., 2005, Dietz, 2002b).

A loss of walking capacity in patients with spastic movement disorder correlates with the behavior of polysynaptic SR in lower limb muscles evoked by a non-noxious tibial nerve stimulation (Hubli et al., 2012b). While healthy subjects and patients (SCI and stroke) with good to moderate walking capacity show dominant early reflex components (60–120 ms latency), in chronic severely affected SCI or stroke patients a late reflex component (120–450 ms latency) dominates. Therefore, the SR might be useful as a marker for the functional state of spinal neuronal circuits associated with locomotor function after a CNS lesion. In addition, these functional states are not fixed since four weeks of intense locomotor training is reflected in both an improvement of locomotor ability and a change in balance of SR components toward the early SR component (Hubli et al., 2012b).

Muscle tone

Muscle tone during functional movements in patients with spastic paresis can hardly be measured under active conditions. Nevertheless, the spastic muscle tone cannot be explained due to an increased activity of motoneurons. The equivalent to the muscle tone assessed clinically is the tension developed at the Achilles tendon during locomotion. Tension development during gait results from muscle stiffness and EMG activity and differs between the affected and non-affected leg in stroke subjects with spastic hemiparesis (Berger et al., 1984): in the non-affected leg, the tension development correlates with the modulation of EMG activity (as in healthy subjects). In the spastic leg, however, the tension development is connected to the stretching of leg extensors which are tonically activated, but on a low level of EMG activity (Berger et al., 1984). This tension development on the spastic side is assumed to occur on a simpler level of organization due to changes in mechanical properties of leg extensor muscles.

Interlimb coordination

Recordings of reflex function reflecting spinal interneuronal activity can be used to acquire insight into the quadrupedal coordination during locomotion (Dietz, 2002b). Recordings of bilateral arm muscle reflex responses to unilateral tibial nerve stimulation allow investigations of the processing of afferent input from the non-affected and the affected legs in stroke patients (Kloter et al., 2011, Michel et al., 2008). In stroke patients, reflex responses in proximal arm muscles are strong on both sides as in healthy subjects, when the nerve of the non-affected leg is stimulated. In contrast, nerve stimulation on the affected leg leads to no responses either on the affected or on the non-affected side (Figure 18.2) (Kloter et al., 2011). An impaired processing of afferent volleys by the corticospinal tract (Lemon, 2008) during locomotion might be responsible for this observation and probably leads to a defective sensorimotor integration and a disturbed inter- and intralimb coupling of stroke patients (Dyer et al., 2009, Finley et al., 2008). The efferent reflex pathway to the muscles of the affected arm seems to be less impaired. Hence, the arm and leg muscle reflex responses are only slightly smaller on the affected side compared to the non-affected side when the nerve of the non-affected leg is stimulated (Figure 18.2). These observations indicate strong sensorimotor interactions between non-affected and affected sides during locomotion in stroke patients. The therapeutic consequences of the findings might have an influence on the functional training in stroke patients by providing additional afferent input from the non-affected side (Dietz, 2011).

Differences between SCI and stroke

The pathophysiology of spasticity varies to a limited extent due to the causative lesion, i.e., independent of whether SCI or stroke, the corticospinal tract is lesioned. The other descending pathways involved by the lesion differ between cerebral and spinal lesions. In both lesion types post-activation depression at the Ia afferents is similarly reduced. However, abnormalities in SCI subjects involve primarily spinal pathways. For example, the decrease of presynaptic inhibition of Ia afferents is more pronounced in patients with SCI than in those suffering from stroke. In stroke patients in particular, ipsilateral non-crossed fibers might be responsible for the preservation of presynaptic inhibition on the spastic side and might lead to minor spastic signs (Faist et al., 1994). Therefore, the gait disorder is frequently more pronounced in SCI than in stroke patients (Pierrot-Deseilligny & Burke, 2005).

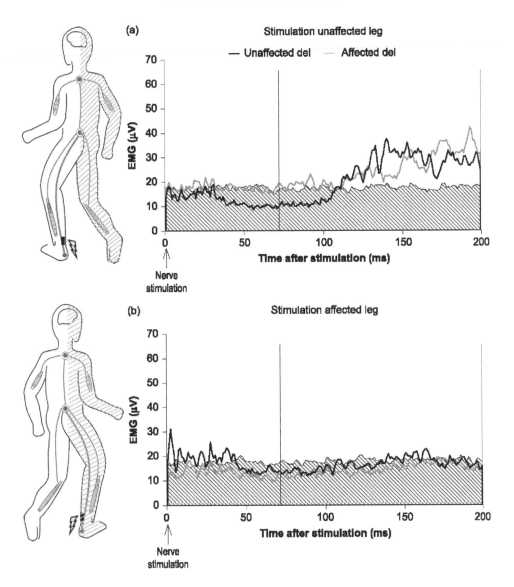

Figure 18.2 Interlimb reflex coupling in stroke patients. Reflex responses in the deltoideus muscle evoked from tibial nerve stimulation. Grand means of the rectified reflex responses in the non-affected (black lines) and affected (grey lines) deltoid muscles to tibial nerve stimulation at the non-affected (a) and affected (b) leg. Background EMG activity is depicted as hatched area. The schematic drawings indicate the side of nerve stimulation (affected side: hatched area) (from Kloter et al. (2011)).

In stroke patients, there is evidence for an increased resistance to stretch in the leg extensors which is mainly due to mechanical rather than reflex activity (Ada et al., 1998, Dietz, 2003a, Sommerfeld et al., 2004). So far it is rather unlikely that the reduction of spasticity by drug therapy will improve gait after stroke or SCI (Landau, 2003). In contrast, the reduction of spastic muscle tone might even be counterproductive as spasticity helps to support the body during gait (Dietz, 2003a, Dietz & Sinkjaer, 2007) (Figure 18.3, schematic drawing).

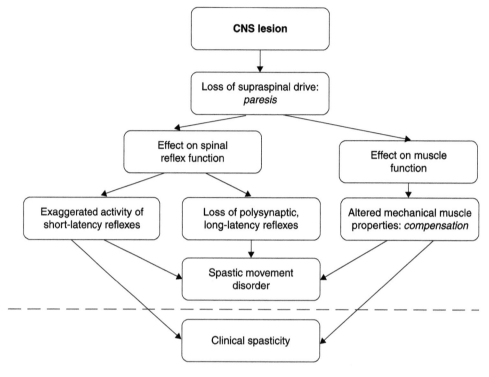

Figure 18.3 Spastic movement disorder: mechanisms involved. An upper motoneuron lesion leads to a loss of supraspinal drive. Consequently, changes in muscle function and reflex behavior occur. Secondary changes to the primary lesion lead to the spastic movement disorder (modified from Dietz et al. (2002b)).

In mobile patients with spinal cord lesions exaggerated stretch reflexes can disrupt movements (Corcos et al., 1986, Dimitrijevic & Nathan, 1967, Mizrahi & Angel, 1979). Nevertheless, training effects can only be achieved when a spastic muscle tone is present (Dietz et al., 1995). In addition, exaggerated stretch reflex activity in the antagonist due to contraction of the agonist can restrict knee movements (Knutsson et al., 1997).

Secondary compensatory mechanisms

A central motor lesion leads to changes in the excitability of SR and a loss of supraspinal drive. The primary deficit, i.e., the paresis, is followed by secondary changes in neuronal control which should not be a priori considered as pathological, but rather as a compensatory strategy of SCI and stroke subjects. The main secondary compensatory mechanisms concern changes in muscle fiber properties in SCI and stroke subjects as well as a partial functional compensation of the hemiplegic by the non-affected limb in stroke subjects.

Changes in mechanical muscle fiber properties

In contrast to healthy subjects, patients with spastic paresis show diminished strength and modulation of EMG activity during functional arm (Ibrahim et al., 1993) or leg (Dietz et al.,

1981) movements. Only weak tonic leg extensor activation is present during locomotion. The decrease of EMG activity is considered to be due to the lack of supraspinal drive leading to paresis and a decreased polysynaptic reflex activity (Kloter et al., 2011) which is important for a proper modulation of EMG activity during functional movements in healthy subjects (Dietz, 2002b). As a consequence, the appropriate muscle tone required to support the body during the stance phase of gait is organized on a lower level. This means that the muscle tone is, not as in healthy subjects, generated through modulation of leg muscle EMG activity, but rather through stretching of the tonically activated leg extensor muscles. Structural changes of muscle fibers (secondary adaptation) by, e.g., loss of sarcomeres (O'Dwyer & Ada, 1996) following a CNS lesion can therefore partially compensate the paresis (primary deficit) by this spastic muscle tone and enable body support during walking (Figure 18.3). However, fast movements can hardly be exerted by this simpler regulation of muscle tone.

Compensation by the non-affected leg

Two findings support the hypothesis that in patients with spastic hemiparesis the non-affected leg functionally compensates in part for the affected, spastic leg which is mainly used as a stick for body support during walking (Dietz & Sinkjaer, 2007). First, the SR evoked by non-noxious tibial nerve stimulation can be used as a marker for the functional state of spinal circuitries associated with locomotor activity in severely affected, incomplete SCI subjects (Hubli et al., 2012b). In SCI subjects, both assessments of general walking ability and muscle strength of lower limbs correlate with SR behavior. In contrast, the SR function of the hemiplegic limb of stroke subjects is related only to assessment of muscle strength of the affected leg, and not to the walking capacity (Hubli et al., 2012a).

Second, the recovery of gait in hemiparetic stroke subjects, e.g., the level of ambulatory independence, body mobility and maximum walking speed, does not depend on an improvement of the EMG patterning of gait-related leg muscle activation in the affected leg (Den Otter et al., 2006). The partial compensation of the functional loss of the affected leg by the non-affected leg might become reduced by training approaches mainly focusing on the affected limb, correspondingly to constraint-induced movement therapy for the affected arm of stroke subjects (Taub et al., 1999).

Conclusion

According to the actual state-of-the-art of spastic movement disorder, the locomotor EMG pattern is preserved after a CNS lesion. In addition, the clinically dominant tendon tap hyperreflexia contributes little to spastic movement disorder. Paresis and attenuation of long-latency (polysynaptic) reflexes hamper walking activity in spastic patients. Secondary to a central lesion, a major alteration in the normal muscle–joint relationship occurs and the muscle tone is regulated on a simpler level to overcome the paresis. These adaptations should be viewed as the optimum for a given state of the sensorimotor system (Latash & Anson, 1996).

Treatment of spasticity is often directed towards a reduction of exaggerated stretch reflex activity and muscle tone, although they are known to be poorly related to the movement disorder following a spinal or cerebral lesion. Consequently, antispastic medications that are directed to reduce the clinical signs of spasticity usually do not improve the movement disorder, and can even increase weakness (Hoogstraten et al., 1988, Latash & Penn, 1996). Therefore, therapeutic interventions in mobile patients with spastic paresis should be focused on the training, relearning and activation of residual motor function (Centonze et al., 2007,

Diserens et al., 2007). In addition, in hemiparetic patients, a better functional outcome might be achieved by facilitating sensorimotor interactions between non-affected and affected sides (Kloter et al., 2011). An additional afferent input from the non-affected side might enhance effects of functional training.

Movement disorder in Parkinson's disease

Clinical presentation

Parkinson's disease (PD) is characterized by deficits in motor control, such as difficulties in movement initiation (Rocchi et al., 2006), scaling movement amplitudes (Jackson et al., 2000) or modulating muscle activity (Flament et al., 2003, Pfann et al., 2001). Movement performance deteriorates with increasing complexity in upper limb movements (Krebs et al., 2001). This is associated with an increased movement time and variability as well as decreased movement velocity in bilateral arm and finger movements (Lazarus & Stelmach, 1992, Shimizu et al., 1987).

In addition, gait and balance difficulties are major clinical problems of PD (Brown & Steiger, 1996). Parkinson disease subjects also have difficulties in gait initiation (Krystkowiak et al., 2006, Rocchi et al., 2006) and in adapting to disturbances (Rogers, 1996). The major deficit in walking in the early stages of PD is a reduction in the speed and amplitude of leg movements leading to the characteristic small and shuffling steps. Furthermore, locomotor hypokinesia is accompanied by a diminished arm swing. As the disease progresses, freezing of stepping becomes more apparent. Patients suffer from disequilibrium and even fall in the more chronic stages of the disease (Brown & Steiger, 1996). In the clinical examination a rigid muscle tone is characteristic of a PD.

Functional movement

Locomotor EMG pattern

Pathophysiologically, an impaired neuronal control of gait associated with rigid and poorly modulated motor performance represents a major deficit of PD (Boonstra et al., 2008, Martin, 1967, Moore et al., 2007). These abnormalities are thought to result from varying combinations of hypokinesia, rigidity, and from deficits of posture and equilibrium. Furthermore, by quantitative gait analysis distinct differences can be evaluated between the gait pattern of patients with vascular or idiopathic PD (Zijlmans et al., 1996) or of those with normal pressure hydrocephalus (Stolze et al., 2001). Furthermore, the characteristic coordination of normal plantigrade gait is lost in PD (Forssberg et al., 1984). Age influences the clinical progression of PD with a faster rate than disease duration and is associated with a decreased levodopa responsiveness (Levy, 2007).

Several studies on gait indicate an impaired programming in patients with PD (for review see Dietz, 1992). The electrophysiological gait analysis of patients with PD, in addition to showing slow and reduced movements, reveals a characteristic pattern of leg muscle activation with a reduced amplitude and little modulated EMG activity in the leg extensor muscles during the stance phase and an increased tibialis anterior activity during swing. Furthermore, the amount of co-activation of antagonistic leg muscles during the support phase of the stride cycle is greater in the patients compared to healthy subjects (Dietz et al., 1981). PD patients are inflexible in adapting and modifying their postural responses to changing support conditions (Schieppati & Nardone, 1991).

Patients with mild to moderate PD also suffer from an impaired acquisition and performance of a high-precision locomotor task, such as obstacle stepping (Michel et al., 2009). During task repetition, improvement is slower and performance level is poorer in PD subjects compared with age-matched healthy subjects. However, PD subjects are able to improve their performance with the repetition of the task (van Hedel et al., 2006). The observation that performance of bilateral obstacle stepping is worse than unilateral obstacle steps (Michel et al., 2009) is in line with findings from upper limb movement tasks, namely that performance is worse in PD compared to control subjects in bilateral compared to unilateral tasks (Shimizu et al., 1987). PD subjects suffer from greater difficulties in daily life situations such as locomotion when attention has to be shared. However, adequate training can improve their adaptive locomotor behavior.

Propriospinal reflexes

A number of electrophysiological studies in PD are concerned with the reflex function during limb displacement in a sitting position. In contrast to the compensatory responses described for perturbations of stance or gait (Dietz, 1997), most of these investigations show an increase in the amplitude of the long-latency EMG response (Cody et al., 1986). An increase of reflex gain at a central site has been postulated (Burke et al., 1977). The discrepant finding of a reduced stretch sensitivity of proprioceptive postural reflexes (Dietz, 1997) may arise primarily from the difference in the motor tasks investigated. The amount of soleus H-reflex inhibition during gait seems to depend on the severity of the disease and might contribute to the problem to initiate a step (Hiraoka et al., 2005).

Several observations have suggested an impaired function of proprioceptive reflexes in PD patients which contributes to the instability during gait (Dietz, 1997). This may be a major reason why these patients rely more on visual input to regulate gait (Bronstein et al., 1990, Prokop et al., 1995). However, the dysfunction of proprioceptive reflexes differs from that observed in spasticity. As in healthy subjects, monosynaptic reflex activity in the gastrocnemius muscle of PD patients is negligible during locomotion. In contrast, the compensatory polysynaptic, or long-latency-reflex EMG responses in the leg extensor muscles are present for the compensation of ground irregularities but significantly smaller than those obtained in age-matched healthy subjects due to a reduced reflex sensitivity (Dietz, 1993, Dietz et al., 1988). The low activation of the leg extensor muscles during conditions of stance and locomotion is assumed to be due to an impaired proprioceptive feedback input from extensor load receptors (Dietz et al., 1993).

This impaired function of polysynaptic reflexes in PD is in line with earlier suggestions of alterations in central responsiveness (Berardelli et al., 1983) and insufficient use of sensory input (Tatton et al., 1984). The consequence of a defective proprioceptive feedback is a poor adaptation to environmental influences (Rogers, 1996). The reduced sensitivity of polysynaptic reflexes in the leg extensor muscles appears to be a direct consequence of the dopamine deficiency in PD patients. It is also observed (without other changes) in young normal subjects following intake of a single dose of haloperidol (Dietz et al., 1990). Pharmacological studies have shown that this effect is due to an impaired function of central dopamine D2 receptors (Dietz et al., 1990). Conversely, only some gait parameters (kinematic) are L-Dopa-sensitive, while others (temporal) are L-Dopa-resistant (Blin et al., 1991).

The diminished gastrocnemius activation following perturbations of locomotion is followed by a stronger tibialis anterior activation in PD patients, which might correspond to the so-called shortening reaction (Berardelli & Hallett, 1984). This may be a result of an impaired proprioceptive feedback control that might be partially compensated for by a greater amount of leg flexor activation which leads to a higher degree of co-activation. Visual input plays a role in the generation of this increased activation (Brouwer & Ashby, 1992, Dietz et al., 1997). The idea of an impaired central regulation is in agreement with the concept of an overcompensated and faulty predictive feedback system (Tatton et al., 1984).

Load receptor function

A basic aspect of the neuronal control of bipedal stance and gait represents the antigravity function of leg extensors. Studies in cats (Duysens & Pearson, 1980, Gossard et al., 1994, Pearson & Collins, 1993) and humans ((Dietz et al., 1989, Dietz et al., 1992, Gollhofer et al., 1989); for review see Dietz, 1992) have indicated that extensor muscle load receptor input has a critical role in the generation of leg extensor activity during gait. Proprioceptive reflexes involved in the maintenance of body equilibrium depend on the presence of contact forces opposing gravity. The reduced sensitivity of the leg extensor EMG activity to body unloading/ loading observed during stance and gait is in agreement with observations in cats (Gossard et al., 1994, Pearson & Collins, 1993) and humans (Dietz et al., 1989, Dietz et al., 1992). These studies provide indirect evidence that input from extensor load receptors plays a major role in leg extensor activation during stance and locomotion. Nevertheless, it cannot be deduced from these experiments whether load information in humans is provided by muscle spindles, cutaneous afferents from the sole of the foot or, as suggested from cat experiments, by group lb fibers originating from Golgi tendon organs (Gossard et al., 1994, Pearson & Collins, 1993).

In humans, evidence for a significant contribution of load receptors to leg muscle activation has come from immersion and unloading experiments ((Dietz et al., 1989); for review see Dietz, 1992). Extensor load receptors are thought to signal changes of the projection of the body's center of mass with respect to the feet. The actual loading of the body obviously serves as a stimulus for extensor load receptors which have been shown to be essential for leg extensor activation during locomotion in cat (Duysens & Pearson, 1980, Mittelstaedt, 1995) and human (Dietz et al., 1992). The negligible effect of body unloading on TA EMG activity is not unexpected as a differential neuronal control of the antagonistic leg muscle occurs (for review, see Dietz, 1992).

Several studies on motor control in patients with PD indicate a defective load control in PD subjects (Burne & Lippold, 1996, Delwaide et al., 1991, Stelmach & Phillips, 1991). Furthermore, when the body becomes unloaded during treadmill locomotion, the leg extensor muscles show load sensitivity in both PD patients and healthy subjects (Figure 18.4) (Dietz & Colombo, 1998, Dietz et al., 1997), i.e., the EMG amplitude decreases with body unloading during stepping in healthy and PD subjects. However, the absolute level of leg extensor EMG amplitude during the stance phase of gait is smaller in patients with PD than in the age-matched healthy subjects. This indicates that in PD the threshold of load receptor reflex activity is maladjusted or biased. The load sensitivity of leg extensors progressively increases from PD patients to age-matched to young healthy subjects. In addition, the absolute level of leg extensor EMG amplitude during the stance phase of stepping becomes progressively larger from PD patients to age-matched to young healthy subjects (Figure 18.4).

While the extensor activation by load receptors is essential for the maintenance of body equilibrium, an attenuated magnitude of this reflex activity exists in PD subjects. It is suggested that decreased sensitivity of extensor load reflex mechanisms contributes to impaired gait already in elderly subjects but is more pronounced in PD patients.

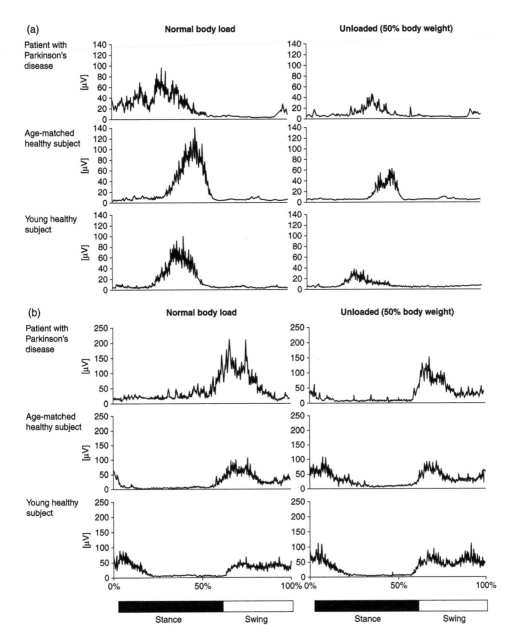

Figure 18.4 Influence of body load on gait pattern: PD vs. healthy subjects. Individual examples of the mean rectified EMG activity (n = 20) (a) gastrocnemius and (b) tibialis anterior muscles of the right leg during locomotion of a patient with PD, an age-matched healthy subject, and a young healthy subject. Two walking conditions are shown: normal body loading (left traces) and 50 percent body unloading (right traces). All recordings are normalized to one step cycle. Stance and swing phases are indicated at the bottom (from Dietz & Colombo (1998)).

Nevertheless, no correlation exists between the impaired load-sensitivity of leg extensors and clinical scores in individual patients with PD (Dietz & Colombo, 1998). This might be because of the fact that extreme (slight and severe) cases were not investigated. Furthermore, gait disability is only partly reflected in the usual clinical scores.

Interlimb coordination

A flexible coupling of thoraco-lumbar and cervical centers allows humans to use the upper limbs for manipulative and skilled movements or, alternatively, for locomotor tasks (Dietz, 2002a). This implicates a functional, task-dependent gating of neuronal pathways between the neuronal circuits controlling lower and upper limb muscles during walking, reflected in the arm swing as a residual function of quadrupedal locomotion.

Unilateral non-noxious leg nerve stimulation during the mid-stance phase of locomotion allows studying the excitability of spinal neuronal circuits underlying the quadrupedal coordination of bipedal gait (Dietz & Michel, 2008). During a precision locomotor task, i.e., obstacle stepping, reflex responses to unilateral tibial nerve stimulation in proximal arm muscles of both sides are enhanced in proximal arm muscles prior to obstacle stepping compared with normal steps, despite a low amplitude background EMG activity in the phase of stimulation (Michel et al., 2008). Corresponding to the reflex activity, the activation of proximal arm muscles is stronger during swing over the obstacle compared to normal swing. Thus, the reflex activity anticipates the following arm muscle activation (Michel et al., 2008).

In PD subjects coordination between the lower limbs as well as between upper and lower limbs is impaired during walking compared with age-matched healthy subjects (Dietz & Colombo, 1998, Dietz & Michel, 2008, Winogrodzka et al., 2005). Defective coordination of upper and lower limbs (Swinnen et al., 1997, Winogrodzka et al., 2005) in combination with reduced arm swing (Carpinella et al., 2007) and abnormal postural reactions to voluntary movements (Rogers et al., 1987) might contribute to the impaired performance of normal stepping and of obstacle avoidance locomotion in PD (Michel et al., 2008).

The spinal interneuronal function underlies the quadrupedal limb coordination during gait and the performance of an obstacle avoidance task was studied in PD subjects (Dietz & Michel, 2008) by the analysis of the quadrupedal distribution of reflex activity to stimulation of a leg nerve. As in healthy subjects, SR responses, evoked by tibial nerve stimulation during mid-stance, are present in all arm and leg muscles investigated in PD subjects. They are larger before execution of obstacle avoidance compared with normal steps in both PD and healthy subjects. However, the SR in the arm muscles prior to normal and obstacle steps is larger in PD compared with age-matched healthy subjects. These observations indicate that quadrupedal limb coordination is basically preserved in PD subjects (Dietz & Michel, 2008). The slightly disturbed interlimb coordination of gait in PD subjects becomes improved by L-Dopa application and subthalamic nucleus stimulation (Carpinella et al., 2007).

Proximal arm muscle activation is stronger during swing over the obstacle compared with normal steps in PD and elderly control subjects, although PD subjects show no relevant arm movements during gait. This mechanism appears to be basically preserved in PD subjects. Despite the slightly worse performance of the obstacle task by PD subjects compared with control subjects, the contribution of upper limb muscle activation to the performance of the precision locomotor task is similar in PD and age-matched healthy subjects.

However, compared with young healthy subjects (27.6 ± 5.8 years) (Michel et al., 2008), the increase of arm muscle EMG during obstacle swing is small in both the PD (65.3 ± 5.9 years) and elderly control subjects (62.0 ± 4.8 years). This attenuated modulation of arm muscle activity (clinically probably reflected in a reduced arm swing) may have contributed to the worse performance of the obstacle task by PD and elderly subjects compared to the young healthy subjects (Michel et al., 2008).

The available data are consistent with the proposal that in PD subjects an enhanced anticipatory spinal neuronal activity (reflected in the SR) in the arm muscles is required to achieve appropriate muscle activation for the automatic control of body equilibrium during the performance of the task. In the tibialis anterior the SR is attenuated presumably because of a stronger voluntary, i.e., cortical, control of leg flexor movements (Dietz & Michel, 2008). In addition, the enhanced anticipatory activation of spinal interneuronal circuits subserving movements of upper limbs might compensate for insufficient lower limb activation. Biomechanical restraints, rather than insufficient upper limb muscle activation, seem to be responsible for the rigid arm position during walking (Dietz & Michel, 2008, Watts et al., 1986).

Secondary compensatory mechanisms

The sensorimotor deficits in PD subjects become partially compensated by secondary changes in neuronal control (Figure 18.5). They mainly concern the role of vision in the control of leg flexor muscles and changes in mechanical muscle properties.

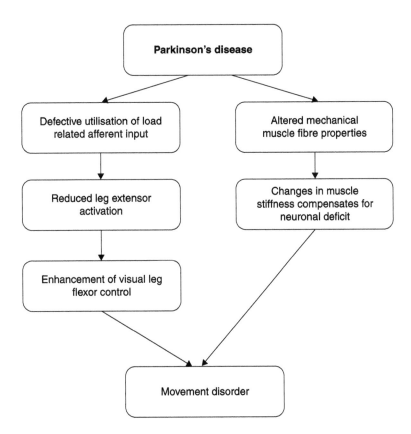

Figure 18.5 Parkinson's disease: mechanisms involved. The movement disorder observed in patients with Parkinson's disease can be mainly ascribed to changes in muscle properties and compensatory strategies of visual feedback.

Changes in mechanical muscle fiber properties

Evidence of changes of inherent muscle stiffness in PD are based on gait studies (Dietz et al., 1988) and perturbation experiments showing a slower ankle joint displacement trajectory in PD patients compared with age-matched healthy subjects (Dietz et al., 1988). This effect may be attributed to altered mechanical properties of the leg extensor muscles in PD and may contribute to rigid muscle tone (Dietz et al., 1988). Such changes in muscle stiffness may be advantageous so far as higher resistance to stretch helps to compensate for small perturbations instead of the impaired reflex function. Changes of inherent mechanical properties of muscle in PD are also present in upper limb muscles (Berardelli et al., 1986, Watts et al., 1986).

These changes also characterize differences between elderly and young normal subjects, although they are more pronounced in PD patients (Dietz et al., 1997). In spasticity such changes in mechanical muscle properties compensate for the paresis in mobile stroke or SCI subjects (for review see Dietz & Sinkjaer (2007)).

Compensation by visual control of leg flexor muscles

The impaired function of proprioceptive reflexes in PD patients, which contributes to the instability of PD patients during gait (for review see Abbruzzese & Berardelli (2003)) may be a major reason why these patients rely more on visual information for the regulation of gait (Bronstein et al., 1990, Schubert et al., 2005). A strong dependency of patients with PD on visual cues during walking is based on the observation of what happens when an optical flow pattern is imposed during stepping on a treadmill. While walking velocity of healthy subjects is influenced only over a short time by the optical flow pattern, patients with PD continuously change their speed sinusoidally with the movements of the optical flow pattern (Prokop et al., 1995). The fact that the control of stance and gait in patients with PD relies more on visual information has the consequence that it is exerted through modulation of leg flexor EMG activity (Brouwer & Ashby, 1992, Dietz et al., 1992), i.e., a stronger TA activation in the patients with PD may in part compensate for impaired leg extensor activity.

When PD subjects walk on a treadmill and have to step repetitively as low as possible over an obstacle without touching it, PD patients perform worse and improve foot clearance more slowly compared with age-matched healthy subjects (Michel et al., 2009, van Hedel et al., 2006). The voluntary command to the prime mover (TA) to overcome the obstacle with a high safety margin might override the automatic control in the lower limb muscles. Nevertheless, in line with an earlier report (van Hedel et al., 2006), subjects with moderate PD perform obstacle stepping almost as well as age-matched healthy subjects, with the exception of a slightly higher foot clearance during obstacle steps and less adaptation of foot clearance.

Restricting vision deteriorates performance in PD and age-matched subjects compared to young healthy subjects, i.e., they are forced to rely more on visual input. Learning a precision locomotor task with restricted vision is impaired in PD more than in age-matched healthy subjects and in both groups in comparison with a group of young healthy subjects. Therefore, the visuo-motor/cortico-spinal control of leg movements plays a major role in PD compared with control subjects. Furthermore, walking in both elderly and PD subjects requires a high attentional demand. When a task becomes more demanding by restricting vision, young healthy subjects use unusual cues (acoustic signals or tactile feedback) to overcome an obstacle. As a consequence, the performance of the obstacle avoidance task deteriorates (Erni & Dietz, 2001).

The increased foot clearance in obstacle steps of PD subjects can be explained by an attempt to securely overcome the obstacle. Compared with young subjects (Dietz & Michel,

2008), elderly healthy subjects show a poorer performance too. This fits with the observation that elderly people also have an increased risk of falls (Dietz & Colombo, 1998, Nieuwenhuijzen et al., 2006). The essential point is that PD and age-matched healthy subjects show similar deficits which are aggravated in PD subjects mainly due to a defective use of proprioceptive feedback (Dietz & Colombo, 1998).

Conclusions

The treatment of PD primarily consists in the application of levodopa and of dopamine agonists in combination with physiotherapy (e.g. treadmill training; (Bello et al., 2008). The disturbed interlimb coordination becomes improved by L-Dopa application and subthalamic nucleus stimulation (Carpinella et al., 2007). Deep brain stimulation can be regarded as a safe treatment of medically refractory movement disorder in PD (Lyoo et al., 2007). Novel deep brain stimulation targets such as the pedunculopontine nucleus might improve treatment of the gait disorder in atypical PD in the near future (Litvan, 2007). However, subthalamic nucleus stimulation seems to be more effective for limb related signs of PD than for the gait disorder. Improvement of gait can also be achieved by a rhythmic auditory facilitation of gait (del Olmo & Cudeiro, 2005) in combination with a pharmacological modulation of both dopamine and norepinephrine pathways (Devos et al., 2007).

During locomotion leg flexor activation by visual control represents a compensatory mechanism for the impaired leg extensor activation due to a defective load receptor input.

An impaired quadrupedal neuronal coordination might also contribute to the locomotor disorder in PD. The goal to treat the gait disorder in PD should also include strengthening the quadrupedal coordination of arm/leg muscle activation during the execution of specific locomotor tasks. The positive effect of task repetition in PD subjects indicates that adequate training can improve adaptive locomotor behavior in PD patients.

General conclusion

An important feature for sensorimotor control in humans is a redundancy of mechanisms involved (e.g. visual, vestibular, proprioceptive input) that enable subjects to solve everyday motor problems and perform complex movements even if they suffer a deficit within the sensorimotor system. The choice of particular movement patterns is based on priorities which might be reconsidered under certain atypical conditions, such as sensorimotor disorders (e.g. stroke, SCI, PD). A new set of priorities will reflect the current state of the sensorimotor system and may select various sources for the generation of motor patterns. The actual motor patterns should not be considered as pathological but rather as adaptive to a primary disorder and may even be viewed as optimal for a given state of the system in the generation of a functional movement. Therefore, therapeutic approaches should not be directed toward restoring the motor patterns to as close to 'normal' as possible but rather toward resolving the original underlying problem (Latash & Anson, 1996) and to guide the spontaneously occurring adaptive changes in a way that an optimal outcome of the individual patient can be achieved. This can be done in a way, for example, to use the development of spastic muscle tone in stroke/SCI subjects to allow stepping but to prevent an overshooting of spasticity/spasms in immobilized subjects. Or to provide additional cues for PD subjects to compensate for the reduced load sensitivity leading to a loss of proprioceptive extensor activation for balance control. By a repetitive functional training locomotor ability can be improved. This training should be adapted individually in such a way that the movements of the subjects become assisted/

supported so far as this is required and by a training speed and duration which is tolerated by the subject.

Dopamine application enhancing motor learning effects (Floel et al., 2005) are controversially discussed. Enhanced training effects were found in stroke subjects (Scheidtmann et al., 2001), but not in SCI subjects (Maric et al., 2008). Furthermore, PD subjects, although suffering a dopamine depletion, show adaptive changes during obstacle stepping similar to age-matched subjects (van Hedel et al., 2006). In any case it has to be acknowledged that after a severe CNS damage the performance of 'normal' movements can never be achieved, independently of the treatment approach. The goal should rather be to obtain an optimum of recovery of function for the individual subject.

In addition, it has to be recognized that elderly subjects also show a recovery of the neurological deficit after a CNS damage. However, they have difficulties in translating this gain in an improvement of function in a hospital setting (Jakob et al., 2009). This should be taken into account in rehabilitation approaches to be applied in elderly people.

For the normal function of the sensorimotor systems we have to be aware that there exists a great redundancy of sources for the normal performance of complex movements, i.e., deficits such as loss of vestibular function do not hamper movement performance under normal conditions. Only when larger or several parts of the systems are damaged (e.g. after stroke or a combination of vestibular loss and polyneuropathy), do complex movements become impaired and automatically regulated on a simpler level of neuronal organization. Thus, the performance of complex movements is still possible, but only within a limited range. In addition, one should be cautious with rehabilitation due to the interference with these beneficial secondary processes occurring after a stroke/SCI.

References

Abbruzzese, G. and Berardelli, A. (2003), 'Sensorimotor integration in movement disorders', *Mov Disord*, 18: 231–40.

Ada, L., Vattanasilp, W., O'Dwyer, N. J. and Crosbie, J. (1998), 'Does spasticity contribute to walking dysfunction after stroke?', *J Neurol Neurosurg Psychiatry*, 64: 628–35.

Bello, O., Sanchez, J. A. and Fernandez-del-Olmo, M. (2008), 'Treadmill walking in Parkinson's disease patients: adaptation and generalization effect', *Mov Disord*, 23: 1243–9.

Berardelli, A., Accornero, N., Argenta, M., Meco, G. and Manfredi, M. (1986), 'Fast complex arm movements in Parkinson's disease', *J Neurol Neurosurg Psychiatry*, 49: 1146–9.

Berardelli, A. and Hallett, M. (1984), 'Shortening reaction of human tibialis anterior', *Neurology*, 34: 242–5.

Berardelli, A., Sabra, A. F. and Hallett, M. (1983), 'Physiological mechanisms of rigidity in Parkinson's disease', *J Neurol Neurosurg Psychiatry*, 46: 45–53.

Berger, W., Horstmann, G. and Dietz, V. (1984), 'Tension development and muscle activation in the leg during gait in spastic hemiparesis: independence of muscle hypertonia and exaggerated stretch reflexes', *J Neurol Neurosurg Psychiatry*, 47: 1029–33.

Blin, O., Ferrandez, A. M., Pailhous, J. and Serratrice, G. (1991), 'Dopa-sensitive and dopa-resistant gait parameters in Parkinson's disease', *J Neurol Sci*, 103: 51–4.

Boonstra, T. A., van der Kooij, H., Munneke, M. and Bloem, B. R. (2008), 'Gait disorders and balance disturbances in Parkinson's disease: clinical update and pathophysiology', *Curr Opin Neurol*, 21: 461–71.

Boorman, G., Hulliger, M., Lee, R. G., Tako, K. and Tanaka, R. (1991), 'Reciprocal Ia inhibition in patients with spinal spasticity', *Neurosci Lett*, 127: 57–60.

Bronstein, A. M., Hood, J. D., Gresty, M. A. and Panagi, C. (1990), 'Visual control of balance in cerebellar and parkinsonian syndromes', *Brain*, 113: 767–79.

Brouwer, B. and Ashby, P. (1992), 'Corticospinal projections to lower limb motoneurons in man', *Exp Brain Res*, 89: 649–54.

Brown, P. and Steiger, M. J. (1996) Basal ganglia gait disorders. In: Bronstein, A.M., Brandt, T. and Woollacott, M.H., editors. Clinical Disorders of Balance, Posture and Gait. London: Arnold. 156–67.

Burke, D. and Ashby, P. (1972), 'Are spinal "presynaptic" inhibitory mechanisms suppressed in spasticity?', *J Neurol Sci*, 15: 321–6.

Burke, D., Hagbarth, K. E. and Wallin, B. G. (1977), 'Reflex mechanisms in Parkinsonian rigidity', *Scand J Rehabil Med*, 9: 15–23.

Burne, J. A., Carleton, V. L. and O'Dwyer, N. J. (2005), 'The spasticity paradox: movement disorder or disorder of resting limbs?', *J Neurol Neurosurg Psychiatry*, 76: 47–54.

Burne, J. A. and Lippold, O. C. (1996), 'Loss of tendon organ inhibition in Parkinson's disease', *Brain*, 119: 1115–21.

Capaday, C. and Stein, R. B. (1987), 'Difference in the amplitude of the human soleus H reflex during walking and running', *J Physiol*, 392: 513–22.

Carpinella, I., Crenna, P., Marzegan, A., Rabuffetti, M., Rizzone, M., Lopiano, L. and Ferrarin, M. (2007), 'Effect of L-dopa and subthalamic nucleus stimulation on arm and leg swing during gait in Parkinson's Disease', *Conf Proc IEEE Eng Med Biol Soc*, 2007: 6665–8.

Carr, L. J., Harrison, L. M., Evans, A. L. and Stephens, J. A. (1993), 'Patterns of central motor reorganization in hemiplegic cerebral palsy', *Brain*, 116: 1223–47.

Cazalets, J. R. and Bertrand, S. (2000), 'Coupling between lumbar and sacral motor networks in the neonatal rat spinal cord', *Eur J Neurosci*, 12: 2993–3002.

Centonze, D., Koch, G., Versace, V., Mori, F., Rossi, S., Brusa, L., Grossi, K., Torelli, F., Prosperetti, C., Cervellino, A., Marfia, G. A., Stanzione, P., Marciani, M. G., Boffa, L. and Bernardi, G. (2007), 'Repetitive transcranial magnetic stimulation of the motor cortex ameliorates spasticity in multiple sclerosis', *Neurology*, 68: 1045–50.

Christensen, L. O., Andersen, J. B., Sinkjaer, T. and Nielsen, J. (2001), 'Transcranial magnetic stimulation and stretch reflexes in the tibialis anterior muscle during human walking', *J Physiol*, 531: 545–57.

Cody, F. W., MacDermott, N., Matthews, P. B. and Richardson, H. C. (1986), 'Observations on the genesis of the stretch reflex in Parkinson's disease', *Brain*, 109: 229–49.

Cole, J. D. and Sedgwick, E. M. (1992), 'The perceptions of force and of movement in a man without large myelinated sensory afferents below the neck', *J Physiol*, 449: 503–15.

Corcos, D. M., Gottlieb, G. L., Penn, R. D., Myklebust, B. and Agarwal, G. C. (1986), 'Movement deficits caused by hyperexcitable stretch reflexes in spastic humans', *Brain*, 109: 1043–58.

Crone, C., Nielsen, J., Petersen, N., Ballegaard, M. and Hultborn, H. (1994), 'Disynaptic reciprocal inhibition of ankle extensors in spastic patients', *Brain*, 117: 1161–8.

del Olmo, M. F. and Cudeiro, J. (2005), 'Temporal variability of gait in Parkinson disease: effects of a rehabilitation programme based on rhythmic sound cues', *Parkinsonism Relat Disord*, 11: 25–33.

Delwaide, P. J., Pepin, J. L. and Maertens de Noordhout, A. (1991), 'Short-latency autogenic inhibition in patients with Parkinsonian rigidity', *Ann Neurol*, 30: 83–9.

Den Otter, A. R., Geurts, A. C., Mulder, T. and Duysens, J. (2006), 'Gait recovery is not associated with changes in the temporal patterning of muscle activity during treadmill walking in patients with post-stroke hemiparesis', *Clin. Neurophysiol.* Jan; 117(1): 4–15. Epub 2005 Dec 5.

Devos, D., Krystkowiak, P., Clement, F., Dujardin, K., Cottencin, O., Waucquier, N., Ajebbar, K., Thielemans, B., Kroumova, M., Duhamel, A., Destee, A., Bordet, R. and Defebvre, L. (2007), 'Improvement of gait by chronic, high doses of methylphenidate in patients with advanced Parkinson's disease', *J Neurol Neurosurg Psychiatry*, 78: 470–5.

Dietrichson, P. (1973), 'The fusimotor system in relation to spasticity and parkinsonian rigidity', *Scand J Rehabil Med*, 5: 174–8.

Dietz, V. (1992), 'Human neuronal control of automatic functional movements: interaction between central programs and afferent input', *Physiol Rev*, 72: 33–69.

Dietz, V. (1993), 'Reflex behavior and programming in Parkinson's disease', *Adv Neurol*, 60: 375–80.

Dietz, V. (1997), 'Neurophysiology of gait disorders: present and future applications', *Electroencephalogr Clin Neurophysiol*, 103: 333–55.

Dietz, V. (2001), 'Spinal cord lesion: effects of and perspectives for treatment', *Neural Plast*, 8: 83–90.

Dietz, V. (2002a), 'Do human bipeds use quadrupedal coordination?', *Trends Neurosci*, 25: 462–7.

Dietz, V. (2002b), 'Proprioception and locomotor disorders', *Nat Rev Neurosci*, 3: 781–90.

Dietz, V. (2003a), 'Spastic movement disorder: what is the impact of research on clinical practice?', *J Neurol Neurosurg Psychiatry*, 74: 820–1.

Dietz, V. (2003b), 'Spinal cord pattern generators for locomotion', *Clin Neurophysiol*, 114: 1379–89.

Dietz, V. (2011), 'Quadrupedal coordination of bipedal gait: implications for movement disorders', *J Neurol*, 258: 1406–12.

Dietz, V. and Berger, W. (1983), 'Normal and impaired regulation of muscle stiffness in gait: a new hypothesis about muscle hypertonia', *Exp Neurol*, 79: 680–7.

Dietz, V. and Berger, W. (1984), 'Interlimb coordination of posture in patients with spastic paresis. Impaired function of spinal reflexes', *Brain*, 107: 965–78.

Dietz, V., Berger, W. and Horstmann, G. A. (1988), 'Posture in Parkinson's disease: impairment of reflexes and programming', *Ann Neurol*, 24: 660–9.

Dietz, V. and Colombo, G. (1998), 'Influence of body load on the gait pattern in Parkinson's disease', *Mov Disord*, 13: 255–61.

Dietz, V., Colombo, G., Jensen, L. and Baumgartner, L. (1995), 'Locomotor capacity of spinal cord in paraplegic patients', *Ann Neurol*, 37: 574–82.

Dietz, V., Feuerstein, T. J. and Berger, W. (1990), 'Significance of dopamine receptor antagonists in human postural control', *Neurosci Lett*, 117: 81–6.

Dietz, V., Fouad, K. and Bastiaanse, C. M. (2001), 'Neuronal coordination of arm and leg movements during human locomotion', *Eur J Neurosci*, 14: 1906–14.

Dietz, V., Horstmann, G. A. and Berger, W. (1989), 'Significance of proprioceptive mechanisms in the regulation of stance', *Prog Brain Res*, 80: 419–23.

Dietz, V., Ketelsen, U. P., Berger, W. and Quintern, J. (1986), 'Motor unit involvement in spastic paresis. Relationship between leg muscle activation and histochemistry', *J Neurol Sci*, 75: 89–103.

Dietz, V., Leenders, K. L. and Colombo, G. (1997), 'Leg muscle activation during gait in Parkinson's disease: influence of body unloading', *Electroencephalogr Clin Neurophysiol*, 105: 400–5.

Dietz, V. and Michel, J. (2008), 'Locomotion in Parkinson's disease: neuronal coupling of upper and lower limbs', *Brain*, 131: 3421–31.

Dietz, V., Quintern, J. and Berger, W. (1981), 'Electrophysiological studies of gait in spasticity and rigidity. Evidence that altered mechanical properties of muscle contribute to hypertonia', *Brain*, 104: 431–49.

Dietz, V., Schubert, M. and Trippel, M. (1992), 'Visually induced destabilization of human stance: neuronal control of leg muscles', *Neuroreport*, 3: 449–52.

Dietz, V. and Sinkjaer, T. (2007), 'Spastic movement disorder: impaired reflex function and altered muscle mechanics', *Lancet Neurol*, 6: 725–33.

Dietz, V., Trippel, M. and Berger, W. (1991), 'Reflex activity and muscle tone during elbow movements in patients with spastic paresis', *Ann Neurol*, 30: 767–79.

Dietz, V., Trippel, M., Ibrahim, I. K. and Berger, W. (1993), 'Human stance on a sinusoidally translating platform: balance control by feedforward and feedback mechanisms', *Exp Brain Res*, 93: 352–62.

Dimitrijevic, M. R. and Nathan, P. W. (1967), 'Studies of spasticity in man. 2. Analysis of stretch reflexes in spasticity', *Brain*, 90: 333–58.

Diserens, K., Perret, N., Chatelain, S., Bashir, S., Ruegg, D., Vuadens, P. and Vingerhoets, F. (2007), 'The effect of repetitive arm cycling on post stroke spasticity and motor control: repetitive arm cycling and spasticity', *J Neurol Sci*, 253: 18–24.

Ditunno, J. F., Little, J. W., Tessler, A. and Burns, A. S. (2004), 'Spinal shock revisited: a four-phase model', *Spinal Cord*, 42: 383–95.

Duysens, J. (2002), 'Human gait as a step in evolution', *Brain*, 125: 2589–90.

Duysens, J. and Pearson, K. G. (1980), 'Inhibition of flexor burst generation by loading ankle extensor muscles in walking cats', *Brain Res*, 187: 321–32.

Dyer, J. O., Maupas, E., de Andrade Melo, S., Bourbonnais, D., Fleury, J. and Forget, R. (2009), 'Transmission in heteronymous spinal pathways is modified after stroke and related to motor incoordination', *PloS One*, 4: e4123.

Edgerton, V. R., Tillakaratne, N. J., Bigbee, A. J., de Leon, R. D. and Roy, R. R. (2004), 'Plasticity of the spinal neural circuitry after injury', *Annu Rev Neurosci*, 27: 145–67.

Elbasiouny, S. M., Moroz, D., Bakr, M. M. and Mushahwar, V. K. (2010), 'Management of spasticity after spinal cord injury: current techniques and future directions', *Neurorehabil Neural Repair*, 24: 23–33.

Erni, T. and Dietz, V. (2001), 'Obstacle avoidance during human walking: learning rate and cross-modal transfer', *J Physiol*, 534: 303–12.

Faist, M., Dietz, V. and Pierrot-Deseilligny, E. (1996), 'Modulation, probably presynaptic in origin, of monosynaptic Ia excitation during human gait', *Exp Brain Res*, 109: 441–9.

Faist, M., Ertel, M., Berger, W. and Dietz, V. (1999), 'Impaired modulation of quadriceps tendon jerk reflex during spastic gait: differences between spinal and cerebral lesions', *Brain*, 122: 567–79.

Faist, M., Mazevet, D., Dietz, V. and Pierrot-Deseilligny, E. (1994), 'A quantitative assessment of presynaptic inhibition of Ia afferents in spastics. Differences in hemiplegics and paraplegics', *Brain*, 117: 1449–55.

Finley, J. M., Perreault, E. J. and Dhaher, Y. Y. (2008), 'Stretch reflex coupling between the hip and knee: implications for impaired gait following stroke', *Exp Brain Res*, 188: 529–40.

Flament, D., Vaillancourt, D. E., Kempf, T., Shannon, K. and Corcos, D. M. (2003), 'EMG remains fractionated in Parkinson's disease, despite practice-related improvements in performance', *Clin Neurophysiol*, 114: 2385–96.

Floel, A., Breitenstein, C., Hummel, F., Celnik, P., Gingert, C., Sawaki, L., Knecht, S. and Cohen, L. G. (2005), 'Dopaminergic influences on formation of a motor memory', *Ann Neurol*, 58: 121–30.

Forssberg, H., Johnels, B. and Steg, G. (1984), 'Is parkinsonian gait caused by a regression to an immature walking pattern?', *Adv Neurol*, 40: 375–9.

Galiana, L., Fung, J. and Kearney, R. (2005), 'Identification of intrinsic and reflex ankle stiffness components in stroke patients', *Exp Brain Res*, 165: 422–34.

Goldberger, M. E. and Murray, M. (1988), 'Patterns of sprouting and implications for recovery of function', *Adv Neurol*, 47: 361–85.

Gollhofer, A., Horstmann, G. A., Berger, W. and Dietz, V. (1989), 'Compensation of translational and rotational perturbations in human posture: stabilization of the centre of gravity', *Neurosci Lett*, 105: 73–8.

Gorassini, M. A., Knash, M. E., Harvey, P. J., Bennett, D. J. and Yang, J. F. (2004), 'Role of motoneurons in the generation of muscle spasms after spinal cord injury', *Brain*, 127: 2247–58.

Gossard, J. P., Brownstone, R. M., Barajon, I. and Hultborn, H. (1994), 'Transmission in a locomotor-related group Ib pathway from hindlimb extensor muscles in the cat', *Exp Brain Res*, 98: 213–28.

Grillner, S. (1985), 'Neurobiological bases of rhythmic motor acts in vertebrates', *Science*, 228: 143–9.

Grillner, S. and Wallen, P. (1985), 'Central pattern generators for locomotion, with special reference to vertebrates', *Annu Rev Neurosci*, 8: 233–61.

Hiersemenzel, L. P., Curt, A. and Dietz, V. (2000), 'From spinal shock to spasticity: neuronal adaptations to a spinal cord injury', *Neurology*, 54: 1574–82.

Hiraoka, K., Matsuo, Y. and Abe, K. (2005), 'Soleus H-reflex inhibition during gait initiation in Parkinson's disease', *Mov Disord*, 20: 858–64.

Hoogstraten, M. C., van der Ploeg, R. J., vd Burg, W., Vreeling, A., van Marle, S. and Minderhoud, J. M. (1988), 'Tizanidine versus baclofen in the treatment of spasticity in multiple sclerosis patients', *Acta Neurol Scand*, 77: 224–30.

Hubli, M., Bolliger, M., Limacher, E., Luft, A. R. and Dietz, V. (2012a), 'Spinal neuronal dysfunction after stroke', *Exp Neurol*, 234: 153–60.

Hubli, M., Dietz, V. and Bolliger, M. (2011), 'Influence of spinal reflexes on the locomotor pattern after spinal cord injury', *Gait Posture*, 34: 409–14.

Hubli, M., Dietz, V. and Bolliger, M. (2012b), 'Spinal reflex activity: a marker for neuronal functionality after spinal cord injury', *Neurorehabil Neural Repair*, 26: 188–96.

Hufschmidt, A. and Mauritz, K. H. (1985), 'Chronic transformation of muscle in spasticity: a peripheral contribution to increased tone', *J Neurol Neurosurg Psychiatry*, 48: 676–85.

Ibrahim, I. K., Berger, W., Trippel, M. and Dietz, V. (1993), 'Stretch-induced electromyographic activity and torque in spastic elbow muscles. Differential modulation of reflex activity in passive and active motor tasks', *Brain*, 116: 971–89.

Jackson, G. M., Jackson, S. R. and Hindle, J. V. (2000), 'The control of bimanual reach-to-grasp movements in hemiparkinsonian patients', *Exp Brain Res*, 132: 390–8.

Jakob, W., Wirz, M., van Hedel, H. J. and Dietz, V. (2009), 'Difficulty of elderly SCI subjects to translate motor recovery–"body function"–into daily living activities', *J Neurotrauma*, 26: 2037–44.

Jones, C. A. and Yang, J. F. (1994), 'Reflex behavior during walking in incomplete spinal-cord-injured subjects', *Exp Neurol*, 128: 239–48.

Kamper, D. G., Schmit, B. D. and Rymer, W. Z. (2001), 'Effect of muscle biomechanics on the quantification of spasticity', *Ann Biomed Eng*, 29: 1122–34.

Kautz, S. A., Patten, C. and Neptune, R. R. (2006), 'Does unilateral pedaling activate a rhythmic locomotor pattern in the nonpedaling leg in post-stroke hemiparesis?', *J Neurophysiol*, 95: 3154–63.

Kloter, E., Wirz, M. and Dietz, V. (2011), 'Locomotion in stroke subjects: interactions between unaffected and affected sides', *Brain*, 134: 721–31.

Knutsson, E., Martensson, A. and Gransberg, L. (1997), 'Influences of muscle stretch reflexes on voluntary, velocity-controlled movements in spastic paraparesis', *Brain*, 120: 1621–33.

Knutsson, E. and Richards, C. (1979), 'Different types of disturbed motor control in gait of hemiparetic patients', *Brain*, 102: 405–30.

Krebs, H. I., Hogan, N., Hening, W., Adamovich, S. V. and Poizner, H. (2001), 'Procedural motor learning in Parkinson's disease', *Exp Brain Res*, 141: 425–37.

Krystkowiak, P., Delval, A., Dujardin, K., Bleuse, S., Blatt, J. L., Bourriez, J. L., Derambure, P., Destee, A. and Defebvre, L. (2006), 'Gait abnormalities induced by acquired bilateral pallidal lesions: a motion analysis study', *J Neurol*, 253: 594–600.

Lajoie, Y., Teasdale, N., Cole, J. D., Burnett, M., Bard, C., Fleury, M., Forget, R., Paillard, J. and Lamarre, Y. (1996), 'Gait of a deafferented subject without large myelinated sensory fibers below the neck', *Neurology*, 47: 109–15.

Landau, W. M. (2003), 'Botulinum toxin for spasticity after stroke', *N Engl J Med*, 348: 258–9; author reply -9.

Latash, M. L. and Anson, J. G. (1996), 'What are "normal movements" in atypic populations?', *Behav Brain Sci*, 19: 55–106.

Latash, M. L. and Penn, R. D. (1996), 'Changes in voluntary motor control induced by intrathecal baclofen in patients with spasticity of different etiology', *Physiother Res Int*, 1: 229–46.

Lazarus, A. and Stelmach, G. E. (1992), 'Interlimb coordination in Parkinson's disease', *Mov Disord*, 7: 159–70.

Lemon, R. N. (2008), 'Descending pathways in motor control', *Annu Rev Neurosci*, 31: 195–218.

Levin, M. F. and Feldman, A. G. (1994), 'The role of stretch reflex threshold regulation in normal and impaired motor control', *Brain Res*, 657: 23–30.

Levy, G. (2007), 'The relationship of Parkinson disease with aging', *Arch Neurol*, 64: 1242–6.

Little, J. W., Ditunno, J. F., Jr., Stiens, S. A. and Harris, R. M. (1999), 'Incomplete spinal cord injury: neuronal mechanisms of motor recovery and hyperreflexia', *Arch Phys Med Rehabil*, 80: 587–99.

Litvan, I. (2007), 'Update of atypical Parkinsonian disorders', *Curr Opin Neurol*, 20: 434–7.

Lyoo, C. H., Aalto, S., Rinne, J. O., Lee, K. O., Oh, S. H., Chang, J. W. and Lee, M. S. (2007), 'Different cerebral cortical areas influence the effect of subthalamic nucleus stimulation on parkinsonian motor deficits and freezing of gait', *Mov Disord*, 22: 2176–82.

Maegele, M., Muller, S., Wernig, A., Edgerton, V. R. and Harkema, S. J. (2002), 'Recruitment of spinal motor pools during voluntary movements versus stepping after human spinal cord injury', *J Neurotrauma*, 19: 1217–29.

Maric, O., Zorner, B. and Dietz, V. (2008), 'Levodopa therapy in incomplete spinal cord injury', *J Neurotrauma*, 25: 1303–7.

Martin, J. P. (1967) The Basal Ganglia and Posture. London: Pitman.

Mazzaro, N., Nielsen, J. F., Grey, M. J. and Sinkjaer, T. (2007), 'Decreased contribution from afferent feedback to the soleus muscle during walking in patients with spastic stroke', *J Stroke Cerebrovasc Dis*, 16: 135–44.

Mendell, L. M. (1984), 'Modifiability of spinal synapses', *Physiol Rev*, 64: 260–324.

Michel, J., Benninger, D., Dietz, V. and van Hedel, H. J. (2009), 'Obstacle stepping in patients with Parkinson's disease. Complexity does influence performance', *J Neurol*, 256: 457–63.

Michel, J., van Hedel, H. J. and Dietz, V. (2008), 'Obstacle stepping involves spinal anticipatory activity associated with quadrupedal limb coordination', *Eur J Neurosci*, 27: 1867–75.

Mirbagheri, M. M., Barbeau, H., Ladouceur, M. and Kearney, R. E. (2001), 'Intrinsic and reflex stiffness in normal and spastic, spinal cord injured subjects', *Exp Brain Res*, 141: 446–59.

Mittelstaedt, H. (1995), 'Evidence of somatic graviception from new and classical investigations', *Acta Otolaryngol Suppl*, 1: 186–7.

Mizrahi, E. M. and Angel, R. W. (1979), 'Impairment of voluntary movement by spasticity', *Ann Neurol*, 5: 594–5.

Moore, O., Peretz, C. and Giladi, N. (2007), 'Freezing of gait affects quality of life of peoples with Parkinson's disease beyond its relationships with mobility and gait', *Mov Disord*, 22: 2192–5.

Nardone, A., Galante, M., Lucas, B. and Schieppati, M. (2001), 'Stance control is not affected by paresis and reflex hyperexcitability: the case of spastic patients', *J Neurol Neurosurg Psychiatry*, 70: 635–43.

Nardone, A. and Schieppati, M. (2005), 'Reflex contribution of spindle group Ia and II afferent input to leg muscle spasticity as revealed by tendon vibration in hemiparesis', *Clin Neurophysiol*, 116: 1370–81.

Nathan, P. W., Smith, M. and Deacon, P. (1996), 'Vestibulospinal, reticulospinal and descending propriospinal nerve fibres in man', *Brain*, 119: 1809–33.

Nielsen, J., Petersen, N., Ballegaard, M., Biering-Sorensen, F. and Kiehn, O. (1993), 'H-reflexes are less depressed following muscle stretch in spastic spinal cord injured patients than in healthy subjects', *Exp Brain Res*, 97: 173–6.

Nielsen, J. F. and Sinkjaer, T. (1996), 'A comparison of clinical and laboratory measures of spasticity', *Mult Scler*, 1: 296–301.

Nieuwenhuijzen, P. H., Horstink, M. W., Bloem, B. R. and Duysens, J. (2006), 'Startle responses in Parkinson patients during human gait', *Exp Brain Res*, 171: 215–24.

O'Dwyer, N. J. and Ada, L. (1996), 'Reflex hyperexcitability and muscle contracture in relation to spastic hypertonia', *Curr Opin Neurol*, 9: 451–5.

O'Dwyer, N. J., Ada, L. and Neilson, P. D. (1996), 'Spasticity and muscle contracture following stroke', *Brain*, 119: 1737–49.

Pearson, K. G. and Collins, D. F. (1993), 'Reversal of the influence of group Ib afferents from plantaris on activity in medial gastrocnemius muscle during locomotor activity', *J Neurophysiol*, 70: 1009–17.

Pfann, K. D., Buchman, A. S., Comella, C. L. and Corcos, D. M. (2001), 'Control of movement distance in Parkinson's disease', *Mov Disord*, 16: 1048–65.

Pierrot-Deseilligny, E. and Burke, D. (2005) The pathophysiology of spasticity and parkinsonian rigidity. In: The Circuitry of the Human Spinal Cord. New York: Cambridge University Press; 2005. 556–601.

Powers, R. K., Campbell, D. L. and Rymer, W. Z. (1989), 'Stretch reflex dynamics in spastic elbow flexor muscles', *Ann Neurol*, 25: 32–42.

Powers, R. K., Marder-Meyer, J. and Rymer, W. Z. (1988), 'Quantitative relations between hypertonia and stretch reflex threshold in spastic hemiparesis', *Ann Neurol*, 23: 115–24.

Prokop, T., Berger, W., Zijlstra, W. and Dietz, V. (1995), 'Adaptational and learning processes during human split-belt locomotion: interaction between central mechanisms and afferent input', *Exp Brain Res*, 106: 449–56.

Remy-Neris, O., Denys, P., Daniel, O., Barbeau, H. and Bussel, B. (2003), 'Effect of intrathecal clonidine on group I and group II oligosynaptic excitation in paraplegics', *Exp Brain Res*, 148: 509–14.

Rocchi, L., Chiari, L., Mancini, M., Carlson-Kuhta, P., Gross, A. and Horak, F. B. (2006), 'Step initiation in Parkinson's disease: influence of initial stance conditions', *Neurosci Lett*, 406: 128–32.

Rogers, M. W. (1996), 'Disorders of posture, balance, and gait in Parkinson's disease', *Clin Geriatr Med*, 12: 825–45.

Rogers, M. W., Kukulka, C. G. and Soderberg, G. L. (1987), 'Postural adjustments preceding rapid arm movements in parkinsonian subjects', *Neurosci Lett*, 75: 246–51.

Rosenfalck, A. and Andreassen, S. (1980), 'Impaired regulation of force and firing pattern of single motor units in patients with spasticity', *J Neurol Neurosurg Psychiatry*, 43: 907–16.

Rushworth, G. (1960), 'Spasticity and rigidity: an experimental study and review', *J Neurol Neurosurg Psychiatry*, 23: 99–118.

Sanes, J. N., Mauritz, K. H., Dalakas, M. C. and Evarts, E. V. (1985), 'Motor control in humans with large-fiber sensory neuropathy', *Hum Neurobiol*, 4: 101–14.

Scheidtmann, K., Fries, W., Muller, F. and Koenig, E. (2001), 'Effect of levodopa in combination with physiotherapy on functional motor recovery after stroke: a prospective, randomised, double-blind study', *Lancet*, 358: 787–90.

Schieppati, M. and Nardone, A. (1991), 'Free and supported stance in Parkinson's disease. The effect of posture and 'postural set' on leg muscle responses to perturbation, and its relation to the severity of the disease', *Brain*, 114: 1227–44.

Schmit, B. D., Benz, E. N. and Rymer, W. Z. (2002), 'Afferent mechanisms for the reflex response to imposed ankle movement in chronic spinal cord injury', *Exp Brain Res*, 145: 40–9.

Schubert, M., Curt, A., Jensen, L. and Dietz, V. (1997), 'Corticospinal input in human gait: modulation of magnetically evoked motor responses', *Exp Brain Res*, 115: 234–46.

Schubert, M., Prokop, T., Brocke, F. and Berger, W. (2005), 'Visual kinesthesia and locomotion in Parkinson's disease', *Mov Disord*, 20: 141–50.

407

Shimizu, N., Yoshida, M. and Nagatsuka, Y. (1987), 'Disturbance of two simultaneous motor acts in patients with parkinsonism and cerebellar ataxia', *Adv Neurol*, 45: 367–70.

Sinkjaer, T., Andersen, J. B. and Nielsen, J. F. (1996), 'Impaired stretch reflex and joint torque modulation during spastic gait in multiple sclerosis patients', *J Neurol*, 243: 566–74.

Sinkjaer, T., Andersen, J. B., Nielsen, J. F. and Hansen, H. J. (1999), 'Soleus long-latency stretch reflexes during walking in healthy and spastic humans', *Clin Neurophysiol*, 110: 951–9.

Sinkjaer, T. and Magnussen, I. (1994), 'Passive, intrinsic and reflex-mediated stiffness in the ankle extensors of hemiparetic patients', *Brain*, 117: 355–63.

Sinkjaer, T., Toft, E. and Hansen, H. J. (1995), 'H-reflex modulation during gait in multiple sclerosis patients with spasticity', *Acta Neurol Scand*, 91: 239–46.

Sinkjaer, T., Toft, E., Larsen, K., Andreassen, S. and Hansen, H. J. (1993), 'Non-reflex and reflex mediated ankle joint stiffness in multiple sclerosis patients with spasticity', *Muscle Nerve*, 16: 69–76.

Sommerfeld, D. K., Eek, E. U., Svensson, A. K., Holmqvist, L. W. and von Arbin, M. H. (2004), 'Spasticity after stroke: its occurrence and association with motor impairments and activity limitations', *Stroke*, 35: 134–9.

Stelmach, G. E. and Phillips, J. G. (1991), 'Movement disorders – limb movement and the basal ganglia', *Phys Ther*, 71: 60–7.

Stolze, H., Kuhtz-Buschbeck, J. P., Drucke, H., Johnk, K., Illert, M. and Deuschl, G. (2001), 'Comparative analysis of the gait disorder of normal pressure hydrocephalus and Parkinson's disease', *J Neurol Neurosurg Psychiatry*, 70: 289–97.

Swinnen, S. P., Van Langendonk, L., Verschueren, S., Peeters, G., Dom, R. and De Weerdt, W. (1997), 'Interlimb coordination deficits in patients with Parkinson's disease during the production of two-joint oscillations in the sagittal plane', *Mov Disord*, 12: 958–68.

Tatton, W. G., Eastman, M. J., Bedingham, W., Verrier, M. C. and Bruce, I. C. (1984), 'Defective utilization of sensory input as the basis for bradykinesia, rigidity and decreased movement repertoire in Parkinson's disease: a hypothesis', *Can J Neurol Sci*, 11: 136–43.

Taub, E., Uswatte, G., and Pidikiti, R. (1999), 'Constraint-Induced Movement Therapy: a new family of techniques with broad application to physical rehabilitation – a clinical review', *J. Rehabil Res Dev*. Jul: 36(3): 237–51. PMID: 10659807.

Thilmann, A. F., Fellows, S. J. and Garms, E. (1990), 'Pathological stretch reflexes on the "good" side of hemiparetic patients', *J Neurol Neurosurg Psychiatry*, 53: 208–14.

Thompson, F. J., Parmer, R. and Reier, P. J. (1998), 'Alteration in rate modulation of reflexes to lumbar motoneurons after midthoracic spinal cord injury in the rat. I. Contusion injury', *J Neurotrauma*, 15: 495–508.

van Hedel, H. J., Waldvogel, D. and Dietz, V. (2006), 'Learning a high-precision locomotor task in patients with Parkinson's disease', *Mov Disord*, 21: 406–11.

Watts, R. L., Wiegner, A. W. and Young, R. R. (1986), 'Elastic properties of muscles measured at the elbow in man: II. Patients with Parkinsonian rigidity', *J Neurol Neurosurg Psychiatry*, 49: 1177–81.

Winogrodzka, A., Wagenaar, R. C., Booij, J. and Wolters, E. C. (2005), 'Rigidity and bradykinesia reduce interlimb coordination in Parkinsonian gait', *Arch Phys Med Rehabil*, 86: 183–9.

Yang, J. F. and Stein, R. B. (1990), 'Phase-dependent reflex reversal in human leg muscles during walking', *J Neurophysiol*, 63: 1109–17.

Zijlmans, J. C., Poels, P. J., Duysens, J., van der Straaten, J., Thien, T., van't Hof, M. A., Thijssen, H. O. and Horstink, M. W. (1996), 'Quantitative gait analysis in patients with vascular parkinsonism', *Mov Disord*, 11: 501–8.

INDEX